Software Engineering

The Aksen Associates Series in Electrical and Computer Engineering

Software Engineering

Stephen R. Schach
Vanderbilt University

Homewood, IL 60430
Boston, MA 02116

Cover designer: Harold Pattek
Compositor: Science Typographers, Inc.
Typeface: 10/12 Times Roman
Printer: R.R. Donnelley and Sons

Printed in the United States of America.

Library of Congress Cataloging-in-Publication Data

Schach, Stephen R.
Software engineering.

Includes bibliographical references.
1. Software engineering. I. Title.
QA76.758.S33 1990 005.1 89-24468
ISBN 0-256-08515-3

1 2 3 4 5 6 7 8 9 0 DO 7 6 5 4 3 2 1 0 9

About the Author

Stephen R. Schach is Associate Professor of Computer Science at Vanderbilt University in Nashville, Tennessee. Dr. Schach received his M.Sc. at the Weizmann Institute of Science in Israel and his Ph.D. at the University of Cape Town, South Africa. He has published over 60 technical papers in a variety of areas, including software engineering, software testing, and computer architecture. Dr. Schach consults for industry and teaches courses in software engineering internationally. He is a member of the Association for Computing Machinery and the IEEE Computer Society.

To
Sharon
David
and
Lauren

Contents

Contents

Preface

The aim of software engineering is the production of quality software, software that is delivered on time and within budget, and that satisfies all its requirements. To this end, a variety of techniques have been developed for performing the various software production tasks, from requirements and specifications to maintenance. An important aspect of software engineering is that software production should be well supported by a set of appropriate automated tools.

Total software expenditure in 1990 in the United States can be conservatively estimated at $120 billion; an additional $240 billion will be spent by the rest of the world. Software pervades almost every aspect of our lives, either directly through such mechanisms as computerized billing systems, airline reservation systems, and hospital records or indirectly because the world's nuclear arsenals are under computer control. It is therefore important that these huge sums of money be well spent, so that the resultant software has a positive influence on our civilization.

Software Engineering is a senior/first-year graduate student textbook, covering all aspects of the production of software. By the end of a course based on this book, a senior will be equipped to start work in the software industry as a software engineer. A graduate student will be qualified to begin a research project in software engineering.

The book has four major themes. The first is analysis and comparison. A variety of techniques are described for each section of the life cycle; these techniques are then carefully contrasted. Because of the plethora of present-day software engineering techniques, it is important to select an appropriate one for the task at hand. The emphasis in this book on analysis and comparison alerts the reader to the need for careful choice and provides him or her with criteria for choosing wisely. A second theme is that the results of experiments in software engineering constitute a powerful tool for determining which techniques are to be preferred for a given situation. The third theme is that maintenance is almost always the most important phase of software production, because maintenance consumes well over half of the total software budget. Maintenance must be planned for throughout the software development life cycle. Fourth, testing is not a phase that is carried out just before the product is delivered to the client, nor is it an activity that is performed at the end of each life-cycle phase. Instead, testing must be done continually throughout the entire software development and maintenance life cycle.

With regard to prerequisites, it is assumed that the reader is familiar with a high-level programming language such as Pascal, C, BASIC, COBOL, or FORTRAN. Although some of the examples are in Ada, no previous knowledge of Ada is needed; even the chapter entitled Ada and Software Engineering has been written for an audience that is assumed to have no previous exposure to Ada. In addition, the reader is expected to have taken a course in data structures and to have acquired the familiarity with computers that is gained through carrying out programming exercises.

How the Book Is Organized

The book is divided into four parts. Each part is prefaced by a summary of the material included in that part. Part One is an introduction to software engineering. The wide scope of software engineering is highlighted, as are the problems of software engineering. In addition, this part provides an overview of the book as a whole.

Part Two is devoted to the software life cycle. Different software life cycles are presented and compared. Also, activities that relate to the life cycle as a whole are described; these include planning a software project and testing.

In contrast, each of the chapters of Part Three is devoted to a separate phase of the life cycle such as specifications, implementation, and maintenance. There are two chapters on design. Theoretical aspects of software design such as objects and abstraction are covered in a chapter entitled Modularity: and Beyond, while practical ways of achieving design goals are covered in a separate chapter. Part Three is the largest part of the book, in order that the reader may be equipped to handle every phase of the life cycle.

Part Four is entitled Major Topics in Software Engineering. Topics covered include CASE (computer-aided software engineering), portability and reusability, the impact of Ada on software engineering, experimentation in software engineering and automatic programming. Of the many possible topics that could have been included in Part Four, these six were chosen because of their relevance both to current software engineering practice and to the way the field is likely to evolve in the future.

Each chapter is concluded with a chapter review, suggestions for further reading in that area, a set of problems, and references for that chapter. In addition, the Bibliography for the book as a whole is at the end of the book.

Software Engineering is designed as a textbook for a one-semester course. The material of the book has been covered in 45 lectures, with ample time for questions and discussion. In addition, this book has been successfully used as a text for short courses for industry.

About the Problem Sets

This book includes three types of problems. First, each chapter has a number of problems intended to highlight key points. These problems are self-contained; there is no need to visit a computer store to determine the specifications of various machines or to contact computer professionals to obtain information such as software productivity data. In general, computer salespeople and software professionals have neither the time nor the inclination to serve as sources of information for solving problems in textbooks, and users of textbooks are not comfortable contacting busy individuals in order to obtain the data they need. For these reasons, the technical information for all of the problems can be found in this book.

Second, there is a software term project. Software engineering is a practical discipline, and purely theoretical courses are not very effective, so a practical project has been included in this book. Because so much software today is produced by teams rather than by individuals, the term project is designed to be solved by students working in teams of 3, the smallest number of team members that cannot confer over a telephone. The term project comprises 13 separate components, each tied in to the relevant chapter. For example, software design is the topic of Chapter 9, so in that chapter the component of the term project is concerned with designing the project software. By breaking up a large project into smaller, well-defined pieces, the instructor will be able to monitor the progress of the class more closely. The project is based on the trials and tribulations of the Plain Vanilla Ice Cream Corporation as described in the Appendix, but the structure of the project is such that the instructor may freely apply the 13 components to any other project of his or her choosing.

Since the book is designed for use by graduate students as well as upperclass undergraduates, the third type of problem is based on research papers in the software engineering literature. In each chapter an important paper has been chosen. The student is asked to read the paper and to answer a question relating to the contents of the paper. Of course, the instructor is free to assign any other research paper; to assist in this regard, the For Further

Reading section at the end of each chapter includes a wide variety of relevant papers.

Acknowledgements

I am indebted to the reviewers of this book, namely Dan Berry, The Technion; David Cheriton, Stanford University; Bob Goldberg, IBM; Bill McCracken, Georgia Institute of Technology; Fred Mowle, Purdue University; and Toby Teorey, University of Michigan, for their careful appraisal of this book and for their many helpful suggestions. In addition to his role as meticulous reviewer, I would like to thank my friend Dan Berry for having introduced me to software engineering many years ago. I should also like to thank software engineering colleagues who commented on a section of the book that pertained to their particular research interests, namely Joyce Blair, Rajeev Gopal, Garry Kampen, Klaas van der Poel, and Pamela Zave.

I would have liked to thank by name all the other individuals who have contributed ideas, but unfortunately that is impossible. This book has evolved over the past 10 years from courses on software engineering I have given to seniors and to graduate students, and also from the many short courses I have given to industry. Lectures on software engineering tend to be exciting, and there are frequently heated classroom discussions. Some of my views on software engineering have resulted from remarks made during those discussions. In the clamor of classroom debate it is sometimes impossible to pin down the source of a comment, let alone to recall it some years later. Nevertheless, I should like to place on record my thanks to the many participants for their stimulating remarks.

I should like to thank Vanderbilt University for granting me sabbatical leave to write this book. In particular, I thank my chairman, Patrick Fischer, for his unfailing support, encouragement, and advice from beginning to end.

Howard Aksen, of Aksen Associates, has been everything that an author could want in an editor. With expertise and warmth he has guided this endeavor from the start. My grateful thanks go to him.

Finally, I should like to thank the three people without whose encouragement I would never have started this book, and without whose love and support I could never have finished it. They accepted without question or complaint the endless evenings and weekends that I spent with "the book" instead of with them. I lovingly dedicate this book to my wife, Sharon, and to my children, David and Lauren.

Stephen R. Schach

Part One

Introduction to Software Engineering

The first two chapters of this book serve a dual role. They introduce the reader to software engineering, and they also provide an overview of the book as a whole.

In Chapter 1, Scope of Software Engineering, it is pointed out that techniques for software production must be cost-effective and must also promote constructive interaction between the members of the software production team. The importance of maintenance, which on average consumes about two-thirds of software expenditure, is stressed throughout the book, starting with this chapter.

Software Production and Its Difficulties is the title of the second chapter. Many problems of software engineering are described, but no solutions are put forward in Chapter 2. Instead, the reader is informed where in the book each problem is tackled. In this way, the chapter serves as a guide to the book as a whole.

Chapter 1 Scope of Software Engineering

1.1 Introduction

A well-known story is told of the executive who received a computer-generated bill for $0.00. After having a good laugh with his friends about "idiot computers," he tossed the bill away. A month later a similar bill arrived, this time marked 30 days. Then came the third bill. The fourth bill arrived a month later, accompanied by a message hinting at possible legal action if the bill for $0.00 was not immediately paid.

The fifth bill, marked 120 days, did not hint at anything—the message was rude and forthright, threatening all manner of legal actions if the bill was not immediately paid. Fearful of his organization's credit rating in the hands of this maniacal machine, the executive called an acquaintance who was a software engineer and related the whole sorry story. Trying not to laugh, the software engineer told the executive to mail in a check for $0.00. This had the desired effect. A receipt for $0.00 was received a few days later, which the executive carefully filed away in case the computer at some future date might allege that $0.00 was still owing.

This well-known story has a less well-known sequel. A few days later the executive was summoned by his bank manager. The banker held up a check and asked, "Is this your check?"

The executive agreed that it was.

"Would you mind telling me why you wrote a check for $0.00?" asked the banker.

So the whole story was retold. When the executive had finished, the banker turned to him and quietly asked, "Have you any idea what your check for $0.00 did to *our* computer system?"

A computer professional can laugh at this story, albeit somewhat nervously. After all, every one of us has designed or implemented a product that, in its original form, would have resulted in the equivalent of sending dunning letters for $0.00. Hopefully, we have always caught this sort of fault during testing. But our laughter has a hollow ring to it, because at the back of our minds there is the fear that someday we will not detect the fault before the product is delivered to the customer.

A decidely less humorous software fault was detected on November 9, 1979. The Strategic Air Command had an alert scramble when the WWMCCS computer system reported that the Russians had launched missiles aimed towards the United States [Neumann, 1980]. What actually happened was that a simulated attack was interpreted as the real thing, just as in the movie *War Games* some 5 years later. While the Department of Defense has understandably not given details about the precise mechanism by which test data were taken for actual data, it seems reasonable to ascribe the problem to a software fault. Either the system as a whole was not designed to differentiate between simulations and reality, or the user interface did not include the necessary checks for ensuring that users of the system would be able to distinguish fact from fiction. In other words, a software fault, if indeed the problem was caused by software, could have brought civilization as we know it to an unpleasant and abrupt end.

Whether we are dealing with billing products or air defense systems, software is being delivered late, over budget, and full of residual faults ("bugs"). Software engineering is an attempt to solve these problems. In other words, software engineering is a discipline whose aim is the production of quality software, software that is delivered on time, within budget, and that satisfies its requirements. In order to achieve this goal, a software engineer has to acquire a broad range of skills, both technical and managerial. These skills have to be applied not just to programming but to every phase of software production, from requirements to maintenance.

The scope of software engineering is very broad. Some aspects of software engineering can be categorized as mathematics or computer science; other aspects fall into the areas of economics, management, or psychology. In order to display the wide-reaching realm of software engineering, four different aspects will now be examined.

1.2 Historical Aspects

It is a fact that electric power generators fail, but far less frequently than payroll products. It is true that bridges sometimes collapse, but considerably less often than operating systems do.

In the belief that software design, implementation, and maintenance could be put on the same footing as traditional engineering disciplines, a NATO study group in 1967 coined the term "software engineering." The claim that building software is similar to other engineering tasks was endorsed by the 1968 NATO Software Engineering Conference. (This endorsement is not very surprising; the very name of the conference reflected the belief that software production should be an engineering-like activity.) A conclusion of the conferees was that software engineering should use the philosophies and paradigms of established engineering disciplines, and that this would solve what they termed the *software crisis*, namely that the quality of software was generally unacceptably low and that deadlines and cost limits were not being met.

The fact that the crisis is still with us, over 20 years later, should tell us two things. First, the software production process is *not* like traditional engineering. Second, the software crisis should rather be termed the *software depression*, in view of its long duration and poor prognosis.

It is certainly true that bridges collapse less frequently than operating systems. Why then cannot bridge-building techniques be used to build operating systems? What the NATO conferees overlooked is that bridges are as different from operating systems as ravens are from writing desks.

A major difference lies in the attitudes of the civil engineering community and the software engineering community to the act of collapsing. When a bridge collapses, as the Tacoma Narrows Bridge did in 1940, this almost always means that the bridge has to be redesigned and rebuilt from scratch. The original design was faulty and posed a threat to human safety; certainly the design has to be drastically changed. Furthermore, the effects of the collapse will in almost every instance have caused so much damage to the bridge fabric that the only reasonable thing to do is to demolish what is left of the faulty bridge and rebuild it.

In contrast, when an operating system crashes it may be possible simply to reboot the system in the hope that the set of circumstances which caused the crash will not recur. This may be the only thing to do, if, as is often the case, there is no evidence as to what caused the crash. The damage caused by the crash will usually be minor: a database partially corrupted, a few files lost. Even when damage to the system is considerable, by using back-up data the system can often be restored to a state not too far removed from the state it was in when the crash occurred.

In those instances when an operating system crash can have significant effects, as is the case with most real-time systems, there is usually some element of fault tolerance, both hardware and software, built into the system as a whole to minimize the effects of a crash. The very concept of fault tolerance highlights this difference between bridges and operating systems. Bridges are engineered in such a way that they will be able to withstand every reasonably anticipated condition such as high winds, flash floods, and so on. An implicit assumption of any software builder, and real-time software builders in particular, is that we cannot hope to anticipate all possible conditions that the software must withstand, so we must design our software in such a way as to try to minimize the damage that an unanticipated condition might cause. In other words, bridges are assumed to be perfectly engineered; operating systems are assumed to be imperfectly engineered. This fundamental difference is why software cannot be "engineered," in the classical sense of the word.

It might be suggested that this difference is only temporary. After all, we have been building bridges for thousands of years, and therefore we have considerable experience and expertise in the types of conditions a bridge must withstand. We have less than 30 years of experience with complex operating systems. Surely with more experience, the argument goes, we will understand operating systems as well as we understand bridges, and so we will be able to construct them using similar principles.

The flaw in this argument is that hardware, and hence the associated operating system, is growing faster in complexity than we can handle it. In the 1960s we had multiprogramming operating systems, virtual memory was a major complicating factor of operating systems of the 1970s, and now we are attempting to come to terms with multiprocessor and distributed (network) operating systems. Until we can handle the complexity caused by the interconnections of the various components of a software product such as an operating system, we cannot hope to understand it fully, and if we do not understand it, we cannot hope to engineer it. And complexity is growing too fast for us to hope to be able to master it.

A second major difference between bridges and operating systems is maintenance. Maintaining a bridge is generally restricted to painting it, repairing minor cracks, resurfacing the road, and so on. A civil engineer if asked to rotate a bridge through 90° or to move it hundreds of miles to another state would consider the requester to be bereft of his or her senses. But we think nothing of asking a software engineer to convert a batch operating system into a time-sharing one or to port it from one machine to another machine with totally different architectural characteristics. It is not unreasonable for 50% of the lines of source code of an operating system to be rewritten over a 5-year period, especially if it is ported to new hardware. But no engineer would consent to half a bridge being replaced; safety requirements would dictate that a new bridge be built.

The area of maintenance is a second fundamental aspect in which software engineering differs from the products of conventional engineering. Further maintenance aspects of software engineering are described in Section 1.4. But first, economics-oriented aspects are presented.

1.3 Economics Aspects

An insight into the relationship between software engineering and computer science can be obtained by comparing and contrasting the relationship between chemistry and chemical engineering. After all, computer science and chemistry are both sciences, and both have a theoretical component and a practical component. In the case of chemistry, the practical component is laboratory work; in the case of computer science, the practical component is programming.

Consider the process of extracting gasoline from coal. During World War II, the Germans used this process to make fuel for their war machine because they were largely cut off from oil supplies. Because of oil embargoes, the government of the Republic of South Africa has poured billions of dollars into SASOL (an Afrikaans acronym standing for "South African coal into oil"). About half of South Africa's liquid fuel needs are met in this way.

From the viewpoint of a chemist there are many possible ways of converting coal into gasoline, and all are equally important. After all, no one chemical reaction is more important than any other. But from the chemical engineer's viewpoint there is, at any one time, exactly one means of synthesizing gasoline from coal that is important, namely that reaction which at that point in time is economically the most feasible. In other words, the chemical engineer evaluates all possible reactions, then rejects all but that one reaction for which the cost per gallon is a minimum.

A similar relationship holds between computer science and software engineering. The computer scientist investigates a variety of ways of producing software, some good and some bad. But the software engineer is interested only in those techniques which make sound economic sense.

For instance, if an organization discovers that a new coding technique CT_2 results in code being produced in nine-tenths of the time needed by the currently used technique CT_1, then common sense seems to dictate that CT_2 is the correct technique to use. In fact, while common sense certainly dictates that if CT_2 is faster than CT_1, then CT_2 is the technique of choice, the preceding aspects of software engineering may in fact imply the opposite. For example, coding technique CT_2 may indeed be 10% faster than CT_1, and the resulting code in both instances may be of comparable quality from the viewpoint of satisfying the client's current needs. But use of technique CT_2 may result in code that is difficult to maintain, so over the life of the product the total software cost may well be higher using CT_2. Of course, if the organization that is doing the coding is in no way responsible for any sort of maintenance, then, from the viewpoint of just that organization, CT_2 is a very attractive proposition. After all, use of CT_2 saves the organization 10% of coding costs. Instead, the client should insist that technique CT_1 be used, and pay the premium in coding costs. In that way over its lifetime the total cost of the software to the client will be lower. But in practice the sole aim of both the client and the coding organization is all too frequently to produce that code as quickly as possible. The long-term effects of using a particular technique are generally ignored in the interests of short-term gain. Applying economic principles to software engineering requires the client to choose those techniques that reduce the client's long-term costs.

The inclusion of maintenance when figuring software costs leads directly to the next section.

1.4 Maintenance Aspects

The life cycle of a software product is the period of time that starts with concept exploration and ends when the product is finally retired (decommissioned). During this period the product goes through a series of phases such as requirements, specifications, design, implementation, installation and checkout, operations and maintenance, and, finally, retirement. Life cycles are discussed in detail in Chapter 3; the topic is being introduced at this point so that the concept of maintenance can be formalized.

Until the end of the 1970s, most organizations were producing software using what is now termed the *waterfall model* as their product life cycle. There are many variations of this model, but by and large the product goes through seven broad phases. These phases probably do not correspond exactly to the phases of any one particular organization, but they are sufficiently close to most practices for the purposes of this book. Similarly, the precise name of each phase varies from organization to organization. The names used here for the various phase are as general as possible in the hope that the reader will feel comfortable with them.

Phase 1. Requirements: The concept is explored and refined, and the client's requirements are ascertained and analyzed.

Phase 2. Specifications: The client's requirements are presented in the form of a document, the *specifications*, "what the product is supposed to do." A plan is drawn up, the *software project management plan*, detailing every aspect of the proposed software development.

Phase 3. Design: The specifications undergo two consecutive design processes. First comes *architectural design* in which the product as a whole is broken down into components, called *modules*. Then each module in turn is designed; this is termed *detailed design*. The resulting document is termed the *design*, "how the product does it."

Phase 4. Implementation: The various components are coded and tested.

Phase 5. Integration: The components of the product are put together and tested as a whole. When the developer is satisfied with the product, it is tested by the client (*acceptance testing*). This phase ends when the product is acceptable to the client and goes into *operations mode*.

Phase 6. Maintenance: Maintenance includes all changes to the product once the client has accepted it as satisfying the specifications. Maintenance includes *corrective maintenance* (or *software repair*), the removal of residual faults but leaving the specifications unchanged, as well as *enhancement* (or *software update*), which consists of changes to the specifications, and the implementation of those changes. There are, in turn, two aspects of enhancement. The first is *perfective maintenance*, changes that the client thinks will improve the effectiveness of the product such as additional functionality or decreased response time. The second is *adaptive maintenance*, changes made in response to changes in the environment in which the product operates such as new government regulations. Studies have indicated that, on average, maintainers spend about 60% of their time on perfective maintenance, 20% on adaptive maintenance, and 20% on corrective maintenance [Lientz, Swanson, and Tompkins, 1978].

Phase 7. Retirement: The product is removed from service.

It is sometimes said that only bad software products undergo maintenance. In fact, the opposite is true; bad products are thrown away, while good products are repaired and enhanced, perhaps for 10 or 15 years. Furthermore, enhancement reflects the fact that a software product is a model of the real world, and the real world is constantly changing.

For instance, if the sales tax rate changes from 6% to 7%, almost every product that deals with buying or selling has to be changed. Suppose the product contains the Ada statement

```
SALES_TAX : constant PERCENTAGE_TYPE := 6.0;
```

declaring that SALES_TAX is a **constant** of type PERCENTAGE_TYPE and is initialized to the value 6.0. In this case maintenance is relatively simple. With the aid of a text editor the value 6.0 is replaced by 7.0, and the product is recompiled and relinked. But if, instead of using the name SALES_TAX, the actual value 6.0 has been used in the product wherever the value of the sales tax is invoked, then such a product will be very difficult indeed to maintain. For example, there may be some instance of the value 6.0 in the source code that should be changed to 7.0 but is overlooked, or an instance of 6.0 that does not refer to sales tax, but will incorrectly be changed to 7.0. In fact, it might be cheaper in the long run to throw the product away and recode it.

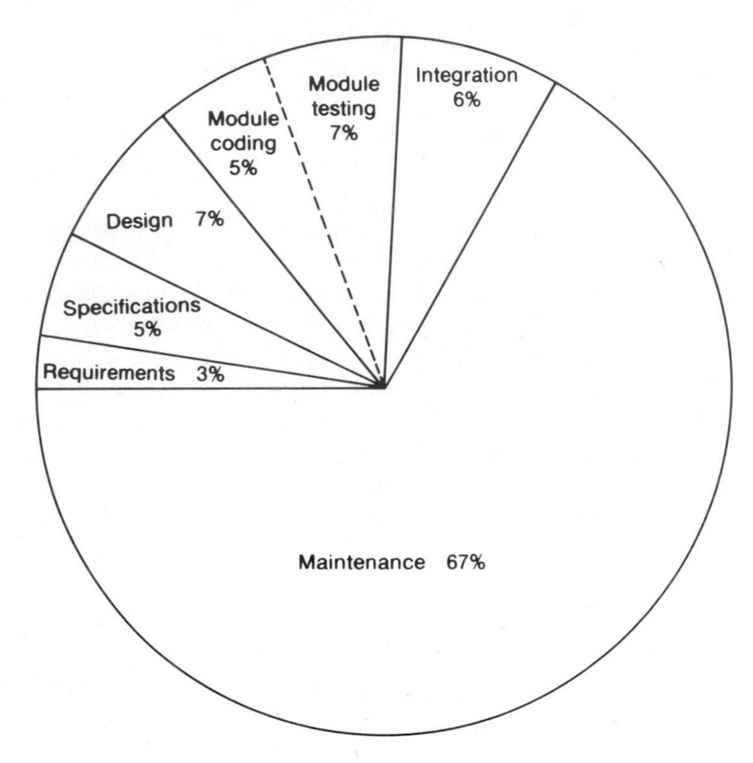

Figure 1.1 Approximate relative costs of life-cycle phases.

The real-time real world is also constantly changing. The missiles with which a jet fighter is armed may be replaced by a new model, so the weapons control component of the avionics system has to be changed. A six-cylinder engine is to be offered as an option in a popular four-cylinder automobile; this implies programming changes to the on-board computer which controls the carburetor, timing, and so on.

Healthy companies change; only dying companies are static. Thus changes to the specifications of products will constantly occur within growing organizations. This means that maintenance in the form of enhancement is a positive part of a company's activities, reflecting that the company is on the move.

But just how much time is devoted to maintenance? The pie chart of Figure 1.1 is obtained by averaging data from various sources, including [Elshoff, 1976], [Daly, 1977], [Zelkowitz, Shaw, and Gannon, 1979], and [Boehm, 1981]. It shows the approximate percentage of time (= money) spent on each phase of the software life cycle. As can be seen from the figure, about

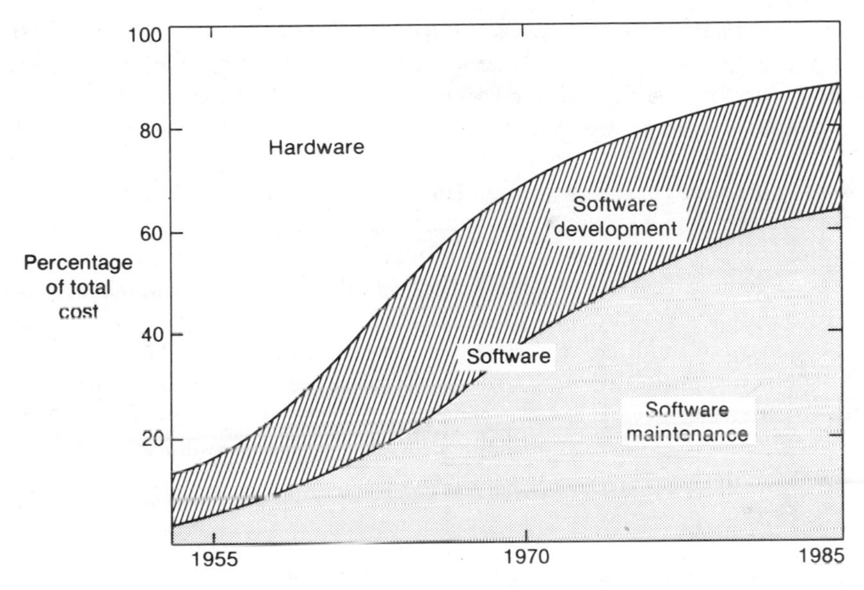

Figure 1.2 Relative hardware/software cost trends [Boehm, 1976]. (© 1976 IEEE.)

two-thirds of total software costs are devoted to maintenance. Thus maintenance is almost always the most important phase of the software life cycle.

Returning to the earlier example, suppose a software organization is currently using coding technique CT_1, but a salesperson promises that CT_2 will reduce coding time by 10%. Even if CT_2 has no adverse effect on maintenance, an astute software manager will think twice before changing coding practices. The entire staff will have to be retrained, new software development tools purchased, and perhaps additional staff hired who are experienced in the new technique. All this expense and disruption will have to be endured for a possible 0.5% decrease in software costs since, from Figure 1.1, module coding consumes only 5% of total software costs.

But suppose a new technique that reduces maintenance by 10% is developed. This should be introduced at once, because the software manager will thereby reduce overall costs by 6.7%. The overheads involved in changing to this technique are a small price to pay for such a large overall saving.

Maintenance constitutes about two-thirds of total software costs. But what fraction is it of total (hardware and software) costs? To answer this question, refer to Figure 1.2, which depicts Boehm's 1973 prediction that by 1980 software costs for large systems would constitute over 80% of total costs, that is, hardware and software costs [Boehm, 1973]. This prediction is depicted by the upper curve in Figure 1.2, the curve separating hardware costs from software costs.

Boehm's prediction, unlike so many in the history of science, not only came true, but was an understatement. For one thing, his prediction came true about 2 years early. Furthermore, while his prediction applied only to large-scale systems, it seems to be true for all systems, from supercomputers down to microcomputers, with one exception. If a microcomputer is used with unmodified off-the-shelf packages, then the cost of the software will be less than the cost of the hardware. This is particularly true when pirated (= stolen) software is being used, as is now unfortunately so often the case.

The lower curve in Figure 1.2 reflects the fact that maintenance costs constitute about two-thirds of total software costs. The lower curve reached the 50% mark around the year 1974. This means that since 1974 out of every $100 spent on hardware, hardware maintenance, software development, and software maintenance, $50 was spent just on software maintenance. Since maintenance is so important, a major aspect of software engineering consists of those techniques, tools, and practices that lead to a reduction in maintenance costs.

1.5 Specification and Design Aspects

The relative costs of fixing a fault at various phases in the software life cycle are shown in Figure 1.3 [Boehm, 1981]. The figure reflects data from IBM [Fagan, 1974], GTE [Daly, 1977], the Safeguard project [Stephenson, 1976], and some smaller TRW projects [Boehm, 1980]. From Figure 1.3, if it costs $10 to fix a fault during implementation, that same fault would have cost only $1 to fix during the specifications phase. But during the maintenance phase, that same fault will cost around $100 to repair. The moral of the story is: We must find faults early, or else it will cost us money. We should therefore employ techniques for detecting faults during the requirements and specifications phases.

There is a further need for such techniques. Other studies have shown [Boehm, 1979] that between 60% and 70% of all faults detected in large-scale projects are specification or design faults. Thus it is important that we improve our specification and design techniques not only in order that faults be found as early as possible, but also because specification and design faults constitute such a large proportion of faults in general. Just as in the example in the previous section which showed that reducing maintenance costs by 10% will reduce overall costs by over 5%, so by reducing specification and design faults by 10% the overall number of faults can be reduced by 6% or 7%.

These ideas highlight another important aspect of software engineering, namely techniques which lead to better specifications and designs.

The next aspect of software engineering considered is the implications of the fact that most software is produced by a team of software engineers, rather than by a single individual who is responsible for every phase of the development and maintenance process.

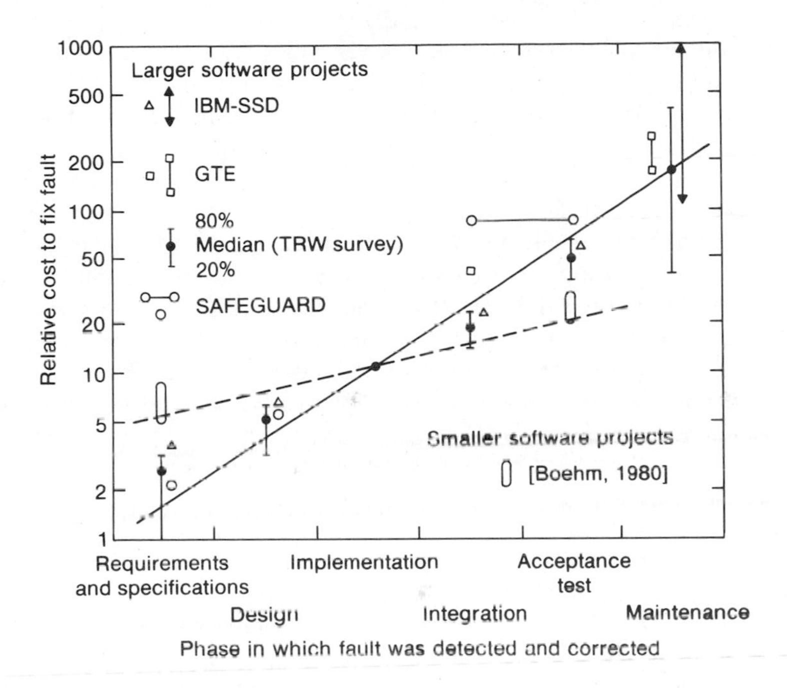

Figure 1.3 Relative cost of fixing a fault at each life-cycle phase (Barry Boehm, *Software Engineering Economics*, © 1981, p. 40. Adapted by permission of Prentice-Hall, Inc., Englewood Cliffs, NJ.)

1.6 Team Programming Aspects

The performance–price factor of a computer is defined as follows:

$$\text{performance–price factor} = \text{time to perform one million additions} \times \text{cost of CPU and main memory} \quad (1.1)$$

This quantity has decreased by an order of magnitude with each succeeding generation of computers. That is to say, the performance–price factor has decreased by an order of magnitude six times over the past 30 years. This decrease has been a consequence of discoveries in electronics, particularly the transistor, and very large-scale integration (VLSI).

The result of these discoveries has been that organizations can easily afford hardware that can run large products, that is to say, products too large to be written by one person within the allowed time constraints. But team programming leads to interface problems between code components and communications problems between team members.

For example, Joe and Freda code modules A and B, respectively, where module A calls module B. When Joe codes A he writes a call to B with five parameters in the argument list. Freda codes B with five parameters, but in a different order from those of Joe. Few linkers other than an Ada linker will detect a type violation, but even if the linker can do this, if the interchanged parameters are both of the same type, then the problem may not be detected for a long period of time. It may be argued that this is a specification problem, and that if the modules had only been properly specified, this would not have happened. That is true, but what happens in practice is that the specifications are often changed once coding commences, but notification of the change is sometimes not distributed to all members of the development team. Thus, when a specification which affects two or more programmers has been changed, poor communications can lead to the interface problems Joe and Freda experienced. This sort of problem will not occur if only one individual is responsible for every aspect of the product, as was the case before powerful computers were affordable.

But interfacing problems are merely the tip of the iceberg when it comes to problems that can be caused by programming in teams. Unless the team is properly organized, an inordinate amount of time can be wasted in conferences between team members. Suppose that a product would take a single programmer 1 year to complete. If the same task were assigned to a team of three programmers, the time for completing the task would generally be closer to 1 year than the expected 4 months, and the quality of the resulting code may well be lower than if the task had been assigned to a single programmer. Since a considerable portion of today's software is being developed and maintained by teams, the scope of software engineering must also include techniques for solving the problems caused by software being developed and maintained by a team.

As has been shown in the preceding five sections, the scope of software engineering is extremely broad. It includes every phase of the software life cycle, from requirements to decommissioning. It also includes human aspects such as programming team organization, economic aspects, and legal aspects such as copyright agreements. All these aspects are implicitly incorporated in the original definition of software engineering given in Section 1.1, namely that software engineering is a discipline whose aim is the production of quality software, software that is delivered on time, within budget, and that satisfies its requirements.

1.7 Terminology

Since the 1970s, the difference between a *program* and a *system* has become blurred. In the "good old days" the distinction was very clear. A program was an autonomous piece of code, generally in the form of a deck of punched cards, that could be executed. A system was a related collection of programs. Thus a system might consist of programs P, Q, R, and S. Magnetic tape T_1 was mounted, then program P was run. It caused a deck of data cards to be read in and produced as output tapes T_2 and T_3. Tape T_2 was then rewound, and program Q was run, producing tape T_4 as output. Program R now merged tapes T_3 and T_4 into tape T_5; T_5 served as input for program S, which printed a series of reports.

Compare that situation with a product performing real-time control of a steel mill running on a machine with a front-end communications processor and a back-end database manager. The single piece of software controlling the steel mill does far more than the old-fashioned system, but in terms of the definitions of classical program and system given previously it is undoubtedly a program. To add to the confusion, the term "system" is now also used to denote the hardware and software together. For example, a flight control system in an aircraft consists of both the in-flight computers and the software running on them. Depending on who is using the term, the flight control system may also include the controls which send commands to the computer such as the joystick, and the parts of the aircraft which are controlled by the computer such as the flaps.

In order to reduce confusion to a minimum, in the rest of this book the term "product" is used to denote a nontrivial piece of software. There are two reasons for this. The first is simply to obviate the program/system confusion by using a third term. The second reason is more important. In this book, software production is viewed as a process, and the end result of a process is termed a product. *Software production* then consists of two phases: *software development* followed by *maintenance*. Finally, the term "system" is used in its modern sense, that is, the combined hardware and software, or as part of universally accepted phrases such as "operating system" and "management information system."

The terms "method" and "technique" are used in this book as follows. A way of carrying out a complete phase such as design or integration is termed a *method*. For example, the finite state machine approach (Section 7.4) is a specifications method. On the other hand, a *technique* is used for a portion of a phase. Black-box testing (Section 6.10) is therefore a technique, because module testing is just one aspect of the implementation phase.

One term that is avoided as far as possible is *bug*. The first use of bug to denote a fault is attributed to Captain (later Rear Admiral) Grace Murray Hopper, the designer of COBOL. One day in the 1940s, a moth apparently flew into a computer and lodged between the contact plates of a relay. Thus there was literally a bug in the system. The term "bug" nowadays is simply a euphemism for *mistake*. While there is generally no real harm in using euphemisms, the word "bug" has overtones that are not conducive to good software production. Specifically, instead of saying, "I made a mistake," a programmer will say, "A bug crept into the code" (not *my* code, but *the* code), thereby transferring responsibility for the mistake from the programmer to the bug. No one blames a programmer for coming down with a case of influenza, because the flu is caused by the flu bug. Referring to a mistake as a bug is a way of casting off responsibility. In contrast, the programmer who says, "I made a mistake," is a computer professional who takes responsibility for his or her actions.

If the reader finds the word "bug" in this book, the author is in no way responsible; it is due solely to the printer's devil!

Chapter Review

Software engineering is defined (Section 1.1) not as a branch of engineering (Section 1.2), but rather as a discipline whose aim is the production of quality software, software that satisfies its requirements, that is delivered on time, and within budget. In order to achieve this goal, appropriate techniques have to be used in all phases of software production, including specifications and design (Section 1.5) and maintenance (Section 1.4). Software engineering addresses all phases of the life cycle and incorporates aspects of many different areas of human knowledge, including economics (Section 1.3) and the social sciences (Section 1.6). In the final section, the terminology used in this book is explained (Section 1.7).

For Further Reading

Barry Boehm is an invaluable source of further information on the scope of software engineering [Boehm, 1976; Boehm and Papaccio, 1988]. Other excellent papers on that topic include [Mills, 1976], [Ramamoorthy, Prakash, Tsai, and Usuda, 1984], [Goldberg, 1986], [Mathis, 1986], and [Card, McGarry, and Page, 1987]. [Vick and Ramamoorthy, 1984] is another useful source.

For an introduction to the importance of economics in software engineering, [Emery, 1974] and [Boehm, 1981] should be consulted. A controversy as to whether software costs really consume 80% of the total computing budget can be found in [Cragon, 1982] and [Boehm, 1983].

With regard to the social sciences and software engineering, [Weinberg, 1971] and [Shneiderman, 1980] contain a wealth of information. Neither book requires prior knowledge of psychology or the behavioral sciences in general. A review of the literature regarding human factors in software engineering can be found in [Laughery and Laughery, 1985]. Other sources of information on human factors include [Rubinstein and Hersh, 1984] and [Norman and Draper, 1986].

Brooks' classic book *The Mythical Man-Month* [Brooks, 1975] is a highly recommended introduction to the realities of software engineering. The book includes aspects of all the topics mentioned previously.

Problems

1.1 You are in charge of developing a new product and ensuring that it will be delivered on time and within budget. Your development budget is $100,000. How much money would you devote to each phase of the life cycle?

1.2 You are a software engineering consultant. The president of a shoe-manufacturing plant wants you to develop a product that will carry out all the accounting functions of the plant and will also provide online information regarding outstanding orders. The product will be used by four accounts clerks and two salesclerks, while two managers must be able to access data when required. The president is willing to pay $25,000 for the hardware and the software together and wants the complete product in four weeks. What do you tell him? Bear in mind that as a consultant you want his business, no matter how unreasonable his request may seem.

1.3 You are the manager of a toy company and decide to call in a software development organization to develop the software to control a robot you have recently purchased to assemble toys. In order to protect your company, what sort of clauses do you insist must be included in the contract with the software developers?

1.4 You are again a software engineer and your job is to supervise the development of the software of Problem 1.3. List ways that your company can fail to satisfy the contract with the toy manufacturer. What are the probable causes of such failures?

1.5 The cost of correcting a software fault in a large air-traffic control system 2 years after delivery was determined to be $5400. The cause of the fault was an ambiguous sentence in a specifications document. Approximately how much would it have cost to have corrected the fault during the specifications phase?

1.6 Suppose that the fault in Problem 1.5 had been detected during the implementation phase. Approximately how much would it have cost to have fixed it then?

1.7 On average, software maintenance consumes over 50% of the total computer budget. What fraction of the total budget is applied to module coding? What fraction of the total budget is applied to perfective maintenance?

1.8 Look up the word "system" in a dictionary. How many different definitions are there? Write down those definitions that can be used in the context of software engineering.

1.9 (Term Project) Suppose that the product described in the Appendix has been implemented exactly as described. In particular, the number of each store is represented by a three-digit integer. Now Plain Vanilla

Ice Cream Corporation (PVIC) expands nationally. Discuss the changes that have to be made to the product and its files when PVIC opens its thousandth store. Also, describe what happens when the number of stores in a state becomes large enough for the state to be split into two regions.

1.10 (Readings in Software Engineering) Your instructor will distribute copies of [Boehm, 1976] and [Boehm and Papaccio, 1988]. Describe the ways that the scope of software engineering has changed between 1976 and 1988.

References

[Boehm, 1973] B. W. BOEHM, "Software and its Impact: A Quantitative Assessment," *Datamation* **19** (May 1973), pp. 48–59.

[Boehm, 1976] B. W. BOEHM, "Software Engineering," *IEEE Transactions on Computers* **C-25** (December 1976), pp. 1226–1241.

[Boehm, 1979] B. W. BOEHM, "Software Engineering, R & D Trends and Defense Needs," in: *Research Directions in Software Technology*, P. Wegner (Editor), The MIT Press, Cambridge, MA, 1979.

[Boehm, 1980] B. W. BOEHM, "Developing Small-Scale Application Software Products: Some Experimental Results," *Proceedings of the Eighth IFIP World Computer Congress*, October 1980, pp. 321–326.

[Boehm, 1981] B. W. BOEHM, *Software Engineering Economics*, Prentice-Hall, Englewood Cliffs, NJ, 1981.

[Boehm, 1983] B. W. BOEHM, "The Hardware/Software Cost Ratio: Is it a Myth?" *IEEE Computer* **16** (March 1983), pp. 78–80.

[Boehm and Papaccio, 1988] B. W. BOEHM AND P. N. PAPACCIO, "Understanding and Controlling Software Costs," *IEEE Transactions on Software Engineering* **14** (October 1988), pp. 1462–1477.

[Brooks, 1975] F. P. BROOKS, JR., *The Mythical Man-Month: Essays on Software Engineering*, Addison-Wesley, Reading, MA, 1975.

[Card, McGarry, and Page, 1987] D. N. CARD, F. E. MCGARRY, AND G. T. PAGE, "Evaluating Software Engineering Technologies," *IEEE Transactions on Software Engineering* **SE-13** (July 1987), pp. 845–851.

[Cragon, 1982] H. G. CRAGON, "The Myth of the Hardware/Software Cost Ratio," *IEEE Computer* **15** (December 1982), pp. 100–101.

[Daly, 1977] E. B. DALY, "Management of Software Development," *IEEE Transactions on Software Engineering* **SE-3** (May 1977), pp. 229–242.

[Elshoff, 1976] J. L. ELSHOFF, "An Analysis of Some Commercial PL/I Programs," *IEEE Transactions on Software Engineering* **SE-2** (June 1976), pp. 113–120.

[Emery, 1974] J. C. EMERY, "Cost/Benefit Analysis of Information Systems," in: *System Analysis Techniques*, J. D. Cougar and R. W. Knapp (Editors), John Wiley and Sons, New York, NY, 1974, pp. 395–425.

[Fagan, 1974] M. E. FAGAN, "Design and Code Inspections and Process Control in the Development of Programs," Technical Report IBM-SSD TR 21.572, IBM Corporation, December 1974.

[Goldberg, 1986] R. GOLDBERG, "Software Engineering: An Emerging Discipline," *IBM Systems Journal* **25** (No. 3/4, 1986), pp. 334–353.

[Laughery and Laughery, 1985] K. R. LAUGHERY, JR., AND K. R. LAUGHERY, SR., "Human Factors in Software Engineering: A Review of the Literature," *Journal of Systems and Software* **5** (February 1985), pp. 3–14.

[Lientz, Swanson, and Tompkins, 1978] B. P. LIENTZ, E. B. SWANSON, AND G. E. TOMPKINS, "Characteristics of Application Software Maintenance," *Communications of the ACM* **21** (June 1978), pp. 466–471.

[Mathis, 1986] R. F. MATHIS, "The Last 10 Percent," *IEEE Transactions on Software Engineering* **SE-12** (June 1986), pp. 705–712.

[Mills, 1976] H. D. MILLS, "Software Development," *IEEE Transactions on Software Engineering* **SE-2** (September 1976), pp. 265–273.

[Neumann, 1980] P. G. NEUMANN, Letter from the Editor, *ACM SIGSOFT Software Engineering Notes* **5** (July 1980), p. 2.

[Norman and Draper, 1986] D. A. NORMAN AND S. W. DRAPER (Editors), *User Centered System Design: New Perspectives on Human–Computer Interaction*, Lawrence Erlbaum Associates, Hillsdale, NJ, 1984.

[Ramamoorthy, Prakash, Tsai, and Usuda, 1984] C. V. RAMAMOORTHY, A. PRAKASH, W.-T. TSAI, AND Y. USUDA, "Software Engineering: Problems and Perspectives," *IEEE Computer* **17** (October 1984), pp. 191–209.

[Rubinstein and Hersh, 1984] R. RUBINSTEIN AND H. M. HERSH, *The Human Factor: Designing Computer Systems for People*, Digital Press, Burlington, MA, 1984.

[Shneiderman, 1980] B. SHNEIDERMAN, *Software Psychology: Human Factors in Computer and Information Systems*, Winthrop Publishers, Cambridge, MA, 1980.

[Stephenson, 1976] W. E. STEPHENSON, "An Analysis of the Resources Used in Safeguard System Software Development," Bell Laboratories, Draft Paper, August 1976.

[Vick and Ramamoorthy, 1984] C. R. VICK AND C. V. RAMAMOORTHY (Editors), *Handbook of Software Engineering*, Van Nostrand Reinhold, New York, NY, 1984.

[Weinberg, 1971] G. M. WEINBERG, *The Psychology of Computer Programming*, Van Nostrand Reinhold, New York, NY, 1971.

[Zelkowitz, Shaw, and Gannon, 1979] M. V. ZELKOWITZ, A. C. SHAW, AND J. D. GANNON, *Principles of Software Engineering and Design*, Prentice-Hall, Englewood Cliffs, NJ, 1979.

Chapter 2 Software Production and Its Difficulties

Different organizations have different ways of producing software. For example, some organizations consider that the software they produce is self-documenting, that is, the product can be understood simply by reading the source code. Other organizations are documentation-intensive. Careful plans are drawn up before any design activities commence, designs are checked and rechecked before coding commences, and detailed descriptions of each module are given to the programmer. Test cases are preplanned, the result of each test run is logged, and the test data are carefully filed away. Once the product enters the maintenance phase, any suggested change must be proposed in writing, with detailed reasons given for making the change. Once written authorization for making the change has been given, the change may be made, but the modified software may not be integrated into the product until the documentation for the product has been updated and the changes to the documentation approved.

Intensity of testing is another measure by which different organizations can be compared. Some organizations devote up to half their software budgets in testing software, while others feel that only the user can thoroughly test a product, so they devote minimal time and effort in testing the product but spend a considerable amount of time fixing faults detected by the user.

Maintenance is a major preoccupation of many software organizations. Some software that is 5 or 10 years old is continually enhanced to meet changing needs; in addition, residual faults continue to appear, even after the software has been in successful operations mode for many years. In contrast, other organizations are essentially concerned with research, leaving development, let alone maintenance, to others. This applies particularly to university computer science departments, where graduate students may build products essentially to prove that a particular design or technique is feasible. The commercial exploitation of the validated concept is then left to other organizations.

But regardless of the exact procedure followed, software production broadly follows the phases outlined in Section 1.4, namely requirements, specifications, design, implementation, integration, maintenance, and, finally, retirement. Some of these phases are known by different names. For example, the *requirements* phase is sometimes called *systems analysis*. Once the client agrees that the product has passed its acceptance test, the product can be used to perform useful work. For this reason, the *maintenance* phase is sometimes referred to as *operations mode*. Furthermore, certain phases may be subdivided; the *design* phase is usually broken down into *architectural design* followed by *detailed design*.

In the preceding list of phases there is no separate testing phase. This omission is deliberate. Testing is not a separate phase, but a process that takes place all the way through software production. The requirements have to be tested, the specifications have to be tested, the design has to be tested, and so on. There are times in the life cycle when testing is carried out to the almost total exclusion of other activities. This occurs towards the end of each phase (*verification*), and is especially true before the product is handed over to the client (*validation*). While there are times when testing predominates, there should never be times when no testing at all is being performed. If testing is treated as a separate phase, then there is a very real danger that testing will not be carried out continually during the entire product development and maintenance life cycle.

In this chapter the phases through which a product passes are described, together with potential difficulties that may arise during each phase. In addition, testing procedures appropriate to each phase are described. Finally, in Section 2.9 inherent difficulties of software production are presented. Solutions to the difficulties associated with the production of software are usually nontrivial, and the rest of this book is devoted to describing suitable techniques. In this chapter, then, only the difficulties are highlighted, but the reader is informed in which sections or chapters solutions are put forward. Thus this chapter is not only an overview of the software production process, but is also a guide to much of the book as a whole.

2.1 Client, Developer, and User

Some preliminary definitions are needed here. The *client* is the individual or organization who wants a product to be developed. The *developer* is the individual or organization who will be responsible for producing the product the client needs. The developer may be responsible for every aspect of the process, from requirements onwards. Alternatively, the developer may be responsible only for the implementation of a product designed by another developer. The term "software development" covers all aspects of software production before the product enters the *maintenance* phase. Any task that is a step towards building a piece of software, including planning, designing, testing, or documenting, constitutes software development. After it has been developed, the software is maintained.

The client and developer may both be part of the same organization. For example, the client may be the head actuary of an insurance company, and the developer the vice-president for management information systems of that insurance company. On the other hand, the client and developer may be two totally independent organizations. For instance, the client may be the Department of Defense, and the developer a major defense contractor. On a much smaller scale, the client may be an attorney in a one-person practice, and the developer a graduate student who earns income to finance his or her studies by writing software.

A third party involved in software production is the *user*. The user is the person or persons on whose behalf the client has commissioned the developer, and who will utilize the software once it is complete. In the insurance company example, the users may be insurance agents who will use the software to select the most appropriate policy for each client. In some instances the client and the user will be the same person, as with the attorney discussed previously.

The seven phases of software production are now presented, and the role played by testing in that phase carefully analyzed. The first phase is the requirements phase.

2.2 Requirements Phase

Software development is an expensive process. A software product will be built only if a client feels a need for that software. The development process therefore begins when the client approaches a developer with regard to a piece of software which, in the opinion of the client, is either essential to the profitability of his or her enterprise or can be economically justified in some other way. At any stage of the process, if the client should no longer believe that developing the software is cost-effective, then development will immediately terminate. Throughout this chapter the assumption is made that the client feels that the product is cost-justifiable.

At an initial meeting between client and developer, the client will outline the product as he or she conceptualizes it. In general, the client will not be a software engineer, so from the viewpoint of the developer the way that the client describes the desired product may be vague, unreasonable, contradictory, or simply impossible to achieve. The task of the developer at this stage is to determine exactly what it is that the client wants and to find out from the client what the constraints on the development process are. Typical constraints are cost and time deadline ("the complete product must cost less than $370,000 and must be completed within 14 months"), but a variety of other constraints are often present such as reliability ("the product must be operational 99% of the time") or the size of the object code ("it has to run on my personal computer"). This preliminary investigation is sometimes called *concept exploration*.

In succeeding meetings between members of the development team and client team, the functionality of the proposed product is successively refined and analyzed from the viewpoint of feasibility and financial justification. This process is sometimes called *systems analysis*. In this book that term is not used, primarily because of the confusion surrounding the word "system" (see Section 1.7).

Up to now, everything seems to be straightforward. Unfortunately, the requirements phase is frequently carried out improperly. When the product is finally delivered to the user, perhaps a year or two after the specifications have been signed off by the client, the client may call the developer and say, "I know that this is what I asked for, but it isn't really what I wanted." What the client asked for, and therefore, what the developer thought the client wanted, was not what the client really *needed*. There can be a number of reasons for this. First, the client may not really understand what is going on in his or her own organization. For example, it is no use asking a software developer for a faster operating system if the slow turnaround is actually being caused by a badly designed database. Or, if the client is running a chain of retail stores that is losing money, the client may ask for a financial management information system that reflects such items as sales, salaries, accounts payable, and accounts receivable. Such a product will be of little use if the real reason for the losses is shrinkage (shoplifting and employee theft). If that is the case, then a stock control product rather than a financial control product is required.

But the major reason that the client frequently asks for the wrong product is that software is complex, and it is difficult for a software professional to visualize a piece of software and its functionality, let alone a client who is barely computer-literate. One way of coping with this second difficulty is prototyping, that is, constructing a working model of the product with which the client can experiment. This topic is discussed in the following section.

2.2.1 Requirements Phase Testing

A *rapid prototype* is a piece of software hurriedly put together by the developer that incorporates much of the functionality of the target product, but that omits those aspects generally invisible to the client, such as file updating or error handling. The client then experiments with the prototype to determine whether the product as perceived by the developer indeed performs what the client needs. If necessary, the prototype is changed until the client is satisfied that it encapsulates the functionality of the target product. Prototyping is discussed in detail in Section 3.3.

2.3 Specifications Phase

Once the developer feels that the client's needs have been understood, then a document is drawn up, the *specifications*. The specifications spell out the functionality of the product, that is, exactly what the product is supposed to do, and list any constraints that the product must satisfy. The specifications document will include the inputs to the product and the required outputs. For example, if the product the client needs is a payroll product, then the inputs will include the pay scales of each employee, data from a time clock, as well as information from personnel files so that taxes can be computed correctly. The outputs will be paychecks and reports such as Social Security deductions. In addition, the specifications will include stipulations that the product must be able to handle correctly a wide range of deductions such as medical insurance payments, union dues, and pension fund contributions.

The specifications of the product constitute a contract. The software developer will be deemed to have completed the contract when a product that satisfies the *acceptance criteria* of the specifications has been delivered. For this reason, the specifications should not include imprecise terms such as "suitable," "convenient," "ample," or "enough," or terms that sound exact but which in practice are equally imprecise, such as "optimal," or "98% complete." Even though there may not be the slightest chance of the specifications forming the basis for a lawsuit, such as when the client and developer are part of the same organization, specifications should always be written as if they will be used as evidence in a trial.

Difficulties can arise in the specifications phase. One possible error that can be made by the specifications team is that the specifications may be *ambiguous*—certain sentences or sections may have more than one possible valid interpretation. Consider the specification, "A part record and a plant record are read from the database. If it contains the letter A followed by the letter Q, then compute the cost of transporting that part to the plant." To what does the "it" in the preceding sentence refer: the part record or the plant record? In fact, the "it" could conceivably even refer to the database!

The specifications may also be *incomplete*, that is to say, some relevant fact may have been omitted. For instance, the specifications may not state what actions are to be taken if the input data contain errors. Furthermore, the specifications may be *contradictory*. For example, in one place in the specifications document for a product to control a fermentation process, it is stated that if the pressure exceeds 35 psi, then valve M17 must immediately be shut. However, in another place it is stated that if the pressure exceeds 35 psi, then the operator must immediately be alerted; only if no remedial action has been taken by the operator within 30 seconds should valve M17 be shut automatically. Software development cannot proceed until the specifications have been corrected. Ways of reducing faults in specifications are described in Chapter 7.

Once the specifications have been completed and checked, the next task is for management to draw up the software project management plan (SPMP), as described in Chapter 4. Major components of the plan are the deliverables (what the client is going to get), the milestones (when the client gets it), and the budget (how much it is going to cost).

2.3.1 Specifications Phase Testing

The software quality assurance (SQA) group must play a role right from the start of the development process. The job of the SQA group is to ensure that the delivered product is what the client ordered, that is, that the product has been correctly built in every way. As pointed out in Chapter 1, a major source of faults in delivered software is faults in specifications that are not detected until the software has gone into operations mode, that is, being used by the client's organization for the purpose for which it was developed. Before the specifications phase can be deemed to be finished, the SQA group must carefully inspect the specifications, looking for contradictions, ambiguities, and incompleteness. In addition, the SQA group must ensure that the specifications are feasible, for example, that any specified hardware component is fast enough, or that the client's current online disk storage capacity is adequate for handling the new product. If a specifications document is to be testable, then one of the properties it must have is that it must be *traceable*. That is to say, every statement in the specifications can be traced back to a statement made by the client team. If the requirements have been methodically presented, properly numbered, cross-referenced, and indexed, the SQA group should not have too much difficulty tracing through the specifications document and ensuring that it is indeed a true reflection of the client's requirements.

An excellent way of checking the specifications is by means of a review. The requirement team is present, as are representatives of the client organization. The meeting is often chaired by a member of the SQA group, who must also be present. The aim of the review is to ensure that the specifications are correct. The reviewers go through the specifications, ensuring that there are no misunderstandings as to what is meant by each statement of the specifications. Walkthroughs and inspections are two types of reviews described in Section 6.6.

A common cause of later difficulties is when both client and developer agree on a statement, but each has understood the statement in a different way. This is another reason that has been put forward for using prototyping; it is difficult to misunderstand a working model. Once the specifications have been approved and signed off by the client and the software project management plan has been completed, the next stage is to design the product.

2.4 Design Phase

The specifications of a product spell out *what* the product is to do. The aim of the design phase is to determine *how* the product is to do it. Starting with the specifications, the design team determines the internal structure of the product. At the design phase, algorithms have to be selected and data structures chosen. The inputs to and outputs from the product are laid down in the specifications, as are all other external aspects of the product. At the design phase, the internal data flows are determined. The design team breaks down the product into *modules*, independent pieces of code with a well-defined interface to the rest of the product. For each module, the designer specifies what it has to do, and how it is to do it. The interface of each module, that is, the parameters passed to the module and the parameters returned by the module, must be specified in detail. For example, a module might measure the water level in a nuclear reactor and cause an alarm to be sounded if the level is too low. A module in an avionics product might take as input two or more sets of coordinates of an incoming enemy missile, compute its trajectory, and advise the pilot as to possible evasive action.

While this decomposition is being performed, the design team must keep a careful record of all design decisions that are taken. This information is essential for two reasons. First, while the product is being designed there will be times when a dead end is reached, and the design team feels the need to backtrack and redesign certain pieces. Having a written record of why specific decisions were taken will assist the team when this occurs and help them to get back on track. The second reason for keeping the design decisions is concerned with maintenance. Ideally, the design of the product should be open-ended, that is to say, it should be done in such a way that no matter how the product is to be enhanced during maintenance, this can be done simply by adding new modules or replacing existing modules; the design as a whole will not be affected. Of course, in practice this is difficult to achieve. Deadline constraints in the real world are such that designers struggle against the clock to complete a design that will satisfy the original specifications, without worrying about any future enhancements. If future enhancements to be added after the product has gone into operations mode are listed in the specifications, then these must be catered for in the design, but this is extremely rare. In general, the specifications will deal only with present requirements, and the design will be appropriately oriented. Another reason why the design cannot possibly cope with all possible future enhancements is that there is no way of determining, when the product is still at the design phase, what all possible enhancements might be. And finally, if the design has to take *all* future possibilities into account, at best it will be unwieldy; at worst it will be so complicated that implementation will be impossible. So the designers have to compromise, putting together a design that can be extended in many reasonable ways without the need for redesign. But in a product that undergoes major enhancement, the time will come when the design simply cannot handle further enhancements. When this stage is reached, the product as a whole has to be redesigned. If the redesign team has at its disposal the reasons for all the design decisions that were originally taken, this can frequently prove to be of great assistance.

The major output from the design phase is the *design* itself. This consists of two parts, *architectural design*, a description in terms of modules of the design as a whole, and *detailed design*, a description of each module. The latter descriptions are handed to the programmers for implementation. Design methodologies are described in Chapter 9, together with ways of describing the design such as graphics, pseudocode, flowcharts, structured English, and so on.

2.4.1 Design Phase Testing

As mentioned in Section 2.3.1, a critical aspect of testability is *traceability*. In the case of the design, this means that every part of the design can be linked to a statement in the specifications. If the design is suitably cross-referenced, then the SQA group will be able to go through the design and have a powerful tool for checking not just that the design is in accordance with the specifications, but that every statement of the specifications is reflected in some part of the design.

The design must be subjected to design reviews, which are similar to the reviews that the specifications undergo, but in this case there is no need for the client to be present. Members of the design team and the SQA group work through the design as a whole as well as each separate module, ensuring that the design is correct. The types of faults to look for include: logic faults, interface faults, lack of exception handling (processing of error conditions), and, most important, nonconformance to the specifications. In addition, the review team should always be aware of the possibility that some specification faults were not picked up during the previous phase. A detailed description of design reviews is given in Section 6.7.

2.5 Implementation Phase

During the implementation phase, the various component modules of the product are coded. Coding is discussed in detail in Chapter 10. The major documentation associated with the coding is the source code itself, suitably commented. But the programmer should always be thinking about future maintenance, so additional documentation should be provided to assist in maintenance. This will include all test cases against which the code was tested, the expected results, and the actual output. This will be used in regression testing, as explained in Section 2.7.1.

2.5.1 Implementation Phase Testing

The modules are tested while they are being implemented (*desk testing*), and after they have been implemented they are run against test cases. This testing is done by the programmer. Thereafter, the quality assurance group tests the modules methodically. Testing techniques are discussed in detail in Chapter 6.

A review is a powerful and successful technique for detecting programming faults. Here the programmer guides the members of the review team through the listing of the module. The review team must include a SQA representative. The procedure is similar to reviews of specifications and designs described previously.

2.6 Integration Phase

The next stage is to put the modules together and determine whether the product as a whole functions correctly. The way in which the modules are integrated (all at once or one at a time) and the specific order (from top to bottom in the module interconnection diagram or bottom to top) can have a critical influence on the quality of the resulting product. For example, suppose the product is integrated bottom-up. If there is a major design fault, then it will show up late, necessitating an expensive rewrite. Conversely, if the modules are integrated top-down, then the lower level modules usually will not receive as thorough a testing as would be the case if the product were integrated bottom-up. These and other problems are discussed in detail in Section 6.14.

2.6.1 Integration Phase Testing

The purpose of *integration testing* is to check that the modules combine together correctly to achieve a product that satisfies its specifications. During integration testing, particular care must be paid to testing the interfaces. It is important that the number, order, and types of formal parameters match the number, order, and types of actual parameters. Of course, this checking is best performed by the linker, but since the majority of linkers cannot do this, the SQA group must. In passing, it should be noted that all Ada linkers perform strong type checking of every interface, that is to say, every formal and actual parameter pair is checked to ensure that the type of the formal parameter is identical to that of the corresponding actual parameter.

When the integration testing has been completed, *product testing* is performed by the SQA group. The functionality of the product as a whole is checked against the specifications. In particular, the constraints listed in the specifications document must be tested. A typical example is whether the response time is fast enough. Not only must the correctness of the product be tested, but also its robustness: Intentionally erroneous input data are submitted to see whether the product will crash, or whether its error handling capabilities are adequate for dealing with bad data. If the product is to be run on a mainframe, together with the client's currently installed software, then tests must also be performed to check that the new product will not have an adverse impact on the client's existing computer operations. Then a check must also be made that the source code and all other types of documentation are complete and consistent with one another. Product testing is discussed in Section 6.15.

The final aspect of integration testing is *acceptance testing*. Here the client reenters the picture. The software is delivered to the client, who tests the software on the actual hardware on which it is to run, using actual data, as opposed to test data. No matter how careful the development team or the SQA group might be, there is a significant difference between test cases, which by their very nature are artificial, and actual data. A software product cannot be considered to satisfy its specifications until the product has passed its acceptance tests. More details about acceptance testing are given in Section 6.16.

2.7 Maintenance Phase

Once the product has been accepted by the client, any changes of any nature constitute maintenance. Maintenance is not an activity that is grudgingly carried out after the product has gone into operations mode. On the contrary, it is an integral part of the life cycle that must be planned for from the beginning. As explained in Section 2.4, the design should, as far as possible, take future enhancements into account. Coding must be performed with a constant eye kept on the implications of future maintenance. After all, as pointed out in Section 1.4, more money is spent on maintenance than on all other software activities combined. It is therefore almost always the most important aspect of software production. Maintenance must never be treated as an afterthought. Instead, the entire software development effort must be carried out in such a way as to minimize the impact of the inevitable future maintenance.

A frequent problem with maintenance is documentation, or rather a lack of it. In the course of developing software against a time deadline, the original specifications and design are frequently not updated, and are consequently almost useless to the maintenance team. Other documentation, such as the database manual or the operating manual, may never have been written, because management decided that getting the product out to the client on time was more important than developing the documentation in parallel with the software. In many instances, the source code is the only documentation available to the maintainer. The high rate of personnel turnover in the industry exacerbates the maintenance situation in that none of the original developers may be working for the organization at the time when maintenance is performed.

Maintenance is usually the most important phase of software production. It is also the most difficult, for the reasons stated previously and for the additional reasons given in Chapter 11.

2.7.1 Maintenance Phase Testing

There are two aspects of the testing of changes to the product. The first is ensuring that the changes have been correctly implemented, that is to say, that the coding which was required is correct. The second aspect is making sure that, in the course of making the required changes to the product, no other inadvertent changes were made. Thus, once the programmer has ensured that the desired changes have been implemented, the product must be tested against previous test cases to make certain that the functionality of the rest of the product has not been compromised. This procedure is called *regression testing*. In order to be able to perform regression testing, it is necessary that all previous test cases be retained, together with the results of running those test cases. Testing during the maintenance phase is discussed in greater detail in Section 6.17.

2.8 Retirement

The final phase in the life cycle of a software product is retirement. After many years of service, a stage is reached when any further maintenance would not be cost-effective. One way that this can happen is that the proposed changes are so drastic that the design as a whole would have to be changed, and it would be cheaper to redesign and recode the entire product from scratch. Another way is that so many changes may have been made to the original design that interdependencies have inadvertently been built into the product, so much so that there is a real danger that even a small change to one minor module might have a drastic effect on the functionality of the product as a whole. Third, the documentation may not have been adequately maintained, thus increasing the risk of a regression fault to the point where it would be safer to recode than to maintain. A fourth possibility is that the hardware on which the product runs is to be replaced by a different machine with a different operating system, and it is cheaper to rewrite from scratch than to modify the product. In each of these instances the current version is replaced by a new version, and the life cycle continues.

True retirement, on the other hand, is a somewhat rare event that occurs when a product has outgrown its usefulness. The client organization no longer requires the functionality provided by the product, and it is finally removed from the computer on which it has been in operations mode for many years.

After this review of the complete life cycle, together with some of the difficulties attendant on each phase, difficulties associated with software production as a whole are now presented.

2.9 Problems with Software Production: Essence and Accidents

Over the past 30 years, hardware has become cheaper and faster. It was mentioned in Section 1.6 that the performance–price factor has decreased by an order of magnitude six times over the past 30 years. Furthermore, hardware has shrunk in size. In the 1950s, companies paid hundreds of thousands of preinflation dollars for a machine as large as a room that was no more powerful than today's desktop personal computers selling for under $1000. It would seem that this trend is inexorable, and computers will continue to become smaller, faster, and cheaper.

Unfortunately, this is not the case. There are a number of physical constraints that must eventually impose limits on the possible future size and speed of hardware. The first of these constraints is the speed of light. The electrons in a digital computer, or more precisely, the electromagnetic waves, simply cannot travel faster than 186,300 miles per second. One way to speed up a computer is to miniaturize its components. In that way, the electrons have smaller distances to travel to accomplish the same tasks. But there are also limits to decreasing the size of a component. An electron travels along a path that can be as narrow as three atoms in width. But if the path along which the electron is to travel is any narrower than that, then the electron can stray onto an adjacent path. In addition, parallel paths may not be too close for the same reason. Thus the speed of light and the nonzero width of an atom result in future physical limits on hardware size and speed. We are nowhere near these limits yet—computers can easily become at least two orders of magnitude faster and smaller without impinging on these physical limits. But there are intrinsic facts of nature that will eventually prevent computers becoming arbitrarily fast or arbitrarily small.

Now what about software? Software is intrinsically conceptual, and therefore nonphysical, although it is of course always stored on some physical medium such as paper or magnetic disk. Superficially, it might appear that with software anything is possible. But Brooks, in a landmark article entitled "No Silver Bullet" [Brooks, 1986], exploded this belief. He argued that, analogous to hardware speed and size limits that cannot physically be exceeded, there are inherent problems with current techniques of software production that can never be solved. To quote Brooks, "... building software will always be hard. There is inherently no silver bullet."

The title of Brooks' article refers to the recommended way of slaying werewolves, otherwise perfectly normal human beings who suddenly turn into wolves. Brooks' line of inquiry is to determine whether a similar silver bullet can be used to solve the problems of software. After all, software usually appears to be innocent and straightforward, but like a werewolf, software can be transformed into something horrifying, in the shape of late deadlines, exceeded budgets, and residual design faults not detected during testing.

Recall that the title of Brooks' article is "*No* Silver Bullet" (author's italics). That is to say, Brooks' theme is that the very nature of software is such that it is highly unlikely that a silver bullet will be discovered which will magically solve all the problems of software production, let alone help to achieve software breakthroughs that are comparable to those that have occurred with unfailing regularity in the hardware field. He divides the difficulties of software into two categories, *essence*, the difficulties inherent in the very nature of software, and *accidents*, the difficulties that are encountered today but are not inherent in software production. That is, essence constitutes those aspects of software production that probably cannot be changed, while accidents are amenable to research breakthroughs, or silver bullets.

What then are the aspects of software production that are inherently difficult? Brooks lists four, which he terms complexity, conformity, changeability, and invisibility. Brooks' use of the word "complexity" is somewhat unfortunate in that the term has many different meanings in computer science in general and software engineering in particular. In the context of his article, Brooks uses the word "complex" in the sense of *complicated* or *intricate*. In fact, the names of all four aspects are used in their nontechnical sense.

Each of the four aspects is now examined in turn.

2.9.1 Complexity

Software is more complex than any other construct made by human beings. Even hardware is almost trivial compared to software. To see this, consider a 16-bit word W in the main memory of a digital computer. Since each of the 16 bits comprising the word can take on exactly two values, namely 0 and 1, word W as a whole can be in any of 2^{16} different states. If we have two words, W_1 and W_2, each 16 bits in length, then the number of possible states of words W_1 and W_2 together is 2^{16} times 2^{16}, or 2^{32}. In general, if a product consists of a number of independent pieces, then the number of possible states of that product is the product of the numbers of possible states of each component.

Now suppose the computer is to be used to run a software product P, and the 16-bit word W is to be used to store the value of an integer X. If the value of X is read in by a statement such as READ(X), then, since the integer X can take on 2^{16} different values, at first sight it might seem that the number of states in which the product could be is the same as the number of states in which the word could be. If the product P consisted only of the single statement READ(X), then the number of states of P would indeed be 2^{16}. But in a realistic, nontrivial software product the value of a variable which is input is later used elsewhere in the product. There is thus an interdependence between the READ(X) statement and statements that use the value of X. The situation is more complex if the flow of control within the product depends on the value of X. For example, X may be the control variable in a **case** statement, or there may be a **for** loop or **while** loop whose termination depends on the value of X. Thus the number of states in any nontrivial product is greater, because of this interaction, than the product of the number of states of each variable. Worse still, this complexity does not grow linearly with the size of the product, but much faster.

Software complexity is an inherent property of software. No matter how a nontrivial piece of software is designed, the pieces of the product will interact. For example, the states of a procedure will depend on the states of its parameters, and the states of global variables (variables that can be accessed by more than one procedure) will also affect the state of the product as a whole. In short, complexity is an essential property, not an accidental one; in Brooks' opinion there is nothing that can be done about it.

Brooks points out that complex phenomena can be described and explained by disciplines such as mathematics and physics because mathematicians and physicists have learned how to abstract the essential features of a complex system, to build a simple model which reflects those essential features alone, and to validate the simple model and to make predictions from it. In contrast, if software is simplified the whole exercise is useless; the simplifying techniques of mathematics and physics work only because the complexities of those systems are accidents, not essence, as is the case with software products.

The consequence of this essential complexity of software is that a product becomes difficult to understand. In fact, it is usually true that no one really understands a large product as a whole. And, if the product as a whole is not understood, then no individual component can be fully understood. This leads to imperfect communication between team members, which, in turn, has as a consequence the time and cost overruns that characterize the development of large-scale software products. In addition, faults in specifications are made simply because of a lack of understanding of all aspects of the product.

This essential complexity affects not only the development process itself, but also the management of the development process. Unless a manager can obtain accurate information regarding the process that he or she is managing, it is difficult to determine personnel needs for the succeeding stages of the project, and budgeting is extremely difficult. Reports to senior management regarding both progress to date and future deadlines are likely to be inaccurate. Drawing up a testing schedule is difficult when neither the manager nor anyone reporting to the manager knows what loose ends still have to be tied. And if a project staffer leaves, trying to train a replacement can be a nightmare.

A further consequence of the complexity of software is that it complicates the maintenance process. As shown in Figures 1.1 and 1.2, about two-thirds of the total software effort is devoted to maintenance. Unless the maintainer really understands the product to be maintained, there is always the danger that either corrective maintenance or enhancement can have the side effect of damaging the product in such a way that further maintenance is required to repair the damage caused by the original maintenance. The possibility of this sort of damage being caused by carelessness is always present, even when the original author makes the change, but it is exacerbated when the maintenance programmer is effectively working in the dark. Poor documentation, or worse, no documentation, or still worse, incorrect documentation, is often a major cause of incorrectly performed maintenance. But no matter how good the documentation may be, software has an inherent complexity that currently transcends all attempts to cope with it, and this complexity will unfavorably impact maintenance.

2.9.2 Conformity

A manually controlled gold refinery is to be computerized. Instead of the plant being controlled by a series of buttons and levers, a computer will send the necessary control signals to the components of the plant. Although the plant is working perfectly, management feels that a computerized control system will increase the yield. The task of the computerization team is to construct a product which will interface with the existing plant. That is to say, the software must conform to the plant, not the plant to the software. This is the first type of conformity identified by Brooks, where software acquires an unnecessary degree of complexity because it has to interface with an existing system.

What if a brand-new computerized gold refinery were to be constructed? At first sight it would appear that the mechanical engineers, metallurgical engineers, and software engineers could sit down together and come up with a plant design in which the machinery and the software fit together in a natural and straightforward manner. In practice, there is generally a perception that it is easier to make the software interface conform to the other components than to change the way the other components have been configured in the past. As a result, even in a new gold refinery the other engineers will insist on designing the machinery as in the past, and the software will be forced to conform to the hardware interfaces. The second type of conformity identified by Brooks which causes unnecessary complexity is that caused by the perceived misconception that the software is the component that is the most conformable.

The problems caused by this forced conformity cannot be removed by redesigning the software, because the complexity is not due to the structure of the software itself. Instead, it is due to the structure of the software caused by the interfaces, to humans or to hardware, imposed on the software designer.

2.9.3 Changeability

As pointed out in Section 1.2, it is considered unreasonable to ask a civil engineer to move a bridge 200 miles or to rotate it through 90°, but it is perfectly acceptable to tell a software engineer to rewrite half an operating system over a 5-year period. Civil engineers know that redesigning half a bridge is expensive and dangerous; it is both cheaper and safer to rebuild from scratch. Software engineers are equally well aware that, in the long run, extensive maintenance is equally unwise, and that rewriting the product will sometimes prove to be cheaper. Nevertheless, major changes to software are frequently demanded.

Brooks points out that there will always be pressures to change software. After all, it *is* easier to change software than, say, the computer on which it runs; that is the reason behind the terms "*soft*ware" and "*hard*ware." In addition, the functionality of a system is embodied in its software, and changes in functionality are achieved through changing the software. It has been suggested that the problems caused by frequent and drastic maintenance are merely problems caused by ignorance, and if the public at large were better educated with regard to the nature of software, then demands for major changes would not occur. But Brooks points out that changeability is a property of the essence of software, an inherent problem that cannot be surmounted. That is to say, the very nature of software is such that, no matter how the public is educated, there will always be pressures for changes in software, and often these changes will be drastic.

There are three reasons why useful software has to undergo change. First, as pointed out in Section 1.4, software is a model of reality, and as the reality changes, so the software must adapt or die. Second, if software is found to be useful, then there are pressures, chiefly from satisfied users, to extend the functionality of the product beyond what is feasible in terms of the original design. Third, successful software survives beyond the lifetime of the hardware for which it was written. In part this is due to the fact that, after 5 years or so, hardware often does not function as well as it did when new. But more significant is the fact that technological change is so rapid that more appropriate hardware components such as larger disks or faster CPUs become available while the software is still viable. The software must therefore be modified in order to run on this new medium.

For these reasons, part of the essence of software is that it has to be changed, and this inexorable and continual change has a deleterious effect on the quality of software.

2.9.4 Invisibility

A major problem with the essence of software is that it is "invisible and unvisualizable" [Brooks, 1986]. Anyone who has been handed a 150-page listing and told to modify the software in some way will know exactly what Brooks is getting at. Unfortunately, there is no acceptable way to represent either a complete product or some sort of overview of the product. In contrast, architects, for example, can provide models or sketches which give an idea of the overall design, as well as blueprints and other detailed diagrams which, to the trained eye, will reflect every detail of the structure to be built. Chemists can build models of molecules, engineers can construct prototypes, and plastic surgeons can use the computer to show potential clients exactly how their faces will look after surgery. Diagrams can be drawn to reflect the structure of silicon chips and other electronic components; the components of a computer can be represented by means of various sorts of schematics, at various levels of abstraction.

Certainly there are ways in which software engineers can represent specific views of their product. For example, a software engineer can draw one directed graph depicting flow of control, another showing flow of data, a third with patterns of dependency, a fourth depicting time sequences. The problem is that few of these graphs are planar, let alone hierarchical. The many crossovers in these graphs are a distinct obstacle to understanding. Parnas [Parnas, 1979] suggests cutting the arcs of the graph until one or more becomes hierarchical. The problem is that the resulting graph, though comprehensible, renders only a subset of the software visualizable, and the arcs that have been cut may be critical from the viewpoint of comprehending the interrelationships between the components of the software.

The result of this inability to represent software visually not only makes software difficult to comprehend, it also severely hinders communication between software professionals—there does not seem to be an alternative to handing a colleague a 150-page listing together with a list of modifications to be made.

It must be pointed out that visualizations of all kinds such as flowcharts, data flow diagrams (Section 7.2), or module interconnection diagrams are extremely useful and powerful ways of visualizing certain aspects of the product. Visual representations are an excellent means of communicating with the client, as well as with fellow software engineers. The problem is that such diagrams can never embody *every* aspect of the product as a whole, nor is there a way of determining what is missing from any one visual representation of the product.

2.9.5 No Silver Bullet?

Brooks' article [Brooks, 1986] is by no means totally gloom-filled. He describes what he considers to be the three major past breakthroughs in software technology, namely high-level languages, time sharing, and software development environments (such as UNIX), but stresses that they solved only accidental, and not essential, difficulties. He evaluates various technical developments that are currently advanced as potential silver bullets, including correctness proofs (Section 6.3), object-oriented design (Section 9.8), Ada

(Chapter 14), and artificial intelligence and expert systems (Chapter 16). Nevertheless, while Brooks feels that some of these approaches may solve remaining accidental difficulties, they are irrelevant from the viewpoint of the essential difficulties.

In order to achieve comparable future breakthroughs, Brooks suggests that we change the way that software is produced. For example, wherever possible software products should be bought off the shelf, rather than custom-built. The use of rapid prototyping (Section 3.3) is for him a major potential source of an order-of-magnitude improvement. For Brooks, the hard part of building software lies in the requirements, specifications, and design phases, not in the implementation phase. Brooks suggests that a prototype be used to provide the client with a working model of the target product which can be tested for correctness, consistency, and usability; other reasons for rapid prototyping are discussed in Chapter 3. The final product is then constructed using the knowledge gained by experimenting with the prototype, and, hopefully, greater use of rapid prototyping may lead to better products. Another suggestion that, in Brooks' opinion, may lead to a major improvement in productivity is greater use of the principle of incremental development, where instead of trying to build the product as a whole it is constructed stagewise. This concept is described in detail in Section 3.4.

But for Brooks, the greatest hope of a major breakthrough in improving software production lies in training and encouraging great designers. As stated previously, in Brooks' opinion one of the hardest aspects of software production is the design phase, and to get great designs, we need great designers. Brooks cites UNIX, APL, Pascal, Modula-2, Smalltalk, and FORTRAN as exciting products of the past. He points out that they have all been the products of one, or a very few, great minds. On the other hand, more prosaic, but useful, products like COBOL, PL/I, ALGOL, MVS/360, and MS-DOS have all been products of committees. Nurturing great designers is for Brooks the most important objective if we wish to improve the current situation with regard to software production.

Parts of Brooks' paper make depressing reading. After all, from the title onwards he states that the inherent nature (or essence) of the current way software is produced makes the finding of a silver bullet a dubious possibility. But he concludes on a note of hope, suggesting that if we change our software production strategies by buying ready-made software wherever possible, using rapid prototyping and incremental building techniques, and attempting to nurture great designers, we may increase software productivity. But an incremental breakthrough, the "silver bullet," is most unlikely.

Brooks' pessimism must be put into perspective. After all, over the past 20 years there has been a steady productivity increase in the software industry of roughly 6% per year. This productivity increase is comparable to what has been observed in many manufacturing industries. What Brooks is looking for is a "silver bullet," a way of rapidly obtaining an order-of-magnitude increase in productivity. It is difficult to disagree with his view that we cannot hope to double productivity overnight. At the same time, the compound growth rate of 6% that has steadily been experienced means that productivity is doubling every 12 years. This improvement may not be as rapid and spectacular as we would like, but the practice of software engineering is steadily improving from year to year.

Chapter Review

The various phases of software production, from requirements (Section 2.2) to retirement (Section 2.8), are outlined, with special attention paid to the problems associated with each phase. Testing is not considered to be a separate phase, because testing must be carried out in parallel with all software production activities. A description of testing activities during each of the production phases is given in Sections 2.2.1 through 2.7.1.

In addition to the problems associated with each separate life-cycle phase, Brooks' views regarding the inherent difficulties of software production as a whole are presented (Section 2.9). Brooks' four aspects of essential software difficulty, namely complexity, conformity, changeability, and invisibility, are described and discussed in Sections 2.9.1 through 2.9.4. Finally, Brooks' viewpoint is analyzed (Section 2.9.5).

For Further Reading

The review articles in the For Further Reading section of Chapter 1, namely [Brooks, 1975], [Boehm, 1976], [Mills, 1976], [Ramamoorthy, Prakash, Tsai, and Usuda, 1984], [Vick and Ramamoorthy, 1984], [Goldberg, 1986], [Mathis, 1986], [Card, McGarry, and Page, 1987], and [Boehm and Papaccio, 1988], also highlight the problems associated with software production. In addition, a number of significant problems in software engineering management are listed in [Thayer, Pyster, and Wood, 1981].

With regard to testing during each of the phases of the life cycle, a good general source is [Myers, 1979]. More specific references are given in Chapter 6 and at the end of that chapter.

No summary such as that presented here can do justice to Brooks' "No Silver Bullet"—the article must be read in the original [Brooks, 1986].

Problems

2.1 Describe a situation in which the client, developer, and user are one and the same person.

2.2 Draw a graph showing the approximate percentage of the working day devoted to testing on a day-to-day basis during the entire life cycle. The horizontal axis must be labeled with the phases, the vertical axis with values from 0% to 100%. Verification and validation activities must be marked as such.

2.3 Consider the requirements phase and the specifications phase. Would it make more sense to combine these two activities into one phase, rather than treating them as two separate phases?

2.4 List the documentation that should be produced during each of the phases of the life cycle.

2.5 The most testing is performed during the integration phase. Would it be better to divide this phase into two separate phases, one incorporating the nontesting aspects, the other all the testing?

2.6 Maintenance is the most important phase of software production and also the most difficult to perform. Nevertheless, it is looked down on by many software professionals, and maintenance programmers are often paid less than developers. Do you think that this is reasonable? If not, how would you change it?

2.7 Why do you think that, as stated in Section 2.8, true retirement is a rare event?

2.8 Brooks says that software production is inherently difficult because of complexity, conformity, changeability, and invisibility. Other areas of human endeavor, including law, medicine, theology, and architecture, are also "difficult." But to what extent is each of these areas affected by complexity, conformity, changeability, and invisibility?

2.9 Brooks considers the most important breakthroughs in software technology to be high-level languages, time sharing, and software development environments. What do *you* consider to be the major software breakthroughs? Give reasons for your answer.

2.10 Draw a detailed flowchart of the software distributed by your instructor. Is it easy to understand? Now make the flowchart planar by deleting all edges that cross over another edge. Is the resulting flowchart easier to understand? How much of the original functionality has been lost? Translate the resulting planar flowchart back into its original programming language and run it. Do you agree with Brooks' statement that software is "invisible and unvisualizable"?

2.11 (Term Project) Discuss how the four essential problems of software production (complexity, conformity, changeability, and invisibility) affect the Plain Vanilla Ice Cream Corporation product described in the Appendix.

2.12 (Readings in Software Engineering) Your instructor will distribute copies of [Brooks, 1986]. Do you agree with Brooks' pessimism, or do you share the optimism of the last paragraph of Section 2.9 of this book?

References

[Boehm, 1976] B. W. BOEHM, "Software Engineering," *IEEE Transactions on Computers* **C-25** (December 1976), pp. 1226–1241.

[Boehm and Papaccio, 1988] B. W. BOEHM AND P. N. PAPACCIO, "Understanding and Controlling Software Costs," *IEEE Transactions on Software Engineering* **14** (October 1988), pp. 1462–1477.

[Brooks, 1975] F. P. BROOKS, JR., *The Mythical Man-Month: Essays on Software Engineering*, Addison-Wesley, Reading, MA, 1975.

[Brooks, 1986] F. P. BROOKS, JR., "No Silver Bullet," in: *Information Processing '86*, H.-J. Kugler (Editor), Elsevier North-Holland, New York, NY, 1986. Reprinted in: *IEEE Computer* **20** (April 1987), pp. 10–19.

[Card, McGarry, and Page, 1987] D. N. CARD, F. E. MCGARRY, AND G. T. PAGE, "Evaluating Software Engineering Technologies," *IEEE Transactions on Software Engineering* **SE-13** (July 1987), pp. 845–851.

[Goldberg, 1986] R. GOLDBERG, "Software Engineering: An Emerging Discipline," *IBM Systems Journal* **25** (No. 3/4, 1986), pp. 334–353.

[Mathis, 1986] R. F. MATHIS, "The Last 10 Percent," *IEEE Transactions on Software Engineering* **SE-12** (June 1986), pp. 705–712.

[Mills, 1976] H. D. MILLS, "Software Development," *IEEE Transactions on Software Engineering* **SE-2** (September 1976), pp. 265–273.

[Myers, 1979] G. J. MEYERS, *The Art of Software Testing*, John Wiley and Sons, New York, NY, 1979.

[Parnas, 1979] D. L. PARNAS, "Designing Software for Ease of Extension and Contraction," *IEEE Transactions on Software Engineering* **SE-5** (March 1979), pp. 128–138.

[Ramamoorthy, Prakash, Tsai, and Usuda, 1984] C. V. RAMAMOORTHY, A. PRAKASH, W.-T. TSAI, AND Y. USUDA, "Software Engineering: Problems and Perspectives," *IEEE Computer* **17** (October 1984), pp. 191–209.

[Thayer, Pyster, and Wood, 1981] R. H. THAYER, A. B. PYSTER, AND R. C. WOOD, "Major Issues in Software Engineering Project Management," *IEEE Transactions on Software Engineering* **SE-7** (July 1981), pp. 333–342.

[Vick and Ramamoorthy, 1984] C. R. VICK AND C. V. RAMAMOORTHY (Editors), *Handbook of Software Engineering*, Van Nostrand Reinhold, New York, NY, 1984.

Part Two

Software Life Cycle

The life cycle is the series of phases, from requirements to maintenance and, finally, retirement, that a software product undergoes. This part of the book is concerned with activities that are carried on throughout the software life cycle.

In Chapter 3 a number of different software life-cycle models are presented, including the waterfall model, the rapid prototyping model, and incremental models. To enable the reader to decide on an appropriate life-cycle model for a specific project, the various life-cycle models are compared and contrasted.

Chapter 4 is entitled Software Planning. Before starting a software project, it is essential to plan the entire operation in detail. Once the project has begun, management must closely monitor progress, noting deviations from the plan and taking corrective action where necessary. In addition to providing a detailed description of a software project management plan, the chapter also addresses the issues of software quality assurance (SQA) and time and cost estimation.

In Chapter 5 the principle of stepwise refinement is described. Stepwise refinement underlies the majority of the techniques described in this book. Since the principle is used in all phases of software production, it is appropriate that it should be presented in Part Two.

One major theme of this book is the importance of maintenance. A second is that testing is not a separate phase that is carried out just before delivering the product to the client or even at the end of each life-cycle phase. Instead, testing must be carried out continually during software development and maintenance. This approach to testing is described in detail in Chapter 6, Testing. A variety of testing methods for each life-cycle phase is presented, then analyzed and compared.

Chapter 3 Software Life-Cycle Models

A software product usually begins as a vague concept such as, "Wouldn't it be nice if the computer could plot our graphs of radioactivity levels," or, "If this corporation doesn't have an exact picture of our cash flow on a daily basis, we will be insolvent in six months," or even, "If we develop and market this new type of spreadsheet we'll make a million dollars!" Once the need for a product has been established, the product goes through a series of development phases. Typically, the product is specified and then designed and implemented. If the client is satisfied, the product is installed, and while it is operational it is maintained. When the product finally comes to the end of its useful life, it is decommissioned. The series of steps through which the product progresses is called the *life cycle*.

The life cycle of every product is different. Some products will spend years in the conceptual stage, perhaps because current hardware is just not fast enough for the product to be viable, or because fundamental research has still to be done before it will be possible to develop an efficient algorithm. Other products will be quickly designed and implemented, and then spend years in the maintenance phase being modified to cater to the users' changing needs. Yet other products will be designed, implemented, and maintained, but after many years of radical maintenance it will be cheaper to develop a completely new product rather than to attempt to patch the current version yet again.

In this chapter a number of different life-cycle models are described. Particular emphasis is placed on the two most widely used models, the waterfall model and the rapid prototyping model. To help shed light on the strengths and weaknesses of these two models, other life-cycle models will be examined, including incremental models and the highly unsatisfactory build-and-fix model presented in the following section.

3.1 Build-and-Fix Model

It is unfortunate that many products are developed using what might be termed the *build-and-fix* approach. The product is constructed without specifications first being drawn up, or any attempt at design. Instead, the developers simply build a product which is reworked as many times as necessary until it satisfies the client. This is shown in Figure 3.1. While this approach may work well on programming exercises 100 or 200 lines long, the build-and-fix approach is totally unsatisfactory for products of any reasonable size. Figure 1.3 (page 13) reflects the fact that the cost of changing a software product is relatively small if the change is made at the requirements, specifications, or design phases, but is unacceptably large if it has to be done when the product has been coded, or worse, if it is already in operations mode. Thus the cost of the build-and-fix approach is far greater than if specifications are drawn up and a design carefully developed.

Instead of the build-and-fix approach, it is essential that before development of a product begins, an overall "game plan" or *life-cycle model*

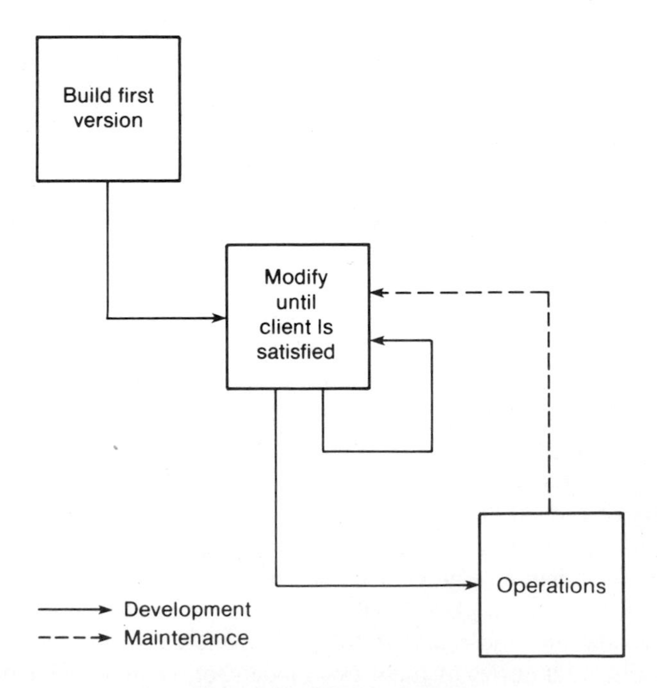

Figure 3.1 Build-and-fix model.

be chosen. The life-cycle model (from now on abbreviated to just "model") includes the various phases of the life cycle, such as requirements, specifications, design, implementation, and maintenance, and the order in which they are to be carried out. As soon as a model has been selected, milestones for the completion of each phase can be fixed. Once the overall plan for developing the product has been agreed to by all parties concerned, building of the product can begin.

Until the early 1980s, the waterfall model was the only widely accepted life-cycle model. This model is now examined in some detail.

3.2 Waterfall Model

The waterfall model was first put forward by Royce [Royce, 1970]. A version of the model appears as Figure 3.2. First, requirements are determined. Then, specifications for the product are drawn up from the requirements, that is to say, a document is produced stating what the product is to do. When the client has signed off the specifications, the design phase begins. In contrast to the specifications which describe *what* the product is to do, the design documents describe *how* the product is to do it.

During the course of developing the design, it sometimes becomes apparent that there is a fault in the specifications. The specifications may be incomplete (some features of the product are simply omitted), contradictory (two or more statements in the specifications define the product in an incompatible way), or ambiguous (the specifications are capable of more than one interpretation). The presence of incompleteness, contradictions, and ambiguities necessitates a revision of the specifications before the software development process can continue. Referring again to Figure 3.2, the arrow from the top of the design box back to the specifications box is a feedback loop. The software production process follows this loop when the developers reevaluate the specifications in the light of information obtained in the course of developing the design. With the client's permission, the necessary changes are then made to the specifications, and the design is adjusted to incorporate these changes. When the developers are finally satisfied, the design documents are handed to the programmers for implementation.

While coding is going on, flaws in the design may appear. For example, the design of a real-time system may prove to be too slow when implemented in code. An example of such a design flaw results from the fact that in FORTRAN the elements of an array B are stored in column-major order, that is to say, in the order B(1, 1), B(2, 1), B(3, 1), . . . , B(N, 1), B(1, 2), B(2, 2), B(3, 2), . . . , B(N, 2), and so on. Suppose a 200 × 200 FORTRAN array B is stored on disk with one column to a block, that is, a 200-word column is read into a buffer in main memory each time a FORTRAN READ statement is executed. The complete array is to be copied from disk into main memory. If the array is read column by column, then exactly 200

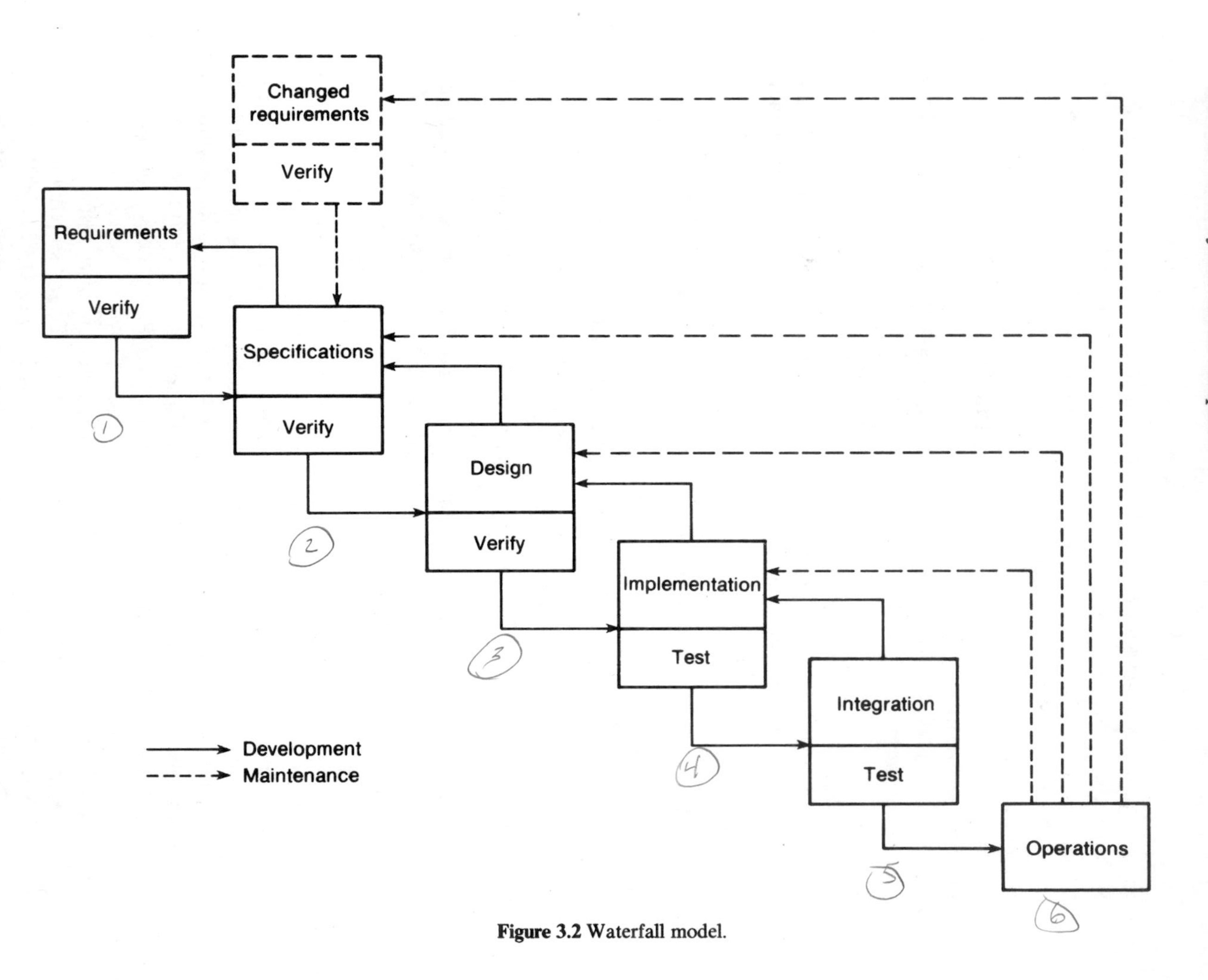

Figure 3.2 Waterfall model.

blocks will have to be transferred from the disk to main memory to copy all 40,000 elements. The first READ statement will cause the first column to be put in the buffer, and the first 200 READs will use the contents of the buffer. Only when the 201st element is required will a second block need to be transferred from disk to main memory. But if the product reads the array row by row, then a fresh block will have to be transferred for every READ, because consecutive READs access different columns, and hence different blocks. Thus 40,000 block transfers would be required, instead of 200 if the array were read in column-major order, and the input/output time for that part of the product would be 200 times longer. Design faults of this type have to be corrected before the team can continue with software development.

The waterfall model with its feedback loops allows for revisions of the design, and even the specifications, at the implementation stage, and, in fact, at every stage of the process. In addition, corrections must be made to all relevant documentation to reflect these revisions.

Once coding is complete, the modules are documented and then integrated to form a complete product. Again, it may be necessary to backtrack and make modifications to the code, preceded perhaps by modifications to the specifications and the design. It is important when changes are made that the documentation be changed to reflect these changes.

When the developers feel that the product has been successfully completed, it is given to the client for acceptance testing. Deliverables at this stage include the user manual and other documentation listed in the contract. And when the client agrees that the product indeed satisfies its specifications, the product is handed over to the client to be put in operations mode.

Once the client has accepted the product, any changes, whether to remove residual faults or to extend the product in some way, constitute maintenance. As can be seen from Figure 3.2, maintenance may require not just implementation changes, but also design and specifications changes. In addition, enhancement is triggered by a change in requirements. This, in turn, is implemented via changes in the specifications, design, and implementation. The waterfall model is a dynamic model, and the feedback loops play an important role in this dynamism. Again, it is vital that the documentation be maintained as meticulously as the code itself.

Up to now, reference has not been made to the lower halves of the boxes in Figure 3.2. An inherent aspect of every phase of the waterfall model is testing. Testing is not a separate phase to be performed only after the product has been constructed; it is not even to be performed only at the end of each phase. Instead, as stated in Chapter 2, testing should proceed continuously throughout the life cycle of the product. Specifically, when the requirements have been drawn up they must be verified, the specifications must be verified, the completed design must be verified, and the code must be tested in a variety of ways. While maintaining the product it is necessary to ensure not only that the modified version of the product still does what the previous version did and that it still does it correctly (regression testing), but in addition that it totally satisfies any further requirements imposed by the client.

The waterfall model has many advantages, not the least of which are the enforced disciplined approach and the requirement that documentation be provided at each stage. Successful termination of each phase of the waterfall model cannot be achieved merely by testing the specifications, design, or code, as the case may be. An essential aspect of the milestone terminating each phase is approval by the SQA group of all the documentation required for that phase.

Specifications documentation, design documentation, code documentation, and other documentation such as the database manual, user manual, and operations manual are essential tools for maintaining the product. As stated in Chapter 1, some 60% or 70% of software budgets are devoted to maintenance, and adherence to the waterfall model with its documentation requirements will make this maintenance easier. The same methodical approach to software production continues during maintenance. Every change that is made must also be reflected in the relevant documentation. Many of the successes of the waterfall model have been due to the fact that it is essentially documentation-driven. But as will be shown, the waterfall model also has its disadvantages.

3.2.1 Analysis of the Waterfall Model

To begin, consider the following two somewhat bizarre scenarios. Joe and Jane Johnson want to build a house. They consult with an architect. Instead of showing them sketches, plans, and perhaps a model, he gives them a 20-page typed document describing the house, written in highly technical terms. Despite the fact that neither Joe nor Jane have any previous architectural experience and hardly understand the document, they enthusiastically sign it and say, "Go right ahead, build the house!"

Another scenario is as follows. Mark Marberry buys his suits by mail order. Instead of mailing him pictures of their suits and samples of available cloths, the company sends Mark a written description of the cut and the cloth of their products. Mark then orders a suit solely on the basis of a written description.

The preceding two scenarios are unlikely in the extreme. But they typify precisely the way software is so often constructed using the waterfall model. The process begins with specifications. In general, specifications are long, detailed, and, quite frankly, boring to read. The client is usually inexperienced in the reading of software specifications, and this difficulty is compounded by the fact that specifications are usually written in a style with which the client is totally unfamiliar. The difficulty is even worse when the specifications are written in a formal specification language like Gist [Balzer,

1985], which is difficult for the untrained client to understand. Nevertheless, the client proceeds to sign off the specifications, whether properly understood or not. In many ways there is little difference between John and Jane contracting to buy a house to be built from a written description that they only partially comprehend and the client signing for a software product described in terms of specifications that are only partially understood.

Mark Marberry and his mail-order suits may seem bizarre in the extreme, but that is precisely what happens when the waterfall model is used in software development. The first time that the client sees a working product is after the entire product has been coded. Small wonder that software developers live in fear of the sentence, "I know this is what I asked for, but it isn't really what I wanted."

What has gone wrong? There is a considerable difference between the way a client understands a product as described by the specifications document and the actual product. The specifications exist only on paper; the client in general therefore cannot determine what the product itself will be like. So the waterfall model, depending as it does so critically on written specifications, can lead to the construction of products which simply do not meet clients' *real* needs.

In fairness, it should be pointed out that, just as an architect can help his or her client to understand what is to be built by providing models, sketches, and plans, so the software engineer can use graphical techniques such as data flow diagrams (Section 7.2) to communicate with the client. The problem is that such graphical aids do not describe how the finished product will work. For example, there is a considerable difference between a flowchart (a diagrammatic description of a product) and the working product itself.

The strength of the next life-cycle model to be examined, the rapid prototyping model, is that it can help to ensure that the client's real needs are met.

3.3 Rapid Prototyping Model

A *prototype* is a working model that is functionally equivalent to a subset of the product. For example, if the target product is to handle accounts payable, accounts receivable, and warehousing, then the prototype might consist of a product that performs the screen handling for data capture and prints the reports, but does no file updating or error handling of any kind. A prototype for a target product that is to determine the concentration of an enzyme in a solution might perform the calculation and print the answer without doing any validation or reasonableness checking of the input data or providing a graphical display of the output data.

The first step in the rapid prototyping life cycle depicted in Figure 3.3 is to build a prototype and to let the client interact with the prototype and experiment with it. Once the client is satisfied that the prototype indeed does most of what is required, the developer can draw up the specifications knowing the client's real needs.

An important aspect of the rapid prototyping model is embodied in the word "rapid." The whole idea is to put the prototype together as quickly as possible. After all, the purpose of the prototype is to provide the client with an understanding of the product, and the sooner, the better. It does not matter if the prototype hardly works, if the product crashes every few minutes, or if the screen layouts are less than perfect. The purpose of the prototype is to enable the client and the developer to agree as quickly as possible on what the product is to do. Thus any imperfections in the prototype may be ignored, provided that they do not seriously impair the functionality of the prototype and hence give a misleading impression of how the product will behave.

A second major aspect of the rapid prototyping model is the fact that the prototype must be built for change. If the first version of the rapid prototype is not what the client needs, then it must be equally rapidly transformed into a second version which hopefully will satisfy the client's requirements. In order to achieve rapidity of development throughout the prototyping process, fourth-generation languages (4GLs) and interpreted languages such as Prolog and BASIC have become popular languages for rapid prototyping purposes. Concerns have been expressed about the maintainability of certain prototyping languages, but from the viewpoint of classical prototyping this is irrelevant. All that counts is: Can a given language be used to produce a prototype quickly? And if necessary, can the prototype be quickly changed? If the answer to both these two questions is yes, then that language is a good candidate for rapid prototyping.

Having constructed the prototype, the life cycle continues as shown in Figure 3.3. A major strength of the rapid prototyping model is that the development of the product is linear, proceeding from the prototype to the delivered product. As a result, the feedback loops of the waterfall model (Figure 3.2) are less likely to be needed in the rapid prototyping model. The members of the development team use the prototype to construct the specifications. Since the working prototype has been validated through interaction with the client, it is reasonable to expect that the resulting specifications will be correct. The design phase is next. Even though the prototype has (correctly) been thrown together, the design team can get insights from it—at worst they will be of the "how not to do it" variety. Again the feedback loops of the waterfall models are less likely to be needed here.

Implementation comes next. In the waterfall model, implementation of the design sometimes leads to design faults coming to light. In the prototyping model, the fact that a working model reflecting at least partial functionality of the complete target product has been built tends to lessen the need to repair the design during or after implementation.

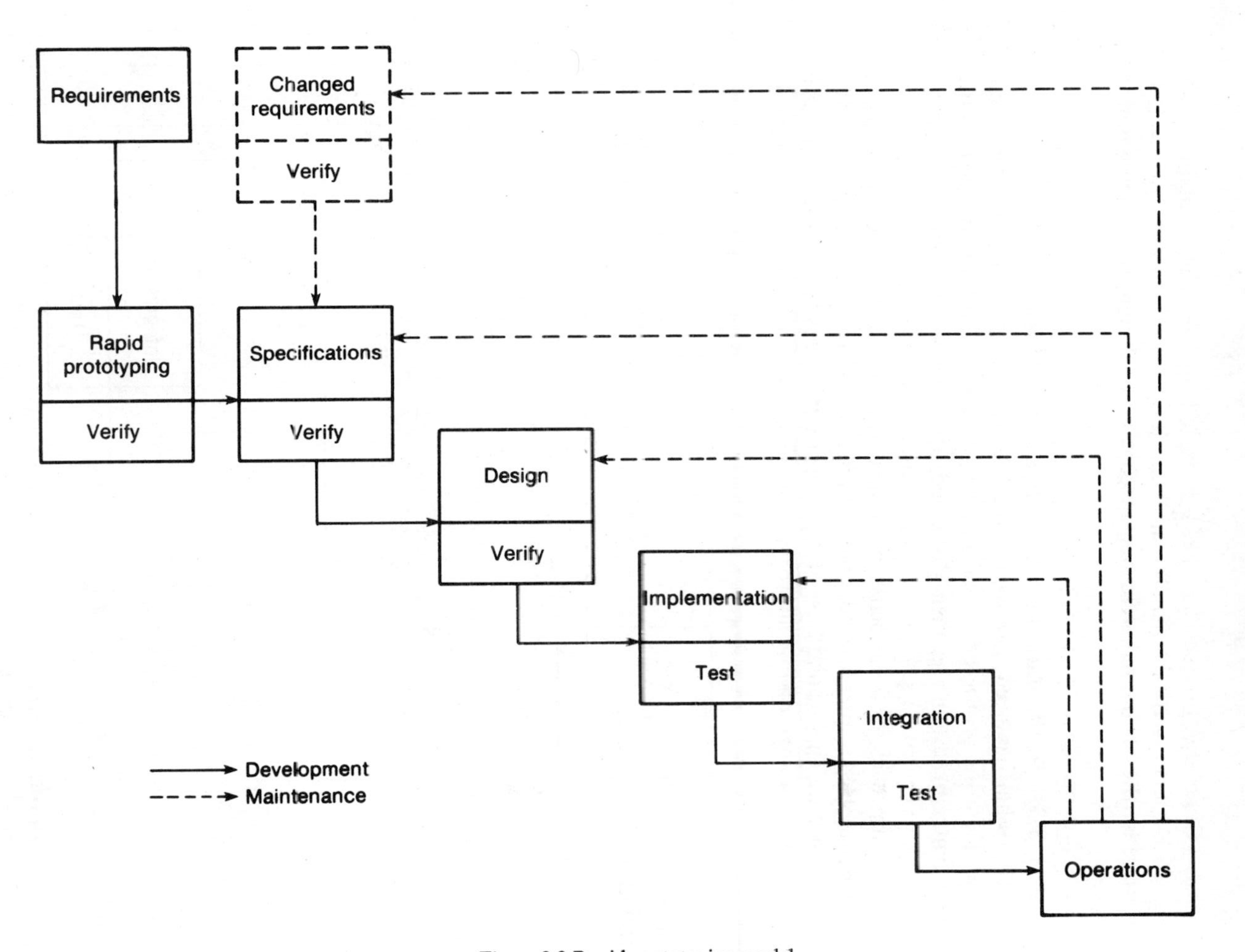

Figure 3.3 Rapid prototyping model.

Once the product has been accepted by the client, maintenance begins. Depending on the maintenance that has to be performed, the cycle is reentered either at the requirements, specifications, design, or implementation phases.

3.3.1 Prototyping as a Specification Technique

What has been described is but one version of the rapid prototyping model. Here the prototype is used solely as a means of accurately determining the client's needs and is then discarded after the specifications have been agreed on. That is to say, prototyping is used as a requirements technique. A second approach is to dispense with specifications as such and to use the prototype itself either as the specifications or as a significant part of the specifications. This second type of rapid prototyping model is shown in Figure 3.4. The advantage of this approach is both speed and accuracy. No time is

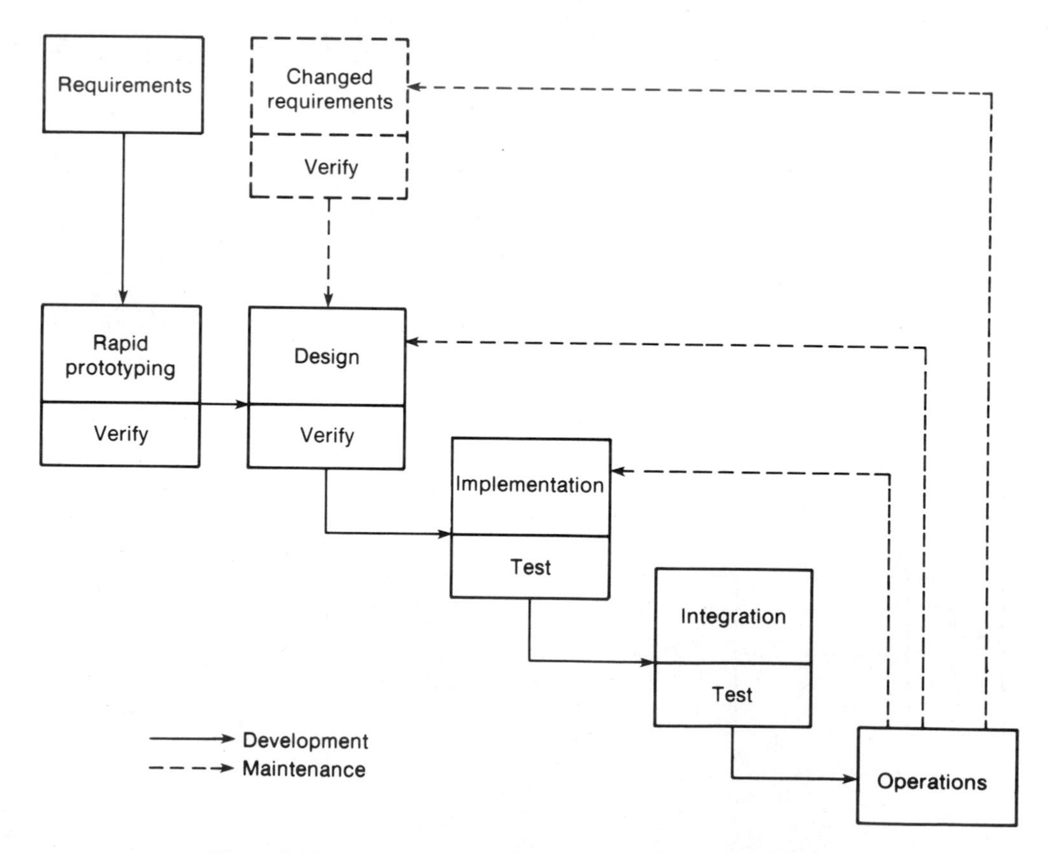

Figure 3.4 Rapid prototyping with prototype serving as specifications.

wasted drawing up written specifications, and the difficulties associated with specifications, such as ambiguities, omissions, and contradictions, cannot arise. Instead, since the prototype is the specifications, all that needs to be done is to state that the product will do what the prototype does, and to list any additional features that the product must support such as file updating, security, and error handling. During the maintenance phase, the current specifications are simply the current version of the product, and changes to the specifications must be made in terms of the functionality of the current version of the product.

This version of the rapid prototyping model can have a major drawback. If there is a disagreement as to whether the developer has satisfactorily discharged his or her obligations, it is unlikely that a prototype will be acceptable in a court of law as the statement of a contract between developer and client. For this reason, using the prototype as the specifications is probably best left to software development within an organization. It is unlikely that, say, the investment management division of a bank will take the data-processing division to court.

3.3.2 Reusing the Prototype

In both versions of the rapid prototyping model discussed previously, the prototype is discarded early in the life cycle. An alternate, but generally unwise, way of proceeding is to develop and refine the prototype until it becomes the product. This is shown in Figure 3.5. In theory, this approach should lead to fast software development; after all, instead of throwing away the code constituting the prototype, as well as the knowledge built into it, the prototype is converted into the final product. But, in practice, the process is very similar to the build-and-fix approach of Figure 3.1. The danger of this form of the rapid prototyping model follows from the fact that the prototype is essentially thrown together, rather than designed and then implemented. The result is that in the process of refining the prototype changes have to be made to a working product, which is an expensive way to proceed, as shown in Figure 1.3 (page 13). Another problem is that the prime objective when constructing a prototype is speed. The resulting code is therefore difficult and expensive to maintain. It might seem wasteful to construct a prototype and then to throw it away and design the product from scratch, but it is far cheaper in both the short term and the long term to do this rather than to try to convert a rapid prototype into production quality software [Brooks, 1975].

One way of ensuring that the prototype is thrown away and the product is properly designed and implemented is to build the prototype in a different language from that of the product. For example, the client may specify that the product must be written in Ada. If the prototype is written, say, in Prolog, then this will ensure that the prototype will be discarded. First,

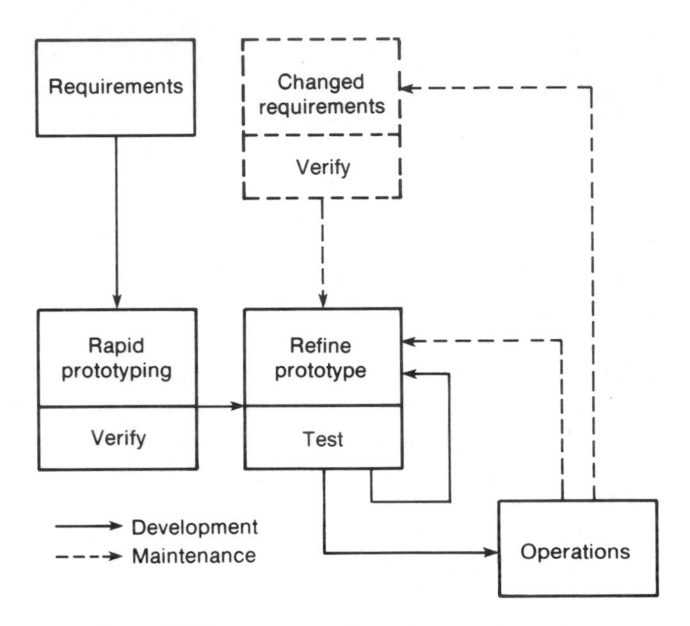

Figure 3.5 Unwise version of rapid prototyping model.

a prototype is written in Prolog and refined until the client is satisfied that it does everything, or almost everything, that the target product is to do. The product is now designed, and the knowledge and skills acquired in constructing the prototype will hopefully go into that design. The design will then be implemented in Ada, and the tested product handed over to the client in the usual way.

3.3.3 Other Uses of Rapid Prototyping

There is generally an element of risk involved in the development of software. Key personnel can resign before the product has been adequately documented. The manufacturer of hardware on which the product is critically dependent can go bankrupt. Too much, or too little, can be invested in testing and quality assurance. After spending hundreds of thousands of dollars on developing a major software product, technological breakthroughs can render the entire product worthless; for example, an organization may research and develop a database management system, but before the product can be marketed, a competing package is announced that does everything that the other product can do, but at a considerably lower price. For obvious reasons, software developers try to minimize the risks involved wherever possible.

One way of minimizing certain types of risk is to prototype. When a software engineer proposes a new design or implementation strategy that could save his or her organization vast sums of money, but it is by no means certain that the new technique is feasible, the obvious way out is to prototype. Instead of using the strategy to develop a complete product, thereby incurring the risk of not merely delaying the completion of the project, but also of conceivably laying the software development organization open to being sued under penalty clauses for late delivery of the product, the new technique is prototyped. For example, a telephone company may come up with a new, apparently highly effective algorithm for routing calls through a communications network. If the product is implemented but does not work the way the theoreticians who invented it claimed it would, not only will the telephone company have wasted the cost of developing the product, but customers may be angered by unnecessary congestion on the network caused by a poor routing algorithm and take their business elsewhere. The solution is to construct a prototype that will handle just the routing of calls and no other part of the control of the communications network. The prototype is then tested on a simulator. In this way the actual system is not disturbed, and for the cost of implementing just the algorithm itself, the telephone company can determine whether it is worthwhile to develop a whole new network controller incorporating the new algorithm.

The use of prototyping as a means of minimizing risk is the idea underlying the *spiral model* [Boehm, 1988]. One way of looking at this life-cycle model is as a waterfall model with each phase preceded by risk analysis. The risks are then resolved by using prototyping of the types described previously. These include prototyping of the requirements to minimize the risk that what is delivered is not what the user really needs, prototyping of the design to minimize the chances of a major design fault, and prototyping of the implementation to ensure that constraints such as storage requirements or response times are met.

Prototyping can be effectively used to provide information with regard to certain classes of risk. For example, timing constraints can generally be tested by constructing a prototype and measuring whether the prototype can achieve the necessary performance. If the prototype is an accurate functional representation of the relevant features of the product, then measurements made on the prototype should give the developer a good idea as to whether the timing constraints can be achieved.

Rapid prototyping is particularly effective when developing the user interface to a product. When rapid prototyping is used in this way it is important that not just the client but also the future users of the product should interact with the prototype user interface, as this will help to achieve user friendliness, a vital objective for all software products. Encouraging users to experiment with the user interface at the prototyping stage greatly reduces the risk of the finished product having to be altered to cater for the users' needs.

Other areas of risk are less amenable to prototyping. For example, there is often a risk that necessary personnel cannot be hired or that key personnel may resign while the project is still incomplete. Another potential risk lies in the fact that a team may not be competent to develop a specific large-scale product. This risk cannot be resolved by seeing how well they perform on a much smaller prototype. There are essential differences between large-scale and small-scale software, and prototyping is of little use in this area. Another area of risk for which prototyping cannot be employed is when the delivery promises of a hardware supplier have to be evaluated. A strategy the developer can adopt is to determine how well other clients of the supplier have been treated in the past, but past performance is by no means a certain determinant of future performance. A penalty clause in the delivery contract is one way of trying to ensure that essential hardware will be delivered on time, but what if the supplier refuses to sign an agreement that includes such a clause? Also, even if the clause is agreed to, nontimely delivery may eventually lead to legal action, but it may take years before the courts can make a ruling on the issue. In the meantime, the software developer may have gone bankrupt through nondelivery of the promised software caused by nondelivery of the promised hardware. In short, while prototyping provides an answer to risk resolution issues in some areas, in other areas it is at best a partial answer, and in certain areas it is no answer at all.

A different use of rapid prototyping is as a means of arriving at a consensus where there is disagreement as to the client's requirements. The author was once called in to advise a committee of six top executives of a major corporation with regard to the development of a management information system (MIS) [Blair, Murphy, Schach, and McDonald, 1988]. A requirement of the product was that one-page reports had to be printed incorporating month-by-month information for a full year. Each member of the committee then insisted that every one-page report had to incorporate data from his or her specific area of responsibility. Since there were six areas in all, this was a clear impossibility. The solution was to construct a rapid prototype. Within 4 person-days a working product was constructed that computed the relevant quantities and printed a one-page report incorporating what the author felt was the most important management information. The prototype was partially acceptable to the management team, the necessary changes were rapidly made, and within 4 weeks of commencing work the complete operational quality product was in place. Without prototyping, consensus would not have been reached as quickly and as amicably. In fact, it is possible that no consensus at all could have been achieved, and the product would never have then been developed.

The management implications of the rapid prototyping model are now considered.

3.3.4 Implications of the Rapid Prototyping Model for Management

One difficulty with rapid prototyping is that the ease with which changes can generally be made to a prototype may encourage the client to request all sorts of major changes to the delivered operational quality version of the product and expect them to be implemented as rapidly as were changes to the prototype. This was the author's experience with the MIS described previously. A related problem is having to explain to the client that the prototype is not of operational quality, and that he or she will have to wait for the operational quality version of the target product to be delivered, notwithstanding the fact that the prototype appears to do everything the client wants. Before prototyping can be used, it is essential that these and other pitfalls be made clear to the managers responsible for the product.

As with the introduction of any new technology, before an organization introduces the rapid prototyping model it is essential that management be aware of the advantages and disadvantages of prototyping. In all fairness, while the case for rapid prototyping is a strong one, it has not yet been proved beyond all doubt. For example, a frequently quoted experiment is that of Boehm, Gray, and Seewaldt who compared seven different versions of a product [Boehm, Gray, and Seewaldt, 1984]. Four versions were specified and three were prototyped with the prototype serving as the specifications. The results were that prototyping and specifying yielded products with roughly equivalent performance, but the prototyped versions comprised about 40% less code and required about 45% less effort. The prototyped versions were rated somewhat lower on functionality and robustness, but higher on ease of use and ease of learning, while specifying produced more coherent designs and software that was easier to integrate.

One important point about this experiment was that it was conducted on seven teams of graduate students, three teams of size 2 and four teams of size 3. The project was only 10 weeks in duration, and no maintenance of the product was performed. In other words, the experiment is not typical of real products in respect to participants, team size, project size, or project life cycle. It is therefore perhaps unwise to treat Boehm's results as proven facts. Instead, they should be taken as indications of the comparative strengths and weaknesses of prototyping as opposed to specifying. In any event, Boehm's results are by no means a blanket endorsement for prototyping. Among other weaknesses pointed out previously, Boehm and his collaborators found that prototyped products are harder to integrate than specified products. Ease of integration is important for large-scale products, especially C^3 software (command, control, and communications). One solution is to use the prototyping model depicted in Figure 3.3, where prototyping is employed as a requirements technique, and not to use the model of Figure 3.4, where prototyping takes the place of specifying.

Two aspects of rapid prototyping must be taken into account by any manager. In Section 3.3.2 it was pointed out that it is short-sighted to turn a prototype into the product, instead of using the prototype solely as a means of determining accurate requirements or specifications. A second important issue is that rapid prototyping under some circumstances can take the place of the specifications phase; it can *never* replace the design phase. A team can certainly use the information and experience gained from the prototype to guide them in fashioning a good design, but the prototype itself is rapidly thrown together, and it is unlikely that a good prototype will have a good design.

A more fundamental issue is that managing the rapid prototyping model requires a major change in outlook of a manager who up to now has managed only the waterfall model. The concept underlying the waterfall model is to do things correctly the first time. Certainly the waterfall model incorporates a number of feedback loops if the team does not accomplish this goal, and the goal is seldom, if ever, reached. However, the ideal for the waterfall model team is not to have to repeat any phase of the development cycle. In contrast, a prototype is specifically built to be changed frequently and then to be thrown away. This concept is diametrically opposed to the approach to which the average manager is accustomed. The rapid prototyping approach of taking several iterations to get it right is probably a more realistic approach than the first-time-right waterfall model expectation, and this is perhaps the line to take when trying to convince a manager to change to the rapid prototyping model.

Another aspect of the rapid prototyping model that requires a different approach on the part of the manager is the increased interaction between the client and the developers, particularly the prototyping team. In the waterfall model, interaction between client and developers is essentially restricted to a series of interviews between the requirements team on the one hand, and the client and his or her employees on the other. When prototyping is used, there is almost continual interaction between the prototyping team and the client's team until the prototype has been accepted.

3.3.5 Integrating the Waterfall and Rapid Prototyping Models

Despite the many successes of the waterfall model, it has a major drawback in that what is delivered to the client may not be what the client really needs. The rapid prototyping model has also had many successes, but it has not yet been proved beyond all doubt. In addition, an experiment to compare the two approaches has demonstrated certain weaknesses of rapid prototyping.

One solution is to combine the two approaches. Comparing the various phases of Figure 3.2 (waterfall model) with those of Figure 3.3 (rapid prototyping model) highlights the fact that prototyping can be used as a requirements technique. In other words, the first step is to build a prototype in order to determine the client's real needs and then to use that prototype as the input to the waterfall model.

This approach has a useful side effect. Some organizations are reluctant to use the rapid prototyping approach because of the risks involved in using any new technology. Introducing prototyping into the organization as a front end to the waterfall model will give management the opportunity to assess the technique while minimizing the associated risk.

A different class of life-cycle model is now examined.

3.4 Incremental Models

Software is built, not written. That is to say, software is constructed step by step, in the same way that a building is constructed. While a software product is in the process of being developed, each step consists of additional pieces that are added to what has gone before. On one day the design is extended, the next day another module is coded. The construction of the complete product proceeds incrementally in this way until the product is finished.

Of course, it is not quite true that progress is made every day. Just as a contractor occasionally has to tear down an incorrectly positioned wall or replace a pane of glass that a careless painter has accidentally cracked, so it is sometimes necessary to respecify, redesign, recode, or at worst, throw away what has been done to date and start again. But the fact that the product sometimes advances in fits and starts does not negate the basic reality that a software product is built piecewise.

The realization that in practice software is engineered incrementally has led to the development of a model that exploits this aspect of software development, the so-called *incremental model* shown in Figure 3.6. The product is designed, implemented, integrated, and tested as a series of incremental *builds*, where a build consists of code pieces from various modules that will interact together to provide a specific functional capability. For example, if the product is to control a nuclear submarine, then the navigation system could constitute a build, as could the weapons control system. In an operating system, the scheduler could be a build, and so could the file management system. At each stage of the incremental model a new build is coded, then integrated into the structure which is tested as a whole. The process stops when the product has the target functionality, that is to say, when the product satisfies all its requirements. The product developer is free to break up the target product into builds as he or she sees fit, subject only to the constraint that as each build is integrated into the existing software the resulting product

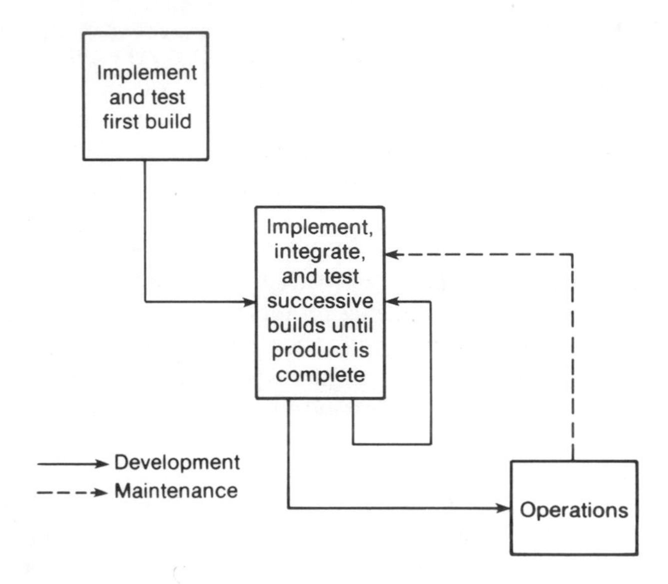

Figure 3.6 Incremental model.

must be testable. If the product is broken into too few builds, then the incremental model degenerates into the build-and-fix approach (Figure 3.1). Conversely, if the product consists of too many builds, then at each stage considerable time is spent in the integration testing of only a small amount of additional functionality. What constitutes an optimal decomposition into builds will vary from product to product and from developer to developer.

The incremental model has had some successes. For example, Wong reports that she used it, together with other modern software development methods, to deliver the software for an air defense system within budget and within 25 months [Wong, 1984].

One incremental model, namely the evolutionary delivery model, is now considered in detail.

3.4.1 Evolutionary Delivery Model

Both the waterfall and prototyping models have as their aim the delivery to the client of a complete operational quality product. That is to say, the client is presented with a product that satisfies *all* of his or her requirements, and that product is of sufficiently high quality for use in operations mode by the client. Correct use of either the waterfall or the prototyping model results in a product that will have been thoroughly tested, and the client should be confident that the product can be utilized for the purpose for which it was designed. Furthermore, the product will come with adequate documentation, so that not only can it be used in operations mode, but all three types of maintenance (adaptive, perfective, and corrective) can be performed as necessary.

With both models there is a projected delivery date, and the intention is to provide the client with the complete product in full working order on or before that date. In contrast, the evolutionary delivery model delivers to the client at each stage an operational quality product, but one which satisfies only a subset of the client's requirements [Gilb, 1985]. The complete product is divided up into builds, and the developer delivers the product build by build. A typical product will usually consist of between 10 and 100 builds. Since at each stage the client has an operational quality product that does a portion of what is required, from the time the first build is delivered the client is able to do useful work and does not have to wait for the final delivery date to be able to use the product. Instead, portions of the total product are available from the beginning.

An important aspect of the evolutionary delivery model is the order in which the builds are developed and delivered. At each stage the client must assign a value to each of the as-yet unimplemented builds. Then the developer estimates the cost of developing each build. The value-to-cost ratio provides the criterion for selecting which build is to be delivered next. The build with the highest value-to-cost ratio is the one which provides the client with the most functionality, in the client's opinion, for the least cost, as charged by the developer. In this way, not only does the client have a *usable* product at all times, but the product is *useful* from the beginning. In addition, while the total cost is the same as for other models, the client spends money in such a way as to maximize the early return on investment. The key functionality is added first, while the bells and whistles are added on later. The complete model is shown in Figure 3.7.

To see how this works, suppose there are three builds that still have to be added to an automated order system. The first is a new way of computing commissions that will provide incentives to the sales force to increase their productivity. The second build on the list is an online query system that management can use to determine the status of every order within seconds, instead of having order queries processed overnight in batch mode. The third build which still remains is the introduction of handheld order-taking devices, which can be used to transmit orders over a telephone line directly into the computer. At present, the sales force has to relay orders to a salesclerk over the telephone, and in the process of entering the order into the computer the clerk occasionally makes mistakes. The third proposed build will obviate such errors. Management decides that the value of the new commission formula to the company, through increased sales, is $50,000. The online query system will increase the productivity of the sales staff, and its value is estimated at $40,000. The order clerks do not make many errors, so the value of the order-taking devices to the corporation is only $20,000. In turn, the developer

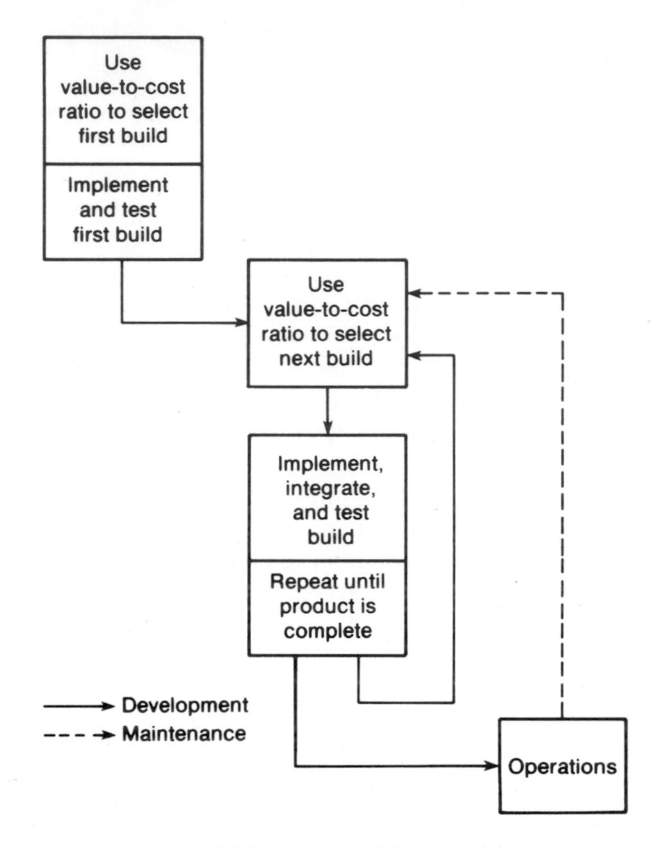

Figure 3.7 Evolutionary delivery model.

tells management that the cost of the build implementing the new commission formula will be $2000, the query system will cost $20,000, while buying the hardware and developing the communications software for the order-taking devices will cost $100,000. Management can now construct Figure 3.8. From the figure it is clear that at this stage the commission formula build has the highest value-to-cost ratio, and hence will provide the highest immediate return on capital. Once this build has been integrated into the product, management and the developer will recompute the value and cost of the two remaining builds. If the numbers shown in Figure 3.8 have not drastically changed, management will instruct the developer to proceed with the online query system and may decide to scrap the order-taking devices.

One aspect that has to be taken into account when computing value-to-cost ratios is that there is sometimes a dependency between builds. For example, a full screen editor cannot be used until a screen driver has been

Part	Value	Cost	Value-to-Cost Ratio
New commission formula	$50,000	$2,000	25.0
Online query system	$40,000	$20,000	2.0
Order taking devices	$20,000	$100,000	0.2

Figure 3.8 Value-to-cost ratios for evolutionary delivery model.

written. The cost of the full screen editor build must therefore include the cost of the prerequisite screen driver build. However, if the value-to-cost ratios dictate that the screen driver should be built before the full screen editor, then once the screen driver has been completed its cost need no longer be included in that of the full screen editor.

Gilb highlights a number of differences between evolutionary delivery on the one hand, and the waterfall and rapid prototyping models on the other [Gilb, 1985]. When his model is used, the first deliverables arrive within weeks, while the client generally waits months or years to receive the product of the waterfall or rapid prototyping models. He points out that the effect of imposing a completely new product can be somewhat traumatic to the client organization, while the gradual introduction of the product via evolutionary delivery permits time to adjust to the new product. Change is an integral part of every growing organization, and since a software product is a model of reality the need for change is also an integral part of delivered software. Change and adaptation are natural to the evolutionary delivery model, while Gilb states, with some justification, that change constitutes a threat to products developed in one large step.

From the client's financial viewpoint, phased delivery does not require a large capital outlay. Instead, there is an excellent cash flow, with the earliest stages delivering high returns on investment. Another advantage of the model is that it is not necessary to complete the product to get a return on investment. Instead, as shown previously, the client can stop development of the product at any time, especially since the order in which the builds of the product are added is determined dynamically so as to maximize the return on investment. What happens with evolutionary delivery is that the client's perception of the complete product is a moving target. As builds are delivered, the client's opinion regarding the values of the remaining builds may change. Just as experience with a prototype alters the client's perception of the product, so the users of the sequence of operational quality builds may view the product in a different light. Future builds previously felt to be essential to the overall product may be deleted, while new builds may be added to the list of additional components that have to be incorporated into the product during some future phase.

Using evolutionary delivery, the design of the product is done in two parts. First, an architecture is designed for the product as a whole into which all projected builds can be incorporated. If a database is to be used, it is designed at this point. Then as each individual build in turn is selected it undergoes detailed design, followed by implementation and integration into the product.

A difficulty with Gilb's model is that each additional build has somehow to be incorporated into the existing structure without destroying what has been built to date. Furthermore, the existing structure must lend itself to extension in this way, and the addition of each succeeding build must be simple and straightforward. Gilb states that the product must always have an *open architecture*, that is, the architecture must be such that additional builds can be easily incorporated. While this is certainly a difficulty in the short term, in the long term it can be a real strength. Every product essentially goes through two phases. First, there is the development phase. This is followed by the maintenance phase as soon as the product enters operations mode. During development it is indeed important to have clear specifications and a coherent and cohesive design. But once the product enters the maintenance phase the requirements change, and radical enhancement can easily destroy a coherent and cohesive design to the extent that further maintenance becomes impossible, and the product must be rebuilt virtually from scratch. From the start Gilb's approach allows for change. The fact that the design must be open-ended is not merely a prerequisite for development using the evolutionary delivery model, but is essential for maintenance, irrespective of the model selected for the development phase. Thus, while the evolutionary delivery model may require more careful design than the holistic waterfall and rapid prototyping models, the payoff is in the maintenance phase. If a design is flexible enough to support the evolutionary delivery model, it will certainly allow virtually any sort of maintenance to be done without falling apart. In fact, the evolutionary delivery model does not distinguish between developing a product and enhancing (maintaining) it; each enhancement is merely an additional build.

A more critical, but related, difficulty is that the evolutionary delivery model can too easily degenerate into the build-and-fix approach. Control of the process as a whole can be lost, and the resulting product, instead of being open-ended, becomes a maintainer's nightmare. In a sense, the evolutionary delivery model is a contradiction in terms, requiring the developer to view the product as a whole in order to begin with a design that will support the entire product, including future enhancements, but simultaneously to view that product as a sequence of builds, each essentially independent of the next. Unless the developer is skilled enough to be able to handle this apparent contradiction, the evolutionary delivery model may lead to an unsatisfactory product.

3.5 Conclusions

Three different classes of software life-cycle models have been examined with special attention paid to some of their strengths and weaknesses. The waterfall model is a known quantity. Its strengths are understood, but so are its weaknesses. The rapid prototyping model was developed as a reaction to a specific perceived weakness in the waterfall model, namely that the delivered product may not be what the client really needs. But less is known about the newer rapid prototyping model than the familiar waterfall model, and in any event the rapid prototyping model may have some problems of its own. One alternative is to combine the good points of both models, as suggested in Section 3.3.5, or to use a variation such as the spiral model (Section 3.3.3). Another is to use a different model, the evolutionary delivery model. This model, notwithstanding its successes, also has some drawbacks.

The best way is for each software development organization to decide on a life-cycle model that is appropriate for that organization, its management, and its employees and to vary the model depending on the features of the specific target software product currently under development. Such a model will incorporate appropriate features from various different life-cycle models. In this way the strengths of the various models can be utilized, and their weaknesses minimized.

Chapter Review

A number of different life-cycle models are described. The build-and-fix model (Section 3.1) should be avoided. The strength of the waterfall model (Section 3.2) is its disciplined, documentation-driven approach. The problem is that the delivered product may not be what the client really needs. The rapid prototyping model (Section 3.3) was developed to solve this problem. The case for the rapid prototyping model is a strong one, but has not been proved beyond all doubt. An incremental model (Section 3.4) such as the evolutionary delivery model (Section 3.4.1) has advantages such as maximizing early return on investments and ease of maintenance, but has difficulties of its own, particularly the need for an open architecture. It is suggested in Section 3.5 that aspects of various life-cycle models should be combined in order to come up with an appropriate life-cycle model. The form of this life-cycle model will depend on the organization, its management, its employees, and the specific product to be built.

For Further Reading

A description of the waterfall model appears in Chapter 4 of [Boehm, 1981]. The same chapter also includes an example of an incremental model different from the one presented here in Section 3.4.1.

For an introduction to prototyping, two suggested books are [Boar, 1984] and [Lantz, 1985]. The application of prototyping to interactive information systems is described in [Mason and Carey, 1983]. Conference papers relating to the prototyping life cycle can be found in [Squires, 1982] and [Budde, Kühlenkamp, Mathiassen, and Züllighoven, 1984]. In addition to the experiment on prototyping by Boehm and his co-workers [Boehm, Gray, and Seewaldt, 1984], another experiment on prototyping is described in [Alavi, 1984].

Gilb's work on the incremental delivery model can be found in [Gilb, 1985] and [Gilb, 1988]. The spiral model is described in [Boehm, 1988], and its application to the TRW Software Productivity System appears in [Boehm, Penedo, Stuckle, Williams, and Pyster, 1984]. Another type of incremental model is described in [Currit, Dyer, and Mills, 1986] and [Selby, Basili, and Baker, 1987].

Many other life-cycle models have been put forward. For example, a collection of papers describing the models of Lehman and Belady can be found in [Lehman and Belady, 1985]. A life cycle that emphasizes human factors is presented in [Mantei and Teorey, 1988].

Problems

3.1 Suppose that you have to write a product about 25 lines long to check that the root of a specific equation has been correctly computed. When the product has been written and tested, it will be thrown away. What life-cycle model would you use?

3.2 You are a software engineering consultant and have been called in by the vice-president for finance of a restaurant chain. He would like your organization to build a product that will monitor the food used by the restaurants, starting with the purchasing in bulk and keeping track of the food as it is distributed to the individual restaurants, cooked, and served to the customers. What criteria would you use in selecting a life-cycle model for the project?

3.3 List the risks involved in developing the software of Problem 3.2. How would you attempt to resolve each risk?

3.4 Your development of the food-tracking product for the restaurant chain is so successful that your organization decides that it must be rewritten as a package that will be sold to a variety of restaurant chains. The new product must therefore be portable and easily adapted to new hardware and/or operating systems. How do the criteria you would use in selecting a life-cycle model for this project differ from those in your answer to Problem 3.2?

3.5 You have just joined Dolphin Software Developers as a software manager. Dolphin has been developing software for years using the waterfall model, usually with some success. On the basis of your experience, you feel that the rapid prototyping model is a far superior way of developing software. Write a report addressed to the vice-president for software development explaining why you feel that the organization should switch to the rapid prototyping model. Remember that vice-presidents do not like reports that are more than one-page in length and that brevity is the way to a vice-president's heart.

3.6 You are the vice-president for software development of Dolphin Software Developers. Reply to the memo of Problem 3.5.

3.7 Describe the sort of product that you feel would be an ideal application for Gilb's evolutionary model.

3.8 Now describe the type of situation where you feel that the evolutionary delivery model might lead to difficulties.

3.9 (Term Project) Because the Plain Vanilla Ice Cream Corporation (PVIC) minicomputer is somewhat overloaded at present, it has been suggested that the product described in the Appendix be implemented on a personal computer. What are the risks involved?

3.10 (Readings in Software Engineering) Your instructor will distribute copies of [Boehm, 1988]. Show that the spiral model incorporates aspects of both the waterfall model and the rapid prototyping model. Which of the two models predominates?

References

[Alavi, 1984] M. ALAVI, "An Assessment of the Prototyping Approach to Information Systems Development," *Communications of the ACM* **27** (June 1984), pp. 556–563.

[Balzer, 1985] R. BALZER, "A 15 Year Perspective on Automatic Programming," *IEEE Transactions on Software Engineering* **SE-11** (November 1985), pp. 1257–1268.

[Blair, Murphy, Schach, and McDonald, 1988] J. A. BLAIR, L. C. MURPHY, S. R. SCHACH, AND C. W. MCDONALD, "Rapid Prototyping, Bottom-Up Design, and Reusable Modules: A Case Study," *ACM Mid-Southeast Summer Meeting*, Nashville, TN, May 1988.

[Boar, 1984] B. H. BOAR, *Application Prototyping: A Requirements Definition Strategy for the '80s*, John Wiley and Sons, New York, NY, 1984.

[Boehm, 1981] B. W. BOEHM, *Software Engineering Economics*, Prentice-Hall, Englewood Cliffs, NJ, 1981.

[Boehm, 1988] B. W. BOEHM, "A Spiral Model of Software Development and Enhancement," *IEEE Computer* **21** (May 1988), pp. 61–72.

[Boehm, Gray, and Seewaldt, 1984] B. W. BOEHM, T. E. GRAY, AND T. SEEWALDT, "Prototyping Versus Specifying: A Multi-Project Experiment," *IEEE Transactions on Software Engineering* **SE-10** (May 1984), pp. 290–303.

[Boehm, Penedo, Stuckle, Williams and Pyster, 1984] B. W. BOEHM, M. H. PENEDO, E. D. STUCKLE, R. D. WILLIAMS, AND A. B. PYSTER, "A Software Development Environment for Improving Productivity," *IEEE Computer* **17** (June 1984), pp. 30–44.

[Brooks, 1975] F. P. BROOKS, JR., *The Mythical Man-Month: Essays on Software Engineering*, Addison-Wesley, Reading, MA, 1975.

[Budde, Kühlenkamp, Mathiassen, and Züllighoven, 1984] R. BUDDE, K. KÜHLENKAMP, L. MATHIASSEN, AND H. ZÜLLIGHOVEN (Editors), *Approaches to Prototyping: Proceedings of the Working Conference on Prototyping,* Namur, Belgium, October 1983, Springer-Verlag, Berlin, West Germany, 1984.

[Currit, Dyer, and Mills, 1986] P. A. CURRIT, M. DYER, AND H. D. MILLS, "Certifying the Reliability of Software," *IEEE Transactions on Software Engineering* **SE-12** (January 1986), pp. 3–11.

[Gilb, 1985] T. GILB, "Evolutionary Delivery versus the 'Waterfall Model'," *ACM SIGSOFT Software Engineering Notes* **10** (July 1985), pp. 49–61.

[Gilb, 1988] T. GILB, *Principles of Software Engineering Management*, Addison-Wesley, Wokingham, UK, 1988.

[Lantz, 1985] K. E. LANTZ, *The Prototyping Methodology*, Prentice-Hall, Englewood Cliffs, NJ, 1985.

[Lehman and Belady, 1985] M. M. LEHMAN AND L. A. BELADY (Editors), *Program Evolution, Processes of Software Change*, Academic Press, London, UK, 1985.

[Mantei and Teorey, 1988] M. M. MANTEI AND T. J. TEOREY, "Cost/Benefit Analysis for Incorporating Human Factors in the Software Development Lifecycle," *Communications of the ACM* **31** (April 1988), pp. 428–439.

[Mason and Carey, 1983] R. E. A. MASON AND T. T. CAREY, "Prototyping Interactive Information Systems," *Communications of the ACM* **26** (May 1983), pp. 347–354.

[Royce, 1970] W. W. ROYCE, "Managing the Development of Large Software Systems: Concepts and Techniques," *Proceedings of WestCon*, August 1970.

[Selby, Basili, and Baker, 1987] R. W. SELBY, V. R. BASILI, AND F. T. BAKER, "Cleanroom Software Development: An Empirical Evaluation," *IEEE Transactions on Software Engineering* **SE-13** (September 1987), pp. 1027–1037.

[Squires, 1982] S. L. SQUIRES, *Working Papers from the ACM SIGSOFT Rapid Prototyping Workshop, ACM SIGSOFT Software Engineering Notes* **7** (December 1982).

[Wong, 1984] C. WONG, "A Successful Software Development," *IEEE Transactions on Software Engineering* **SE-10** (November 1984), pp. 714–727.

Chapter 4 Software Planning

Software has to be built, and there is no easy panacea for the difficulties of constructing a software product. A large software product takes time and resources to put together. And like any other large construction project, careful planning at the beginning of the project is perhaps the single most important factor that distinguishes success from failure.

Software is constructed in order to satisfy a need. For example, a corporation may wish to automate its personnel records, an air force might feel the need to modernize by implementing electronic warfare systems on its tactical aircraft, or a programmer in an organization that builds software packages for personal computers may suddenly come up with a wonderful idea for a new type of video game that will make millions of dollars. But before coding of any of these projects can be started, it is necessary to plan in detail the entire software development effort.

4.1 Components of a Software Project Management Plan

A software project management plan has three main components: the work to be done, the resources with which to do it, and the money to pay for it all. In this section these three ingredients of the plan are discussed. The terminology is taken from [IEEE 1058, 1987], which is discussed in greater detail in Section 4.3.

Software development requires *resources*. The major resources required are the people who will develop the software, the hardware on which the software will be run, and the support software such as operating systems, text editors, and version control software. (Version control software enables the development team to have different versions of various modules in differ-

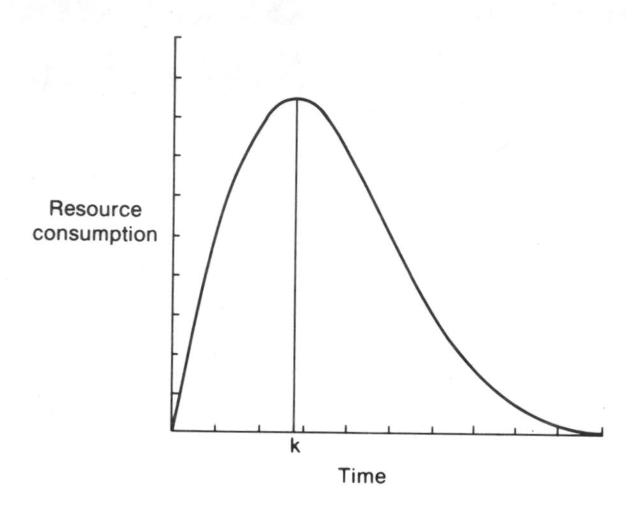

Figure 4.1 Rayleigh curve showing how resource consumption varies with time.

ent stages of development, and to maintain control over which module is in what stage of development, and which should be linked to specific other modules.)

Use of resources such as personnel varies with time. Norden has shown [Norden, 1958] that the Rayleigh distribution

$$R_C = \frac{t}{k^2} e^{-t^2/2k^2} \qquad 0 \leq t < \infty \tag{4.1}$$

is a good approximation of the way that resource consumption R_C varies with time t. Parameter k is a constant, the time at which consumption is at its peak, and $e = 2.71828\ldots$, the base of Naperian logarithms. A typical Rayleigh curve is shown in Figure 4.1. Putnam investigated the applicability of Norden's results to software development and found that personnel and other resource consumption was modeled with some degree of accuracy by the Rayleigh distribution [Putnam, 1978].

It is therefore insufficient in a software plan to state that three senior programmers with at least 5 years of experience are required. What is needed is something like the following:

> Three senior programmers with at least 5 years of experience in real-time programming are needed, two to start 3 months after the project commences, the third to start 6 months after that. Two will be phased out when integration testing commences, the third when product maintenance begins.

This time dependence of resource needs applies not only to personnel but also to computer time, support software, computer hardware, office facilities, and even travel. Thus the software project management plan will be a function of time.

The work to be done falls into two categories. First, there is work which is carried on throughout the project and which does not relate to any specific phase of software development. Examples of such work are project management and quality control. Work that is carried on right through the project is termed a *project function*. The second kind of work is that which relates to a specific phase in the development of the product. An *activity* is a major unit of work that has precise beginning and ending dates, consumes *resources* such as computer time or person-days, and results in *work products* such as a budget, design documents, schedules, source code, or user's manual. An activity, in turn, comprises a set of *tasks*, a task being the smallest unit of work subject to management accountability. There are thus three levels of units of work: project functions that are carried on throughout the project, activities (major units of work), and tasks (minor units of work).

A critical aspect of the plan is concerned with completion of work products. The date on which a work product is deemed to be completed is termed a *milestone*. In order to determine whether a work product has reached a milestone, it must first pass a series of *reviews* performed by fellow team members, management, and/or the client. A typical milestone is the date on which coding is completed and passes review. Once a work product has been reviewed and agreed upon, it becomes a *baseline* and can be changed only through formal procedures. In reality, there is more to a work product than merely the product itself. A *work package* defines not just the work product, but also the staffing requirements, duration, resources, name of the responsible individual, and acceptance criteria for the work product. *Money* is of course a vital component of the plan. A detailed budget must be worked out and the money allocated, as a function of time, to the project functions and activities.

The issue of how to draw up a plan for software production is now addressed.

4.2 How to Plan Software Production

There are three main phases to the planning. First, the problem must be clearly stated; the preliminary concept must be refined until there is absolutely no doubt as to why the product is required, and what exactly the product has to do. Second, alternative solution strategies must be compared until an optimal solution strategy has been determined. Third, a project management plan for the product as a whole must be developed. These three phases are now examined in detail.

4.2.1 Defining the Problem

When a software development organization decides to risk significant amounts of time and money in developing software for a client, the chances of the product being developed on time and within budget are pretty slim unless there is complete unanimity among the members of the team responsible for developing the software as to what the software product will do. The first step in achieving the unanimity is to determine precisely what the client's current product does. For example, it is inadequate to say, "They need a computer-aided design system because they claim their manual design system is lousy." Unless the development team knows exactly what is wrong with the current manual system, there is a high probability that aspects of the new computerized system will be found to be equally "lousy." Similarly, if a personal computer manufacturer is contemplating implementation of a new operating system, the first step is to evaluate the operating system currently running on the company's hardware and to analyze carefully exactly why it is unsatisfactory. To take an extreme example, it is vital to know whether the problem exists only in the mind of the sales manager who is attempting to blame the lack of success of the sales team on the operating system, or whether users of the operating system are thoroughly disenchanted with both the functionality and the reliability of what they have purchased. Only after a clear picture of the present situation has been gained can the team attempt to answer the critical question: What must the new product be able to do? The process of answering this question is carried out during the requirements phase.

Once the development team believes that it understands what the product should do, the user's requirements will then be concretely expressed in the form of the specifications document. It is important that this document be intelligible to the client. After all, the client is paying for the product, and unless the client believes that he or she really understands what the new product will be like, there is a good chance that the client will either decide not to authorize the development of the product or will ask some other development organization to build it. This, in turn, implies that the specifications document should be in a form that the client and his or her staff can easily comprehend. One solution to the problem of difficult-to-understand specifications is to use rapid prototyping and let the prototype either take the place of the specifications document or form part of the specifications; this approach was described in Section 3.3.1.

The specifications usually incorporate constraints that the product has to satisfy. Cost is almost always a constraint and so is a time deadline. A common stipulation is that "the product shall be installed in such a way that it can run in parallel with the existing product" until the client is satisfied that the new product indeed satisfies every aspect of the specifications. Other constraints might include portability, namely that the product be constructed in such a way that it can also run on other hardware under the same operating system, or perhaps that the product can run under a variety of different operating systems. Reliability may be another constraint. If the product has to monitor patients in an intensive care unit, it is of paramount importance that it should be fully operational 24 hours a day. Rapid response time may be another requirement; a typical constraint in this category might be "99% of all queries of type-4 shall be answered within 0.75 seconds." Response-time constraints have to be expressed in probabilistic terms because the response time will depend on the current load on the computer.

A vital component of the specifications is the set of acceptance criteria. It is important both from the viewpoint of the client and the developer to spell out a series of tests such that if the product passes these tests, it will be deemed to satisfy its specifications, and the developer's job is done. Some of the acceptance criteria may be restatements of the constraints, while others will address different issues. For example, the client might supply the developer with a description of the data that the product must be able to handle. An acceptance criterion would then be that the product correctly process data of this type and that its error-processing capabilities must be able to filter out data that do not conform.

At this stage the developers know what the proposed product has to be able to do, and what the constraints are. Now the second phase of the planning process can be started, namely to determine feasible alternative development strategies and from these to select an optimal strategy.

4.2.2 Determining an Optimal Solution Strategy

The specifications document is a contract between client and developer. It specifies precisely what the product must do and the constraints on the product. Once the development team fully understands the problem, possible solution strategies can be suggested. A solution strategy is a general approach to building the product. For example, one possible solution strategy for a product is to use an online database; another is to use conventional flat files and to extract the information using overnight batch runs. When determining solution strategies it is often a good idea to come up with strategies without worrying about the constraints that were determined during the first phase. Then the various solution strategies can be evaluated in the light of the constraints, and, if necessary, modifications can be made. There are a number of ways of determining whether a specific solution strategy will satisfy the client's constraints. An obvious one is prototyping, which can be a good technique for resolving issues relating to user interfaces and timing constraints, as previously discussed in Section 3.3.3. Other techniques for determining whether constraints will be satisfied include analytic network modeling [MacNair and Sauer, 1985] and simulation [Bratley, Fox, and Schrage, 1987].

During this process a number of solution strategies will be put forward and then discarded. It is important that a written record be kept of all discarded strategies, and why they were rejected. This document will assist the development team if they are ever called upon to justify their chosen strategy. But more important, there is an ever-present danger during the maintenance phase that the process of enhancement will be accompanied by an attempt to come up with a new and unwise solution strategy. Having a record of why certain strategies were rejected during the development phase can be very helpful in this respect.

By this point the development team will have determined one or more possible solution strategies that satisfy the constraints. A two-stage decision now has to be taken. First, whether the client should be advised to computerize, and if so, which of the viable solution strategies should be adopted. The answer to the first question can best be decided by determining whether the benefits to the client of the target product will outweigh its costs. If the client decides to proceed with the project, then the client must inform the development team as to the optimization criterion to be used, such as minimizing the total cost to the client or maximizing the return on investment. The developers will then advise the client as to which of the viable solution strategies will best satisfy the optimization criterion.

4.2.3 Cost – Benefit Analysis

The technique of comparing estimated future benefits against projected future costs is termed *cost–benefit analysis*. To see how this can be done, consider how Krag Central Electric Company (KCEC) decided in 1965 whether or not to computerize their billing system. Billing was then being done manually by 80 clerks who mailed bills every 2 months to KCEC customers. Computerization would require KCEC to buy or lease the software and hardware, including data capture equipment.

One advantage to KCEC would be that bills would be mailed monthly instead of every 2 months, thereby improving the company's cash flow considerably. Furthermore the 80 billing clerks would be replaced by 11 data capture clerks. On the other hand, a complete data-processing department would have to be set up, staffed by well-paid computer professionals. Over a 7-year period, costs were estimated as follows. Salary savings were estimated to be $1,575,000, and improved cash flow was projected to be worth $875,000. The total benefits were therefore estimated at $2,450,000. The cost of hardware and software, including maintenance, was estimated to be $1,250,000. In the first year there would be a conversion cost of $350,000, and the cost of explaining the new computerized system to customers was estimated to be a further $125,000. Thus the total costs were estimated at $1,725,000, about $750,000 less than the estimated benefits. KCEC immediately decided to computerize.

Cost–benefit analysis is not always straightforward. On the one hand, salary savings can be estimated by a management consultant, a CPA can project cash flow improvements, and the costs of hardware, software, and conversion can be estimated by a software engineering consultant. But how is the cost of dealing with customers trying to adjust to computerization to be determined? Or, returning to the example of the three projected builds in Figure 3.8, how did management determine that increased sales as a result of implementing the new commission formula would be worth $50,000, that the productivity boost from the online query system would result in a profit increase of $40,000, and that the value of the errors obviated by handheld order-taking devices would be $20,000?

The point is that tangible benefits are easy to measure, but intangible benefits can be hard to quantify directly. A practical way of assigning a dollar value to intangible benefits is to make assumptions in order to be able to estimate the benefits. These assumptions must then be stated in conjunction with the resulting estimates of the benefits. After all, managers have to make decisions. If no data are available, then making assumptions from which such data can be determined is usually the best that can be done under the circumstances. This approach has the further advantage that if someone else reviewing the data and the underlying assumptions can come up with better assumptions, then better data can be provided, and the associated intangible benefits can be computed more accurately. The same technique can be used for intangible costs.

Once the solution strategy has been selected, the final stage in planning the project is to draw up the project management plan, the controlling document for managing the software development process.

4.2.4 Software Project Management Plan

At this stage the plan itself can be drawn up. As discussed in Section 4.1, this involves determining the various work units, estimating the resources required and the budget, and coming up with a detailed timetable showing what is to be done when, what resources will be required to do it, and what it will cost, all as a function of time. To do this requires a number of tools such as cost models which will be discussed in Section 4.5. At this stage, only the framework for presenting the plan will be discussed; the rest will follow.

Figure 4.2 shows the software project management plan (SPMP) prescribed in IEEE Draft Standard 1058 [IEEE 1058, 1987]. While there are many other ways of drawing up an SPMP, there will be distinct advantages to following a future IEEE standard. First, it is a standard drawn up by representatives of many different major organizations involved in software. There is input from both industry and universities, and the members of the working group and the reviewing teams between them have literally millions of hours of experience in drawing up project management plans. The draft standard incorporates this experience. A second advantage is that the IEEE SPMP is designed for use with all types of software product, irrespective of

1 Introduction
- 1.1 Project Overview
- 1.2 Project Deliverables
- 1.3 Evolution of the Software Project Management Plan
- 1.4 Reference Materials
- 1.5 Definitions and Acronyms

2 Project Organization
- 2.1 Process Model
- 2.2 Organizational Structure
- 2.3 Organizational Boundaries and Interfaces
- 2.4 Project Responsibilities

3 Managerial Process
- 3.1 Management Objectives and Priorities
- 3.2 Assumptions, Dependencies, and Constraints
- 3.3 Risk Management
- 3.4 Monitoring and Controlling Mechanisms
- 3.5 Staffing Plan

4 Technical Process
- 4.1 Methods, Tools, and Techniques
- 4.2 Software Documentation
- 4.3 Project Support Functions

5 Work Packages, Schedule, and Budget
- 5.1 Work Packages
- 5.2 Dependencies
- 5.3 Resources Requirements
- 5.4 Budget and Resource Allocation
- 5.5 Schedule

Additional Components

Figure 4.2 Components of IEEE software project management plan [IEEE 1058, 1987]. (© 1987 IEEE.)

size. It does not impose a specific life-cycle model, nor are specific techniques prescribed. The plan is essentially a framework, the contents of which will be tailored in a different way by each organization for a particular application area, development team, or method. By adhering to this framework on an industry-wide basis, the advantages of standardization will accrue. Eventually, when an organization hires new personnel, they will not require training in the format used for the organization's SPMP, because the last firm they worked for will also be using the same SPMP format, namely the IEEE Software Project Management Plan described in detail in the following section.

4.3 IEEE Software Project Management Plan

The plan framework itself is now described in detail. The numbers and headings in the text correspond to the entries in Figure 4.2. The various terms used have been defined in Section 4.1.

1. Introduction: The five subsections comprising this section of the SPMP provide an overview of the project and of the product to be developed.

1.1. Project Overview: In this subsection a brief description is given of the project objectives, the product to be delivered, the activities, and their resulting work products. In addition, the milestones are listed, as are the resources required. The master schedule and master budget are also given in this subsection.

1.2. Project Deliverables: All the items to be delivered to the client are listed here, together with the delivery dates.

1.3. Evolution of the Software Project Management Plan: No plan can be cast in concrete. The SPMP, like any other plan, requires continual updating in the light of experience and of change within both the client organization and the software development organization. In this section the formal procedures and mechanisms for changing the plan are described.

1.4. Reference Materials: All documents referenced in the SPMP are listed here.

1.5. Definitions and Acronyms: This information is needed in order to ensure that the SPMP is unambiguously understood by every reader.

2. Project Organization: In the four subsections comprising the project organization section the developers specify how the product is to be developed, both from the viewpoint of the life cycle of the product and the organizational structure of the developers.

2.1. Process Model: In this subsection the life-cycle model is specified in terms of the activities, such as designing the product or performing integration testing, and project functions, such as project management or configuration management. Key aspects here include specifications of milestones, baselines, reviews, work products, and deliverables.

2.2. Organizational Structure: The management structure of the development organization is described in this section. It is important to demarcate the lines of authority and responsibility within the organization.

2.3. Organizational Boundaries and Interfaces: No project is constructed in a vacuum. The project members have to interact with the client organization and with other members of their own organization. In addition, subcontractors may be involved in a large project. Administrative and managerial boundaries between the project itself and these other entities must be laid down. In addition, many software development organizations are divided into two types of groups, development groups who work on a single project and support groups who provide support functions such as configuration management and SQA on an organizational-wide basis. Administrative and managerial boundaries between the project group and the support groups must also be clearly defined.

2.4. Project Responsibilities: For each project function such as SQA and for each activity such as integration testing, the responsible individual must be identified.

3. Managerial Process: In the five subsections which comprise this section of the SPMP a description is given of how the project is to be managed.

3.1. Management Objectives and Priorities: In this subsection the philosophy, goals, and priorities for management are described. The types of items that appear in this section of the plan may include frequency and mechanism of reporting, the relative priorities among requirements, schedule and budget for the project, and risk management procedures.

3.2. Assumptions, Dependencies, and Constraints: The need for these items was described in Section 4.2.1.

3.3. Risk Management: The various risk factors associated with the project are listed in this subsection, as well as the mechanisms to be used for tracking these risk factors.

3.4. Monitoring and Controlling Mechanisms: In this subsection reporting mechanisms for the project are described in detail, including review and audit mechanisms.

3.5. Staffing Plan: The personnel who will be staffing the project constitute an important resource. In this subsection the numbers and types of personnel required are listed, together with the durations for which they will be needed.

4. Technical Process: Technical aspects of the project are specified in the three subsections constituting this section.

4.1. Methods, Tools, and Techniques: Technical aspects of hardware and software are described in detail in this subsection. Items that should be covered include the computing systems (hardware, operating systems, and software) to be used for developing the product, as well as the target systems on which the product will be run. Other necessary aspects include development methods, team structure, programming language(s), tools, and techniques to be employed. In addition, technical standards such as documentation standards and coding standards are included, perhaps by reference to other documents, as well as procedures for developing and modifying work products.

4.2. Software Documentation: This subsection contains the documentation requirements, namely the milestones, baselines, and reviews for software documentation.

4.3. Project Support Functions: In this subsection appear the detailed plans for the supporting functions such as configuration management and quality assurance.

5. Work Packages, Schedule, and Budget: In the five subsections comprising this section the emphasis is on the work packages, their interdependencies, resource requirements, and associated budgeting allocations.

5.1. Work Packages: In this subsection the work packages are specified, with their associated work products broken down into activities and tasks.

5.2. Dependencies: Module coding follows design and precedes integration testing. In general, there will be interdependencies among the work packages and dependencies on external events. In this subsection these dependencies are specified.

5.3. Resource Requirements: To complete the project, a wide variety of resources will be required. In this subsection the total resources are presented as a function of time.

5.4. Budget and Resource Allocation: The various resources listed previously cost money. Just as the resource needs vary with time, so does the allocation of the budget to the various component resources. The allocation of resources and budget to the various project functions, activities, and tasks is presented in this subsection.

5.5. Schedule: In the final subsection a detailed schedule is given for each component of the project. This master plan will then be followed, hopefully resulting in a project that will be completed on time and within budget.

Additional Components: For certain projects, additional components may need to appear in the plan. In terms of the IEEE framework they appear at the end of the plan. Such additional components may include subcontractor management plans, security plans, independent verification and validation plans (Section 4.4.1), training plans, hardware procurement plans, installation plans, and the product maintenance plan.

4.4 Quality Issues

The term "software quality" is frequently misunderstood within the software context. After all, the word "quality" implies excellence of some sort, but this is unfortunately not what is implied in the software context. To put it bluntly, the state of the art in software development is such that merely getting the software to function correctly is enough—excellence is an order of magnitude more than what will totally satisfy almost all clients. Informally then, the quality of software is the extent to which the product satisfies its requirements.

4.4.1 Independent Verification and Validation

Before describing software quality assurance, the issue of independence must be raised. It is important that software quality be determined by an external group, personnel outside the software development group. In the late 1950s, the Atlas Missile Program made use of an independent software tester to ensure unbiased evaluation of the software. This outside evaluator is now termed a *verification and validation* (V & V) contractor. The two terms, verification and validation, do not mean exactly the same thing. Verification applies to a single activity of software development. It is the process of determining whether the work products of that activity satisfy their requirements. That is to say, *verification* is the process of determining whether a particular activity has been completed correctly. *Validation*, on the other hand, consists of determining whether the product as a whole satisfies its requirements. To add to the confusion, the term "verification" is also used by some software engineers to mean proving a product correct using mathematical techniques as described in Section 6.3. For the sake of clarity, the abbreviation V & V will be used to denote verification and validation as described in this paragraph.

A more important abbreviation is IV & V, *independent* verification and validation. It is absolutely essential that the group responsible for V & V be independent of the team who develop the software. At best, the IV & V group should be employees from a different organization, but if the IV & V group is from the same organization as the development team, then it is mandatory that the two be managerially independent. That is to say, the software development team should report to one manager, the IV & V group to another. If the development manager wishes to deliver a particular product to the client, but the IV & V manager feels that the product has not yet been adequately tested, then neither manager can overrule the other. The decision whether to release the product on time with the risk that it is still full of faults, or to release it late with the risk of antagonizing the client, cannot be taken either by the manager responsible for development or by the manager responsible for IV & V. Instead, the senior manager to whom both these managers report should decide whether the development organization should risk the consequences of delivering a substandard product on time or of delivering a quality product past its deadline. The existence of an independent V & V group ensures that if quality is to be compromised it must be done for good reason, not merely to ensure that a deadline is met.

4.4.2 Software Quality Assurance

Software quality assurance (SQA) goes further than IV & V. As mentioned previously, verification takes place when an activity has been completed, and validation occurs when the product as a whole is complete. In contrast, the term "quality assurance" applies to every aspect of the software development process. Development of the various standards to which the software must conform fall within the ambit of SQA, as do the monitoring procedures for assuring compliance with those standards. The SQA group is responsible for performing the IV & V or for ensuring that it is adequately performed by an independent group.

At first sight, having a separate SQA group would appear to add to the cost of software development. But this is not so. If there is no SQA group, then every member of the software development organization is involved to some extent in verification, validation, and other quality assurance activities. Suppose that there are 100 software professionals in the organization, and that they devote, on average, 30% of their time to quality assurance activities. The 100 individuals should be divided into two groups. Software development is then performed by 70 individuals, and the other 30 are responsible for SQA. The same amount of time is then devoted to SQA as before, and the only additional expense is a manager to lead the SQA group. In return for this expenditure, SQA can be performed by an independent group. In addition, they will specialize in SQA, leading to products of higher quality than when quality assurance activities are performed throughout the organization.

Like every other component of the software development process, SQA activities must be planned. Once the SQA plan has been drawn up, the details should appear in section 3.4 of the SPMP.

4.5 Estimating Time and Cost

The budget is an integral part of any software project management plan. The client will want to know how much he or she will have to pay for the product before development commences. If the development team underestimates the actual cost, then building the product will cause the development organization to lose money. On the other hand, if the development team overestimates, then the client may decide that on the basis of cost–benefit analysis or return on investment there is no point in having the product built. Alternatively, the client may give the job to another development organization whose estimate is more reasonable. Either way, it is clear that accurate cost estimation is critical.

Another important part of any plan is timing. The client will certainly want to know when the complete product will be delivered. If the development organization is unable to keep to its schedule, then at best the organization loses credibility, at worst penalty clauses will be invoked. In all cases, the managers responsible for the software project management plan have a lot of explaining to do. Conversely, if the development organization overestimates the time that will be needed to build the product, there is a good chance that the client will go elsewhere.

Unfortunately, it is by no means easy to estimate cost and time requirements accurately. There are just too many variables to be able to get an accurate handle on either cost or time. One big difficulty is the human factor. Over twenty years ago Sackman and co-workers showed differences of up to 28 to 1 between pairs of programmers [Sackman, Erikson, and Grant, 1968]. It is easy to try to brush off Sackman's results by saying that experienced programmers will always outperform beginners, but that is *not* what was done. Instead, Sackman and his colleagues compared matched pairs of programmers. That is to say, they observed, for example, two programmers with 10 years of experience on similar types of projects and measured the time it took them to perform tasks like coding and debugging. Then they repeated this with, say, two beginners who had been in the profession for the same short length of time and who had similar educational backgrounds. Comparing worst and best performances, they observed differences of 6 to 1 in product size, 8 to 1 in product execution time, 9 to 1 in development time, 18 to 1 in coding time, and 28 to 1 in debugging time. A particularly alarming observation is that the best and worst performances on one product were by two programmers, each of whom had 11 years of experience. Even when the best and worst cases were removed from Sackman's sample, observed differences were still of the order of 5 to 1. On the basis of these results, it is clear that we cannot hope to estimate software cost or completion time with any degree of accuracy. It has been argued that in a large project these differences will tend to cancel out, but this is perhaps wishful thinking; the presence of one or two very good, or very bad, team members will cause marked deviations from planned milestones and impact the budget.

Another human factor that can affect estimation is that in a free country there is no way of ensuring that a critical staff member will not resign during the project and take a position at some other organization. Time and money are then spent hiring a replacement and attempting to fit him or her into the vacated position, or in reorganizing the remaining team members to try to compensate. Either way schedules slip, and estimates come unstuck.

Underlying the cost estimation problem is the issue: How is the size of a product to be measured?

4.5.1 Estimating the Size of a Product

The most common measure of the size of a product is the number of lines of code. Two different units are commonly used, namely lines of code (LOC) and thousand delivered source instructions (KDSI). There are many problems associated with the use of lines of code [van der Poel and Schach, 1983]. First, creation of source code is only a small part of the total software development effort. It seems somewhat far-fetched that the time required for specifying, planning, designing, implementing, integrating, and testing can be expressed solely as a function of the number of lines of code of the final product. Second, implementing the same product in two different languages will result in versions with different lengths of code. Also, with languages such as Lisp or with many nonprocedural fourth-generation languages (4GLs), the concept of LOC or KDSI is not defined. Third, it is often not clear exactly how to count lines of code. Should only executable lines of code be counted, or data definitions as well? Should comments be counted? If not, there is a danger that programmers will be reluctant to spend time on "nonproductive" commenting, but if comments are counted, then there is the opposite danger that programmers will write reams of comments in an attempt to boost their apparent productivity. Also, what about counting job control language statements? And how are changed lines or deleted lines to be counted? In short, the apparently straightforward measure of lines of code (LOC) is anything but straightforward to count. Fourth, not all the code written is delivered to the client. It is not uncommon for half the code to consist of tools needed to support the development effort. Fifth, the number of lines of code in the final product can be determined only when the product is completely finished. Thus basing cost estimation on lines of code is doubly dangerous. To start the estimation process, the number of lines of code in the finished product must be estimated. Then this estimate is used to arrive at the cost of the product. Not only is there uncertainty in every costing technique, but if the input to an uncertain cost estimator is itself uncertain, namely the number of lines of code in a product which has not yet been built, then the reliability of the resulting cost estimate is unlikely to be too high.

Since lines of code are so unreliable, other metrics must be considered. Software Science [Halstead, 1977; Shen, Conte, and Dunsmore, 1983] provides a variety of measures of product size. These are derived from the fundamental metrics of Software Science, namely the number of operands and operators in the software product and the number of unique operands and unique operators. As with lines of code, these numbers can be determined only after the product has been completed, thus severely reducing the predictive power of the metrics. Also, studies have cast doubt on the validity of Software Science [Hamer and Frewin, 1982; Coulter, 1983; Shen, Conte, and Dunsmore, 1983].

An alternative approach to sizing has been the use of metrics based on measurable quantities that can be determined early in the software life cycle. For example, van der Poel and Schach put forward the FFP metric for cost estimation of medium-scale data-processing products, that is to say, products that take between 2 and 10 person-years to complete [van der Poel and

Schach, 1983]. The three basic structural elements of a data-processing product are its files, flows, and processes; the name FFP is an acronym formed from the initial letters of those three elements. A file is defined to be a collection of logically or physically related records that is permanently resident in the product; transaction and temporary files are excluded. A flow is a data interface between the product and the environment such as a screen or a report. A process is a functionally defined logical or arithmetic manipulation of data. Examples are sorting, validating, or updating. Given the number of files Fi, flows Fl, and processes Pr in a product, its size S and cost C are given by

$$S = Fi + Fl + Pr \tag{4.2}$$

$$C = b \times S \tag{4.3}$$

where b is a constant that will vary from organization to organization and measures the efficiency (productivity) of the software development process within that organization. That is to say, the size of a product is simply the sum of the number of files, flows, and processes, a quantity that can be determined once the architectural design is complete. The cost is then proportional to the size, the constant of proportionality being determined by a least-squares fit to cost data relating to products previously developed by that organization. Unlike metrics based on lines of code, the cost can be estimated before coding commences.

The validity and reliability of the metric were demonstrated using a purposive sample which covered a range of medium-scale data-processing applications. Unfortunately, the metric was never extended to include databases, an essential component of many data-processing products.

A similar, but independently developed, metric for the size of a product is one developed by Albrecht based on function points [Albrecht, 1979; Albrecht and Gaffney, 1983]. Albrecht's metric is based on the number of inputs Inp, outputs Out, inquiries Inq, master files Maf, and interfaces Inf. In its simplest form the number of function points FP is given by

$$FP = 4 \times Inp + 5 \times Out + 4 \times Inq + 10 \times Maf + 7 \times Inf \tag{4.4}$$

Because this is a measure of the product's size, it can be used for cost estimation and productivity estimation.

Equation (4.4) is an oversimplification of a three-step calculation. First, each of the components of a product, namely Inp, Out, Inq, Maf, and Inf, must be classified as simple, average, or complex. Each component is then assigned a number of function points depending on its level. For example, an average input is assigned 4 function points, as reflected in Equation (4.4), but a simple input is assigned only 3, while a complex input is assigned 6 function points. The function points assigned to each component are then summed, yielding the unadjusted function points (UFP).

Second, the technical complexity factor (TCF) is computed. This is a measure of the effect of 14 technical factors such as transaction rates, performance, and online updating. Each of these 14 factors is assigned a value from 0 ("not present or no influence") to 5 ("strong influence throughout"). The resulting 14 numbers are summed, yielding the total degree of influence (DI). The TCF is then given by

$$TCF = 0.65 + 0.01 \times DI \tag{4.5}$$

Since DI can vary from 0 to 70, TCF will vary from 0.65 to 1.35.

Finally, FP, the number of function points, is given by

$$FP = UFP \times TCF \tag{4.6}$$

Experiments by Albrecht and others [Albrecht, 1979; Albrecht and Gaffney, 1983; Behrens, 1983; Jones, 1987] to measure software productivity rates have shown a better fit using function points than using KDSI. Jones states that he has observed errors in excess of 800% counting KDSI, but *only* (author's italics) 200% in counting function points [Jones, 1987], a most revealing remark.

To show the superiority of function points over lines of code, Jones cites the example shown in Figure 4.3 [Jones, 1987]. The same product was coded both in assembler and in Ada, and the results compared. First, consider KDSI per person-month. This measure tells us that coding in assembler is apparently 60% more efficient than coding in Ada, a fact which is patently false. Third-generation languages like Ada have superseded assembler simply because it is much more efficient to code in a third-generation language like

	Assembler version	Ada version
Source code size	70 KDSI	25 KDSI
Development costs	$1,043,000	$590,000
KDSI per person-month	0.335	0.211
Cost per source statement	$14.90	$23.60
Function points per person-month	1.65	2.92
Cost per function point	$3023	$1710

Figure 4.3 Comparison of assembler and Ada products [Jones, 1987]. (© 1987 IEEE.)

Ada. Now consider the second measure, cost per source statement. Note that one Ada statement in this product is equivalent to 2.8 assembler statements. Use of cost per source statement as a measure of efficiency again implies that it is more efficient to code in assembler than in Ada, the statement rejected previously as being nonsense. However, when function points per person-month is taken as the metric of programming efficiency, the superiority of Ada over assembler is clearly reflected.

On the other hand, both function points and the FFP metric of Equations (4.2) and (4.3) suffer from the same disadvantage: Product maintenance is often inaccurately measured. When a product is maintained, major changes to the product can be made without changing the number of files, flows, and processes or the number of inputs, outputs, inquiries, master files, and interfaces.

4.5.2 Methods of Cost Estimation

Notwithstanding the difficulties with estimating cost, it is essential that software developers simply do the best they can to obtain accurate estimates of both project time and project cost, while taking into account as many of the factors that can affect their estimates as possible. These include the skill levels of the personnel, the complexity of the project, the size of the project (cost increases with size, but much faster than linearly), familiarity of the development team with the application area, the hardware on which the product is to run, and availability of software tools. Another factor is the deadline effect. If a project has to be completed within a certain time, the effort, in person-years, is greater than if there is no constraint on completion time.

From the preceding list, which is by no means complete, it is clear that estimation is a difficult problem. There are a number of approaches that have been used, with greater or lesser success.

1. Expert Judgement by Analogy: In this technique, a number of experts are consulted. An expert arrives at an estimate by comparing the target product to be constructed to completed products with which the expert was actively involved and noting the similarities and differences. For example, an expert may compare the target product to a similar product developed 2 years ago for which the data were input in batch mode, while the target product is to have online data capture. Since the organization has familiarity with the type of product to be developed, the expert feels that this will reduce development time and cost by 15%. But the online front end is somewhat complex; this will increase time and costs by 25%. Finally, the target product has to be developed in a language with which most of the team members are unfamiliar, thus increasing time by 15% and cost by 20%. Combining these three figures, the expert decides that the target product will take 25% more time and cost 30%

more than the previous one. Thus, since the previous product took 12 months to develop and cost $800,000, the target product will take 15 months and cost $1,040,000.

Two other experts within the organization compare the same products. One comes to the conclusion that on the basis of her judgement the target product will take 13.5 months and cost $1,100,000. The other comes up with the figures of 16 months and $980,000. How can the predictions of these three experts be reconciled? One technique is the Delphi technique, which allows experts to arrive at a consensus without having group meetings, which can have the undesirable side effect of one persuasive member swaying the group. In this technique [Helmer-Hirschberg, 1966] the experts work independently. Each produces a cost estimate together with a rationale for that estimate. These estimates and rationales are then distributed to all the experts, who now produce a second estimate. This process of estimation and distribution continues until the experts hopefully agree to within an accepted tolerance. No group meetings take place during the iteration process.

Valuation of real estate is frequently done on the basis of expert judgement by analogy. An appraiser will compare a house with similar houses which have recently been sold to arrive at a valuation. Suppose that house A is to be valued, house B next door has just been sold for $150,000, and house C on the next street was sold 3 months ago for $165,000. The appraiser may reason as follows: House A has one more bathroom than house B, and the yard is 10,000 square feet larger. House C is approximately the same size as house A, but the roof is in poor condition. On the other hand, it has a jacuzzi. After careful thought, the appraiser may arrive at a figure of $168,000 for house A. In the case of software products, expert judgement by analogy is less precise than real estate valuation. Recall that our first software expert said that using an unfamiliar language would increase time by 15% and cost by 20%. From where do these figures come? Unless the expert has some validated data from which the effect of each and every difference can be determined (a highly unlikely possibility), errors induced by what can only be described as guesses will result in hopelessly incorrect cost estimates. In addition, unless the experts are blessed with total recall, their recollections of completed products may be sufficiently inaccurate as to invalidate their predictions. Finally, experts are human, and therefore have biases. There is always a danger that these biases will affect their predictions. At the same time, the results of estimation by a group of experts should reflect their collective experience; if this is broad enough, the result may well be accurate.

2. Bottom-Up Approach: One way of trying to reduce the errors resulting from evaluating a product as a whole is to break up the product into smaller components. Estimates of cost and time are made for each component separately, and these are then combined to provide an overall figure. This approach has the advantage that estimating costs for several smaller components is generally quicker and more accurate than for one large one. In addition, the estimation process is likely to be more detailed than with one large monolithic product. The disadvantage of this is that a product is often more than the sum of its components; there may be product-level costs involved. But more important, how are the costs of the separate components to be derived? All that has been achieved by breaking down the product is to replace the single problem of estimating time and cost for a large product by the problems of estimating these quantities for a number of smaller problems (plus the difficulty of determining the product-level costs, if any).

3. Algorithmic Models: In this approach a model such as function points or the FFP metric is used to determine product cost. All that the estimator has to do is to measure the parameters of the model, and cost and time estimates can be computed using the algorithm. On the surface an algorithmic model is superior to expert opinion, because a human expert is, as pointed out previously, subject to biases and may overlook certain aspects of the completed and target products. In contrast, an algorithmic model is unbiased; every product is treated the same way. The danger with an algorithmic model is that its estimates are only as good as the underlying assumptions. For example, underlying the function-point model is the assumption that every aspect of a product is embodied in the five quantities on the right-hand side of Equation (4.4) and the 14 technical factors.

Many algorithmic models have been proposed. Some are based on mathematical theories as to how software is developed. For example, underlying the SLIM model is the assumption that personnel utilization during software development is given by the Rayleigh distribution of Equation (4.1) [Putnam, 1978]. Other models are statistically based; large numbers of projects are studied and empirical rules determined from that data. Then there are hybrid models incorporating mathematical equations, statistical modeling, and expert judgement. Once such model is the RCA PRICE S model [Freiman and Park, 1979]. The most important hybrid model is Boehm's COnstructive COst MOdel (COCOMO), which is described in detail in the next section.

4.5.3 COCOMO Model

The COCOMO model is actually a series of three models, ranging from a macroestimation model, which treats the product as a whole, down to a microestimation model, which treats the product in detail. In this section a description is given of Intermediate COCOMO, which has a middle level of complexity and detail. The COCOMO models are described in detail in [Boehm, 1981]; an overview is presented in [Boehm, 1984a].

Computing the development time using Intermediate COCOMO is done in two stages. First, a rough estimate is provided of the development effort. Two parameters have to be estimated, the length of the product in KDSI (thousand delivered source instructions) and the product's development mode, the intrinsic level of difficulty of developing the product. From these two parameters the *nominal effort* can be computed. For example, if the project is adjudged to be essentially straightforward (*organic mode*), then the nominal effort (in person-months) is given by

$$\text{Nominal effort} = 3.2 \times (\text{KDSI})^{1.05} \text{ person-months} \qquad (4.7)$$

For example, if the product to be built is organic and is estimated to be 12,000 delivered source statements (12 KDSI), then the nominal effort is

$$3.2 \times (12)^{1.05} = 43 \text{ person-months}$$

Next, this nominal value must be multiplied by 15 *software development effort multipliers*. These multipliers and their values are given in Figure 4.4. Each of these multipliers can have up to six values. For example, the product complexity multiplier is assigned the values 0.75, 0.85, 1.00, 1.15, 1.30, or 1.65, according to whether the developers rate the project complexity as very low, low, nominal (average), high, very high, or extra high. As can be seen from Figure 4.4, all 15 multipliers take on the value 1.00 when the corresponding parameter is nominal. Boehm has provided guidelines to help the developer determine whether the parameter should indeed be rated nominal or whether the rating is low or high. For example, consider again the module complexity multiplier. If the control operations of the module essentially consist of a sequence of the operations of structured programming (such as **if-then-else, do-while, case**), then the complexity is rated *very low*. If these operators are nested, then the rating is *low*. Adding intermodule control and decision tables increases the rating to *nominal*. If the operators are highly nested, with compound predicates, and if there are queues and stacks, then the rating is *high*. The presence of reentrant and recursive coding and fixed-priority interrupt handling pushes the rating to *very high*. Finally, multiple resource scheduling with dynamically changing priorities and microcode-level control ensures that the rating is *extra high*. These ratings apply to control operations. A module also has to be evaluated from the viewpoint of computational operations, device-dependent operations, and data management operations. For details of the criteria for computing each of the 15 multipliers, the reader should refer to [Boehm, 1981].

To see how this works, in [Boehm, 1984a] the example is given of microprocessor-based communications processing software for a highly ambi-

	Rating					
Cost Drivers	Very Low	Low	Nominal	High	Very High	Extra High
Product Attributes						
Required software reliability	0.75	0.88	1.00	1.15	1.40	
Data base size		0.94	1.00	1.08	1.16	
Product complexity	0.70	0.85	1.00	1.15	1.30	1.65
Computer Attributes						
Execution time constraint			1.00	1.11	1.30	1.66
Main storage constraint			1.00	1.06	1.21	1.56
Virtual machine volatility*		0.87	1.00	1.15	1.30	
Computer turnaround time		0.87	1.00	1.07	1.15	
Personnel Attributes						
Analyst capabilities	1.46	1.19	1.00	0.86	0.71	
Applications experience	1.29	1.13	1.00	0.91	0.82	
Programmer capability	1.42	1.17	1.00	0.86	0.70	
Virtual machine experience*	1.21	1.10	1.00	0.90		
Programming language experience	1.14	1.07	1.00	0.95		
Project Attributes						
Use of modern programming practices	1.24	1.10	1.00	0.91	0.82	
Use of software tools	1.24	1.10	1.00	0.91	0.83	
Required development schedule	1.23	1.08	1.00	1.04	1.10	

*For a given software product, the underlying virtual machine is the complex of hardware and software (operating system, database management system) it calls on to accomplish its task.

Figure 4.4 Intermediate COCOMO software development effort multipliers [Boehm, 1984a]. (© 1984 IEEE.)

tious new electronic funds transfer network with high reliability, performance, development schedule, and interface requirements. This product fits the complex ("embedded") mode and is estimated to be 10,000 delivered source instructions (10 KDSI) in length, so the nominal development effort is given by

$$\text{Nominal effort} = 2.8 \times (\text{KDSI})^{1.20} \quad (4.8)$$

Since the project is estimated to be 10 KDSI in length, the nominal effort is

$$2.8 \times (10)^{1.20} = 44 \text{ person-months}$$

The estimated development effort is obtained by multiplying the nominal effort by the 15 multipliers. The ratings of these multipliers and their values are given in Figure 4.5. Using these values, the product of the multipli-

Cost Drivers	Situation	Rating	Effort Multiplier
Required software reliability	Serious financial consequences of software faults	High	1.15
Data base size	20,000 bytes	Low	0.94
Product complexity	Communications processing	Very High	1.30
Execution time constraint	Will use 70% of available time	High	1.11
Main storage constraint	45K of 64K store (70%)	High	1.06
Virtual machine volatility	Based on commercial microprocessor hardware	Nominal	1.00
Computer turnaround time	Two-hour average turnaround time	Nominal	1.00
Analyst capabilities	Good senior analysts	High	0.86
Applications experience	Three years	Nominal	1.00
Programmer capability	Good senior programmers	High	0.86
Virtual machine experience	Six months	Low	1.10
Programming language experience	Twelve months	Nominal	1.00
Use of modern programming practices	Most techniques in use over one year	High	0.91
Use of software tools	At basic minicomputer tool level	Low	1.10
Required development schedule	Nine months	Nominal	1.00

Figure 4.5 Intermediate COCOMO effort multiplier ratings for microprocessor communications software [Boehm, 1984a]. (© 1984 IEEE.)

ers is found to be 1.35, so the estimated effort for the project is

$$1.35 \times 44 = 59 \text{ person-months}$$

From this figure, additional factors can be used to determine dollar costs, development schedules, phase and activity distributions, computer costs, annual maintenance costs, and other related items—for details, see [Boehm, 1981]. In other words, Intermediate COCOMO is a complete algorithmic cost estimation model, giving the user virtually every conceivable assistance in project planning.

The COCOMO model has been validated with respect to a broad sample of 63 projects covering a wide variety of application areas. The results of applying Intermediate COCOMO to this sample are that the actual values come within 20% of the predicted values about 68% of the time. Although attempts have been made to improve upon this accuracy, this makes little sense because the data are generally in the $\pm 20\%$ range. Nevertheless, the accuracy obtained places Intermediate COCOMO at the cutting edge of cost estimation research; no other technique is consistently as accurate.

The major problem with the COCOMO model is that its most important input is the number of lines of code in the target product. If this estimate is incorrect, then it affects every single prediction of the model.

Because of the possibility that the predictions of the COCOMO model or any other estimation technique may be inaccurate, it is necessary for management to monitor all predictions continually during software development.

4.5.4 Tracking Time and Cost Estimates

While the product is being developed, the actual development effort must be constantly compared against the predictions made. For example, suppose that the estimation model used by the software developers predicted that the specification phase would last 3 months and require 7 person-months of effort. However, 4 months have gone by and 10 person-months of effort have been expended, but the specifications document is by no means complete. Deviations of this kind can serve as an early warning that something has gone wrong and that corrective action must be taken. The problem could be that the size of the product was seriously underestimated, or that the development team is not as competent as it was thought to be. Whatever the reason, there are going to be serious time and cost overruns, and management must take appropriate action now to minimize the effects of this.

Careful tracking of predictions must be done throughout the development life cycle, irrespective of the methods by which the predictions were made. Deviations could be due to faulty predictions, inefficient software development, a combination of both, or some other reason. The important thing is to detect deviations early and to take immediate corrective action.

4.6 Training Requirements

When the subject of training is raised during the planning phase, a common response is, "We don't need to worry about training until the product is finished, and then we can train the user." This is a somewhat unfortunate remark, implying as it does that only users require training. In fact, training may also be needed by members of the development team, starting with training in software planning. When new software development methods are used such as new design techniques or testing procedures, training must be provided to every member of the team involved to a lesser or greater extent in using the method. The introduction of hardware or software tools of any sort such as workstations or an integrated project support environment (IPSE) (see Section 12.4) requires similar training. Programmers may need training in the operating system of the machine to be used for product development, as well as in the implementation language. Documentation preparation training is frequently overlooked, as evidenced by the poor quality of so much documentation. Computer operators will certainly require some sort of training to be able to run the new product and may require additional training if new hardware is to be utilized.

The required training can be obtained in a number of ways. The easiest and least disruptive is in-house training, either by fellow employees or by consultants. There are many commercial establishments offering a variety of training courses; these courses may or may not be exactly what is required. Often, colleges offer training courses in the evenings. Videotape-based courses for self-instruction are another alternative.

Once the training needs have been determined and the training plan drawn up, the plan must be incorporated into the SPMP.

4.7 Documentation Standards

The development of a software product is accompanied by a wide variety of documentation. Jones has found that 28 pages of documentation were generated per 1000 instructions (KDSI) for an IBM internal commercial product around 50 KDSI in size, and about 66 pages per KDSI for a commercial software product of the same size. Operating system IMS/360 Version 2.3 was about 166 KDSI in size, and 157 pages of documentation per KDSI were produced. The documentation was of various types, including planning, control, financial, and technical [Jones, 1986]. In addition to these types of documentation, source code itself is a form of documentation, and comments within the code constitute further documentation.

A considerable portion of the software development effort is absorbed by documentation. A survey of 63 development projects and 25 maintenance projects showed that for every 100 hours spent on activities related to code, 150 hours were spent on activities related to documentation [Boehm, 1981]. For large TRW products, the proportion of time devoted to documentation-related activities rose to 200 hours [Boehm, Penedo, Stuckle, Williams, and Pyster, 1984].

Standards are needed for every type of documentation. Uniformity in, say, design documentation within an organization reduces misunderstandings between team members and aids the quality assurance group. While new employees have to be trained in the documentation standards, no further training will be needed when existing employees move from project to project within the organization. From the viewpoint of product maintenance, uniform coding standards assist maintenance programmers in understanding source code. Standardization is even more important for user manuals, because these have to be read by a variety of personnel, few of whom are computer experts. The IEEE is currently drafting a standard for user manuals (IEEE Draft Standard P1063 for Software User Documentation).

As part of the planning process, standards must be set up for all documentation to be produced during software production. These standards are incorporated in the SPMP. Where a preexisting standard is to be used, such as the IEEE Standard for Software Test Documentation [ANSI/IEEE 829, 1983], the standard is listed in section 1.4 of the SPMP (Reference Materials). If a standard is specially written for the development effort, then it appears in section 4.1 (Methods, Tools, and Techniques).

Documentation is an essential aspect of the software production effort; in a very real sense the product *is* the documentation. Planning the documentation effort in every detail, and then ensuring that the plan is adhered to, is a critical component of successful software production.

Chapter Review

The main theme of this chapter is the importance of planning in the software production process. The three major components of a software project management plan, namely the work to be done, the resources with which to do it, and the money to pay for it, are outlined in Section 4.1. In Section 4.2 the planning process is described. One particular software project management plan, namely the IEEE draft standard, is described in detail in Section 4.3. Quality issues are highlighted in Section 4.4; emphasis is placed on *independent* software quality assurance (Section 4.4.2). A vital component of any software project management plan is estimating the time and the cost (Section 4.5). Several measures are put forward for estimating the size of a product, including function points (Section 4.5.1). Various methods of cost estimation are then described, especially the Intermediate COCOMO model (Section 4.5.3). The chapter concludes with sections on training requirements and documentation standards (Sections 4.6 and 4.7).

For Further Reading

An introduction to software management can be found in [DeMarco, 1982]. A somewhat more advanced text is [Jones, 1986]. Two IEEE Computer Society tutorials, namely [Reifer, 1986] and [Thayer, 1988], consist of papers that are good sources for more information on software management in general, as well as planning in particular.

There are many excellent books on cost–benefit analysis, including [Dasgupta and Pearce, 1972], [Mishan, 1982], and [Pearce, 1971]. For information on cost–benefit analysis as applied to information systems, the reader should consult [King and Schrems, 1978].

Books on software quality assurance include [Dunn and Ullman, 1982] and [Schulmeyer and McManus, 1987]. [Beizer, 1984] has two chapters on SQA. There is also an IEEE Computer Society tutorial containing a number of useful papers on SQA [Chow, 1985]. Practical experiences with SQA are described in [Gustafson and Kerr, 1982] and [Poston and Bruen, 1987].

For further information on IEEE Draft Standard 1058 for Software Project Management Plans, the draft standard itself should be carefully read [IEEE 1058, 1987].

Sackman's work is described in [Sackman, Erikson, and Grant, 1968]. A more detailed source is [Sackman, 1970].

Many critical reviews of Software Science have been published, including [Hamer and Frewin, 1982], [Coulter, 1983], and [Shen, Conte, and Dunsmore, 1983]. A further source of information on Software Science is [Conte, Dunsmore, and Shen, 1986]; the book also contains analyses of a wide number of other metrics, including lines of code, Putnam's SLIM model, COCOMO, and function points. Information on function points can also be found in [Albrecht, 1979], [Albrecht and Gaffney, 1983], [Behrens, 1983], and [Jones, 1987]. A careful analysis of function points, as well as suggested improvements, appears in [Symons, 1988].

For the reader who is interested in metrics in general, important books include [Conte, Dunsmore, and Shen, 1986] and [Musa, Iannino, and Okumoto, 1987]. Over 120 papers on software metrics published between 1980 and 1988 are listed in [Côté, Bourque, Oligny, and Rivard, 1988]. Data on the validity of function points as well as other metrics can be found in [Kemerer, 1987].

The theoretical justification for the COCOMO model, together with full details for implementing it, appears in [Boehm, 1981]; a shorter version is to be found in [Boehm, 1984a].

Standards of various types are described in [Branstad and Powell, 1984]. A complete listing of IEEE standards can be obtained from IEEE, 345 East 47th Street, New York, NY 10017.

Problems

4.1 Show that the Rayleigh distribution [Equation (4.1)] attains its maximum value when $t = k$. Find the corresponding resource consumption.

4.2 Why should the following constraints not appear in a specifications document:
- (i) The communications network must be set up at a reasonable cost.
- (ii) Response time for a type-3 query should not exceed 4.5 seconds.
- (iii) The product must significantly increase productivity within the sales division.

4.3 Why do you think that some cynical software organizations refer to *milestones* as *millstones*? (*Hint*: Look up the figurative meaning of "millstone" in your dictionary.)

4.4 A new form of gastrointestinal disease is sweeping the country of Rojbari. Like histoplasmosis, it is transmitted as an airborne fungus. Although the disease is almost never fatal, an attack is very painful, and the sufferer is unable to work for about 2 weeks. The government of Rojbari wishes to determine how much money, if any, to spend on attempting to eradicate the disease. The committee charged with advising the Secretary for Health is considering four aspects of the problem, namely health care costs (Rojbari provides free health care for all its citizens), loss of earnings (and hence loss of taxes), pain and discomfort, and gratitude towards the government. Explain how cost–benefit analysis can be of assistance to the committee. For each benefit or cost, suggest how a dollar estimate for that benefit or cost could be obtained.

4.5 A software development organization currently employs 70 software professionals, including 12 managers, all of whom do both development and verification and validation of software. Latest figures show that 29% of their time is spent on verification and validation. The average annual cost to the company of a manager is $112,000, while nonmanagerial professionals cost $83,000 a year on average; both figures include overheads. Use cost–benefit analysis to determine whether a separate SQA group should be set up within the organization.

4.6 Repeat the cost–benefit analysis of Problem 4.5 for a firm with only seven software professionals, including two managers. Assume that the other figures remain unchanged.

4.7 You are a software specialist at PWV Software Developers. A year ago your manager announced that the product on which you would be working would comprise 12 files, 84 flows, and 113 processes.

(i) Using the FFP metric, determine its size.

(ii) For PWV Software, the constant b in Equation (4.3) has been determined to be $802. What was the cost estimate for the project predicted by the FFP metric?

(iii) The product was recently completed, and today it was announced that the project cost $162,000. What does this tell you about the productivity of your development team?

4.8 A target product will have 6 simple inputs, 3 average inputs, and 9 complex inputs. There are 17 average outputs, 51 average inquiries, 10 average master files, and 22 average interfaces. Determine the *UFP* (unadjusted function points).

4.9 If the total degree of influence for the product of Problem 4.8 is 39, determine the number of function points.

4.10 You are in charge of developing a 43-KDSI embedded product that is nominal except that the database size is rated very high and the use of software tools is low. Using the Intermediate COCOMO model, what is the estimated effort in person-months?

4.11 You are in charge of developing two 25-KDSI organic mode products. Both are nominal in every respect except that product P_1 has extra high complexity and product P_2 has extra low complexity. To develop the product, you have two teams at your disposal. Team A has very high analyst capabilities, applications experience, and programmer capability, and high virtual machine experience and programming language experience. Team B is rated very low on all five attributes.

(i) What is the total effort (in person-months) if team A develops product P_1 and team B develops product P_2?

(ii) What is the total effort (in person-months) if team B develops product P_1 and team A develops product P_2?

(iii) Which of the two preceding staffing assignments makes more sense? Is your intuition backed by the predictions of the COCOMO model?

4.12 You are in charge of developing a 28-KDSI organic mode product.

(i) Assuming a cost of $7500 per person-month, how much is the project estimated to cost?

(ii) Your entire development personnel resigns at the start of the project. You are fortunate enough to be able to replace them by very highly experienced and capable personnel, but the cost per person-month will rise to $8500. How much money do you expect to gain (or lose) as a result of the personnel change?

4.13 (Term Project) Draw up a software project management plan for the Plain Vanilla Ice Cream Corporation product described in the Appendix.

4.14 (Readings in Software Engineering) Your instructor will distribute copies of [Boehm, 1984a]. What types of forecasts do you feel are necessary for managing the software development process adequately but are not supplied by the Intermediate COCOMO model?

References

[Albrecht, 1979] A. J. ALBRECHT, "Measuring Application Development Productivity," *Proceedings of the IBM SHARE / GUIDE Applications Development Symposium*, Monterey, CA, October 1979, pp. 83–92.

[Albrecht and Gaffney, 1983] A. J. ALBRECHT AND J. E. GAFFNEY, JR., "Software Function, Source Lines of Code, and Development Effort Prediction: A Software Science Validation," *IEEE Transactions on Software Engineering* **SE-9** (November 1983), pp. 639–648.

[ANSI/IEEE 829, 1983] "Software Test Documentation," ANSI/IEEE 829-1983, American National Standards Institute, Inc., Institute of Electrical and Electronic Engineers, Inc., 1983.

[Behrens, 1983] C. A. BEHRENS, "Measuring the Productivity of Computer Systems Development Activities with Function Points," *IEEE Transactions on Software Engineering* **SE-9** (November 1983), pp. 648–652.

[Beizer, 1984] B. BEIZER, *Software System Testing and Quality Assurance*, Van Nostrand Reinhold, New York, NY, 1984.

[Boehm, 1981] B. W. BOEHM, *Software Engineering Economics*, Prentice-Hall, Englewood Cliffs, NJ, 1981.

[Boehm, 1984a] B. W. BOEHM, "Software Engineering Economics," *IEEE Transactions on Software Engineering* **SE-10** (January 1984), pp. 4–21.

[Boehm, Penedo, Stuckle, Williams, and Pyster, 1984] B. W. BOEHM, M. H. PENEDO, E. D. STUCKLE, R. D. WILLIAMS, AND A. B. PYSTER, "A Software Development Environment for Improving Productivity," *IEEE Computer* **17** (June 1984), pp. 30–44.

[Brandstad and Powell, 1984] M. BRANSTAD AND P. B. POWELL, "Software Engineering Project Standards," *IEEE Transactions on Software Engineering* **SE-10** (January 1984), pp. 73–78.

[Bratley, Fox, and Schrage, 1987] P. BRATLEY, B. L. FOX, AND L. E. SCHRAGE, *Guide to Simulation*, Second Edition, Springer-Verlag, New York, NY, 1987.

[Chow, 1985] T. S. CHOW (Editor), *Tutorial: Software Quality Assurance: A Practical Approach*, IEEE Computer Society Press, Washington, DC, 1985.

[Conte, Dunsmore, and Shen, 1986] S. D. CONTE, H. E. DUNSMORE, AND V. Y. SHEN, *Software Engineering Metrics and Models*, Benjamin/Cummings, Menlo Park, CA, 1986.

[Côté, Bourque, Oligny, and Rivard, 1988] V. CÔTÉ, P. BOURQUE, S. OLIGNY, AND N. RIVARD, "Software Metrics: An Overview of Recent Results," *Journal of Systems and Software* **8** (March 1988), pp. 121–131.

[Coulter, 1983] N. S. COULTER, "Software Science and Cognitive Psychology," *IEEE Transactions on Software Engineering* **SE-9** (March 1983), pp. 166–171.

[Dasgupta and Pearce, 1972] A. K. DASGUPTA AND D. W. PEARCE, *Cost–Benefit Analysis*, MacMillan, London, UK, 1972.

[DeMarco, 1982] T. DEMARCO, *Controlling Software Projects: Management, Measurement, and Estimation*, Yourdon Press, New York, NY, 1982.

[Dunn and Ullman, 1982] R. DUNN AND R. ULLMAN, *Quality Assurance for Computer Software*, McGraw-Hill, New York, NY, 1982.

[Freiman and Park, 1979] F. R. FREIMAN AND R. E. PARK, "PRICE Software Model—Version 3: An Overview," *Proceedings of the IEEE-PINY Workshop on Quantitative Software Models*, October 1979, pp. 32–41.

[Gustafson and Kerr, 1982] G. G. GUSTAFSON AND R. J. KERR, "Some Practical Experience with a Software Quality Assurance Program," *Communications of the ACM* **25** (January 1982), pp. 4–12.

[Halstead, 1977] M. H. HALSTEAD, *Elements of Software Science*, Elsevier North-Holland, New York, NY, 1977.

[Hamer and Frewin, 1982] P. G. HAMER AND G. D. FREWIN, "M. H. Halstead's Software Science—A Critical Examination," *Proceedings of the IEEE Sixth International Conference on Software Engineering*, Tokyo, Japan, 1982, pp. 197–205.

[Helmer-Hirschberg, 1966] O. HELMER-HIRSCHBERG, *Social Technology*, Basic Books, New York, 1966.

[IEEE 1058, 1987] "Draft Standard for Software Project Management Plans," IEEE 1058-1987, Institute of Electrical and Electronic Engineers, Inc., 1987.

[Jones, 1986] C. JONES, *Programming Productivity*, McGraw-Hill, New York, NY, 1986.

[Jones, 1987] C. JONES, Letter to the Editor, *IEEE Computer* **20** (December 1987), p. 4.

[Kemerer, 1987] C. F. KEMERER, "An Empirical Validation of Software Cost Estimation Models," *Communications of the ACM* **30** (May 1987), pp. 416–429.

[King and Schrems, 1978] J. L. KING AND E. L. SCHREMS, "Cost–Benefit Analysis in Information Systems Development and Operation," *ACM Computing Surveys* **10** (March 1978), pp. 19–34.

[MacNair and Sauer, 1985] E. A. MACNAIR AND C. H. SAUER, *Elements of Practical Performance Modeling*, Prentice-Hall, Englewood Cliffs, NJ, 1985.

[Mishan, 1982] E. J. MISHAN, *Cost–Benefit Analysis: An Informal Introduction*, Third Edition, George Allen & Unwin, London, UK, 1982.

[Musa, Iannino, and Okumoto, 1987] J. D. MUSA, A. IANNINO, AND K. OKUMOTO, *Software Reliability: Measurement, Prediction, Application*, McGraw-Hill, New York, NY, 1987.

[Norden, 1958] P. V. NORDEN, "Curve Fitting for a Model of Applied Research and Development Scheduling," *IBM Journal of Research and Development* **2** (July 1958), pp. 232–248.

[Pearce, 1971] D. W. PEARCE, *Cost–Benefit Analysis*, Macmillan, London, UK, 1971.

[Poston and Bruen, 1987] R. M. POSTON AND M. W. BRUEN, "Counting Down to Zero Software Failures," *IEEE Software* **4** (September 1987), pp. 54–61.

[Putnam, 1978] L. N. PUTNAM, "A General Empirical Solution to the Macro Software Sizing and Estimating Problem," *IEEE Transactions on Software Engineering* **SE-4** (July 1978), pp. 345–361.

[Reifer, 1986] D. J. REIFER (Editor), *Tutorial: Software Management*, Third Edition, IEEE Computer Society Press, Washington, DC, 1986.

[Sackman, 1970] H. SACKMAN, *Man–Computer Problem Solving: Experimental Evaluation of Time-Sharing and Batch Processing*, Auerbach, Princeton, NJ, 1970.

[Sackman, Erikson, and Grant, 1968] H. SACKMAN, W. J. ERIKSON, AND E. E. GRANT, "Exploratory Experimental Studies Comparing Online and Offline Programming Performance," *Communications of the ACM* **11** (January 1968), pp. 3–11.

[Schulmeyer and McManus, 1987] G. G. SCHULMEYER AND J. I. MCMANUS (Editors), *Handbook of Software Quality Assurance*, Van Nostrand Reinhold, New York, NY, 1987.

[Shen, Conte, and Dunsmore, 1983] V. Y. SHEN, S. D. CONTE, AND H. E. DUNSMORE, "Software Science Revisited: A Critical Analysis of the Theory and its Empirical Support," *IEEE Transactions on Software Engineering* **SE-9** (March 1983), pp. 155–165.

[Symons, 1988] C. R. SYMONS, "Function Point Analysis: Difficulties and Improvements," *IEEE Transactions on Software Engineering* **14** (January 1988), pp. 2–11.

[Thayer, 1988] R. H. THAYER (Editor), *Tutorial: Software Engineering Management*, IEEE Computer Society Press, Washington, DC, 1988.

[van der Poel and Schach, 1983] K. G. VAN DER POEL AND S. R. SCHACH, "A Software Metric for Cost Estimation and Efficiency Measurement in Data Processing System Development," *Journal of Systems and Software* **3** (September 1983), pp. 187–191.

Chapter 5 Stepwise Refinement: A Basic Software Engineering Technique

5.1 Miller's Law

Stepwise refinement underlies many software engineering techniques. It can be defined as "postpone decisions as to details as late as possible in order to be able to concentrate on the important issues."

As will be seen during the course of this book, stepwise refinement underlies many specifications methods, design and implementation methods, even testing and integration methods. The reason why stepwise refinement is so important is because of Miller's law, which states that [Miller, 1956] at any one time a human being can concentrate on 7 ± 2 *chunks* (quanta of information).

The problem when developing software is that we need to concentrate on many more than seven chunks at a time. For example, a module usually has considerably more than seven variables, and a client has more than seven requirements. Stepwise refinement enables the software engineer to concentrate on those seven chunks that are the most relevant at the current phase of development.

To see how this works, a case study is now presented that illustrates how stepwise refinement can be used in the design of a product.

5.2 Stepwise Refinement Case Study

The case study presented in this section seems almost trivial, in that it involves the updating of a sequential master file, a common operation in many application areas. This choice of a familiar example is deliberate, in order to enable the reader to concentrate on stepwise refinement rather than the

Transaction Type	Name	Address
3	BROWN	
1	HARRIS	2 Oak Lane, Townsville
2	JONES	Box 345, Tarrytown
3	JONES	
1	SMITH	1304 Elm Avenue, Oak City

Figure 5.1 Input transaction records for sequential master file update.

problem in hand, which may be stated as follows:

> Design a product to update the sequential master file containing name and address data for the monthly magazine, *True Life Software Disasters*. There are three types of transactions, namely insertions, modifications, and deletions, with transaction codes 1, 2, and 3, respectively. Thus the transaction types are:
>
> Type 1: INSERT (a new subscriber into the master file)
> Type 2: MODIFY (an existing subscriber record)
> Type 3: DELETE (an existing subscriber record)
>
> Transactions are sorted into alphabetical order and by transaction code within alphabetical order.

The first step in designing a solution is to set up a typical file of input transactions such as that shown in Figure 5.1. The file contains five records, namely DELETE BROWN, INSERT HARRIS, MODIFY JONES, DELETE JONES, and INSERT SMITH. Note that it is not unusual to have to perform both a modification and a deletion of the same subscriber in one run.

The problem may be represented as shown in Figure 5.2. There are two input files:

1. Old master file name and address records.
2. Transaction records.

There are three output files:

3. New master file name and address records.
4. Exception report.
5. Summary and end-of-job message.

To begin the design process, the starting point is the single box UPDATE MASTER FILE shown in Figure 5.3. This can be broken down into

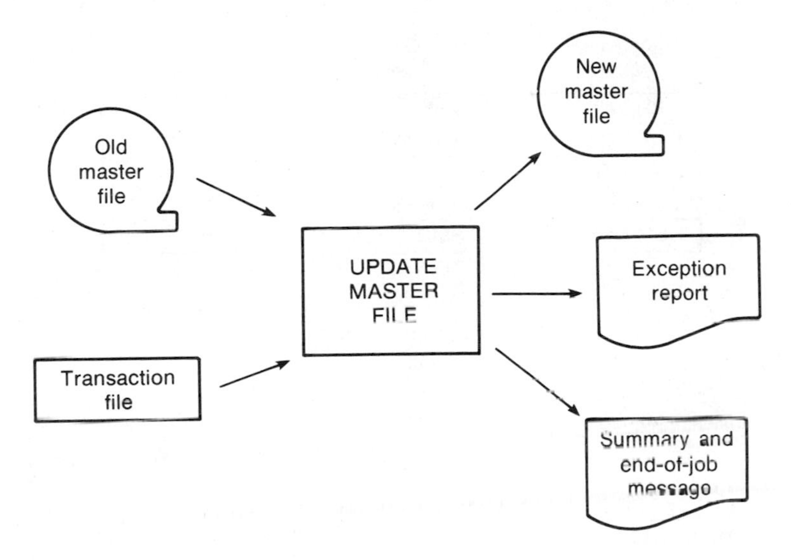

Figure 5.2 Sequential master file update.

three boxes, labeled INPUT, PROCESS, and OUTPUT. The assumption is that when PROCESS requires a record, our level of competence is such that the correct record can be produced at the right time. Similarly, we are capable of writing the correct record to the correct file at the correct time. Therefore the technique is to separate out the INPUT and OUTPUT aspects and concentrate on PROCESS.

What is this PROCESS? To determine what it does, consider the example shown in Figure 5.4. The key of the first transaction record (BROWN) is compared with the key of the first old master file record (ABEL). Since BROWN comes after ABEL, the ABEL record is written to the new master file, and the next old master file record (BROWN) is read. Now the key of the transaction record matches the key of the old master file record, and since the

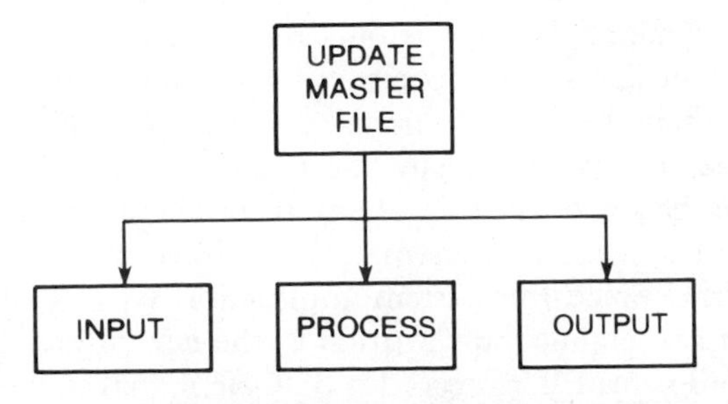

Figure 5.3 First refinement of design.

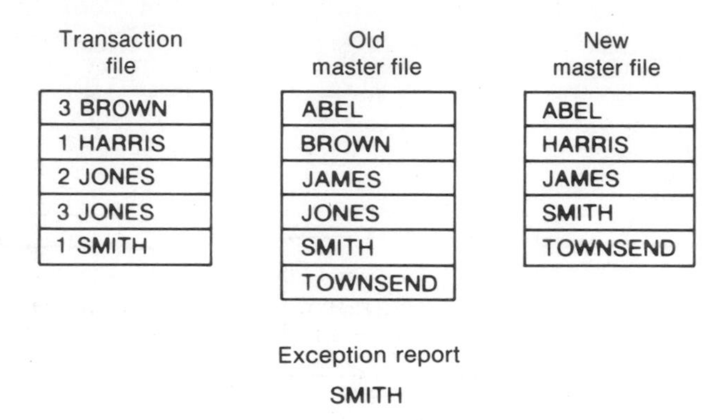

Figure 5.4 Transaction file, old master file, new master file, and exception report.

transaction type is 3 (delete), what is required is to delete the BROWN record. This is implemented by not copying the BROWN record onto the new master file. Thus the next transaction record (HARRIS) and old master file record (JAMES) are read, overwriting the BROWN records in their respective buffers. HARRIS comes before JAMES and is inserted into the new master file; the next transaction record (JONES) is read. Since JONES comes after JAMES, the JAMES record is written to the new master file, and the next old master file record is read; this is JONES. As can be seen from the transaction input file, the JONES record is to be modified and then deleted, so the next transaction record (SMITH) and the next old master file record (also SMITH) are read. Unfortunately, the transaction type is 1 (insert), but SMITH is already on the master file. So there is an error of some sort in the data, and the SMITH record is written to the exception report. To be more precise, the SMITH transaction record is written to the exception report, the SMITH old master file record is written to the new master file.

Now that PROCESS is fully understood, it may be represented as in Figure 5.5. UPDATE MASTER FILE of Figure 5.3 may now be refined, resulting in the second refinement shown in Figure 5.6.

The next step is to refine the INPUT and OUTPUT boxes of Figure 5.6, resulting in Figure 5.7. There is only one problem—the design of Figure 5.7 has a major fault. To see this, consider the situation with regard to the data of Figure 5.4 when the current transaction is 2 JONES, that is, modify JONES, and the current old master file record is JONES. In the design of Figure 5.7, since the key of the transaction record is the same as the key of the old master file record, the leftmost path is followed to the Test transaction type decision box. Since the current transaction type is MODIFY, the old master file record is modified and written to the new master file, and the next transaction record is read. This record is 3 JONES, that is, delete JONES. But the modified JONES record has already been written to the new master file.

transaction record key = old master file record key	1. INSERT: Print error message 2. MODIFY: Change master file record 3. DELETE: *Delete master file record
transaction record key > old master file record key	Copy old master file record onto new master file
transaction record key < old master file record key	1. INSERT: Write new subscriber to new master file 2. MODIFY: Print error message 3. DELETE: Print error message

*Deletion of a master file record is implemented by not copying the record onto the new master file.

Figure 5.5 Diagrammatic representation of PROCESS.

The reader may be wondering why an incorrect refinement has deliberately been presented. The point is that when using stepwise refinement it is necessary to desk-check each successive refinement before proceeding to the next. If a particular refinement turns out to be in error, it is not necessary to restart the process from the beginning, but merely to go back to the previous refinement and proceed from there. In this instance, the second

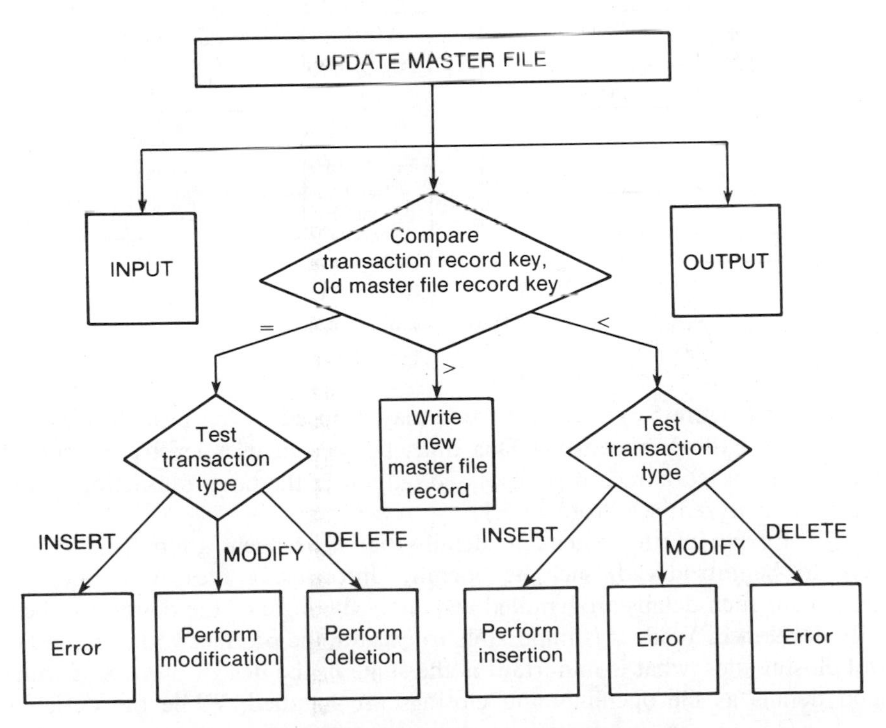

Figure 5.6 Second refinement of design.

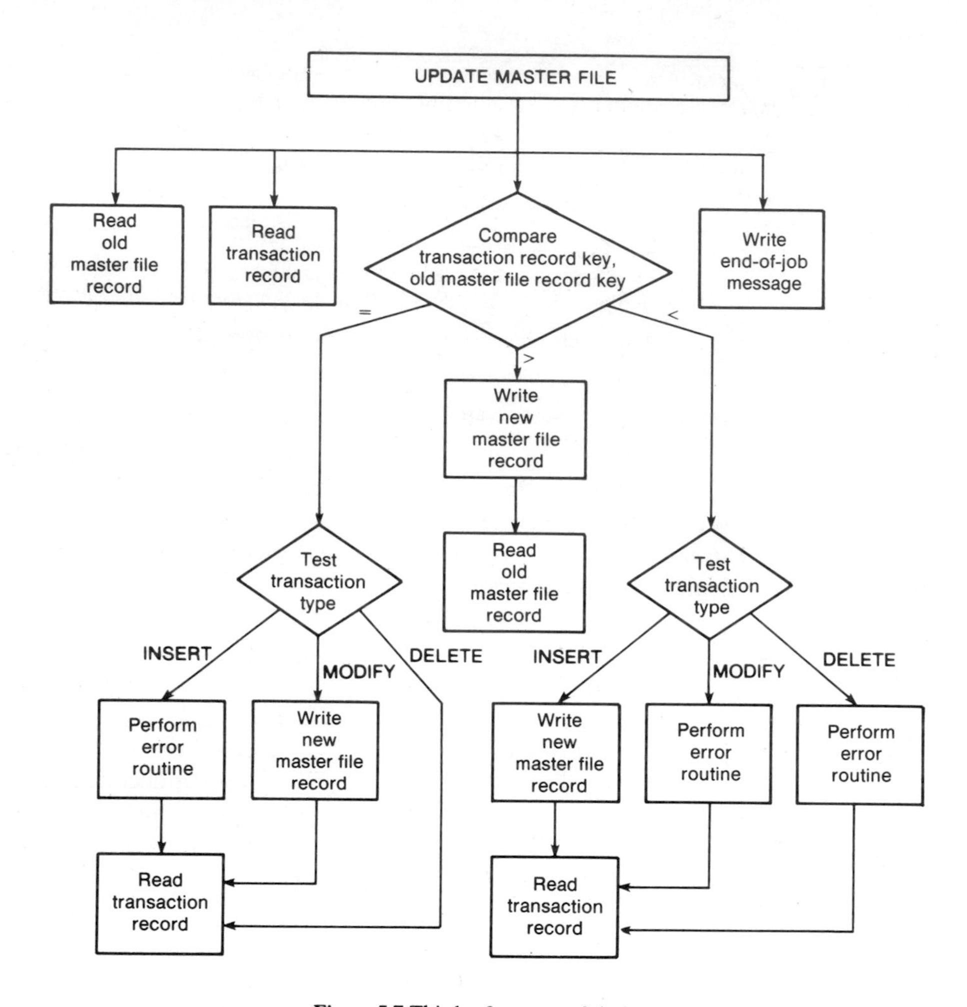

Figure 5.7 Third refinement of design.

refinement (Figure 5.6) is correct, so it may be used as the basis for another attempt at a third refinement. This time, the design uses level-1 lookahead, that is, a transaction record is processed only after the next transaction record has been analyzed; see Problem 5.1.

In the fourth refinement, details that have been ignored up to now have to be introduced, such as opening and closing files. With stepwise refinement such details are handled last, after the logic of the design has been fully developed. While it is impossible to execute the product without opening and closing files, what is important is the stage in the design process at which such details as file openings and closings are handled. While the design is being developed, the 7 ± 2 chunks on which the designer can concentrate at once should *not* include details like opening and closing files. File openings and closings have nothing to do with the problem itself; they are merely necessary aspects of the implementation of any design.

The term "stepwise refinement" was first introduced by Wirth [Wirth, 1971]. In the preceding case study stepwise refinement was applied to a flowchart, while Wirth applied the method to pseudocode. The representation to which stepwise refinement is applied is not important; stepwise refinement is a general technique that can be used in every phase of software development and with almost every representation.

The power of stepwise refinement is that it helps the software engineer to concentrate on the relevant aspects of the current development phase, namely design in the preceding case study, and to ignore details that, though necessary in the overall scheme of things, need not be considered, and in fact should be ignored, until later. Miller's law is a fundamental restriction on the mental powers of human beings. Since we cannot fight our nature, we must live with it, accepting our limitations and doing the best we can under the circumstances.

Chapter Review

Miller's law (Section 5.1) states that human beings can concentrate at any one time on only 7 ± 2 chunks of information. Stepwise refinement is a technique that allows the software engineer to concentrate on the currently relevant seven chunks. In Section 5.2 the application of stepwise refinement to design is illustrated by means of a case study.

For Further Reading

For further information regarding Miller's law and for his theory of how the brain operates on chunks, the reader should consult [Tracz, 1979] and [Moran, 1981], as well as Miller's original paper [Miller, 1956]. An analysis of Miller's law from the viewpoint of cognitive psychology and Software Science is to be found in [Coulter, 1983]. Wirth's paper on stepwise refinement is a classic of its kind and deserves careful study [Wirth, 1971]. Equally significant from the viewpoint of stepwise refinement are the books by Dijkstra [Dijkstra, 1976] and Wirth [Wirth, 1975].

Mills has applied stepwise refinement to box-structured design, a technique for producing a design from a specification [Mills, Linger, and Hevner, 1987; Mills, 1988]. Rajlich has extended stepwise refinement to large-scale products [Rajlich, 1985a].

Problems

5.1 Consider the effect of introducing *lookahead* to the design of the third refinement of the sequential master file update problem. That is to say, before processing a transaction the next transaction must be read. If both transactions apply to the same master file record, then the decision regarding the processing of the current transaction will depend on the type of the next transaction. Draw up a 3×3 table with the rows labeled with the type of the current transaction and the columns labeled with the type of the next transaction, and fill in the action to be taken in each instance. For example, two successive insertions of the same record are clearly an error. But two modifications may be perfectly valid—for example, a subscriber can change his or her address more than once in a given month. Now develop a flowchart for the third refinement that incorporates lookahead.

5.2 Check whether your solution to Problem 5.1 solves the problem of a modification transaction followed by a deletion transaction, both transactions being applied to the same master file record. If not, modify your solution.

5.3 Check whether your solution to Problem 5.1 also solves the problem of an insertion followed by a modification followed by a deletion, all applied to the same master file record. If not, modify your solution.

5.4 Check whether your solution also solves the problem of n insertions, modifications, or deletions, $n > 2$, all applied to the same master file record. If not, modify your solution.

5.5 The last transaction record does not have a successor. Check whether your flowchart for Problem 5.1 takes this into account and processes the last transaction record. If not, modify your solution.

5.6 In some applications an alternative to lookahead can be achieved by careful ordering of the transactions. For example, the original problem caused by a modification followed by a deletion of the same master file record could have been solved by processing a deletion before a modification. This would have resulted in the master file being correctly written and an error message appearing in the exception report. Investigate whether there is an ordering of the transactions that can solve all of the difficulties listed in Problems 5.2, 5.3, and 5.4.

5.7 (Term Project) Design the input routines of the Plain Vanilla Ice Cream Corporation product described in the Appendix using stepwise refinement.

5.8 (Readings in Software Engineering) Your instructor will distribute copies of [Wirth, 1971]. List the differences between Wirth's approach and the approach to stepwise refinement presented in this chapter.

References

[Coulter, 1983] N. S. Coulter, "Software Science and Cognitive Psychology," *IEEE Transactions on Software Engineering* **SE-9** (March 1983), pp. 166–171.

[Dijkstra, 1976] E. W. Dijkstra, *A Discipline of Programming*, Prentice-Hall, Englewood Cliffs, NJ, 1976.

[Miller, 1956] G. A. Miller, "The Magical Number Seven, Plus or Minus Two: Some Limits on our Capacity for Processing Information," *The Psychological Review* **63** (March 1956), pp. 81–97.

[Mills, 1988] H. D. Mills, "Stepwise Refinement and Verification in Box-Structured Systems," *IEEE Computer* **21** (June 1988), pp. 23–36.

[Mills, Linger, and Hevner, 1987] H. D. Mills, R. C. Linger, and A. R. Hevner, "Box Structured Information Systems," *IBM Systems Journal* **26** (No. 4, 1987), pp. 395–413.

[Moran, 1981] T. P. Moran (Editor), Special Issue: The Psychology of Human–Computer Interaction, *ACM Computing Surveys* **13** (March 1981).

[Rajlich, 1985a] V. Rajlich, "Stepwise Refinement Revisited," *Journal of Systems and Software* **5** (February 1985), pp. 81–88.

[Tracz, 1979] W. J. Tracz, "Computer Programming and the Human Thought Process," *Software—Practice and Experience* **9** (February 1979), pp. 127–137.

[Wirth, 1971] N. Wirth, "Program Development by Stepwise Refinement," *Communications of the ACM* **14** (April 1971), pp. 221–227.

[Wirth, 1975] N. Wirth, *Algorithms + Data Structures = Programs*, Prentice-Hall, Englewood Cliffs, NJ, 1975.

Chapter 6 Testing

The software development cycle all too frequently includes testing as a separate phase, after implementation and before maintenance. Nothing could be more dangerous from the viewpoint of trying to achieve high-quality software. On the contrary, testing must be an integral part of every aspect of software production. In order to stress the point, this chapter is placed before the chapters on the individual life-cycle phases.

6.1 What Is Testing?

It has been claimed that testing is a demonstration that faults are not present. (A *fault* is the IEEE standard terminology for what is popularly called a "bug," while a *failure* is the incorrect behavior of the product as a consequence of the fault [ANSI/IEEE 729, 1983].) Despite the fact that some organizations spend up to 50% of their software budget on testing, delivered *tested* software is notoriously unreliable.

The reason for this contradiction is simple. As Dijkstra put it, "Program testing can be a very effective way to show the presence of bugs, but it is hopelessly inadequate for showing their absence" [Dijkstra, 1972]. What Dijkstra is saying is that if a product is run with test data and the output is wrong, then the product definitely contains a fault. But if the output is correct, then there still may be a fault in the product; all that particular test has shown is that the product runs correctly on that particular set of test data.

6.2 Definition of Testing

What then is testing? According to Goodenough, testing is a process of inferring certain behavioral properties of a product based, in part, on the results of executing the product in a known environment with selected inputs [Goodenough, 1979].

This definition has three troubling implications. First, the definition states that testing is an inferential process. The tester takes the product, runs it with known input data, and examines the output. From this the tester has to infer what, if anything, is wrong with the product. From this viewpoint, testing is comparable to trying to find the proverbial black cat in a dark room. All the tester has to help find any faults is perhaps 10 or 20 sets of inputs and corresponding outputs, possibly a user fault report, and thousands of lines of code. From this the tester has to deduce if there is a fault, and what it is.

A second implication of the definition arises from the phrase "in a known environment." We can never really know our environment, either the hardware or the software. We can never be certain that the operating system is functioning correctly or that the run-time routines are correct. There may be an intermittent hardware fault in the main memory of the computer. So what is observed as the behavior of the product may in fact be a correct product interacting with a faulty compiler, or faulty hardware, or some other faulty component of the environment.

The third part of the definition of testing that has worrisome implications is the phrase "with selected inputs." In the case of a real-time system, there is frequently no control over the inputs to the system. Consider avionics software. The flight control system has two types of inputs. The first type of input is what the pilot wants the aircraft to do. Thus if the pilot pulls back on the joystick in order to climb or opens the throttle in order to increase the speed of the aircraft, these mechanical motions are transformed into digital signals which are sent to the flight control computer. The second type of input is the current physical state of the aircraft such as its altitude, speed, and the elevations of the flaps. The flight control system uses the values of those quantities to compute what signals should be sent to the components of the aircraft such as the flaps and the engines in order to implement the pilot's intentions. While the pilot's inputs can easily be set to any desired values simply by setting the aircraft's controls appropriately, the inputs corresponding to the current physical state of the aircraft cannot be so easily manipulated. In fact, there is no way that one can force the aircraft to provide "selected inputs." How then can such a real-time system be tested? The answer is to use a simulator.

A simulator is a working model of the environment in which the product, in this case the flight control software, executes. The flight control software can be tested by causing the simulator to send selected inputs to the flight control software. That is to say, the simulator has controls that allow the operator to set an input variable to any selected value. Thus, if the purpose of the test is to determine how the flight control software performs if one engine catches fire, then the controls of the simulator are set in such a way that the inputs sent to the flight control software are indistinguishable from what the inputs would be if an engine of the actual aircraft were on fire. The output can then be analyzed by examining the output signals sent from the flight control software to the simulator. But a simulator can at best be a good approximation of a faithful model of some aspect of the system; it can never be the system itself. Using a simulator means that while there is indeed a "known environment," there is little likelihood that this known environment is in every way identical to the actual environment in which the testing must be performed.

The preceding definition of testing speaks of "behavioral properties." What behavioral properties must be tested? An obvious answer is: Test whether the product functions correctly. But as will be shown, correctness is neither necessary nor sufficient. Before discussing correctness, four other behavioral properties will be considered, namely *utility*, *reliability*, *robustness*, and *performance* [Goodenough, 1979].

6.2.1 Utility

Utility is the extent to which a user's needs are met when a correct product is used under conditions permitted by its specifications. In other words, a product that, as far as is known, is functioning correctly is subjected to inputs which are valid in terms of the specifications. The user may test, for example, how easy the product is to use, whether the product performs useful functions, and whether the product is cost-effective. Irrespective of whether the product is correct or not, these are vital issues that have to be tested. If the product is not cost-effective, there is no point in buying it. And unless it is easy to use, either it will not be used at all, or it will be used incorrectly. Thus, when considering buying an existing product, the utility of the product should be tested first, and if the product fails on that score, then there is no point in further testing.

6.2.2 Reliability

Another aspect of a product that must be tested is its reliability. *Reliability* is a measure of the frequency and criticality of product failure, where failure is an unacceptable effect or behavior occurring under permissible operating conditions [Goodenough, 1979]. In other words, it is necessary to know how often the product fails (*mean time between failures*) and how bad the effects of that failure can be. When a product fails, an important issue is how long it takes, on average, to repair it (*mean time to repair*). But often more important is how long it takes to repair the *results* of the failure. This last point is frequently overlooked. Suppose that the software running on a communications front end fails, on average, only once every 6 months, but when it fails it completely wipes out a database. At best the database can be reinitialized to its status when the last checkpoint dump was taken, and the audit trail can then be used to put the database into a state which is virtually up to date. But if this recovery process takes the better part of 2 days, during which time the database and communications front end are inoperative, then the reliability of the product is low, notwithstanding the fact that the mean time between failures is 6 months.

6.2.3 Robustness

Another aspect of every product that requires testing is its *robustness*. While it is difficult to come up with a precise definition of robustness, robustness is essentially a function of a number of factors such as range of operating conditions, the possibility of unacceptable effects on valid input, and the acceptability of effects when the product is given invalid input. With regard to the first factor, a product with the wide range of permissible operating conditions is more robust than a product that is more restrictive. The second factor is the possibility of unacceptable effects when the product satisfies its specifications. That is to say, suppose the user gives a valid command: What is the possibility that this command has disastrous consequences? The third factor is the acceptability of effects and behavior when the product is *not* used under permissible operating conditions. When testing for utility, the test data have to satisfy the input specifications of the product; for this third aspect of robustness, test data that do not satisfy the input specifications are deliberately input, and the tester determines how badly the product reacts. For example, when the product solicits a name, the tester may reply with a stream of unacceptable characters such as control-A escape-% ?$#@. If the computer responds with a message such as Incorrect Data—Try Again, or better, informs the user as to why the data do not conform to what was expected, then it is more robust than a product that crashes whenever the data deviate even slightly from what is required.

6.2.4 Performance

Performance is another aspect of the product that must be tested. It is essential to know the extent to which the product meets its constraints with regard to response time or space requirements. For an embedded computer system such as an on-board computer in a handheld antiaircraft missile, the space constraints of the system may be such that there is no way that more than say, 128 kilobytes (kb) of main memory can be provided. No matter how excellent the software may be, if it cannot operate without 256 kb of memory, then it is no use at all.

Real-time software is characterized by hard time constraints, that is to say, time constraints of such a nature that if a constraint is not met, information is lost. For example, a nuclear reactor control system may have to sample the temperature of the core and process the data every tenth of a second. If the system is not fast enough to be able to handle interrupts from

Procedure Sort

Input specification:	A: **array** (1 .. N) **of** INTEGER;
Output specification:	B: **array** (1 .. N) **of** INTEGER such that $B(1) \leq B(2) \leq \ldots \leq B(N)$;

Figure 6.1 Specifications for sort procedure.

the temperature sensor every tenth of a second, then data will be lost, and there is no way of recovering the data; the next time that the system receives temperature data it will be the current temperature, not the reading that was missed. If the reactor is on the point of a meltdown, it is critical that all relevant information be both received and processed as laid down in the specifications. With all real-time systems, the performance must meet every time constraint listed in the specifications.

6.2.5 Correctness of Specifications

Finally, a definition of correctness can be given. A product is *correct* if it satisfies its output specifications, independent of its use of computing resources, when operated under permitted conditions [Goodenough, 1979]. In other words, if input satisfying the input specifications is provided and the product is given all the resources it needs, then the product is correct if the output satisfies the output specifications.

This definition of correctness, like the definition of testing itself, has worrisome implications. Suppose that a product has been successfully tested against a broad variety of test data. Does this mean that the product is acceptable? Unfortunately, it does not. If a product is correct, all that it means is that it satisfies its specifications. But what if the specifications themselves are incorrect? To illustrate this difficulty, consider the specifications shown in Figure 6.1. The specifications state that the input to the sort is an array A of N integers, while the output is another array B which is sorted in nondecreasing order. Superficially, the specifications seem to be perfectly correct. But consider **procedure** TRICK_SORT shown in Figure 6.2. In that procedure, all N elements of array B are set to 0. The procedure satisfies the specifications of Figure 6.1 and is therefore correct!

```
procedure TRICK_SORT;
begin
    B := 0;
end TRICK_SORT;
```

Figure 6.2 Procedure satisfying specifications of Figure 6.1.

Procedure Sort

Input specification:	A: **array** (1 .. N) **of** INTEGER:
Output specification:	B: **array** (1 .. N) **of** INTEGER such that $B(1) \le B(2) \le \ldots \le B(N)$ and the elements of array B are a permutation of the elements of array A

Figure 6.3 Corrected specifications for sort procedure.

What has happened? Unfortunately, the specifications of Figure 6.1 are wrong. What has been omitted is a statement that the elements of B, the output array, are a permutation (rearrangement) of the elements of the input array A. After all, an intrinsic aspect of sorting is that it is a rearrangement process. And the procedure of Figure 6.2 capitalizes on this specification fault. In other words, **procedure** TRICK_SORT is correct, but the specifications of Figure 6.1 are wrong. Corrected specifications appear in Figure 6.3.

From the preceding example, it is clear that the consequences of specification faults are nontrivial. After all, the correctness of a product is meaningless if its specifications are incorrect.

The fact that a product is correct is not *sufficient*, because the specifications in terms of which it was shown to be correct may themselves be wrong. But is it *necessary*? Consider the following example. A software organization has acquired a superb new C compiler. The new compiler can translate twice as many lines of source code per minute as the old compiler the organization was using, the object code runs nearly 45% faster, and the size of the object code is about 20% smaller. In addition, the error messages are much clearer, and the cost of annual maintenance and updates is less than half of those of the old compiler. There is one problem, however: The first time that a **while** statement appears in any routine, the compiler prints a spurious error message. The compiler is therefore not correct, because the specifications for a compiler implicitly or explicitly require that error messages be printed if, and only if, there is a fault in the source code. It is certainly possible to use the compiler—in fact, in every way but one the compiler is absolutely ideal. Furthermore, it is reasonable to expect that this minor fault will be corrected in the next release. In the meantime, the programmers will learn to ignore the spurious error message. Not only can the organization live with the incorrect compiler, but if anyone were to suggest replacing it by the old correct compiler there would be an outcry. Thus the correctness of a product is neither necessary nor sufficient.

Both the preceding examples are admittedly somewhat artificial. But they make the point that correctness simply means that the product is a correct implementation of its specifications. In other words, there is more to testing than just showing that the product is correct.

```
I := 1;
S := 0;
while I ≤ N loop
    S := S + Y(I);
    I := I + 1;
end loop;
```

Figure 6.4 Code fragment to be proved correct.

With all the difficulties associated with testing, computer scientists have tried to come up with other ways of ensuring that the product does what it is supposed to do. One such alternative that has received considerable attention for more than 20 years is correctness proving.

6.3 Testing versus Correctness Proofs

It is important to distinguish between testing and correctness proofs. *Testing* is performed by executing the product in a known environment using selected test data. A *correctness proof* is a formal mathematical verification that the product is correct; the product is not executed on a computer. For this reason, the process is frequently termed *verification*. Unfortunately, the term "verification" has another meaning in the software engineering context. As defined in Section 4.4.1, verification is the process of determining whether the work products of an activity satisfy their requirements. That is to say, verification is the process of determining whether a particular activity has been completed correctly. Throughout this book, the term "verification" is used in this sense, while the formal mathematical process will be termed *correctness proving*, to remind readers that what is involved is a mathematical proof process.

6.3.1 Example of a Correctness Proof

To see how correctness proving is performed, consider the code fragment shown in Figure 6.4. The flowchart equivalent to the code is given in Figure 6.5. It will now be shown that the code fragment is correct in that after the code has been executed, the variable S will contain the sum of the N elements of the array Y. In Figure 6.6, lines of the flowchart are labeled with the letters A through H. In addition, *assertions* have been associated with each of the places, that is to say, claims have been made that certain mathematical properties hold at each place. The correctness of the assertions will be proved in the following discussion.

The input specification, the condition that holds at A before the code is executed, is that the variable N is a natural number, that is,

$$A: \mathrm{N} \in \{1, 2, 3, \ldots\} \tag{6.1}$$

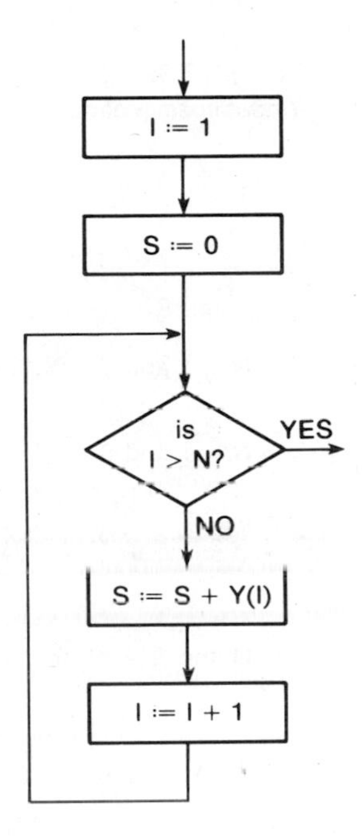

Figure 6.5 Flowchart of Figure 6.4.

An obvious output specification is that if control reaches point H, the value of S contains the sum of the N values stored in array Y, that is,

$$H\text{: S} = \text{Y(1)} + \text{Y(2)} + \ldots + \text{Y(N)} \tag{6.2}$$

In fact, the code fragment can be proved to be correct with respect to a stronger output specification, namely

$$H\text{: I} = \text{N} + 1 \text{ and S} = \text{Y(1)} + \text{Y(2)} + \ldots + \text{Y(N)} \tag{6.3}$$

A natural reaction to the last sentence is to ask: Where did output specification (6.3) come from? By the end of the proof, the reader will hopefully have the answer to that question; also see Problems 6.6 and 6.7.

A third component of the proof process is to provide a loop invariant for the loop. That is to say, a mathematical expression must be provided which holds at point D, irrespective of whether the loop has been executed 0, 1, or many times. The loop invariant that will be proved to hold is

$$D\text{: I} \leq \text{N} + 1 \text{ and S} = \text{Y(1)} + \text{Y(2)} + \ldots + \text{Y(I} - 1) \tag{6.4}$$

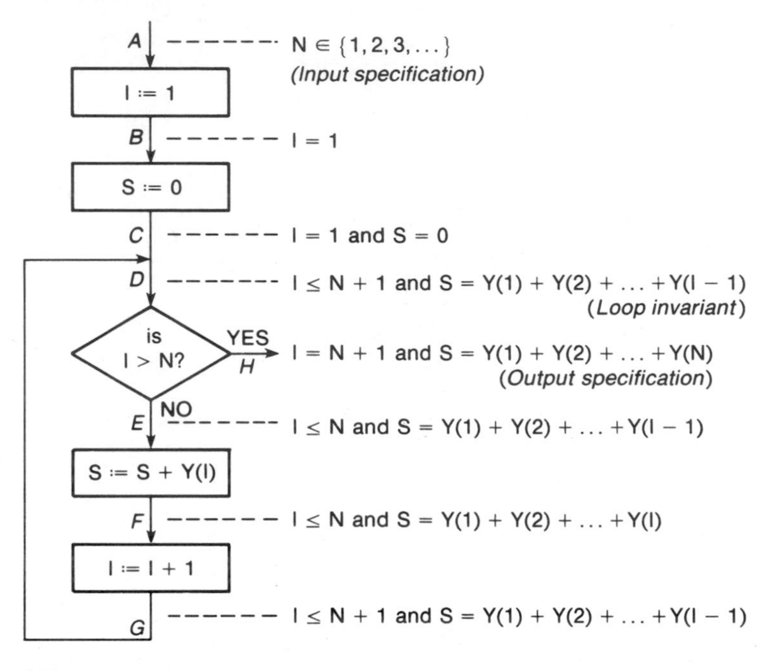

Figure 6.6 Figure 6.5 with input specification, output specification, loop invariant, and assertions.

Now it will be shown that if specification (6.1) holds at point *A*, then specification (6.3) will hold at point *H*, that is to say, the code fragment will be proved to be correct.

First, the assignment statement I := 1 is executed. Control is now at point *B*, where the following assertion holds:

$$B\colon \mathrm{I} = 1 \tag{6.5}$$

To be more precise, at point *B* the assertion should read I = 1 and N ∈ {1, 2, 3, ...}. But the input specification (6.1) holds at all points in the flowchart. For brevity, the and N ∈ {1, 2, 3, ...} will be omitted from now on.

At point *C*, as a consequence of the second assignment statement, namely S := 0, the following assertion is true:

$$C\colon \mathrm{I} = 1 \text{ and } \mathrm{S} = 0 \tag{6.6}$$

Now the loop is entered. It will be proved by induction that the loop invariant (6.4) is indeed correct. Just before the loop is executed for the first time, assertion (6.6) holds, that is, the value of I is equal to 1, and S is equal to

0. Now consider loop invariant (6.4). Since $I = 1$ by assertion (6.6), and $N \geq 1$ from input specification (6.1), it follows that $I \leq N + 1$, as required. Furthermore, since $I = 1$ it follows that $I - 1 = 0$, so the sum in (6.4) is empty and $S = 0$, as required. Loop invariant (6.4) is therefore clearly true just before the first time the loop is entered.

Now the inductive hypothesis step is performed. Assume that at some stage during the execution of the code fragment the loop invariant holds. That is, for I equal to some value I_0, $1 \leq I_0 \leq N + 1$, execution is at point D, and the assertion that holds is

$$D\colon I_0 \leq N + 1 \text{ and } S = Y(1) + Y(2) + \ldots + Y(I_0 - 1) \qquad (6.7)$$

Control now passes to the test box. If $I_0 > N$, then since $I_0 < N + 1$ by hypothesis (6.7), it follows that $I_0 = N + 1$. By inductive hypothesis (6.7), this implies that

$$H\colon I_0 = N + 1 \text{ and } S = Y(1) + Y(2) + \ldots + Y(N) \qquad (6.8)$$

which is precisely the output specification (6.3).

On the other hand, if the test is $I_0 > N$? fails, then control passes from point D to point E. Since I_0 is not greater than N, $I_0 \leq N$, so (6.7) becomes

$$E\colon I_0 \leq N \text{ and } S = Y(1) + Y(2) + \ldots + Y(I_0 - 1) \qquad (6.9)$$

The statement $S := S + Y(I_0)$ is now executed, so from assertion (6.9), at point F the following assertion must hold:

$$F\colon I_0 \leq N \text{ and } S = Y(1) + Y(2) + \ldots + Y(I_0 - 1) + Y(1_0)$$

$$= Y(1) + Y(2) + \ldots + Y(I_0) \qquad (6.10)$$

The next statement to be executed is $I_0 := I_0 + 1$. To see the effect of this statement, suppose that the value of I_0 before executing this statement is, say, 17. Then the last term in the sum in (6.10) is $Y(17)$. Now the value of I_0 is increased by 1 to 18. The sum S is unchanged, so the last term in the sum is still $Y(17)$, which is now $Y(I_0 - 1)$. Also, at point F, $I_0 \leq N$. Increasing the value of I_0 by 1 means that if the inequality is to hold at point G, then $I_0 \leq N + 1$. Thus, the effect of increasing I_0 by 1 is that the following assertion holds at point G:

$$G\colon I_0 \leq N + 1 \text{ and } S = Y(1) + Y(2) + \ldots + Y(I_0 - 1) \qquad (6.11)$$

Assertion (6.11) which holds at point *G* is identical to assertion (6.7) which, by assumption, holds at point *D*. But point *D* is topologically identical to point *G*. In other words, if (6.7) holds at *D* for $I = I_0$, then it will again hold at *D* with $I = I_0 + 1$. It has been shown that the loop invariant holds for $I = 1$. By induction it follows that loop invariant (6.4) holds for all values of I, $1 \leq I \leq N + 1$.

All that remains is to prove that the loop terminates. Initially, by assertion (6.6), the value of I is equal to 1. Each time the loop is iterated the value of I is increased by 1 by the statement I := I + 1. Eventually, I must reach the value N + 1, at which time the loop is exited, and the value of S is given by assertion (6.8), thus satisfying output specification (6.3).

To review, given the input specification (6.1), it was proved that loop invariant (6.4) holds whether the loop has been executed 0, 1 or more times. Furthermore, it was proved that after N + 1 iterations the loop terminates, and when it does so the values of I and S satisfy the output specification (6.3). In other words, the code fragment has been mathematically shown to be correct.

6.3.2 Correctness Proof Case Study

An important aspect of correctness proofs is that they should be done in conjunction with design and coding. As Dijkstra put it, "The programmer should let the program proof and program grow hand in hand" [Dijkstra, 1972]. For example, when a loop is incorporated into the design a loop invariant is put forward, and as the design is refined, so is the invariant. Developing a product in this way gives the programmer confidence that the product is correct and tends to reduce the number of faults. Quoting Dijkstra again, "The only effective way to raise the confidence level of a product significantly is to give a convincing proof of its correctness" [Dijkstra, 1972]. But even if a product is proved to be correct, it must be thoroughly tested as well. To illustrate the necessity for testing in conjunction with correctness proving, consider the following.

In 1969, Naur published a paper on a technique for constructing and proving a product [Naur, 1969]. The technique was illustrated by what Naur termed a "line-editing problem," but today it would be considered a text-processing problem. It may be stated as follows:

> Given a text consisting of words separated by BLANK characters or by NL (new line) characters, convert it to line-by-line form in accordance with the following rules:
>
> **1.** line breaks must be made only where the given text has BLANK or NL;
> **2.** each line is filled as far as possible, as long as
> **3.** no line will contain more than MAXPOS characters.

Naur constructed a procedure using his technique and informally proved its correctness. The procedure was about 25 lines of ALGOL 60. The paper was then reviewed by Leavenworth in *Computing Reviews* [Leavenworth, 1970]. The reviewer pointed out that in the output of the procedure published by Naur, the first word of the first line is preceded by a blank unless the first word is exactly MAXPOS characters long. Although this may seem a trivial fault, it is a fault that would surely have been detected had the procedure been tested, that is, run with test data, rather than only proved. But worse was to come. London detected three additional faults in Naur's procedure [London, 1971]. One was that the procedure does not terminate unless a word longer than MAXPOS characters is encountered. Again, this fault is likely to have been detected if the procedure had been tested. London then presented a corrected version of the procedure and proved formally that the resulting procedure was correct; recall that Naur had used only informal proof techniques.

The next episode in this saga was that Goodenough and Gerhart found three further faults that London had not detected, despite his formal proof [Goodenough and Gerhart, 1975]. These included the fact that the last word will not be output unless it is followed by a BLANK or NL. Yet again, reasonable choice of test data would have caused this fault to be detected without much difficulty. In fact, of the total of seven faults collectively detected by Leavenworth, London, and Goodenough and Gerhart, four could have been detected simply by running the procedure on test data, such as the illustrations given in Naur's original paper.

The lesson from this saga is clear: Even if a product has been proved correct, it must still be tested thoroughly. The example in Section 6.3.1 showed that proving the correctness of even a small code fragment can be a lengthy process. Furthermore, the case study of this section showed that it is a difficult and error-prone process, even for a 25-line procedure. The following question must therefore be answered: Is correctness proving an interesting research idea, or is it a powerful software engineering technique whose time has come? This question is answered in the next section.

6.3.3 Correctness Proofs and Software Engineering

A number of reasons have been put forward as to why correctness proving should not be viewed as a standard software engineering technique. First, it is claimed that software engineers do not have adequate mathematical training. Second, it is suggested that proving is too expensive to be practical, and third, that proving is too hard. Each of these reasons will be shown to be an oversimplification.

Although the proof given in Section 6.3.1 can be understood with hardly more than high school algebra, nontrivial proofs require that input specifications, output specifications, and loop invariants be expressed in first- or second-order predicate calculus, or equivalent. Not only does this make the proof process simpler for a mathematician, but it allows correctness proving to be done by a computer. To complicate matters, predicate calculus is now somewhat outdated. To prove the correctness of concurrent products, techniques using temporal or other modal logics are required [Lamport, 1980; Manna and Pnueli, 1981]. There is no doubt that correctness proving requires a training in mathematical logic. Fortunately, most computer science majors today either take courses in the requisite material or are given the background to enable them to learn correctness proving techniques on the job. Thus universities and colleges are now turning out sufficient computer science graduates with the mathematical skills for correctness proving. The claim that practicing software engineers do not have the necessary mathematical training may have been true in the past, but no longer applies in the light of the thousands of computer science majors joining the industry each year.

The assertion that proving is too expensive a technique to be used in software development is also not correct. On the contrary, the economic viability of correctness proving can be determined on a project-by-project basis using cost–benefit analysis (Section 4.2.3). For example, consider the software for the NASA Space Station. Human lives are at stake, and if something goes wrong a space shuttle rescue mission may not arrive in time. The cost of proving life-critical space station software correct is large. But the potential cost of a software fault that is overlooked because correctness proving is not performed, is even larger.

The third assertion is that proving is too hard. Despite this claim, many nontrivial products have been successfully proved to be correct, including operating system kernels, compilers, and communications systems [Landwehr, 1983; Berry and Wing, 1985]. Furthermore, many tools such as theorem provers exist to assist in correctness proving. A theorem prover takes as input a product, its input and output specifications, and loop invariants. The theorem prover then attempts to prove mathematically that the product, when given input data satisfying the input specifications, will produce output data satisfying the output specifications.

At the same time, there are some difficulties with correctness proving. For example, how can we be sure that a theorem prover is correct; if the theorem prover prints out `This product is correct`, can we believe it? To take an extreme case, consider the so-called theorem prover shown in Figure 6.7. No matter what code is submitted to this theorem prover it will print out `This product is correct`. In other words, what reliability can be placed on the output of a theorem prover? One suggestion is to submit a theorem prover to itself and see whether it is correct. Apart from the philosophical implications, a simple way of seeing that this will not work is to consider what would happen if the theorem prover of Figure 6.7 were submitted to itself for

```
procedure THEOREM_PROVER;
begin
   PUT ("This product is correct");
end THEOREM_PROVER;
```

Figure 6.7 "Theorem prover."

proving. It will, as always, print out This product is correct, thereby "proving" its own correctness.

A further difficulty is finding the input and output specifications, and especially the loop invariants, or their equivalents, in other logics such as modal logic. Suppose that a product is correct. Unless a suitable invariant for each loop can be found, there is no way of proving the product correct. Yes, tools do exist to assist in this task [Tamir, 1980], but even with state-of-the-art tools, a software engineer may simply not be able to come up with a correctness proof.

Worse than not being able to find loop invariants, what if the specifications themselves are incorrect? An example of this is **procedure** TRICK_SORT (Figure 6.2). A good theorem prover, when given the incorrect specifications of Figure 6.1, will undoubtedly declare that the procedure of Figure 6.2 is correct.

Manna and Waldinger have stated that, "We can never be sure that the specifications are correct," and, "We can never be certain that a verification system is correct" [Manna and Waldinger, 1978]. These statements from two of the leading experts in the field encapsulate the various points made previously.

Does all this mean that there is no place for correctness proofs in software engineering? Not at all. Proving products correct is an important, and sometimes vital, software engineering tool, where appropriate. Proofs are appropriate where human lives are at stake or where otherwise indicated by cost–benefit analysis: If the cost of verifying software is less than the probable cost if the product fails, then the product should be proved. But, as the text editor case study shows, proving alone is not enough. Instead, correctness proving should be viewed as an important component of the set of techniques that may be utilized to check that a product is correct. Since the aim of software engineering is the production of quality software, correctness proving is indeed a software engineering technique.

6.4 Overview of Remainder of Chapter 6

Testing is not a process that is applied only when the coding is done. On the contrary, testing is an inherent component of the product life cycle. The requirements must be checked. During the specifications phase the specifications must be checked, as must the software production management plan. The design phase requires careful validation at every stage. Each module must certainly be tested, and the product as a whole needs testing at the integration phase. After acceptance testing the product goes into operations mode and maintenance begins. And hand in hand with maintenance goes repeated checking of modified versions of the product.

Testing essentially falls into two broad classes, testing of the product as a whole and testing of individual modules. The order of the material of the remaining sections of this chapter follows the order of the product life cycle, starting with testing at the requirements phase and ending with maintenance testing.

6.5 Testing during the Requirements Phase

The aim of the requirements phase is to determine the client's real needs. It is essential to verify that this has been performed correctly. Frequently, the best way to do this is to build a prototype and to let the client experiment with it. This technique is described in detail in Section 3.3.

Once the requirements have been satisfactorily checked, the specifications phase commences.

6.6 Testing during the Specifications Phase

During the specifications phase the functionality of the proposed product is encapsulated into the specifications document. It is vital to verify that the specifications are correct. The best way to do this is by means of a walkthrough or by an inspection of the specifications.

It is not a good idea for the person responsible for drawing up the specifications to be solely responsible for reviewing them. Almost everyone has blind spots which allow faults to creep into the specifications document, and those same blind spots will prevent the faults from being detected on review. Even assigning the review task to one person other than the author of the specifications does not help much; we have all had the experience of reading through a document many times, but failing to detect, say, a blatant spelling error that a second reader picks up almost immediately. A *walkthrough* is a review performed together by a team of software professionals with a broad range of skills. The advantage of a review by a team of experts is that the different skills of the participants increase the chances of a fault being highlighted. In addition, when there is a team of skilled individuals working together there is often a synergistic effect.

6.6.1 Preparation for Walkthroughs

The first step is to decide on the participants of the walkthrough. Ideally, the group should consist of between three and six individuals. There should be at least one representative from the team responsible for drawing up the specifications. The manager responsible for the specifications should also be present. There should also be a client representative. It is a good idea to have a representative of the design team who will perform the next phase of the development, because they are the immediate users of the specifications document [Dunn, 1984]. Last, but definitely not least, there must be a representative of the software quality assurance group. For reasons that will be explained, the SQA group member should chair the walkthrough.

Once a time and a venue for the walkthrough have been arranged, the material for the walkthrough must be distributed to the participants well in advance of the walkthrough itself in order to allow for careful preparation. Each reviewer should study the material and come up with two lists, a list of items that he or she does not understand, and a list of items the reviewer believes are incorrect.

6.6.2 Managing Walkthroughs

The session should be chaired by the SQA representative. There are two reasons for this. First, there must be a written record kept of the proceedings, and the SQA personnel need to take minutes in any event as part of their job. Then consider the motivations of the participants. The representative responsible for the specifications may be anxious to have the specifications approved as quickly as possible in order to start some other task. The client representative may decide that any defects not detected at the review will probably show up during acceptance testing and will then be fixed at no cost to the client organization. But the SQA representative has the most at stake. The quality of the product is a direct reflection of the professional competence of the SQA group. So the SQA representative has the most to lose if the walkthrough is poorly performed and specification faults are allowed to slip through. For these reasons, the walkthrough should be chaired by the SQA representative.

The person leading the walkthrough guides the other members of the walkthrough team through the document in order to uncover any faults. It is not the task of the team to correct faults, but merely to record them for later correction. One reason for this is that specifications produced by a committee, namely the walkthrough team, within the time constraints of the walkthrough are likely to be inferior in quality to what one individual trained in the techniques of drawing up specifications can do. Second, a correction produced by a walkthrough team of five individuals will take at least as much time as if it were done by one person, and therefore costs five times as much from the viewpoint of the salaries of the five participants. The third reason is that there simply is not enough time in a walkthrough both to detect and to correct faults. No walkthrough should last longer than 2 hours. The time should be spent detecting and recording faults, not correcting them.

There are two ways of conducting a walkthrough. The first is participant-driven. Each participant in turn goes through his or her list of unclear items and items which appear to be incorrect. The representative of the specifications team must respond to each query, clarifying what is unclear to the reviewer. With regard to suspected faults detected by a reviewer, the specifier will either agree that a fault has been made or will explain why the reviewer was mistaken. The second way of conducting a review is document-driven. The person who drew up the specifications walks the participants through the document, with the reviewers interrupting either with their prepared comments or with comments triggered by the presentation. This second approach is likely to be more thorough. In addition, it generally leads to the detection of more faults for the following reason. Although many faults are highlighted by the team as a whole or by individuals as a result of the interplay between the participants, in practice it happens that the majority of faults at a document-driven walkthrough are spontaneously detected by the review leader. Time after time the presenter will pause in the middle of a sentence, his or her face will light up, and a fault, one that has lain dormant under many rereadings of the specifications, will suddenly be detected. A fruitful field for research by a psychologist would be to determine why verbalization so often leads to fault detection during specifications walkthroughs. Similar spontaneous fault detection takes place during design and code walkthroughs as well. Not surprisingly, the more thorough document-driven review is the technique prescribed in the IEEE Draft Standard for Software Reviews and Audits [IEEE 1028, 1986].

The primary role of the walkthrough leader is to evoke questions and discussion. A walkthrough is an interactive process; it is not supposed to be one-sided instruction by the presenter. It is also essential that the walkthrough not be used as a means of evaluating the participants. If this happens, the major thrust of the walkthrough will not be to detect faults, but it will degenerate into a point-scoring session, no matter how well the session leader tries to run the walkthrough. It has been suggested that the manager responsible for the specifications be a member of the walkthrough team. If this manager is also responsible for the annual evaluations of the members of the walkthrough team and particularly the presenter, the fault detection capabilities of the team will be sharply reduced as the primary motive of the presenter will be to impress his or her manager by trying to minimize the number of faults that show up. To prevent this conflict of interests, the management structure of the development team should be set up in such a way that the person responsible for the specifications is not also directly responsible for evaluating any members of a specifications walkthrough team.

The second major product of the specifications phase is the software project management plan (SPMP), and this, too, must be carefully verified by means of a management review [IEEE 1028, 1986].

6.6.3 Specifications Inspections

In addition to the specifications walkthrough, another powerful mechanism for detecting faults in specifications is the specifications inspection. Also conducted by a team, the inspectors review the specifications against a checklist. Typical items on a specifications inspection checklist include: Have the hardware resources required been specified? Have the acceptance criteria been specified?

Inspections were first suggested by Fagan in the context of testing the design and the code [Fagan, 1976]. Fagan's work is described in detail in Section 6.7.2. However, inspections have also proved to be of use in the testing of specifications.

6.7 Testing during the Design Phase

Not only must the design be correct, but if it does not faithfully reflect the client's requirements, then it is of no use. The goal of testing at the design phase is therefore to verify that the specifications have been accurately and completely incorporated into the design, as well as to ensure the correctness of the design. For example, the design must not have any logic faults, and all interfaces must be correctly defined. It is important that any faults in the design be detected before coding commences, otherwise the cost of fixing the faults will be 10 to 100 times higher. Detection of design faults can be achieved by means of design walkthroughs, as well as design inspections.

6.7.1 Design Walkthroughs

A design walkthrough is similar to the walkthrough at the specifications phase except that there is no need for the representatives of the client organization to be present at a design walkthrough. Where the product is transaction-oriented, the design walkthrough should reflect this fact [Beizer, 1984]. (A *transaction* is an operation from the viewpoint of the user of the product, such as "process a request" or "print a list of today's orders.") Walkthroughs that will include all possible transaction types should be scheduled. The reviewer should relate each transaction to the specification, showing how the transaction arises from the specification document. For example, if the application is an automatic teller machine (ATM) product, a transaction will correspond to each operation that the customer can perform, such as deposit to or withdraw from a credit card account. In other instances, the correspondence between specifications and transactions will not necessarily be one to one. In a traffic-light control system, for example, if an automobile driving over a sensor pad results in the system deciding to change a particular light from red to green in 15 seconds, then further impulses from that sensor pad may be ignored. Conversely, in order to promote traffic flow, a single impulse may cause a whole series of lights to be changed from red to green.

Restricting reviews to transaction-driven walkthroughs will not detect cases where the designers have overlooked instances of transactions required by the specifications. To take an extreme example, the specifications for the traffic-light controller may stipulate that between 12:00 P.M. and 6:00 A.M. all lights are to flash orange in one direction and red in the other direction. If the designers overlooked this stipulation, then clock-generated transactions at 12:00 P.M. and 6:00 A.M. would not be included in the design, and if these transactions were overlooked, they could not be tested in a design walkthrough based on transactions. Since it is not adequate just to schedule design walkthroughs that are transaction-driven, specification-driven walkthroughs are also essential to ensure that no statement of the specifications has either been overlooked or misinterpreted.

6.7.2 Design Inspections

As mentioned in Section 6.6.3, inspections were first put forward by Fagan for the testing of designs and code [Fagan, 1976]. For Fagan, an inspection goes far beyond a walkthrough. An inspection goes through five formal steps. First, an *overview* of the design is given by the designer. At the end of the session, design documentation is distributed to the participants. Second, the participants *prepare* individually for the inspection, trying to understand the documentation in detail. In this they should be aided by having at their disposal lists of fault types found in recent inspections, with the fault types ranked by frequency. These lists of faults will help them to concentrate on those areas where the chances of detecting a fault are higher. The third step is the *inspection*. To begin, one participant walks the inspection team through the design, ensuring that every piece of logic is covered at least once, and every branch is taken at least once. Then fault finding as such commences, with the participants aided by checklists. As with walkthroughs, the purpose is only to find and note down the faults, not to correct them. Within one day, the leader of the inspection team (the *moderator*) must produce a written report of the inspection, to ensure that everything discovered will be followed through. The fourth stage is the *rework* in which the designer resolves all faults and problems noted in the written report. The final stage is the *follow-up*. The moderator must ensure that every single issue raised has been satisfactorily resolved, either by fixing the design or by clarifying items that were incorrectly flagged as faults. All fixes must be checked to ensure that no new faults have been introduced as a result of a fix [Fagan, 1986]. If more than 5% of the material inspected has been reworked, then the team reconvenes for a 100% reinspection.

The design inspection should be conducted by a team of four, consisting of the moderator, designer, implementor, and tester. The moderator is manager and leader of the inspection team. The designer is a member of the team that produced the design, while the implementor will be responsible, either individually or as part of a team, for translating the design into code. Fagan states that the tester is any programmer responsible for setting up test cases; it is of course preferable that the tester should be a member of the SQA group. In the IEEE standard the team consists of between three and six participants [IEEE 1028, 1986]. Special roles are played by the *moderator*, the *reader* who leads the team through the design, and the *recorder* who is responsible for producing a written report of the faults as detected.

The checklist for a design inspection should include items such as: Is each item of the specifications adequately and correctly addressed? For each interface, do the actual and formal parameters correspond? Have error-handling mechanisms been adequately identified? Is the design compatible with the hardware resources, or does it require more hardware than is actually available? Is the design compatible with the software resources—for example, does the operating system stipulated in the specifications have the functionality required by the design?

An important component of the inspection procedure is the record of fault statistics. Faults must be recorded by severity (major or minor) and by fault type, such as interface fault, or logic fault. This information can be used in a number of useful ways. First, the number of faults in a given product can be compared with averages of faults detected at the same stage of development in comparable products, giving management an early warning that something is amiss and allowing timely corrective action to be taken. Second, if testing the design of two or three modules results in a disproportionate number of faults of a particular type coming to light, this information can be used to alert management that the same situation is probably present in the remaining modules. Management can then test the other modules for this fault and remedy the situation. Third, if the inspection of a particular module reveals far more faults than were found in any other module in the product, then a case might be made for redesigning that module from scratch. Finally, information regarding the number and type of faults detected at the design inspection will aid the team performing the code inspection of the same module at a later stage.

Fagan's first experiment was performed on a systems product [Fagan, 1976]. One hundred person-hours were devoted to inspections, at a rate of two 2-hour inspections per day by a 4-person team. Of all the faults that were found during the development of the product, no fewer than 67% were located by inspections before module testing was started. Furthermore, during the first 7 months of the operational phase, 38% fewer faults were detected in the inspected product than in a comparable product reviewed using informal walkthroughs.

Fagan then conducted a further experiment, this time on an applications product [Fagan, 1976]. He found that 82% of all faults were found during design and code inspections. A useful side effect of the inspections was that programmer productivity rose because less time had to be spent on module testing. Using an automated estimating model, Fagan determined that as a result of the inspection process, the savings on programmer resources were 25% despite the fact that time had to be devoted to the inspections. In a different experiment Jones has found that over 70% of faults could be removed by having design and code inspections [Jones, 1978].

A risk of the inspection process is that, like the walkthrough, it might be used for performance appraisal. The danger is particularly acute in the case of inspections because of the detailed fault information that is available. Fagan dismisses this fear by stating that over a period of 3 years he knew of no IBM manager who used such information against a programmer, or as he put it, no manager has tried to "kill the goose that lays the golden eggs" [Fagan, 1976]. However, if inspections are not conducted properly, they may not prove to be as wildly successful as they have been at IBM. Unless top management is made aware of the problem, misuse of inspection information is a possibility.

6.7.3 Comparison of Inspections and Walkthroughs

Superficially, the difference between an inspection and a walkthrough is that the inspection team uses a checklist of queries to aid it in finding the faults. But the difference goes deeper than that. A walkthrough is a two-step process: preparation, followed by group analysis of the design. An inspection is a five-step process, namely overview, preparation, inspection, rework, and follow-up, and the procedure to be followed in each of those steps is formalized. An example of that formalization is the careful categorization of faults and the use of that information in code inspections when the product has been implemented as well as in design inspections of future products.

The inspection process takes much longer than a walkthrough. The question is then: Is it worth the additional time and effort? The key point is not only that inspections are excellent detectors of faults, but that faults are detected early in the life cycle. Design faults are usually detected before implementation commences, that is, before they become expensive to fix. Similarly, coding faults are found before integration of the product. The data of Fagan clearly indicate that inspections are a powerful and cost-effective tool for fault detection [Fagan, 1976]. Unfortunately, not much research has been done subsequent to the pioneering studies of Fagan investigating the fault-detecting capabilities of inspections during the design phase or the comparative value of inspections as opposed to other design review techniques [Jones, 1986]. Nevertheless, the data of Fagan are convincing, and in the absence of conflicting evidence, the case for inspections is a strong one.

Testing during the implementation phase is now considered.

6.8 Testing during the Implementation Phase

There is a fundamental difference between testing the specifications and design and testing the individual modules. The difference is that the specifications and design exist only on paper, while modules can be run with test data on a computer. In fact, the term "test" is infrequently used in the context of specifications and design; instead the term "verify" is employed. This highlights the difference between executing code with test data (testing) and reading through documentation at a walkthrough or inspection (verifying).

6.8.1 Who Should Perform Module Testing?

Suppose a programmer is asked to test a module that he or she has written. Testing has been described by Myers as the process of executing a product with the intention of finding faults [Myers, 1979]. Testing is thus a destructive process. On the other hand, the programmer who is doing the testing will ordinarily not wish to destroy his or her own work. If the fundamental attitude of the programmer towards the code is the usual protective one, then the chances of that programmer using test data which will highlight faults is considerably lower than if the major motivation were truly destructive.

A successful test is one that finds faults. This, too, poses a difficulty. It means that if the module passes the test, then the test has failed. Conversely, if the module does not perform according to specifications, then the test succeeds. When a programmer is asked to test a module he or she has written, the programmer is being asked to execute the module in such a way that a failure (incorrect behavior) ensues. This goes against the creative instincts of programmers.

An inescapable conclusion of this is that a programmer should not test his or her own module. After a programmer has been *con*structive and has built a module, for the programmer to test that module requires that programmer to be *de*structive and to attempt to destroy the very thing that he or she has created. A second reason why testing should be done by someone else is that the programmer may have misunderstood some aspect of the design or specifications. If testing is done by someone else, the resulting faults may be discovered. Nevertheless, debugging (finding the cause of the failure and correcting the fault) must be done by the original programmer, the person who is most familiar with the code.

At the same time, the statement that a programmer should not test his or her own code must not be taken too far. Consider the programming process. The programmer begins by reading the module specifications and then designs the module. The module design may be in the form of a flowchart or perhaps psuedocode. But whatever technique is used, the programmer must certainly desk-check the module before entering it into the computer. That is to say, the programmer must try out the flowchart or pseudocode with various test cases and trace through the detailed design to check that each test case is correctly executed. Only when the programmer is satisfied that the detailed design is correct should the text editor be invoked and the module coded.

Once the module is in machine-readable form, it undergoes a series of tests. First, unless a syntax-directed editor has been used (Section 10.10), the programmer attempts to compile the module. Once this has been successfully achieved, the next step is to link and load it. Then the programmer attempts to execute the module. If the module executes, then test data are used to determine that the module works successfully, probably the same test data that were used to desk-check the code. Next, if the module executes correctly when correct test data are used, the programmer then tries out incorrect data to test the robustness of the module. When the programmer is satisfied that the module is operating correctly, systematic testing commences. It is this systematic testing that should *not* be performed by the programmer.

If the programmer is not to test his or her module, who is to do it? As stated in Section 4.4.1, independent verification and validation (IV & V) must be performed by the software quality assurance (SQA) group. The key word here is *independent*. Only if the SQA group is truly independent of the development team can its members fulfill their mission of ensuring that the product indeed satisfies its specifications, without software development managers applying pressures such as product deadlines that might hamper their work. SQA personnel must report to their own managers and thus protect their independence.

How is systematic testing to be performed? An essential part of a test case is the definition of the expected output before the test is run. It is a complete waste of time for the tester to sit at a terminal, execute the module, enter arbitrary test data, and then peer at the screen and say, "I guess that looks right." Equally futile is for the tester to plan out test cases with great care and then run each test case in turn, look at the output, and say, "Yes, that certainly looks right." It is far too easy to be fooled by plausible results. If a programmer is allowed to test his or her own code, then there is always the danger that the programmer will see what he or she wants to see. The same danger can occur even when the testing is done by someone else. The solution is for management to insist that before a test is performed, both the test data and the expected results of that test be recorded. After the test has been performed, the actual results obtained should be recorded and compared with the expected results.

Even in small organizations and with small products, it is important that this recording be done in machine-readable form, because test cases should never be thrown away. The reason for this maintenance. While the product is being maintained, regression testing must be performed. Stored test cases that the product has previously executed successfully must be rerun to ensure that the modifications made to add new functionality to the product have not destroyed the product's existing functionality. This is discussed further in Section 6.17.

6.8.2 When to Rewrite rather than Debug a Module

When a member of the SQA group detects a failure (erroneous output), the module must, as stated previously, be returned to the original programmer for debugging, that is, detection of the fault and correction of the code. There will be occasions when it is preferable for the module to be thrown away and redesigned and recoded from scratch, either by the original programmer or by another, possibly more senior, member of the development team. To see why this may be necessary, consider Figure 6.8. The graph shows the apparently nonsensical fact that the probability of the existence of more faults in a module is proportional to the number of faults already found in that module [Myers, 1979]. To see why this should be so, consider two modules, M_A and M_B. Suppose that both modules are approximately the same length and that both have been tested for the same number of hours. Suppose further that only 2 faults were detected in M_A, but that 47 faults were detected in M_B. It is likely that more faults remain to be rooted out of M_B than out of M_A. Furthermore, additional testing and debugging of M_B is likely to be a lengthy process, and there will always be the suspicion that M_B is still not perfect. In both the short run and the long run, it is preferable to discard M_B, redesign it, and then recode it.

The distribution of faults in modules is certainly not uniform. Myers cites the example of faults found by users in OS/370 [Myers, 1979]. It was found that 47% of the faults were associated with only 4% of the modules. An

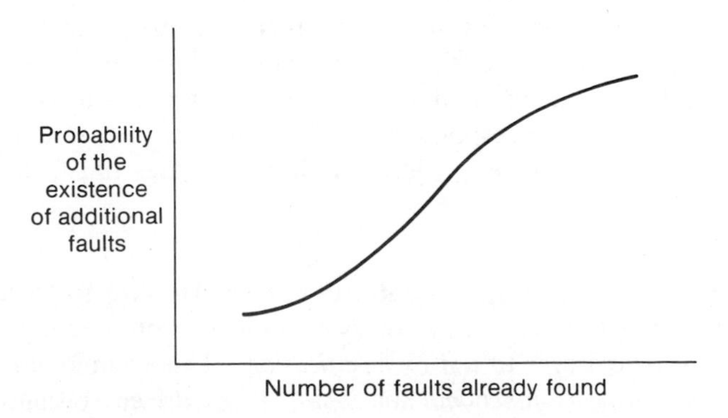

Figure 6.8 Graph showing probability that there are faults still to be found is proportional to number of faults already detected.

earlier study by Endres regarding internal tests of DOS/VS (Release 28) at IBM Laboratories, Böblingen, Germany, showed similar nonuniformity [Endres, 1975]. Of the total of 512 faults detected in 202 modules, only 1 fault was detected in each of 112 of the modules. On the other hand, there were modules with 14, 15, 19, and 28 faults, respectively. Endres points out that the latter three modules were three of the largest modules in the product, each comprising over 3000 lines of DOS macro assembler language. But the module with 14 faults was a relatively small module which was previously known to be very unstable. This type of module is a prime candidate for being discarded and recoded.

The way for management to cope with this sort of situation is to predetermine the maximum number of faults that will be permitted in a given module, and when that maximum is reached the module must be thrown away. This maximum will vary from application area to application area and from module to module. After all, a module which reads in a record from a database and checks that the part number is valid should be permitted to contain far fewer faults than a complex module from a tank weapons control system that must coordinate data from a variety of sensors and then direct the aim of the main gun towards the intended target. One way to decide on the maximum fault figure for a specific module is to examine fault data on similar modules that have required corrective maintenance. But whatever estimation technique is used, management must ensure that the module is scrapped if that figure is exceeded.

Techniques for selecting test cases for each module are now presented.

6.9 Module Test Case Selection

The worst way to test a module is to use random test data. The tester sits at a terminal, and whenever the module requests input, the tester responds with arbitrary data. As will be shown, there is never time to test more than the tiniest fraction of all possible tests cases, which can easily number many more than 10^{100}. The few test cases that can be run, perhaps of the order of 1000, are too valuable to waste on random data. Worse, there is a tendency when the machine solicits input to respond more than once with the same data, thus wasting even more test cases. It is clear that test cases must be constructed systematically.

6.9.1 Testing to Specifications versus Testing to Code

There are two basic ways of systematically constructing test data to test a module. The first is to *test to specifications*. This technique is also called *black-box*, *data-driven*, *functional*, or *input/output-driven* testing. In this approach, the code itself is ignored; the only information used in drawing up test cases is the specifications document. The other extreme is to *test to code* and to ignore the specifications when selecting test cases. Other names for this technique are *glass-box*, *logic-driven*, or *path-oriented* testing.

First, the feasibility of each of these two approaches will be considered separately.

6.9.2 Feasibility of Testing to Specifications

Consider the following example. Suppose that the specifications for a certain data-processing product state that five types of commission and seven types of discount must be provided. Testing just commission and discount requires 35 test cases. It is no use saying that commission and discount are computed in two entirely separate modules, and hence may be tested independently—in black-box testing the product is treated as a black box and its internal structure is completely irrelevant.

Thus testing to specifications is impossible in practice because of the combinatorial explosion. There are simply too many test cases to consider. Testing to code is therefore now examined.

6.9.3 Feasibility of Testing to Code

The most common form of testing to code requires that each path through the module must be executed at least once. To see the infeasibility of this, consider the flowchart depicted in Figure 6.9. Despite the fact that the flowchart appears to be almost trivial, there are over 10^{12} different paths through the flowchart. There are five possible paths through the diamond in the center, and the total number of possible paths through the flowchart is therefore

$$5^1 + 5^2 + 5^3 + \ldots + 5^{18} = 4.77 \times 10^{12} \qquad (6.12)$$

To see the implications of over a trillion paths, consider how long it would take to test each path. If a team of programmers could be found who

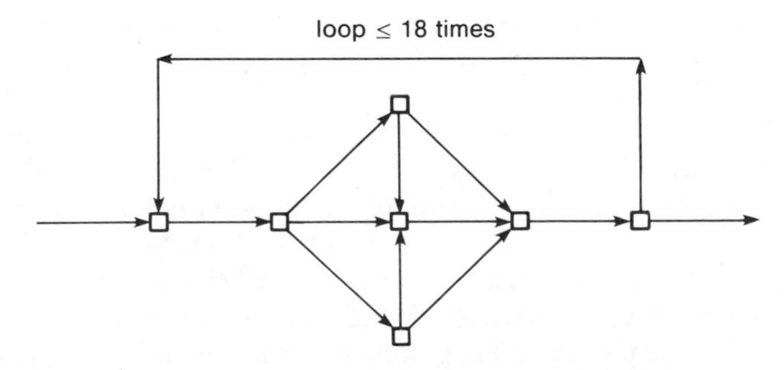

Figure 6.9 Flowchart with over 10^{12} possible paths.

```
if (X + Y + Z)/3 = X then
  PUT ("X, Y, Z are equal in value");
else
  PUT ("X, Y, Z are unequal");
end if;

Test case 1: X = 1, Y = 2, Z = 3
Test case 2: X = Y = Z = 2
```

Figure 6.10 Incorrect code fragment for determining if three integers are equal, together with two test cases.

could generate, run, and examine test cases at an average rate of one every 7 seconds, then it would take more than a million years to test the flowchart exhaustively. And if there are this many paths through a simple flowchart containing a loop, it is not difficult to imagine the total number of different paths in a module of reasonable size and complexity, let alone in a large module with many loops. In short, the combinatorial explosion renders testing to code as infeasible as testing to specifications.

There are additional reasons why testing to code is problematic. Testing to code requires the tester to exercise every path. It is possible to exercise every path without detecting every fault in the product, that is to say, testing to code is not reliable. To see this, consider the code fragment shown in Figure 6.10 [Myers, 1976]. The code was written to test the equality of three integers, X, Y, and Z, using the totally fallacious assumption that if the average of three numbers is equal to the first number, then the three numbers are equal. Two test cases are also shown in the figure. In the first test case the value of the average of the three numbers is 6 / 3 or 2, which is not equal to 1. The product therefore correctly informs the tester that X, Y, and Z are unequal. The integers X, Y, and Z are all equal to 2 in the second test case, so the product computes their average as 2, which is equal to the value of X, and the product correctly concludes that the three numbers are equal. Thus both paths through the product have been exercised without the fault being detected. Of course, the fault would come to light if test data such as X = 2, Y = 1, Z = 3 were used.

A third difficulty with path testing is that a path can be tested only if it is present. Consider the code fragment shown in Figure 6.11. It is clear that there are two paths to be tested, corresponding to the cases D = 0 and D ≠ 0. But now consider the single statement of Figure 6.11(b). Now there is only one path, and this path can be tested without the fault being detected. In fact, if a programmer omits checking whether D = 0 in his or her code, it is likely that the programmer is unaware of the potential danger, and that the case D = 0 will not be included in the programmer's test data. This is an additional

```
(a)  if D = 0 then
       ZERO_DIVISION_ROUTINE;
     else
       X := N / D;
     end if;

(b)  X := N / D;
```

Figure 6.11 Two code fragments for computing quotient.

argument for having an independent software quality assurance group whose job includes detecting faults of this type.

The examples show conclusively that the criterion "exercise all paths in the product" is not *reliable*, as there exist products for which some data exercising a given path will detect a fault and some data exercising the same path will not. However, path-oriented testing is *valid*, since it does not inherently preclude selecting test data that might reveal the fault.

Because of the combinatorial explosion of possible paths in products containing loops, less comprehensive glass-box criteria than exercising all paths have been proposed. One such weaker criterion is exercising both the true branch and the false branch of all conditional statements. However, the coverage of this criterion is lower than that of exercising all paths, and hence the criterion results in lower reliability. A still weaker criterion is to execute every statement.

Because of the combinatorial explosion, neither testing to specifications nor testing to code is feasible. A compromise is needed, using techniques that will highlight as many faults as possible, while accepting that there is no way to guarantee that all faults have been removed. A reasonable way to proceed is to use black-box test cases first (testing to specifications) and then develop additional test cases using glass-box techniques (testing to code).

6.10 Black-Box Module-Testing Techniques

Exhaustive black-box testing generally requires literally billions and billions of test cases. The art of testing is to set up a small, manageable set of test cases in such a way as to maximize the chances of detecting a fault while minimizing the chances of wasting a test case by having the same fault detected by more than one test case. Every test case must be chosen so as to detect a previously undetected fault. One such technique is equivalence testing combined with boundary value analysis.

6.10.1 Equivalence Testing and Boundary Value Analysis

Suppose the specifications for a database management product state that the product must be able to handle any number of records between 1 and

16,383 ($2^{14}-1$). If the product can handle 34 records and 14,870 records, then the chances are good that it will work fine for 8252 records. In fact, the chances of detecting a fault, if present, are likely to be equally good if any test case between 1 and 16,383 records is selected. Conversely, if the product works correctly for any one test case in the range between 1 and 16,383, then it will probably work for any other test case in the range. The range between 1 and 16,383 constitutes an *equivalence class*, that is to say, a set of test cases such that any one member of the class is as good a test case as any other member of the class. To be more precise, the specified range of numbers of records that the product must be able to handle defines three different equivalence classes:

Equivalence class 1: Less than 1 record.
Equivalence class 2: Between 1 and 16,383 records.
Equivalence class 3: More than 16,383 records.

Testing the database management product using the technique of equivalence classes then requires that one test case from each equivalence class be selected. The test case from equivalence class 2 should be correctly handled, while error messages should be printed for the test cases from class 1 and class 3.

A successful test case is one that detects a previously undetected fault. In order to maximize the chances of finding such a fault, a high-payoff technique is *boundary value analysis*. Experience has shown that when a test case that is on or just to one side of the boundary of an equivalence class is selected, the probability of detecting a fault increases. Thus, when testing the database product, seven test cases should be selected:

Test case 1: 0 records	Member of equivalence class 1 and adjacent to boundary value
Test case 2: 1 record	Boundary value
Test case 3: 2 records	Adjacent to boundary value
Test case 4: 723 records	Member of equivalence class 2
Test case 5: 16,382 records	Adjacent to boundary value
Test case 6: 16,383 records	Boundary value
Test case 7: 16,384 records	Member of equivalence class 3 and adjacent to boundary value

The preceding example applies to the input specifications. An equally powerful technique is to examine the output specifications. For example, in 1989 the minimum Social Security (FICA) deduction from any one paycheck permitted by the tax code was $0.00, and the maximum was $3379.50, corresponding to gross earnings of $45,000. Thus, when testing a payroll product, the test cases for the Social Security deduction from paychecks should include input data that are expected to result in deductions of exactly

$0.00 and $3379.50. In addition, test data should be set up that might result in deductions of less than $0.00 or more than $3379.50.

In general, for each range (R_1, R_2) listed in either the input or the output specifications, five test cases should be set up, corresponding to values less than R_1, equal to R_1, greater than R_1 but less than R_2, equal to R_2, and greater than R_2. Where it is specified that an item has to be a member of a certain set (e.g., "the input must be a letter"), two equivalence classes must be tested, namely a member of the specified set and a nonmember of the set. Where the specifications lay down a precise value (e.g., "the response must be followed by a $ sign"), then there are again two equivalence classes, namely the specified value and anything else.

The use of equivalence classes, together with boundary value analysis, to test both the input specifications and the output specifications is a valuable technique for generating a relatively small set of test data with the potential of uncovering a number of faults that might well remain hidden if less powerful techniques for test data selection were used.

6.10.2 Functional Testing

An alternative form of black-box testing is to base the test data on the functionality of the module. In *functional testing* [Howden, 1987], each of the functions implemented in the module is identified. Typical functions in a module for a computerized warehouse product might be "get next database record," or "determine whether quantity on hand is below the reorder point." In a weapons control system, a module might include the function "compute trajectory." In a module of an operating system, one function might be "determine whether file is empty."

After determining all the functions of the module, test data are set up to test each function separately. Now the functional testing is taken a step further. If the module consists of a hierarchy of lower-level functions, connected together by the control structures of structured programming (Section 10.3), then functional testing proceeds recursively. For example, if a higher-level function is of the form

⟨*higher-level function*⟩ ::= **if** ⟨*conditional expression*⟩ **then**
        ⟨*lower level function 1*⟩;
    else
        ⟨*lower level function 2*⟩;
    end if;   (6.13)

then, since ⟨*conditional expression*⟩, ⟨*lower-level function* 1⟩, and ⟨*lower-level function* 2⟩ have been subjected to functional testing, ⟨*higher-level function*⟩ can be tested using branch coverage, a glass-box technique described in

Section 6.11.1. Note that this form of structural testing is a hybrid technique —the lower-level functions are tested using a black-box technique, but the higher-level functions are tested using a glass-box technique.

In practice, however, higher-level functions are not constructed in such a structured fashion from lower-level functions. Instead, the lower-level functions are usually intertwined in some way. To determine faults in this situation, *functional analysis* is required, a somewhat complex procedure; for details, see [Howden, 1987]. A further complicating factor is that functionality frequently does not coincide with module boundaries. Thus the distinction between module testing and integration testing becomes blurred; one module cannot be tested without, at the same time, testing the other modules whose functionality it uses. The random interrelationships between modules from the viewpoint of functional testing may have unacceptable consequences from the viewpoint of management. For example, milestones and deadlines can become somewhat ill-defined, making it difficult to determine the status of the product with respect to the software project management plan.

6.11 Glass-Box Module-Testing Techniques

6.11.1 Structural Testing: Statement, Branch, and Path Coverage

In glass-box techniques test cases are selected on the basis of examination of the code, rather than the specifications. The simplest form of glass-box testing is *statement coverage*, that is, running a series of test cases which will ensure that every statement is correctly executed. To keep track of which statements are still to be executed, an automated tool is required which keeps a record of how many times each statement has been executed over the series of tests. A weakness of this approach is that there is no guarantee that all outcomes of branches are properly tested. To see this, consider the code fragment of Figure 6.12. The programmer has made a mistake; the compound conditional A > 1 **and** B = 0 should have read A > 1 **or** B = 0. The test data shown in the figure allow the statement X := 9 to be executed without the fault being highlighted.

An improvement over statement coverage is *branch coverage*, that is, running a series of tests to ensure that all branches are tested at least once. Again, a tool is usually needed to help the tester keep track of which branches

```
if A > 1 and B = 0 then
   X := 9;
end if;

Test case: A = 2, B = 0
```

Figure 6.12 Code fragment with test data.

have or have not been tested. Techniques such as statement or branch coverage are termed *structural tests*.

The most powerful form of structural testing is *path coverage*, that is, testing all paths. As shown previously, in a product with loops the number of paths can be very large; it may in fact be infinite. As a result, researchers have been investigating ways of reducing the number of paths to be examined while still being able to uncover more faults than would be possible using branch coverage. One criterion for selecting paths is to restrict test cases to *linear code sequences* [Woodward, Hedley, and Hennell, 1980]. To do this, first identify the set of points L from which control flow may jump. The set L includes entry and exit points and statements such as an **if** statement or a **goto** statement. Then linear code sequences are those paths which begin at an element of L and end at an element of L. The technique has been successful in that it has uncovered many faults without having to test every path.

Another way of reducing the number of paths to test is *all-definition-use-path coverage* [Rapps and Weyuker, 1985]. In this technique, each occurrence of a variable Z, say, in the source code is labeled either as a *definition* of the variable, such as Z := 1 or READ(Z), or a *use* of the variable, such as Y := Z + 3 or **if** Z < 9 **then** ERROR_B **end if**. All paths between the definition of a variable and the use of that definition are identified. This can be done by an automatic tool. A test case is now set up for each such path. While all-definition-use-path coverage is an excellent test technique, it does have the disadvantage that the upper bound on the number of paths is 2^d, where d is the number of decision statements (branches) in the product. Examples can be constructed exhibiting the upper bound. However, it has been shown that for real products, as opposed to artificial examples, this upper bound is not reached, and the actual number of paths is proportional to d [Weyuker, 1988]. In other words, the number of test cases needed for all-definition-use-path coverage is generally much smaller than the theoretical upper bound. Thus all-definition-use-path coverage is a practical test case selection technique.

When using structural testing, the situation can arise in which the tester simply cannot come up with a test case that will exercise a specific statement, branch, or path. What may have happened is that there is an infeasible path ("dead code") in the module, that is, a path that cannot possibly be executed for any input data. Figure 6.13 shows two examples of infeasible paths. In Figure 6.13(a) the programmer has omitted a minus sign. If K is less than 2, then K cannot possibly be greater than or equal to 3, so the statement X := X*K cannot be reached. Similarly, in Figure 6.13(b) the range 1 through 0 is empty, so the statement TOTAL := TOTAL + VALUE(J) can never be reached; the programmer had intended the range to be 1 through 10, but made a typing mistake. A tester using statement coverage would soon realize that neither statement could be reached, and the faults would be found.

```
(a)  if K < 2 then
         if K > 3 then        [should be: K > -3]
              ↑
            X := X*K;
         end if;
     end if;

(b)  for J in 1 .. 0 loop     [should be: 1 .. 10]
                   ↑
         TOTAL := TOTAL + VALUE(J);
     end loop;
```

Figure 6.13 Two examples of infeasible paths.

6.11.2 Measures of Complexity

The quality assurance viewpoint provides an additional approach to glass-box testing. Suppose that a manager is told that module M_1 is more complex than module M_2. Irrespective of the precise way in which the term "complex" is defined, the manager will intuitively feel that M_1 is likely to have more faults than M_2. Following this idea, computer scientists have come up with a number of metrics of software complexity as an aid in determining which modules are most likely to have faults. If the complexity of a module is measured and found to be unreasonably high, a manager may direct that the module be redesigned and reimplemented on the grounds that it will probably be cheaper and faster to start from scratch than to attempt to debug a fault-prone module.

The simplest measure of complexity is lines of code. The underlying assumption behind this metric is that there is a constant probability p that a line of code contains a fault. Thus, if a tester believes that, on average, a line of code has a 2% chance of containing a fault and the module under test is 100 lines long, then this implies that the module is expected to contain two faults, while a module that is twice as long is twice as fault-prone. Basili and Hutchens as well as Takahashi and Kamayachi showed that the number of faults is indeed related to the size of the product as a whole [Basili and Hutchens, 1983; Takahashi and Kamayachi, 1985].

Attempts have been made to find more accurate predictors of faults based on measures of product complexity. A typical contender is McCabe's measure of cyclomatic complexity [McCabe, 1976]. The cyclomatic complexity is essentially the number of decisions in the module. More precisely, it is the number of branches plus 1. Cyclomatic complexity is an additive metric; the complexity of a product consisting of N modules is the sum of the complexities of the individual modules. McCabe's metric can be computed almost as easily as lines of code. It has been shown to be a surprisingly good measure of

```
if K < 2 then
  if K > 3 then
     X := X*K;
  end if;
end if;
```

Distinct operators:

if < **then** > := * ; **end if**

Distinct operands:

K 2 3 X

Number of distinct operators $n_1 = 8$
Number of distinct operands $n_2 = 4$
Total number of operators $N_1 = 13$
Total number of operands $N_2 = 7$

Figure 6.14 Software Science metrics applied to code fragment of Figure 6.13(a).

faults. For example, Walsh analyzed 276 procedures in the Aegis system, a shipboard combat system [Walsh, 1979]. Walsh measured the cyclomatic complexity M and found that 23% of the procedures with M greater than or equal to 10 had 53% of the faults detected. Also, the procedures with M greater than or equal to 10 had 21% more faults per line of code than the procedures with smaller M values.

Halstead's Software Science metrics [Halstead, 1977] have also been used for fault prediction. Two of the four basic elements of Software Science are n_1, the number of distinct operators in the module, and n_2, the number of distinct operands. Typical operators include +, *, **if, goto**; operands are the user-defined variable names. The other two basic elements are N_1, the total number of operators, and N_2, the total number of operands. The code fragment of Figure 6.13(a) is reproduced in Figure 6.14. From the latter figure it is clear that the code fragment has eight distinct operators and four distinct operands. The total number of operators and operands are 13 and 7, respectively. These four basic elements then serve as input for fault prediction metrics [Ottenstein, 1979].

Musa, Iannino, and Okumoto have analyzed the data available on fault densities and have come to a number of conclusions [Musa, Iannino, and Okumoto, 1987]. First, most measures of complexity, including Halstead's and McCabe's, show a high correlation with the number of lines of code, or more precisely, the number of deliverable, executable source instructions. In other words, when researchers measure what they believe to be the complexity of a module or product, the result they obtain may largely be a reflection of the number of lines of code, a measure that correlates strongly with the number of faults. Second, complexity metrics provide little improvement over lines of code for predicting fault rates. These conclusions are supported by Basili and Hutchens and also by Gremillion [Basili and Hutchens, 1983; Gremillion, 1984].

At first sight, the fact that lines of code is an effective metric for predicting fault rates is in direct contradiction to the many arguments and experimental results negating the validity of that metric [van der Poel and Schach, 1983; Jones, 1986]. In fact, there is no contradiction. The work of Jones and others is within the context of software productivity, and in that area there is no question that lines of code is a poor metric. The results of Basili and others referenced in the previous paragraph relate specifically to fault rates and have nothing to do with productivity. Thus, when performing module testing, the larger modules should indeed be given special attention, whether from the viewpoint of test cases or walkthroughs and inspections as described in the next section.

6.11.3 Code Walkthroughs and Inspections

In Sections 6.7.1 and 6.7.2 a strong case was made for the use of design walkthroughs and design inspections. The same arguments hold equally well for code walkthroughs and code inspections. In brief, the fault-detecting power of these two nonexecution-based techniques leads to rapid and thorough fault detection. The additional time required for walkthroughs or inspections is more than repaid by increased productivity due to the presence of fewer faults in the code at the integration-testing phase. Furthermore, code inspections have led to a 95% reduction in corrective maintenance costs [Crossman, 1982]. Further arguments in favor of walkthroughs and inspections are given in the next section.

6.12 Comparison of Module-Testing Techniques

A number of studies have been performed comparing strategies for module testing. Myers compared black-box testing, a combination of black-box and glass-box testing, and three-person code walkthroughs [Myers, 1978a]. The experiment was performed using 59 highly experienced programmers testing the same product. All three techniques were equally effective in finding faults, but code walkthroughs proved to be less cost-effective than the other two techniques. Hwang compared black-box testing, glass-box testing, and code reading by one person [Hwang, 1981]. All three techniques were found to be equally effective, with each technique having its own strengths and weaknesses.

A major experiment was conducted by Basili and Selby [Basili and Selby, 1987]. The techniques compared were the same as in Hwang's experiment, namely black-box testing, glass-box testing, and one-person code reading. The subjects were 32 professional programmers and 42 advanced students. Each tested three products, using each testing technique once. Fractional factorial design [Basili and Weiss, 1984] was used to compensate for the fact that the products were tested in different ways by different participants; no participant tested the same product in more than one way. Different results were obtained from the two groups of participants. With regard to the professional programmers, code reading detected more faults than the other two techniques, and the fault detection rate was faster. Two different groups of advanced students participated. In one group there was no significant difference between the three techniques; in the other, code reading and black-box testing were equally good, and both outperformed glass-box testing. However, the rates at which students detected faults were the same for all techniques. Overall, code reading led to the detection of more interface faults than did the other two techniques, while black-box testing was most successful at finding control faults. The main conclusion that can be drawn from this experiment is that code inspection is at least as successful at detecting faults as glass-box and black-box testing.

The Cleanroom software development method incorporates a number of different techniques, including an incremental life cycle (Section 3.4), formal techniques for specification and design (Section 7.7), nonexecution based module-testing techniques such as code reading [Mills, Dyer, and Linger, 1987; Linger, Mills, and Witt, 1979; Selby, Basili, and Baker, 1987] and code walkthroughs and inspections (Section 6.11.3). An experiment conducted on teams of graduate students was undertaken to compare the Cleanroom method with the team software approach currently in use in many organizations. The data relating to testing showed that the products developed using Cleanroom had a higher percentage of successful test cases than the non-Cleanroom products. This is a further indication of the power of nonexecution-based testing techniques.

Some management implications of module testing are now considered.

6.13 Management Aspects of Module Testing

An important decision that must be made during the development of every module is how much time, and therefore money, to spend on testing that module. As with so many other economic issues in software engineering, cost–benefit analysis (Section 4.2.3) can play a useful role. For example, the decision as to whether or not the cost of correctness proving exceeds the benefit of the assurance that a specific product satisfies its specifications can be decided on the basis of cost–benefit analysis. Cost–benefit analysis can also be used to compare the cost of running additional test cases against the cost of failure of the delivered product caused by inadequate testing.

There is another approach for determining whether testing of a specific module should continue or whether it is likely that virtually all the faults have been removed. The techniques of reliability analysis can be used to provide statistical estimates of how many faults are still remaining. A number of different techniques have been put forward for determining statistical estimates of the number of remaining faults. The basic idea underlying these techniques is the following. Suppose that a product is tested for 1 week. On Monday 23 faults are found, and 7 more on Tuesday. On Wednesday 5 more faults are found, 2 on Thursday, and 0 on Friday. Since the rate of fault detection is steadily decreasing from 23 faults per day to none, it seems likely that most faults have been found, and testing of that module could well be halted. Determining the probability that there are no more faults in the code requires a level of mathematics and statistics beyond that expected for readers of this book. Details are therefore not given here; the reader who is interested in reliability analysis should consult a text such as [Shooman, 1983].

6.14 Testing during the Integration Phase

During the implementation phase the individual modules are coded and tested. The next step is to combine them into a single product which is then tested; this is termed *integration*. As will be shown, following this prescription to the letter can lead to an unsatisfactory product. Instead, implementation and integration should be interlaced in a methodical way, allowing management to maintain control of the development process at all times.

First, the effects of separate implementation and integration phases are examined.

6.14.1 Separate Development

Consider the product depicted in Figure 6.15. One approach in developing the product is to code and test each module separately. Then all 13 modules are linked together, and the product as a whole is tested. There are two difficulties with this approach. First, consider module A. When it has been coded it cannot be tested on its own, because it calls modules B, C, and D. Thus, in order to test module A, modules B, C, and D must be coded as *stubs*. In its simplest form, a stub is an empty module. A more effective stub prints a message such as Procedure DISPLAY RADAR PATTERN called. Best of all, a stub should return precooked values corresponding to preplanned test cases.

Now consider module H. To test it on its own requires a driver, a module which calls it (a) once, (b) several times, or (c) many times, checking each time the value returned by the module under test. Similarly, testing module D requires a driver and two stubs. Thus one problem with separate development is that effort has to be put into constructing stubs and drivers, all of which are thrown away after module testing is completed.

The second, and much more important, difficulty with this approach to integration is its lack of fault isolation. If the product as a whole is tested against a specific test case and the product fails, then the fault could lie in any of the 13 modules or 13 interfaces. In a large product with, say, 103 modules

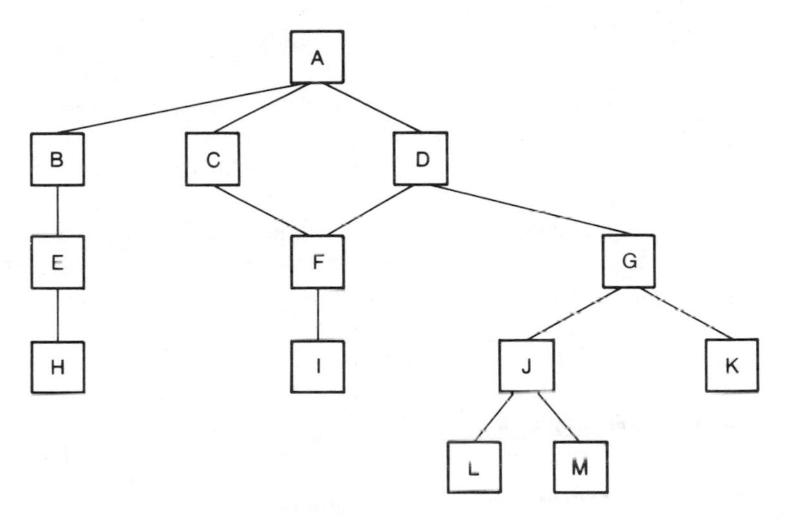

Figure 6.15 Typical module interconnection diagram.

and 108 interfaces, there are no fewer than 211 places where the fault might lie.

The solution to both these difficulties is to combine module and integration testing.

6.14.2 Top-Down Development

When the product shown in Figure 6.15 is developed top-down, a module is coded and integrated after all the modules that call that module have been coded and integrated. One possible top-down ordering is A, B, C, D, E, F, G, H, I, J, K, L, and M. First, module A is coded and tested with B, C, and D implemented as stubs. Now stub B is expanded into module B, linked to module A, and tested, with module E implemented as a stub. Development proceeds in this way. Another possible top-down ordering is A, B, E, H, C, D, F, I, G, J, K, L, and M. With this latter ordering, portions of the development can proceed in parallel in the following way. After A has been coded and tested, one team can use module A to develop B, E, and H, while another team can use it to work in parallel on C, D, F, and I. Once D and F are completed, a third team can start work on G, J, K, L, and M.

Suppose that module A by itself executes correctly on a specific test case, but when the same test data are submitted after B has been coded and integrated into the product, now consisting of modules A and B linked together, then the test fails. The fault can be in one of only two places, namely in module B or the interface between modules A and B. In general, whenever a module M_NEW is added to what has been tested so far and a previously successful test case fails, the fault almost certainly will lie either in M_NEW or in the interface(s) between M_NEW and the rest of the product. Thus top-down development provides fault isolation. A further advantage of this scheme is that little extra work is involved in writing stubs. Each stub is simply expanded into the corresponding complete module at the appropriate step in the development of the product.

A big advantage of top-down development is that major design flaws show up early. The modules of a product can be divided in two groups, *logic modules* and *functional modules*. (There is no connection between the term "functional module," the term "functional cohesion" defined in Section 8.2.7, and the term "functional testing" explained in Section 6.10.2.) Logic modules essentially incorporate the decision-making flow of control aspects of the product. The logic modules are generally those situated close to the root in the module interconnection diagram. For example, in Figure 6.15 it is reasonable to expect modules A, B, C, D, and perhaps G and J to be logic modules. The functional modules, on the other hand, perform the actual operations of the product. For example, a functional module may be named GET LINE FROM TERMINAL or MEASURE TEMPERATURE OF REACTOR CORE. The functional modules are generally found in the lower levels, close to the leaves, of the module interconnection diagram. In Figure 6.15 modules E, F, H, I, K, L, and M turn out to be functional modules; this is shown in Figure 6.16.

For any product it is important that the logic modules be coded and tested before the functional modules. This will ensure that, if there are any major design faults, they will show up early. Suppose that the whole product is completed before a major fault is detected. Large parts of the product will have to be rewritten, especially the logic modules which embody the flow of control. Many of the functional modules will probably be reusable in the rebuilt product; for example, a module like GET LINE FROM TERMINAL or MEASURE TEMPERATURE OF REACTOR CORE will be needed no matter how the product is restructured. However, the way the functional modules will be interconnected to the other modules in the product may have to be changed, thereby necessitating further unnecessary work. Thus the earlier a design fault is detected, the quicker and cheaper it will be to get the product corrected and back on development schedule. The order in which modules are developed using the top-down strategy essentially ensures that logic modules are indeed developed before functional modules, because logic modules are almost always the ancestors of functional modules in the module interconnection diagram. This is a major strength of top-down development.

Nevertheless, there is at least one difficulty associated with top-down development. The lower-level modules in the module interconnection diagram are not tested as frequently as the upper-level modules. For example, if there are 184 modules altogether, the root module will be tested 184 times, while the

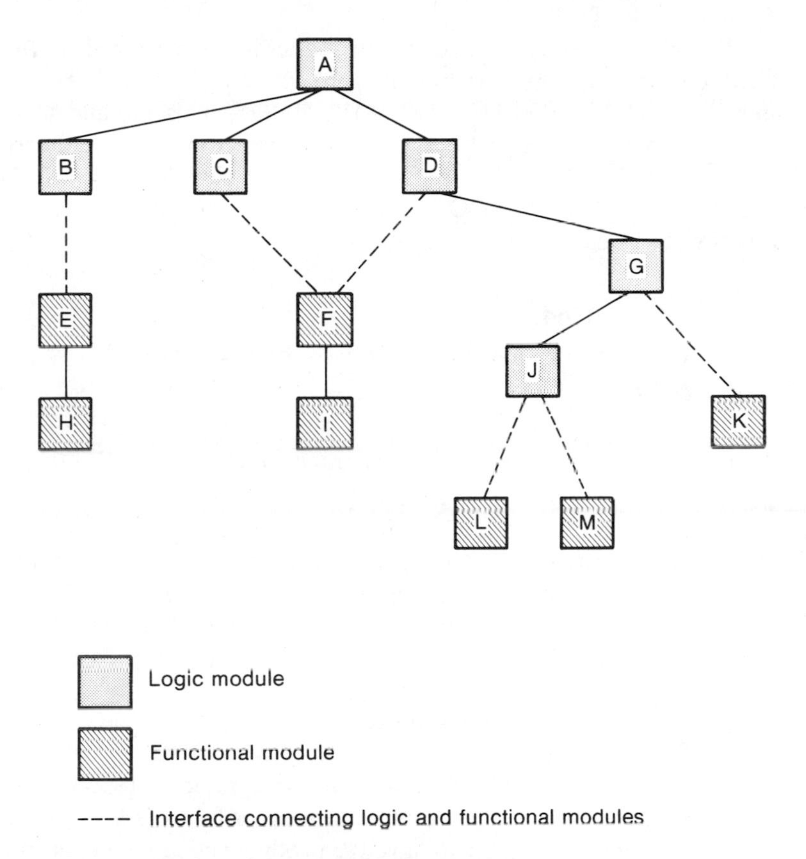

Figure 6.16 Product of Figure 6.15 implemented using sandwich development.

last module to be integrated into the product will be tested only once. This is relevant because of the need to reuse existing modules and thereby reduce development time and cost of future products. The logic modules are likely to be somewhat problem-specific, and hence unusable in another context, but the functional modules, particularly if they have informational or functional cohesion (Section 8.2), are probably reusable in other products. For this reason it is vital that these functional modules be thoroughly tested. Reusing a module that is thought, incorrectly, to have been properly tested is likely to be less cost-effective than rewriting that module from scratch, because the inherent assumption that a module is correct can lead to false deductions being made when a fault is detected in the product. Instead of suspecting the improperly tested module, the tester may think that the fault lies elsewhere, resulting in a waste of effort. Thus reusability of functional modules may be incompatible with top-down development.

The situation is aggravated if the product is well designed; in fact, the better the design, the worse the modules are likely to be tested. To see this, consider the module COMPUTE SQUARE ROOT. This module takes two arguments, a floating-point number X whose square root is to be determined and an ERROR_FLAG which is set to **true** if X is negative. Suppose further that COMPUTE SQUARE ROOT is called by module M_3, and that M_3 contains the statement

```
if X >= 0 then
  COMPUTE SQUARE ROOT (X, ERROR_FLAG);
end if;
```

In other words, COMPUTE SQUARE ROOT is never called unless the value of X is nonnegative, and therefore the module can never be tested with negative values of X to see if it functions correctly. The type of design where the calling module includes a safety check and the called module does as well is called *defensive programming* and is good programming practice. At the same time, as a result of defensive programming, subordinate functional modules are unlikely to be thoroughly tested if developed top-down.

6.14.3 Bottom-Up Development

In bottom-up development, if module M_ABOVE calls module M_BELOW, then M_BELOW is coded and integrated before M_ABOVE. Returning to Figure 6.15, one possible bottom-up ordering is L, M, H, I, J, K, E, F, G, B, C, D, and A. In order to have the product coded by a team, a better bottom-up ordering is as follows: H, E, and B are given to one team, and I, F, and C to another. The third team starts with L, M, J, K, and G, then codes D and integrates their work with the work of the second team. Finally, when B, C, and D have been successfully integrated, A can be coded and integrated.

The functional modules are thus thoroughly tested. In addition, the testing is done with the aid of drivers, rather than by fault shielding, defensively programmed calling modules. While bottom-up development solves the major difficulty of top-down development and shares with top-down development the advantage of fault isolation, it unfortunately has a difficulty of its own. Specifically, major design faults will be detected late in the integration cycle. The logic modules are integrated last, and hence if there is a major design fault, it will be picked up at the end of the integration process with the resulting huge costs of redesigning and recoding large portions of the product.

Thus both top-down and bottom-up development have their strengths and weaknesses. The solution for product development is to combine the two development strategies in such a way as to make use of their strengths and to minimize their weaknesses. This leads to the idea of sandwich development.

6.14.4 Sandwich Development

Consider the module interconnection diagram shown in Figure 6.16. Six of the modules, namely A, B, C, D, G, and J, are logic modules, and therefore should be developed top-down. Seven are functional modules, namely E, F, H, I, K, L, and M, and should be developed bottom-up. Since neither top-down nor bottom-up development is suitable for all the modules, the solution is to partition them. The six logic modules are given to one team and are developed top-down. In this way, any major design faults will be caught early. The seven functional modules are developed bottom-up. They thus receive a thorough testing, unshielded by defensively programmed calling modules, and can therefore be reused with confidence in other products. When all modules have been appropriately integrated, the interfaces between the two groups of modules are tested one by one. At all times during this process there is fault isolation. In passing, the term "sandwich development" [Myers, 1979] comes from the fact that the logic modules and functional modules can be considered as the top and bottom of a sandwich, the interfaces that connect them as the sandwich filling.

6.14.5 Management Implications of Integration Testing

A difficulty for management is that it is sometimes discovered at integration time that pieces of the modules simply do not fit together. For example, suppose that team 1 coded module M_1, while team 2 coded module M_2. In the version of the design documentation used by team 1, module M_1 calls module M_2 passing four parameters, but the version of the design documentation used by team 2 states clearly that M_2 has only three parameters. A problem like this can arise when a change is made to one copy of the design document, but this change is not propagated to all members of the development group. Both teams know that they are in the right, and neither is prepared to compromise, because the team that gives in will have to recode large portions of the product.

In order to solve these and similar problems of incompatibility, the entire integration process should be run by the SQA group. Furthermore, as with testing at other phases, the SQA group has the most to lose if the integration testing is improperly performed and is therefore the most likely to ensure that testing is performed thoroughly. Thus the manager of the SQA group should have responsibility for all aspects of integration testing. He or she must decide which modules will be developed top-down and which will be developed bottom-up and will assign the integration-testing tasks to the appropriate individuals or teams. The SQA group, which will have drawn up the integration test plan in the SPMP, is responsible for implementing that plan.

At the end of the integration phase all the modules will have been tested and combined into a single product.

6.15 Product Testing

The fact that the last module has been successfully integrated into the product does not mean that the task of the developer is complete. The SQA group still has a number of testing tasks that must now be performed in order to be certain that the product will not fail the acceptance test, the final hurdle that the product development team must overcome. The failure of a product to pass its acceptance test is almost always a poor reflection of the management capabilities of the development organization. The client may conclude that the developers are incompetent, which all but guarantees that the client will do everything in his or her power to avoid employing those developers again. Worse, the client may believe that the developers are dishonest and deliberately handed over substandard software in order to finish the contract and be paid as quickly as possible. If the client genuinely believes this and tells other potential clients, then the developers have a major public relations problem on their hands. It is up to the SQA group to make sure that the product passes the acceptance test with flying colors.

To ensure a successful acceptance test, the SQA group must perform product testing. This essentially consists of tests that the SQA group believes will closely approximate the forthcoming acceptance tests. First, black-box test cases for the product as a whole must be run. Up to now, test cases will have been set up on a module-by-module basis, ensuring that each module individually satisfies its specifications. Now test cases that treat the complete product as a black box must be run. Second, the robustness of the product as a whole must be tested. Again, the robustness of individual modules will have been tested during integration, but now productwide robustness is the issue for which test cases must be set up and run. In addition, the product must be subjected to *stress testing*, that is, making sure that it functions correctly when operating under a peak load, such as all terminals trying to log on at the same time or customers operating all the automatic teller machines simultaneously. It must also be subjected to *volume testing*, for example, making sure that it can handle large input files. Third, the SQA group must check that the product satisfies all its constraints. For example, if the specifications state that the response time for 95% of queries when the product is working under full load must be under 3 seconds, then it is the responsibility of the SQA group to verify that this is indeed the case. There is no question that this will be checked during acceptance testing, and if the product fails to meet a major constraint at acceptance testing, then the development organization will lose a considerable amount of credibility. Similarly, storage constraints and security constraints must be checked. A common constraint is that the new product must be compatible with the existing one; if this constraint is in the specifications, then it, too, must be carefully checked. Fourth, the SQA group must review all documentation that in terms of the contract is to be handed over to the client together with the code. The SQA group must check that the documentation conforms to the standards laid down in the SPMP. In addition, the documentation must be verified against the product. For instance, the SQA group has to determine that the user manual indeed reflects the correct way of using the product, and that if the operator follows the instructions listed in the operator's manual, then the product will function as specified.

Once the SQA group assures management that the product can handle anything the acceptance testers can throw at it, the product, that is, the software plus the documentation, is handed over to the client organization for acceptance testing.

6.16 Acceptance Testing

The purpose of acceptance testing is for the client to determine whether the product indeed satisfies its specifications as claimed by the developer. Acceptance testing is done either by the client organization, or by the SQA group in the presence of client representatives, or by an independent SQA group hired by the client for this purpose. Acceptance testing naturally includes correctness testing, but in addition it is necessary to test performance and robustness. The four major components of acceptance testing, namely testing correctness, robustness, performance, and documentation, are exactly what is done by the developer during product testing; this is not surprising, because product testing is a rehearsal for the acceptance test.

A key aspect of acceptance testing is that it must be performed on actual data, rather than test data. No matter how well test cases are set up, by their very nature they are artificial. More importantly, test data should be a true reflection of the corresponding actual data, but in practice this is not always the case. For example, the member of the specifications team responsible for characterizing the actual data may perform this task incorrectly. Alternatively, even if the data are correctly specified, the SQA group member who uses that data specification may misunderstand it or misinterpret it. The resulting test cases will not be a true reflection of the actual data, leading to an inadequately tested product. For these reasons, acceptance testing must be performed on actual data. Furthermore, since the development team endeavors to ensure that the product testing will duplicate every aspect of the acceptance testing, as much of the product testing as possible should also be performed on actual data.

When a new product is to replace an existing product, the specifications will almost always include a clause to the effect that the new product must be installed in such a way that it can run in parallel with the existing product. The reason for this is that there is a very real possibility that the new product may be faulty in some way. The existing product works correctly, but is inadequate in some respects. If the existing product is replaced by a new product that works incorrectly, then the client is in trouble. To prevent this from occurring, both products must run in parallel until the client is satisfied that the new product can take over the functions of the existing product. Successful parallel running then brings the acceptance testing to a favorable conclusion, and the existing product can be retired.

When the product has passed its acceptance test the task of the developers is complete. Any changes now made to that product constitute maintenance.

6.17 Testing during the Maintenance Phase

Recall that there are two types of maintenance. *Corrective maintenance* consists of changes that are made to the product to correct residual faults, that is to say, changes to the code, and perhaps to the design, without changing the specifications in any way. *Enhancement* consists of changes to the specifications and their implementation through changes to the design, the code, and the documentation.

Perhaps the most important aspect of maintenance is to ensure that a change made in order to achieve a specific objective, whether corrective or enhancement, does not have unacceptable side effects, that is to say, it must not damage another part of the product. When a change made to one area affects what appears to be a completely unrelated area, this is termed a *regression fault*. At development time many members of the development team have a broad overview of the product as a whole, but as a result of the rapid personnel turnover in the computer industry it is unlikely that members of the maintenance team will previously have been involved in the original development. Thus the maintainer tends to see the product as a set of loosely related modules and is generally not conscious of the fact that a change to one module may seriously impact one or more other modules, and hence the product as a whole. Even if the maintainer wished to understand every aspect of the product, the pressures to fix or to extend the product are generally such that there is no time for the detailed study needed to achieve this. Furthermore, in many cases there is little or no documentation to assist in gaining that understanding. One way of trying to minimize this difficulty is to use regression testing, that is, testing the changed product against previous test cases to ensure that it still works correctly.

For this reason it is vital to store all test cases, together with their expected outcomes, in machine-readable form. As a result of changes made to the product, certain stored test cases may have to be modified. For example, if, as a consequence of tax legislation, salary withholding percentages change, then the correct output from a payroll product for each test case involving withholding will change. Similarly, if satellite observations lead to corrections in the latitude and longitude of an island, then the correct output from a product that calculates the position of an aircraft using the coordinates of that island will correspondingly change. Depending on the maintenance performed, some valid test cases will become invalid and vice versa. But the computations that need to be made to correct the stored test cases are essentially the same computations that would have to be made in order to set up new test data for checking that the maintenance has been correctly performed. No additional work is therefore involved in maintaining the file of test cases and their expected outcomes.

It can be argued that regression testing is a waste of time because regression testing requires the complete product to be retested against a whole host of test cases, most of which apparently have nothing to do with the modules that were modified as a result of product maintenance. The word "apparently" is the key word in the previous sentence. The dangers of unwitting side effects of maintenance are too great for that argument to hold water; regression testing is an essential aspect of maintenance in all situations.

In the next two sections the testing of two somewhat specialized types of software is discussed, namely distributed software and real-time software.

6.18 Testing Distributed Software

The difficulties of testing are considerably exacerbated when the product under test is implemented as a distributed product running on two or more different pieces of hardware. Examples of distributed computer systems are a hypercube [Seitz, 1985] and a network of computers connected by Ethernet [Metcalfe and Boggs, 1976]. In such systems the various component computers are loosely coupled, with the individual CPUs each having their own memory and communicating with one another by message passing.

The process of testing a product running on a uniprocessor is well understood. A vast literature exists on the subject, and tools have been constructed to aid the tester. Implicit in this uniprocessor testing process are a number of assumptions. It is assumed that there is a global environment and that the execution of the product within that environment is deterministic. Furthermore, it is assumed that the instructions of the product are executed sequentially and that the addition of debugging statements between product statements will not modify the execution of the product.

When attempting to test a product running on a distributed system, the familiar techniques of testing on a uniprocessor frequently are not very effective because the previous assumptions simply do not hold. On a distributed system there is no global environment. Instead, the instructions are executed in parallel on different processors. Furthermore, as a consequence of timing considerations, product execution behavior may not be reproducible from one execution to the next. Finally, adding debugging statements in one process could affect the behavior of other processes executing in parallel, resulting in timing-dependent faults, or so-called Heisenbugs. For example, suppose that process A is running on processor P_A and process B is running on processor P_B. At some point process A sends a message to process B, that is to say, somewhere in process A there is a **transmit** command and a corresponding **receive** command in process B. Now suppose that the programmer has coded B incorrectly, so that if B reaches the **receive** instruction and the message from A is not waiting in the buffer, then process B fails. On the other hand, if the message from A is waiting, then process B works correctly. The programmer tests the product and it crashes, because process B is faster than process A, and the message is not waiting for B at the appropriate time. So the programmer attempts to debug process B by inserting a few write statements which will indicate to the programmer how much of process B has been executed before the failure. However, the additional output statements slow down process B, and process A has a chance to transmit the message before process B needs it. As a result, the product works perfectly. However, the moment the programmer removes the debugging statements from process B it runs at its normal speed and reaches the **receive** before A can **transmit** and therefore fails as it did before. Thus debugging causes the fault to disappear and to reappear when the process has apparently been successfully debugged.

Testing distributed software requires special tools, especially distributed debuggers to try to locate faults [Wahl and Schach, 1988]. Timing problems that arise when a distributed product is run on distributed hardware cannot be sorted out if the software is run on a uniprocessor. The distributed debugger must maintain history files, that is, a chronological record of events such as data inputs or messages transmitted from one process to another. The need for history files is a consequence of timing considerations. Suppose a distributed product is tested and fails. The programmer analyzes the output from the test run and concludes that a particular statement is in error. But now the test case must be rerun to determine that the fault has in fact been correctly fixed. Because the modified product is now different from the original, if only slightly so, the processes will run at different relative rates, and the messages between processors may be sent in a different order. As a result, the test case run after the product has been modified may not be the same as the one in which the fault was detected, and the programmer will not necessarily be able to determine if the fault has in fact been fixed. History files solve this problem in that the execution record they contain can be used by a distributed debugger to reproduce the exact sequence of events which caused the failure.

6.19 Testing Real-Time Software

There are a number of difficulties specific to the testing of real-time software. It is a characteristic of real-time systems that they are critically dependent on the timing of inputs and the order of inputs, while the developer has no control either over the timing or over the order in which the inputs arrive. The reason for this is that inputs to a real-time system, such as the temperature of a computer-controlled nuclear reactor, the speed of an aircraft in an avionics system, or the heart rate of a patient in a computer-controlled intensive care unit, come from the real world. From the testing viewpoint, this implies that the number of test cases that must be considered to be reasonably sure that the software is functioning correctly is generally much larger than for products that are not subject to real-time constraints. Another difficulty is that inputs to the system may well occur in parallel, and many real-time systems are therefore implemented as parallel systems running on a number of processors, with the attendant problems of testing distributed software described in the previous section.

Real-time software, and especially embedded real-time software, frequently must run without the supervision of computer operators. For example, the computers in a deep space probe or in a satellite in Earth orbit must run untended for the entire duration of the mission. The software must be able to cope with a wide variety of emergencies and exceptional conditions. If the worst happens and the software fails, it must be able to reinitialize itself and restart its functions. But embedded real-time systems on Earth must be equally robust. Consider, for example, the software for a computer controlling the engine of an automobile; if it cannot recover by itself from an unexpected event, the engine will be immobilized until a mechanic trained in computers can restart the computer system. A similar situation can occur with a system for monitoring the patients in a hospital intensive care unit. This robustness requirement implies that real-time software must incorporate a great number of exception-handling routines which will enable the software to recover without human intervention from a wide variety of ordinarily unexpected circumstances. From the testing viewpoint this means that real-time software has to be far more complicated than would be the case if a human operator were available to restart the product and perform the necessary recovery routines in the event of a failure.

Yet another difficulty with testing real-time software arises in setting up test cases. The inputs to most real-time systems come from the real world. The timing and sequence of inputs constitute a major aspect of real-time software that must be tested, but it is often impossible to organize a specific set of inputs occurring in a specific order at a specific time. For example, it is impossible to arrange for all the patients in a 10-bed intensive care unit to develop cardiac failure at the same time or for a set of medical emergencies to occur in a given order during a certain time interval. In other instances, the test cases may theoretically be possible to arrange, but are too hazardous to risk. Examples of this are determining whether the software controlling a nuclear power plant can successfully prevent accidents of the types that occurred at Three Mile Island and at Chernobyl, and whether a missile detection system would function correctly if an unfriendly power were to launch thousands of intercontinental ballistic missiles and submarine-launched ballistic missiles all at the same time.

Real-time software is frequently more complex than most people realize, even the developers. As a result, there are sometimes subtle interactions between components that even the most skilled of testers would not usually bring to light. As a result, an apparently minor change can have major consequences. A famous example of this is the fault that delayed the first space shuttle orbital flight in April, 1981 [Garman, 1981]. The space shuttle avionics are controlled by four identical synchronized computers. There is also an independent fifth computer for back-up in case the set of four computers fails. A change had been made 2 years earlier to the module that performs bus initialization before the avionics computers are synchronized. An unfortunate side effect of this change was that a record containing a time just slightly later than the current time was erroneously sent to the data area used for synchronization of the avionics computers. The time that was sent was sufficiently close to the actual time for this fault not to be detected. About 1 year later the time difference was slightly increased, just enough to cause a 1 in 67 chance of causing a synchronization failure. Then on the day of the first shuttle launch the synchronization failure occurred, and three of the four identical avionics computers were synchronized one cycle late relative to the first computer. A fail-safe device which prevents the independent fifth computer from receiving information from the other four computers unless they are in agreement had the unanticipated consequence of preventing initialization of the fifth computer, and the launch had to be postponed. An all too familiar aspect of this incident was that the fault was in the bus initialization module, a module that apparently had no connection whatsoever with the synchronization routines.

There are five main approaches to testing real-time software, namely structure analysis, correctness proofs, systematic testing, statistical testing, and simulation [Quirk, 1985]. *Structure analysis techniques* can be used to investigate control flow and data flow without executing the product. These techniques can then be used to prove that all parts of the product can be reached and that there is a path leading to termination from every node in the graph, to check that variables are assigned a value before they are used, and whether there are variables that are never referenced. These techniques are not perfect, however; for example, it is impossible to check that all elements of an array are assigned a value before they are used, because the indices of array elements may be determined only at run time. Structure analysis can assist in deadlock detection and can also be used for certain classes of timing checks [Quirk, 1983]. *Correctness proofs* have been discussed in Section 6.3. Real-time systems were mentioned as one of the areas where correctness proving may be an economically viable technique. A number of systems have been constructed which have been used to prove real-time systems correct. *Systematic testing* is carried out by running sets of test cases consisting of the same input data arranged in all possible orderings, thereby detecting faults caused by overlooking certain orderings of the inputs. Thus if 5 different inputs that can occur are being tested, these 5 inputs can be arranged in 5! different ways, yielding 120 different test cases. Clearly the combinatorial explosion makes this technique infeasible when the number of different inputs is even moderately large. As a result, *statistical techniques* are employed to select a manageable sample from the huge number of possible test cases in such a way that probabilistic statements can be made about the reliability of the product, for example, that the probability of software failure is below 0.05%.

The most important testing technique for real-time software is *simulation*. A simulator is "a device which calculates, emulates or predicts the behavior of another device, or some aspect of the behavior of the world" [Bologna, Quirk, and Taylor, 1985]. In real-time testing, simulators can be used to simulate the real world by providing input to the product, and then they can evaluate the outputs from the product. In this way, a simulator can be considered as a test-bed on which the product can be run. The programmer arranges for the simulator to provide selected inputs to the product, and then the simulator assists the programmer in analyzing the outputs from the product to determine whether the product is functioning correctly.

Simulation is particularly important when it is impossible or too dangerous to test a product against suitable sets of test cases. Simulation can be used to check that a product can adequately handle situations like an aircraft stalling and about to crash, the pile of a nuclear reactor about to go critical, or a large number of incoming nuclear missiles. Another major use for a simulator is in the training environment. Perhaps the most common instance of this is a flight simulator which is used to train pilots. But a flight simulator can also be used to test avionics software without putting a human pilot at risk as would be the case if an actual aircraft were flown with new and untested flight control software.

6.20 When Testing Stops

After a product has been successfully maintained for many years it may eventually lose its usefulness and be superseded by a totally different product, in much the same way that electronic valves were replaced by transistors. Alternatively, a product may still be useful, but the cost of porting it to new hardware or of running it under a new operating system may be larger than the cost of constructing a new product, using the old one as a prototype. Thus, finally, the software product is decommissioned and removed from service.

Only at that point, when the software has been irrevocably discarded, can testing stop.

Chapter Review

A key theme of this chapter is that testing must be carried out in parallel with all life-cycle activities. The chapter begins with a definition of testing (Sections 6.1 and 6.2) and a discussion of behavioral properties of a product that must be tested, including utility, reliability, robustness, performance, and correctness (Sections 6.2.1 through 6.2.5). In Section 6.3 testing is carefully contrasted with proofs of correctness, and an example of such a proof is given in Section 6.3.1. The role of correctness proofs in software engineering is then analyzed (Sections 6.3.2 and 6.3.3), and it is pointed out that cost–benefit analysis can be used to decide whether or not to use correctness proving for a given project. A description is then given of testing during each of the phases of the software life cycle, including requirements (Section 6.5), specifications (Section 6.6), design (Section 6.7), and implementation (Section 6.8). The importance of walkthroughs and inspections is stressed. Another key issue is that module testing must be performed by the independent SQA group, and not by the programmer (Section 6.8.1). Test cases must be selected systematically (Section 6.9). Various black-box and glass-box testing techniques are described (Sections 6.10 and 6.11) and then compared (Section 6.12); the power of nonexecution-based testing is highlighted. This is followed by a discussion of the managerial implications of module testing (Section 6.13). Integration phase testing is treated next (Section 6.14); the need to interlace implementation and integration methodically is underlined throughout. The next topics described are product testing (Section 6.15) and acceptance testing (Section 6.16). During the maintenance phase, regression testing is essential (Section 6.17). Then the problems of testing of distributed software and real-time software are presented (Sections 6.18 and 6.19). Finally, the issue of when testing can finally stop is discussed in Section 6.20.

For Further Reading

The attitude of software producers to the testing process has changed over the years, from viewing testing as a means of showing that a product runs correctly, to the modern attitude that testing should be used to prevent requirement, specification, design, and implementation faults. This progression is described in [Gelperin and Hetzel, 1988].

A good introduction to proving programs correct is Chapter 3 of [Manna, 1974]. One of the standard techniques of correctness proving is using the so-called Hoare logic, as described in [Hoare, 1969]. An alternative approach to ensuring that products satisfy their specifications is to construct the product stepwise, checking that each step preserves correctness. This is described in [Dijkstra, 1968c], [Wirth, 1971], [Dahl, Dijkstra, and Hoare, 1972], and [Dijkstra, 1976]. Functional correctness is another approach. It has been developed by Mills and co-workers [Linger, Mills, and Witt, 1979; Mills, Basili, Gannon, and Hamlet, 1987] and is used in the IBM Cleanroom software development method [Currit, Dyer, and Mills, 1986; Selby, Basili, and Baker, 1987]. Practical uses of correctness proofs are described in [Musser, 1980] and [Good, 1983]. A collection of research papers on verification can be found in the August 1985 Special Issue of *ACM SIGSOFT Software Engineering Notes* devoted to the proceedings of the Third Formal Verification Workshop. An important article regarding acceptance of correctness proofs by the software engineering community is [De Millo, Lipton, and Perlis, 1979].

With regard to reviews during the life cycle, useful checklists for inspections can be found in [Fagan, 1976], [Perry, 1983], and [Dunn, 1984]. The original paper on design inspections is [Fagan, 1976]; detailed information can be obtained from that paper. Later advances in review techniques are described in [Weinberg and Freedman, 1984] and [Fagan, 1986]. The IEEE Draft Standard for Software Reviews and Audits [IEEE 1028, 1986] is another excellent source of information on nonexecution-based testing. A review of techniques for validating specifications and designs can be found in [Boehm, 1984b]. A new approach to design reviews is presented in [Parnas and Weiss, 1987].

The classic work on execution-based testing is [Myers, 1979], a work that has had a significant impact on the field of testing. Issues in test case selection are discussed in [Petschenik, 1985]. Functional testing is described in [Howden, 1987], while structural techniques are compared in [Clarke, Podgurski, Richardson, and Zeil, 1985]. [Munoz, 1988] describes an approach to product testing. Statistical techniques are described in [Shooman, 1983] and also in [Goel, 1985]. The December 1985 and January 1986 Special Issues of *IEEE Transactions on Software Engineering* contain a wide variety of articles on software reliability.

Approaches to testing distributed software are presented in [Garcia-Molina, Germano, and Kohler, 1984] and [Wahl and Schach, 1988]. A wide-ranging source of material on testing real-time systems is [Quirk, 1985]. Glass has written a number of works on real-time systems, including [Glass, 1982] and [Glass, 1983]. Checking that real-time constraints have been satisfied is described in [Dasarathy, 1985].

Problems

6.1 You have been testing a module for 4 days and have found one fault. What does this tell you about the existence of other faults?

6.2 You are the SQA manager for Hardy Hardware, a chain of 650 hardware stores in all 50 states. Your organization is considering buying a word-processing package for use throughout the organization. Before authorizing the purchase of the package, you decide to test it thoroughly. What properties of this package do you investigate?

6.3 All 650 stores in the Hardy Hardware chain are now to be connected by a communications network. A salesman is offering you a 4-week free trial to experiment with the package he is trying to sell you. What sort of software tests would you perform, and why?

6.4 You are the general in charge of developing the software for controlling a new handheld antitank missile. The software has been delivered to you for acceptance testing. What properties of the software do you test?

6.5 How are the terms *testing*, *correctness proving*, *validation*, and *verification* used in this book?

6.6 What happens to the correctness proof of Section 6.3.1 if the loop invariant

$$D\text{: S} = \text{Y}(1) + \text{Y}(2) + \ldots + \text{Y}(\text{I} - 1)$$

is used instead of (6.4)?

6.7 Assume that you have some experience with loop invariants and that you know that invariant (6.4) is the correct invariant for the loop of Figure 6.6. Show that output specification (6.3) is a natural consequence of the loop invariant.

6.8 Consider the following code fragment:

```
I := 0;
F := 1;
while I < N loop
   I := I + 1;
   F := F * I;
end loop;
```

Prove that this code fragment correctly computes N! if $\text{N} \in \{1, 2, 3, \ldots\}$.

6.9 Can correctness proving solve the problem that the product as delivered to the client may not be what the client really needs? Give reasons for your answer.

6.10 What are the similarities between a design walkthrough and a design inspection? What are the differences?

6.11 Design and implement a solution to the Naur text-processing problem (Section 6.3.2) using a language specified by your instructor. Run it against test data and write down the number of faults that you find and the cause of the fault (e.g., logic fault, loop counter fault). Do not correct any of the faults you detected. Now exchange products with a fellow student and see how many faults each of you finds in the other's product and whether or not they are new faults. Again record the cause of each fault and compare the fault types found by each of you. Tabulate the results for the class as a whole.

6.12 As SQA manager for a software development organization you are responsible for determining the maximum number of faults that may be found in a given module. If this maximum is exceeded, then the module must be redesigned and recoded. What criteria would you use to determine the maximum for a given module?

6.13 Set up black-box test cases for Naur's text-processing problem (Section 6.3.2).

6.14 Using your product for Problem 6.11 (or code distributed by your instructor), set up statement coverage test cases.

6.15 Repeat Problem 6.14 for branch coverage.

6.16 Repeat Problem 6.14 for all-definition-use-path coverage.

6.17 Repeat Problem 6.14 for path coverage.

6.18 Repeat Problem 6.14 for linear code sequences.

6.19 Draw a flowchart of your product of Problem 6.11 (or code distributed by your instructor). Determine its cyclomatic complexity. If you are unable to determine the number of branches, consider the flowchart as a directed graph. Determine the number of edges e, nodes n, and connected components c (the main program and any procedure or function each constitute a connected component). The cyclomatic complexity V is then given by [McCabe, 1976]

$$V = e - n + c$$

6.20 Determine Halstead's four basic metrics for the code fragment of Figure 6.13(b).

6.21 (Term Project) Draw up a set of black-box test cases for the Plain Vanilla Ice Cream Corporation product described in the Appendix.

6.22 (Readings in Software Engineering) Your instructor will distribute copies of [Gelperin and Hetzel, 1988]. The authors view the period since 1988 as the "prevention-oriented period." To what extent do you agree with this assertion?

References

[ANSI/IEEE 729, 1983] "A Standard Glossary of Software Engineering Terminology," ANSI/IEEE 729-1983, American National Standards Institute, Inc., Institute of Electrical and Electronic Engineers, Inc., 1983.

[Basili and Hutchens, 1983] V. R. BASILI AND D. H. HUTCHENS, "An Empirical Study of a Syntactic Complexity Family," *IEEE Transactions on Software Engineering* **SE-9** (November 1983), pp. 664–672.

[Basili and Selby, 1987] V. R. BASILI AND R. W. SELBY, "Comparing the Effectiveness of Software Testing Strategies," *IEEE Transactions on Software Engineering* **SE-13** (December 1987), pp. 1278–1296.

[Basili and Weiss, 1984] V. R. BASILI AND D. M. WEISS, "A Methodology for Collecting Valid Software Engineering Data," *IEEE Transactions on Software Engineering* **SE-10** (November 1984), pp. 728–738.

[Beizer, 1984] B. BEIZER, *Software System Testing and Quality Assurance*, Van Nostrand Reinhold, New York, NY, 1984.

[Berry and Wing, 1985] D. M. BERRY AND J. M. WING, "Specifying and Prototyping: Some Thoughts on Why They Are Successful," in: *Formal Methods and Software Development*, *Proceedings of the International Joint Conference on Theory and Practice of Software Development*, *Volume 2*, Springer-Verlag, Berlin, West Germany, 1985, pp. 117–128.

[Boehm, 1984b] B. W. BOEHM, "Verifying and Validating Software Requirements and Design Specifications," *IEEE Software* **1** (January 1984), pp. 75–88.

[Bologna, Quirk, and Taylor, 1985] S. BOLOGNA, W. J. QUIRK, AND J. R. TAYLOR, "Simulation and System Validation," in: *Verification and Validation of Real-Time Software*, W. J. Quirk (Editor), Springer-Verlag, Berlin, West Germany, 1985, pp. 179–201.

[Clarke, Podgurski, Richardson, and Zeil, 1985] L. A. CLARKE, A. PODGURSKI, D. J. RICHARDSON, AND S. J. ZEIL, "A Comparison of Data Flow Path Selection Criteria," *Proceedings of the Eighth International Conference on Software Engineering*, London, UK, August 1985, pp. 244–251.

[Crossman, 1982] T. D. CROSSMAN, "Inspection Teams, Are They Worth It?" *Proceedings of the Second National Symposium on EDP Quality Assurance*, Chicago, IL, November 1982.

[Currit, Dyer, and Mills, 1986] P. A. CURRIT, M. DYER, AND H. D. MILLS, "Certifying the Reliability of Software," *IEEE Transactions on Software Engineering* **SE-12** (January 1986), pp. 3–11.

[Dahl, Dijkstra, and Hoare, 1972] O.-J. DAHL, E. W. DIJKSTRA, AND C. A. R. HOARE, *Structured Programming*, Academic Press, New York, NY, 1972.

[Dasarathy, 1985] B. DASARATHY, "Timing Constraints of Real-Time Systems: Constructs for Expressing Them, Methods of Validating Them," *IEEE Transactions on Software Engineering* **SE-11** (January 1985), pp. 80–86.

[De Millo, Lipton, and Perlis, 1979] R. A. DE MILLO, R. J. LIPTON, AND A. J. PERLIS, "Social Processes and Proofs of Theorems and Programs," *Communications of the ACM* **22** (May 1979), pp. 271–280.

[Dijkstra, 1968c] E. W. DIJKSTRA, "A Constructive Approach to the Problem of Program Correctness," *BIT* **8** (No. 3, 1968), pp. 174–186.

[Dijkstra, 1972] E. W. DIJKSTRA, "The Humble Programmer," *Communications of the ACM* **15** (October 1972), pp. 859–866.

[Dijkstra, 1976] E. W. DIJKSTRA, *A Discipline of Programming*, Prentice-Hall, Englewood Cliffs, NJ, 1976.

[Dunn, 1984] R. H. DUNN, *Software Defect Removal*, McGraw-Hill, New York, NY, 1984.

[Endres, 1975] A. ENDRES, "An Analysis of Errors and their Causes in System Programs," *IEEE Transactions on Software Engineering* **SE-1** (June 1975), pp. 140–149.

[Fagan, 1976] M. E. FAGAN, "Design and Code Inspections to Reduce Errors in Program Development," *IBM Systems Journal* **15** (No. 3, 1976), pp. 182–211.

[Fagan, 1986] M. E. FAGAN, "Advances in Software Inspections," *IEEE Transactions on Software Engineering* **SE-12** (July 1986), pp. 744–751.

[Garcia-Molina, Germano, and Kohler, 1984] H. GARCIA-MOLINA, F. GERMANO, JR., AND W. H. KOHLER, "Debugging a Distributed Computer System," *IEEE Transactions on Software Engineering* **SE-10** (March 1984), pp. 210–219.

[Garman, 1981] J. R. GARMAN, "The 'Bug' Heard 'Round the World," *ACM SIGSOFT Software Engineering Notes* **6** (October 1981), pp. 3–10.

[Gelperin and Hetzel, 1988] D. GELPERIN AND B. HETZEL, "The Growth of Software Testing," *Communications of the ACM* **31** (June 1988), pp. 687–695.

[Glass, 1982] R. L. GLASS, "Real-Time Checkout: The 'Source Error First' Approach," *Software—Practice and Experience* **12** (January 1982), pp. 77–83.

[Glass, 1983] R. L. GLASS (Editor), *Real-Time Software*, Prentice-Hall, Englewood Cliffs, NJ, 1983.

[Goel, 1985] A. L. GOEL, "Software Reliability Models: Assumptions, Limitations, and Applicability," *IEEE Transactions on Software Engineering* **SE-11** (December 1985), pp. 1411–1423.

[Good, 1983] D. I. GOOD, "The Proof of a Distributed System in GYPSY," Technical Report ICSCA-CMP-30, University of Texas at Austin, 1983.

[Goodenough, 1979] J. B. GOODENOUGH, "A Survey of Program Testing Issues," in: *Research Directions in Software Technology*, P. Wegner (Editor), The MIT Press, Cambridge, MA, 1979, pp. 316–340.

[Goodenough and Gerhart, 1975] J. B. GOODENOUGH AND S. L. GERHART, "Toward a Theory of Test Data Selection," *IEEE Transactions on Software Engineering* **SE-1** (June 1975), pp. 156–173.

[Gremillion, 1984] L. L. GREMILLION, "Determinants of Program Repair Maintenance Requirements," *Communications of the ACM* **27** (August 1984), pp. 826–832.

[Halstead, 1977] M. H. HALSTEAD, *Elements of Software Science*, Elsevier North-Holland, New York, NY, 1977.

[Hoare, 1969] C. A. R. HOARE, "An Axiomatic Basis for Computer Programming," *Communications of the ACM* **12** (October 1969), pp. 576–583.

[Howden, 1987] W. E. HOWDEN, *Functional Program Testing and Analysis*, McGraw-Hill, New York, NY, 1987.

[Hwang, 1981] S.-S. V. HWANG, "An Empirical Study in Functional Testing, Structural Testing, and Code Reading Inspection," Scholarly Paper 362, Department of Computer Science, University of Maryland, College Park, MD, 1981.

[IEEE 1028, 1986] "Draft Standard for Software Reviews and Audits," IEEE P1028, Institute of Electrical and Electronic Engineers, Inc., 1986.

[Jones, 1978] T. C. JONES, "Measuring Programming Quality and Productivity," *IBM Systems Journal* **17** (No. 1, 1978), pp. 39–63.

[Jones, 1986] C. JONES, *Programming Productivity*, McGraw-Hill, New York, NY, 1986.

[Lamport, 1980] L. LAMPORT, "'Sometime' is Sometimes 'Not Never': On the Temporal Logic of Programs," *Proceedings of the Seventh Annual ACM Symposium on Principles of Programming Languages*, Las Vegas, NV, 1980, pp. 174–185.

[Landwehr, 1983] C. E. LANDWEHR, "The Best Available Technologies for Computer Security," *IEEE Computer* **16** (July 1983), pp. 86–100.

[Leavenworth, 1970] B. LEAVENWORTH, Review #19420, *Computing Reviews* **11** (July 1970), pp. 396–397.

[Linger, Mills, and Witt, 1979] R. C. LINGER, H. D. MILLS, AND B. I. WITT, *Structured Programming: Theory and Practice*, Addison-Wesley, Reading, MA, 1979.

[London, 1971] R. L. LONDON, "Software Reliability through Proving Programs Correct," *Proceedings of the IEEE International Symposium on Fault-Tolerant Computing*, March, 1971.

[Manna, 1974] Z. MANNA, *Mathematical Theory of Computation*, McGraw-Hill, New York, NY, 1974.

[Manna and Pnueli, 1981] Z. MANNA AND A. PNUELI, "Verification of Concurrent Programs, Part I: The Temporal Framework," Report No. STAN-CS-81-836, Department of Computer Science, Stanford University, Stanford, CA, June 1981.

[Manna and Waldinger, 1978] Z. MANNA AND R. WALDINGER, "The Logic of Computer Programming," *IEEE Transactions on Software Engineering* **SE-4** (1978), pp. 199–229.

[McCabe, 1976] T. J. McCabe, "A Complexity Measure," *IEEE Transactions on Software Engineering* **SE-2** (December 1976), pp. 308–320.

[Metcalfe and Boggs, 1976] R. M. Metcalfe and D. R. Boggs, "Ethernet: Distributed Packet Switching for Local Computer Networks," *Communications of the ACM* **19** (July 1976), pp. 395–404.

[Mills, Basili, Gannon, and Hamlet, 1987] H. D. Mills, V. R. Basili, J. D. Gannon, and R. G. Hamlet, *Principles of Computer Programming: A Mathematical Approach*, Allyn and Bacon, Newton, MA, 1987.

[Mills, Dyer, and Linger, 1987] H. D. Mills, M. Dyer, and R. C. Linger, "Cleanroom Software Engineering," *IEEE Software* **4** (September 1987), pp. 19–25.

[Munoz, 1988] C. U. Munoz, "An Approach to Software Product Testing," *IEEE Transactions on Software Engineering* **14** (November 1988), pp. 1589–1596.

[Musa, Iannino, and Okumoto, 1987] J. D. Musa, A. Iannino, and K. Okumoto, *Software Reliability: Measurement, Prediction, Application*, McGraw-Hill, New York, NY, 1987.

[Musser, 1980] D. R. Musser, "Abstract Data Type Specification in the AFFIRM System," *IEEE Transactions on Software Engineering* **SE-6** (January 1980), pp. 24–32.

[Myers, 1976] G. J. Myers, *Software Reliability: Principles and Practices*, Wiley-Interscience, New York, NY, 1976.

[Myers, 1978a] G. J. Myers, "A Controlled Experiment in Program Testing and Code Walkthroughs/Inspections," *Communications of the ACM* **21** (September 1978), pp. 760–768.

[Myers, 1979] G. J. Myers, *The Art of Software Testing*, John Wiley and Sons, New York, NY, 1979.

[Naur, 1969] P. Naur, "Programming by Action Clusters," *BIT* **9** (No. 3, 1969), pp. 250–258.

[Ottenstein, 1979] L. M. Ottenstein, "Quantitative Estimates of Debugging Requirements," *IEEE Transactions on Software Engineering* **SE-5** (September 1979), pp. 504–514.

[Parnas and Weiss, 1987] D. L. Parnas and D. M. Weiss, "Active Design Reviews: Principles and Practices," *Journal of Systems and Software* **7** (December 1987), pp. 259–265.

[Perry, 1983] W. E. Perry, *A Structured Approach to Systems Testing*, Prentice-Hall, Englewood Cliffs, NJ, 1983.

[Petschenik, 1985] N. H. Petschenik, "Practical Priorities in System Testing," *IEEE Software* **2** (September 1985), pp. 18–23.

[Quirk, 1983] W. J. Quirk, "Recent Developments in the SPECK Specification System," Report CSS.146, Harwell, UK, 1983.

[Quirk, 1985] W. J. Quirk (Editor), *Verification and Validation of Real-Time Software*, Springer-Verlag, Berlin, West Germany, 1985.

[Rapps and Weyuker, 1985] S. RAPPS AND E. J. WEYUKER, "Selecting Software Test Data Using Data Flow Information," *IEEE Transactions on Software Engineering* **SE-11** (April 1985), pp. 367–375.

[Seitz, 1985] C. L. SEITZ, "The Cosmic Cube," *Communications of the ACM* **28** (January 1985), pp. 22–33.

[Selby, Basili, and Baker, 1987] R. W. SELBY, V. R. BASILI, AND F. T. BAKER, "Cleanroom Software Development: An Empirical Evaluation," *IEEE Transactions on Software Engineering* **SE-13** (September 1987), pp. 1027–1037.

[Shooman, 1983] M. L. SHOOMAN, *Software Engineering: Design, Reliability, and Management*, McGraw-Hill, New York, NY, 1983.

[Takahashi and Kamayachi, 1985] M. TAKAHASHI AND Y. KAMAYACHI, "An Empirical Study of a Model for Program Error Prediction," *Proceedings of the Eighth International Conference on Software Engineering*, London, UK, 1985, pp. 330–336.

[Tamir, 1980] M. TAMIR, "ADI: Automatic Derivation of Invariants," *IEEE Transactions on Software Engineering* **SE-6** (January 1980), pp. 40–48.

[van der Poel and Schach, 1983] K. G. VAN DER POEL AND S. R. SCHACH, "A Software Metric for Cost Estimation and Efficiency Measurement in Data Processing System Development," *Journal of Systems and Software* **3** (September 1983), pp. 187–191.

[Wahl and Schach, 1988] N. J. WAHL AND S. R. SCHACH, "A Methodology and Distributed Tool for Debugging Dataflow Programs," *Proceedings of the Second Workshop on Software Testing, Verification, and Analysis*, Banff, Canada, July 1988, pp. 98–105.

[Walsh, 1979] T. J. WALSH, "A Software Reliability Study Using a Complexity Measure," *Proceedings of the AFIPS National Computer Conference*, New York, NY, 1979, pp. 761–768.

[Weinberg and Freedman, 1984] G. M. WEINBERG AND D. P. FREEDMAN, "Reviews, Walkthroughs, and Inspections," *IEEE Transactions on Software Engineering* **SE-10** (January 1984), pp. 68–72.

[Weyuker, 1988] E. J. WEYUKER, "An Empirical Study of the Complexity of Data Flow Testing," *Proceedings of the Second Workshop on Software Testing, Verification, and Analysis*, Banff, Canada, July 1988, pp. 188–195.

[Wirth, 1971] N. WIRTH, "Program Development by Stepwise Refinement," *Communications of the ACM* **14** (April 1971), pp. 221–227.

[Woodward, Hedley, and Hennell, 1980] M. R. WOODWARD, D. HEDLEY, AND M. A. HENNELL, "Experience with Path Analysis and Testing of Programs," *IEEE Transactions on Software Engineering* **SE-6** (May 1980), pp. 278–286.

Part Three Phases of the Life Cycle

In Part Three the various life-cycle phases are described in depth. In Chapter 7, Specification Methods, three approaches to specification, namely informal, semiformal, and formal, are presented. Instances of each approach are described. Methods that are described in depth and illustrated by case studies include structured systems analysis, finite state machines, Petri nets, and PAISLey. The chapter concludes with a comparison of the various approaches, and with suggestions regarding which would be appropriate for a given project.

The next two chapters are devoted to software design. Chapter 8 is entitled Modularity: and Beyond. Here the theoretical ideas underlying a good design are described. Modularity, data encapsulation, abstract data types, information hiding, and objects are treated as a natural progression.

These ideas are then utilized in Chapter 9, Design Methods. Three classes of design method are presented, namely data-oriented, procedure-oriented, and hybrid. Instances of each are described in detail, including data flow analysis, transaction analysis, Jackson system development, and object-oriented design, and case studies are presented. Again, the emphasis is on comparison and contrast.

Implementation is the subject of Chapter 10. Areas covered include the choice of programming language, fourth-generation languages (4GLs), good programming practice, and programming standards. The issue of programming team organization is handled in depth. An analysis is made of the advantages and disadvantages of democratic teams, chief programmer teams, and hybrid teams. The chapter concludes with a description of essential coding tools.

Part Three concludes with Chapter 11, Maintenance. Topics covered in this chapter include the importance of maintenance and the difficulties of maintenance. The problem of multiple versions is addressed, and configuration control suggested as a solution. The management of maintenance is considered in some detail.

By the end of Part Three, the reader should have a clear understanding of every phase of the life cycle, the difficulties associated with each phase, and how to go about tackling those difficulties.

Chapter 7 Specification Methods

Before a product can be built it is necessary to draw up a set of specifications, that is, a document which states what the product has to do. In theory, this is a straightforward, almost mechanical task. All that is needed is for the requirements team to talk to the client, find out what the client wants, and then for the specifications team to write it all down in the form of the specifications.

In practice, of course, this is nonsense for two reasons. First, what the client thinks he or she wants may not be what really is required, or as a prominent politician once put it, "I know you believe you understood what you think I said, but I am not sure you realize that what you heard is not what I meant!" The problems of the requirements phase were discussed in Chapter 3. Second, after the specifications document has been signed off by the client, it is virtually the sole source of information available to the design team for drawing up the design. That is to say, even if the client's needs have been accurately determined during the requirements phase, if the specifications document contains faults such as omissions, contradictions, and ambiguities, the inevitable result will be faults in the design that will be carried over into the implementation. What is needed, therefore, are techniques for representing the client's needs in a format that will result in a fault-free product being delivered to the client at the end of the development cycle. Such specification methods form the subject of this chapter.

7.1 Informal Specifications

In many development projects, the specifications consist of page after page of English, or some other natural language such as French or Japanese. A typical paragraph of such a specifications document reads:

> BV.4.2.5. If the sales for the current month are below the target sales, then a report is to be printed, unless the difference between target sales and actual sales is less than half of the difference between target sales and actual sales in the previous month, or if the difference between target sales and actual sales for the current month is under 5%.

The background leading up to that paragraph is as follows. The management of a retail chain sets a target sales figure for each store for each month, and if a store does not meet this target, then a report is to be printed. Suppose that for one particular store the sales target for January was $100,000, but actual sales were only $64,000, that is, 36% below target. In this case, a report must be printed. Now suppose further that for February the target figure was $120,000, and that actual sales were only $100,000, 16.7% below target. Although the store has not made the target figure, the percentage difference for February, namely 16.7%, is less than half of the previous month's percentage difference, namely 36%, so management feels that an improvement has been made, and no report is to be printed. Then in March the target was again $100,000 but the store made $98,000, only 2% below target. Since the percentage difference is small, namely less than 5%, no report should be printed.

Careful rereading of the preceding specifications paragraph shows some divergence from what the retail chain's management actually asked for. Paragraph BV.4.2.5 speaks of the "difference between target sales and actual sales"; percentage difference is not mentioned. The difference in January was $36,000 and in February it was $20,000. The *percentage* difference, which is what management wanted, dropped from 36% in January to 16.7% in February, less than half of the January percentage difference. However, the *actual* difference dropped from $36,000 to $20,000, an improvement of $16,000, which is less than half of $36,000. So if the development team had faithfully implemented the specifications, the report would have been printed, which is not what management wanted. Then the last clause speaks of a "difference ... [of] 5%." What is meant, of course, is a *percentage* difference of 5%, only the word "percentage" does not occur anywhere in the paragraph.

There are thus a number of faults in the specifications. First, the wishes of the client have been ignored. Second, there is ambiguity—should the last clause read "percentage difference ... [of] 5%," or "difference ... [of] $5000," or something else entirely? In addition, the style is poor. What the paragraph says is, "If something happens, print a report. However, if something else happens, don't print it after all. And if a third thing happens, don't print it either." It would have been much clearer if the specifications had simply stated when the report is to be printed. All in all, paragraph BV.4.2.5 is not a very good example of how to write a specifications document.

In fact, paragraph BV.4.2.5 is fictitious, but it is, unfortunately, typical of all too many specifications documents. The reader may feel that the example is unfair, and that this sort of problem cannot arise if specifications are written with care by professional specification writers. To refute this charge, the case study of Chapter 6 is now resumed.

7.1.1 Informal Specifications Case Study

Recall from Section 6.3.2 that in 1969 Naur published a paper on correctness proving [Naur, 1969]. He illustrated his technique by means of a text-processing problem. Using his technique, Naur constructed a procedure to solve the problem and informally proved the correctness of his procedure. A reviewer of Naur's paper pointed out one fault in the procedure [Leavenworth, 1970]. London then detected three additional faults in Naur's procedure [London, 1971]. He presented a corrected version of the procedure and proved its correctness formally. Goodenough and Gerhart then found three further faults that London had not detected [Goodenough and Gerhart, 1975]. Of the total of seven faults collectively detected by the reviewer, London, and Goodenough and Gerhart, two can be considered as specification faults. For example, Naur's specifications do not state what happens if the input includes two successive adjacent breaks (BLANK or NL characters). For this reason, Goodenough and Gerhart then produced a new set of specifications. These specifications were about four times longer than Naur's, which are given in Section 6.3.2.

In 1985, Meyer wrote an article on formal specification techniques [Meyer, 1985]. The main thrust of his article is that when specifications are written in a natural language such as English they tend to have contradictions, ambiguities, and omissions, and that the solution to this is to express specifications formally using mathematical terminology. Meyer detected some 12 faults in Goodenough and Gerhart's specifications and then developed a mathematical set of specifications to correct all the problems. Meyer then paraphrased these mathematical specifications and constructed English specifications. In the author's opinion, Meyer's English specifications contain a fault. Meyer points out in his paper that if the maximum number of characters per line is, say, 10, and the input is, say, WHO WHAT WHEN, then, in terms of both Naur's and Goodenough and Gerhart's specifications, there are two equally valid outputs, namely WHO WHAT on the first line and WHEN on the second, or WHO on the first line and WHAT WHEN on the second. In fact, Meyer's paraphrased English specifications also contain this ambiguity.

The key point is that Goodenough and Gerhart's specifications were constructed with the greatest of care. After all, the reason they were constructed was to correct Naur's specifications. Furthermore, Goodenough and

Gerhart's paper went through two versions, the first of which was published in the proceedings of a refereed conference and the second in a refereed journal [Goodenough and Gerhart, 1975]. Finally, both Goodenough and Gerhart are experts in software engineering in general, and specifications in particular. Thus if two experts with as much time as they needed carefully produced specifications in which Meyer detected 12 faults, what chance does an ordinary computer professional working under time pressure have of producing fault-free specifications? Worse still, the text-processing problem can be coded in 25 or 30 lines, while real-world products can consist of hundreds of thousands of lines of source code.

It is clear that natural language is not a good way of specifying a product. In this chapter better alternatives will be described. The order in which the specification methods will be presented will be from the informal to the more formal.

7.2 Structured Systems Analysis

The use of graphics to specify software was an important technique of the 1970s. Two methods became particularly popular, namely those of DeMarco, and Gane and Sarsen [DeMarco, 1978; Gane and Sarsen, 1979]. The two methods are both equally good and are similar in many ways. Gane and Sarsen's approach is presented here because it is currently used in object-oriented design, described in Section 9.8.

As an aid to understanding the method, consider the following case study.

7.2.1 Structured Systems Analysis Case Study

> Sally's Software Store buys software from various suppliers and sells it to the public. Popular software packages are kept in stock, but the rest must be ordered as required. Institutions and corporations are given credit facilities, as are some members of the public. Sally's Software is doing well, with a monthly turnover of 300 packages at an average retail cost of $250 each. Despite her business success, Sally has been advised to computerize. Should she?

The question is wrongly put. It should read: What sections of her business, namely accounts payable, accounts receivable, and inventory, should be computerized, *if any*? Even this is inadequate—is the system to be batch or online? Is there to be an in-house computer, or is a service bureau to be used? But even if the question is refined further, it still misses the fundamental issue: What is Sally's objective in computerizing her business?

Only when this has been answered is there any point in looking further. For example, if she wishes to computerize simply because she sells software, then she needs an in-house system with a variety of sound and light effects that ostentatiously shows off the possibilities of a computer. But if she uses her business to launder "hot" money, then she needs a product that keeps four or five different sets of books and does not have an audit trail.

For the purposes of this case study, the assumption will be made that Sally wishes to computerize "in order to make more money." This does not help very much, but it is clear that cost–benefit analysis can be used to determine whether or not to computerize each of the three sections of her business.

The main danger of many standard approaches is that one is tempted to come up with the solution first, for example, a Lime III microcomputer with two disk drives and a printer, and then to find out what the problem is! Instead, with Gane and Sarsen's method the client's needs are analyzed using a nine-step method. An important point is that stepwise refinement is used in many of those steps; this will be indicated as the method is demonstrated.

The first step is to determine the *logical data flow*, as opposed to the physical data flow. This is done by drawing a data flow diagram (DFD). The DFD shows what happens, not how it happens. It uses the four basic symbols shown in Figure 7.1.

Step 1. Draw the DFD: The DFD for any nontrivial product is likely to be large. The DFD is a pictorial representation of all aspects of the logical

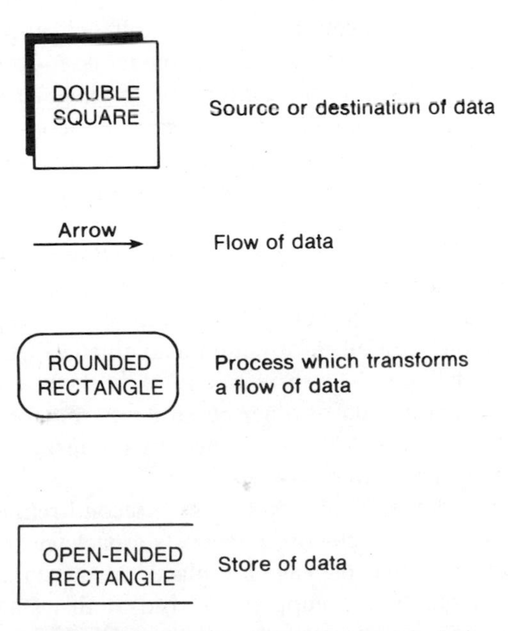

Figure 7.1 Symbols of structured systems analysis.

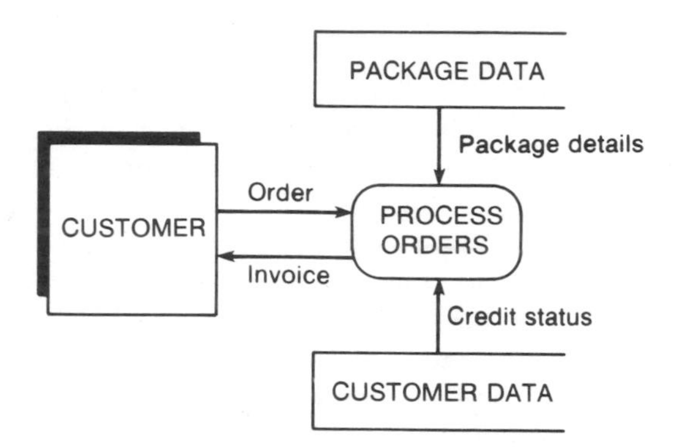

Figure 7.2 Data flow diagram: first refinement.

data flow and as such is guaranteed to contain considerably more than 7 ± 2 elements. For this reason, the DFD must be developed by stepwise refinement (Chapter 5). Returning to the case study, the first refinement is shown in Figure 7.2.

This diagram of *logical data flow* can have many interpretations. Two possible scenarios are as follows:

Scenario 1: PACKAGE DATA consists of some 900 floppy disks and diskettes on shelves, as well as a number of catalogs in a desk drawer. CUSTOMER DATA is a collection of 5 × 7 cards held together by a rubber band, plus a list of customers whose payments are overdue. PROCESS ORDERS is Sally looking for the appropriate package from the shelves, if necessary looking it up in a catalog, then finding the correct 5 × 7 card and checking that the customer's name is not on the list of defaulters. This scenario is totally manual and corresponds to the way Sally is currently conducting her business.

Scenario 2: PACKAGE DATA and CUSTOMER DATA are computer files and PROCESS ORDERS is Sally entering the customer's name and the name of the package at a terminal. This scenario corresponds to a fully computerized solution, with all information available online.

The DFD of Figure 7.2 represents not only the preceding two scenarios, but also literally an infinity of other possibilities. The key point is that the DFD represents a flow of information—the actual package that the customer wants does not enter into it in any way.

The DFD is now refined stepwise. The second refinement is depicted in Figure 7.3. Now when the customer requests a package that Sally does not have on hand, details of that package are placed in the data store PENDING ORDERS, which might be a computer file, but at this stage of the process could equally well be a manila folder. PENDING ORDERS is scanned daily,

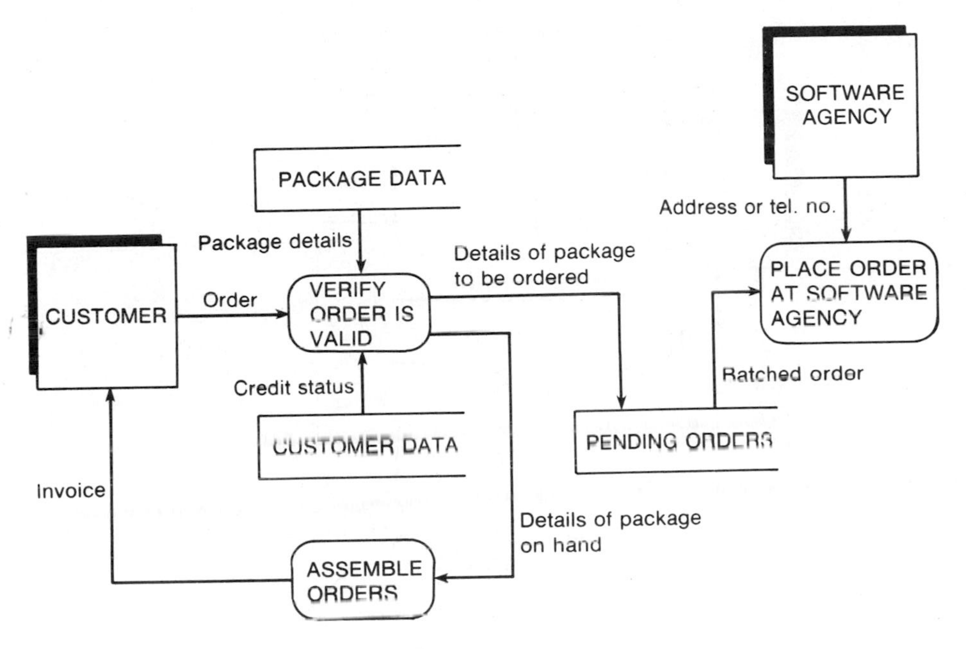

Figure 7.3 Data flow diagram: second refinement.

by the computer or by Sally, and if there are sufficient orders for one agency, then a batched order is placed. Also, if an order has been waiting for, say, 5 working days, it is ordered, irrespective of how many packages are waiting to be ordered from the relevant agency. This DFD does not show the logical flow of data when the software package arrives from the agency, nor does it show financial functions such as accounts payable and accounts receivable. These will be added in the third refinement.

Only a portion of the third refinement is shown in Figure 7.4 because the DFD is starting to get large. The rest of the DFD relates to accounts payable and to the software agencies. The final DFD will be larger still, stretching over perhaps six legal-size sheets. But it will be easily understood by the client who will sign it off, confirming that it is an accurate representation of the logical flow of data in her business.

Of course, for a larger product the DFD will be larger. After a certain point, it becomes impractical to have just one DFD, and a hierarchy of DFDs is needed. A single box at one level is expanded into a complete DFD at a lower level. A difficulty with this approach frequently arises as a consequence of the positioning of sources and destinations of data. A particular process P may be reflected at level L, say, and expanded at level L + 1. The correct place for the sources and destinations of data for process P is level L + 1, because

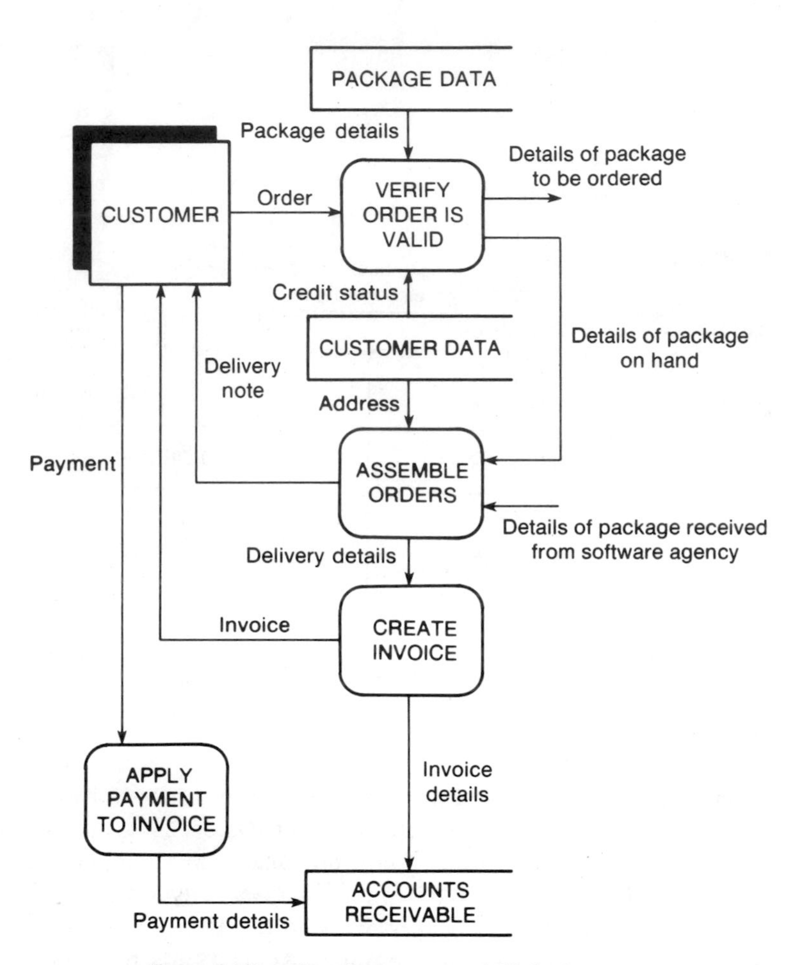

Figure 7.4 Data flow diagram: part of third refinement.

that is where they are used. But a client, even one who is experienced in reading a DFD, frequently cannot understand the DFD at level L because the sources and destinations of data relating to process P appear to be missing. In order to prevent this confusion, it is sometimes advisable to draw the correct DFD and then modify it by moving sources and destinations of data one or more levels up so that the client will not have this difficulty in comprehending it.

Step 2. Decide what sections to computerize, and how (batch or online): The choice of what to automate often depends on how much the client is prepared to spend. Obviously, it would be nice to automate the entire operation, but the cost of this may be prohibitive. In order to determine which sections to automate, cost–benefit analysis is applied to the various possible strategies for computerizing each section. For example, for each section of the DFD a decision has to be taken as to whether that group of operations should be performed in batch or online. With large volumes to process and tight controls required, batch is often the answer, but with small volumes and an in-house microcomputer, online would appear to be better. Returning to the case study, one alternative would be to automate accounts payable in batch and to perform validation of orders online. A second alternative would be to automate everything, with the editing of the software agency consignment notes against orders being done online or batch, the rest of the operations online. A key point is that the DFD corresponds to all the preceding possibilities.

The next three stages of Gane and Sarsen's method are, respectively, the stepwise refinement of the flows of data (arrows), processes (rounded rectangles), and data stores (open rectangles).

Step 3. Put in the details of the data flows: First, decide what data items must go into the various data flows. Then refine each flow stepwise.

In the case study, there is a data flow Order. This can be refined as follows:

```
Order:
    Order_identification
    Customer_details
    Package_details
```

Now, each of the preceding components of Order is refined further. In the case of a larger product, a data dictionary is needed to keep track of the various data elements involved.

Step 4. Define the logic of the processes: Now that the data elements within the product have been determined, what happens within each process can be investigated further. In the case study, suppose that there is a process GIVE EDUCATIONAL DISCOUNT. Sally must provide the software developers with details of what discount she gives to educational institutions, for example, 10% on up to four packages, 15% on five or more. In order to cope with the difficulties of natural language specifications, this should be translated from English into a decision tree. This is shown in Figure 7.5.

An advantage of a decision tree is that in more complex cases it is easy to check that all possibilities have been taken into account. An example of this is shown in Figure 7.6. From the figure it is immediately obvious that

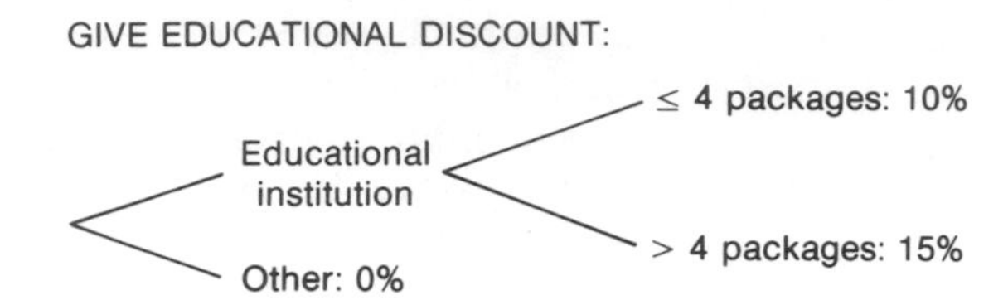

Figure 7.5 Decision tree depicting Sally's Software Store educational discount policy.

the cost of a seat behind the end zone for an alumnus has not been specified. Another good way of representing processes is using decision tables [Pollack, Hicks, and Harrison, 1971]. Decision tables have a further advantage in that tools exist which allow the contents of a decision table to be entered automatically into a computer, thereby obviating the need to code that part of the product.

Step 5. Define the data stores: At this stage it is necessary to define the exact contents of each store and its representation (format). Thus, if the product is to be implemented in COBOL, this information must be provided down to the **pic** level; if Ada is to be used, the **digits** or **delta** must be specified. In addition, it is necessary to specify where immediate access is required.

The issue of immediate access depends on what queries are going to be put to the product. For example, suppose that in the case study it is decided to validate orders online. A customer may order a package by name ("Do you have Lotus 1-2-3 in stock?"), by function ("What accounting packages do you have in stock?"), or by machine ("Do you have anything new for the Macintosh II?"), but rarely by price ("What do you have for $149.50?"). Thus immediate access to PACKAGE DATA is required by name, function, and machine. This is depicted in the data immediate-access diagram (DIAD) of Figure 7.7.

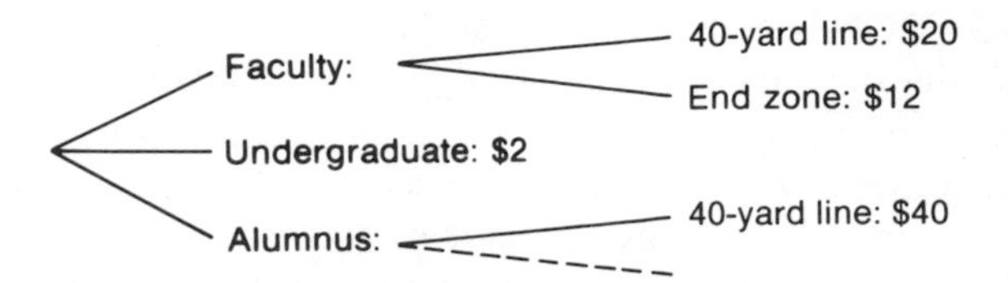

Figure 7.6 Decision tree describing seating prices for college football games.

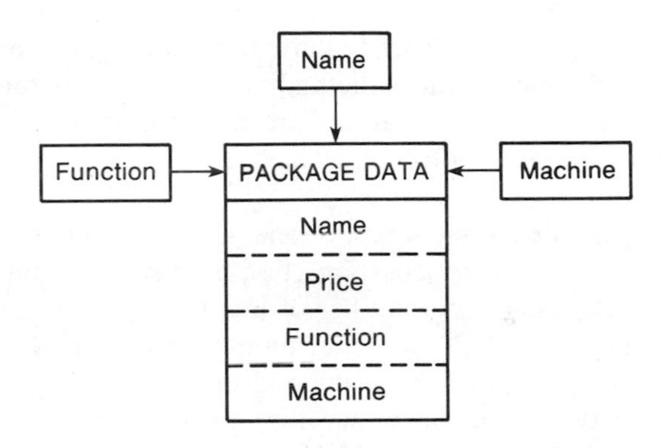

Figure 7.7 Data immediate-access diagram (DIAD) for PACKAGE DATA.

Step 6. Define the physical resources: Now that the developers know what is required online and the representation (format) of each element, a decision can be made regarding blocking factors. In addition, for each file the following can be specified: file name; organization (sequential, indexed, etc.); storage medium; and records, down to the field level.

Step 7. Determine the input/output specifications: The input forms must be specified, at least with respect to components, if not detailed layout. Input screens must similarly be decided on. The printed output must also be specified, where possible in detail, otherwise just estimated length.

Step 8. Perform sizing: At this step numerical data that will be used in step 9 to determine the hardware requirements must be computed. This includes the volume of input (daily or hourly), the frequency of each printed report and its deadline, the size and number of records of each type that are to pass between CPU and mass storage, and the size of each file.

Step 9. Determine the hardware requirements: From the information on disk files determined previously, mass storage requirements can be computed. In addition, mass storage requirements for back-up purposes can be determined. From knowledge of input volumes, the needs in this area can be determined; since the number of lines and frequency of printed reports is known, output devices can be specified. If the client already has hardware, it can be determined whether this hardware will be adequate or if additional hardware will have to be purchased. On the other hand, if the client does not have suitable hardware, a recommendation can be made as to what should be acquired and whether it should be purchased or leased.

Determining the hardware requirements is the ninth and final step of Gane and Sarsen's specification method, and the case study is therefore concluded. The resulting specifications are now handed to the design team, and the product life cycle continues.

Despite its many strengths, Gane and Sarsen's method does not provide the answer to every question. For example, it cannot be used to determine response times. The number of input/output channels can at best be roughly gauged. Also, CPU size and timing cannot be estimated with any degree of accuracy. Nonetheless, at the end of the specifications phase hardware decisions have to be made whether or not accurate information is available. These are distinct drawbacks of Gane and Sarsen's method and, to be fair, virtually every other method for either specifications or design. Nevertheless, the situation is considerably better than what used to be done before methodical approaches to specifying were put forward, which was to make decisions regarding hardware right at the beginning of the software development process. Gane and Sarsen's method has led to major improvements in the ways that products are specified, and the fact that Gane and Sarsen and the authors of most competing methods essentially ignore time as a variable should not detract from the advantages that these methods have brought to the software industry.

7.3 Other Semiformal Methods

Gane and Sarsen's method is clearly more formal than writing a specification in a natural language. At the same time, it is less formal than many of the methods presented in the following discussion such as Petri nets (Section 7.5) and executable specifications (Section 7.6). Dart and her coworkers classify specification and design methods as informal, semiformal, or formal [Dart, Ellison, Feiler, and Habermann, 1987]. In terms of this classification Gane and Sarsen's structured systems analysis is a semiformal method, while the other two methods mentioned in this paragraph are formal methods.

Gane and Sarsen's method has been included in this book for two reasons. First, it is widely used; there is a good chance that the reader may be employed at some future date by an organization that uses structured systems analysis or some variant of it. Second, data flow diagrams are used in object-oriented design (Section 9.8), one of the more promising of today's design methods. But there are many other good semiformal methods; see, for example, the proceedings of the various International Workshops on Software Specifications and Design. Because of space limitations, all that will be given here is a brief description of a few well-known methods.

PSL/PSA [Teichroew and Hershey, 1977] is a computer-aided method for specifications of information-processing products. The name comes from the two components of the method, namely the problem statement language

(PSL) that is used to describe the product and the problem statement analyzer (PSA) that enters the PSL description into a database and produces reports on request. PSL/PSA is widely used, particularly for documenting products.

SADT [Ross, 1985] consists of two interrelated components, a box-and-arrow diagramming language termed structural analysis (SA) and a design technique (DT)—hence the name SADT. Stepwise refinement underlies SADT to a greater extent than with Gane and Sarsen's method; a conscious effort has been made to adhere to Miller's law. As Ross puts it, "Everything worth saying, about anything worth saying something about, must be expressed in six or fewer pieces" [Ross, 1985]. SADT has had a great many successes in specifying a wide variety of products, especially complex, large-scale projects. Like many other similar semiformal methods, its applicability to real-time systems is less clear.

On the other hand, SREM (software requirements engineering method, pronounced "Shrem") was explicitly designed for specifying the conditions under which specific actions are to occur [Alford, 1985]. For this reason SREM has been particularly useful for specifying real-time systems and has now been extended to distributed systems. SREM consists of a number of components. RSL is a specifications language. REVS is a set of tools which perform a variety of specifications-related tasks such as translating the RSL specifications into an automated database, automatically checking for data flow consistency (ensuring that no data item is used before it has been assigned a value), and generating simulators from the specifications that can be used to ensure that the specifications are correct. In addition, SREM now has a design method named DCDS, distributed computing design system.

The power of SREM comes from the fact that the model underlying the whole method is a finite state machine (FSM), described in Section 7.4. A textual representation has been developed for specifying products, resulting in a method that is semiformal and hence easier to use than the more formal FSM method described in the following discussion. As a result of this formal model underlying SREM, it is possible to perform the consistency checking mentioned previously and also to verify that performance constraints on the product as a whole can be met, given the performance of individual components. SREM has been used by the U.S. Air Force to specify two C^3I systems (command, control, communications, and intelligence) [Scheffer, Stone, and Rzepka, 1985]. While SREM proved to be of great use in the specifications phase, it appears that the REVS tools employed later in the development cycle were considered to be less useful.

Formal methods are described in the next four sections. The underlying theme is that formal methods lead to more precise specifications than semiformal or informal methods can. However, the use of formal methods in general requires lengthy training, and software engineers using formal methods need to have been exposed to the relevant mathematics. The following sections have been written with the mathematical content kept to a minimum. Furthermore, wherever possible, mathematical formulations are preceded by informal presentations of the same material. Nevertheless, the level of Sections 7.4 through 7.7 is higher than that of the rest of the book and may be omitted at first reading.

7.4 Finite State Machines

Consider the following example, originally due to the M202 team at the Open University, U.K. [Brady, 1977]. A safe has a combination lock that can be in one of three positions, labeled 1, 2, and 3. The dial can be turned left or right (L or R). Thus at any time there are six possible dial movements, namely 1L, 1R, 2L, 2R, 3L, and 3R. The combination to the safe is 1L, 3R, 2L; any other dial movement will cause the alarm to go off. The situation is depicted in Figure 7.8. There is one initial state, namely Safe Locked. If the input is 1L, then the next state is A, but if any other dial movement, 1R, say, or 3L, is made, then the next state is Sound Alarm, one of the two final states. If the correct combination is chosen, then the sequence of transitions is from Safe Locked to A to B to Safe Unlocked, the other final state. What is shown in Figure 7.8 is a finite state machine (FSM). An FSM need not necessarily be depicted graphically. The same information is shown in tabular form in Figure 7.9. For each state other than the two final states, the transition to the next state is indicated, depending on the way the dial is moved.

A finite state machine consists of five parts: a set of states J, a set of inputs K, the transition function T which specifies the next state given the current state and the current input, the initial state S, and the set of final states F. In the case of the combination lock on the safe:

The set of states J is {Safe Locked, A, B, Sound Alarm}.
The set of inputs K is {1L, 1R, 2L, 2R, 3L, 3R}.
The transition function T is depicted in Figure 7.9.
The initial state S is Safe Locked.
The set of final states F is {Safe Locked, Sound Alarm}.

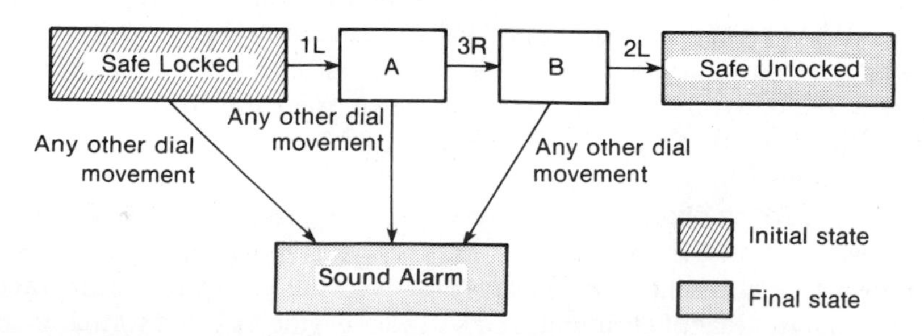

Figure 7.8 Finite state machine representation of combination safe.

Table of next states

Current state / Dial movement	Safe Locked	A	B
1L	A	Sound Alarm	Sound Alarm
1R	Sound Alarm	Sound Alarm	Sound Alarm
2L	Sound Alarm	Sound Alarm	Safe Unlocked
2R	Sound Alarm	Sound Alarm	Sound Alarm
3L	Sound Alarm	Sound Alarm	Sound Alarm
3R	Sound Alarm	B	Sound Alarm

Figure 7.9 Transition function of finite state machine.

In more formal terms, a finite state machine is a 5-tuple (J, K, T, S, F), where

J is a finite, nonempty set of states.
K is a finite, nonempty set of inputs.
T is a function from J ~ F × K into J called the transition function.
S ∈ J is the initial state.
F is the set of final states, J ⊇ F.

Use of the finite state machine approach is widespread in computing applications. For example, every menu-driven product is an implementation of a finite state machine. The display of a menu corresponds to a state, and an input entered at the keyboard or an icon selected with the mouse is an event that causes the product to go into some other state. For example, entering V when the main menu appears on the screen might cause a volumetric analysis to be performed on the current data set. A new menu then appears, and the user may enter G, P, or X. Selecting G causes the results of the calculation to be graphed, P causes them to be printed, while X causes a return to the main menu. Each transition has the form

current state [menu] and **event** [option selected] ⇒ **next state** (7.1)

For the purposes of specifying a product, a useful extension of FSMs is to add a sixth component to the preceding 5-tuple, namely a set of predicates P, where each predicate is a function of the global state X of the product [Kampen, 1987]. More formally, the transition function T is now a function from J ~ F × K × P into J. Transition rules now have the form

current state and **event** and **predicate** ⇒ **next state** (7.2)

7.4.1 Finite State Machine Case Study

To see how this formalism works in practice, the method will now be applied to a modified version of the so-called elevator problem, attributed to Davis [IWSSD, 1986].

> An n elevator product is to be installed in a building with m floors. The problem concerns the logic to move elevators between floors according to the following constraints:
>
> C_1: Each elevator has a set of buttons, one for each floor. These illuminate when pressed and cause the elevator to visit the corresponding floor. The illumination is canceled when the corresponding floor is visited by the elevator.
>
> C_2: Each floor, except the first floor and top floor, has two buttons, one to request an up-elevator and one to request a down-elevator. These buttons illuminate when pressed. The illumination is canceled when an elevator visits the floor and moves in the desired direction.
>
> C_3: When an elevator has no requests to service, it should remain at its current floor with its doors closed.

The product will now be specified using an extended finite state machine [Kampen, 1987]. There are two sets of buttons in the problem: In each of the n elevators there is a set of m buttons, one for each floor. Since these n × m buttons are inside the elevators, they will be referred to as elevator buttons. Then on each floor there are two buttons, one to request an up-elevator and one to request a down-elevator. These will be referred to as floor buttons.

The FSM for an elevator button is shown in Figure 7.10. Let EB(e, f) denote the button in elevator e that is pressed to request floor f. EB(e, f) can be in two states, with the button on (illuminated) or off. More precisely, the states are:

EBON(e, f):	<u>E</u>levator <u>B</u>utton (e, f) <u>ON</u>	(7.3)
EBOFF(e, f):	<u>E</u>levator <u>B</u>utton (e, f) <u>OFF</u>	

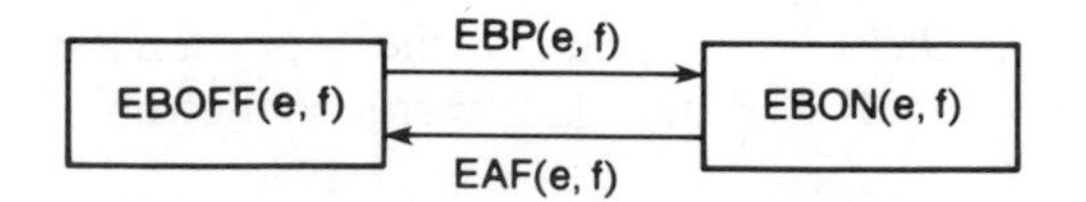

Figure 7.10 FSM for elevator button [Kampen, 1987]. (© 1987 IEEE.)

If the button is on and the elevator arrives at floor f, then the light is turned off. Conversely, if the light is off and a button is pressed, then the light comes on. There are thus two events involves, namely

EBP(e, f):	Elevator Button (e, f) Pressed	(7.4)
EAF(e, f):	Elevator e Arrrives at Floor f	

In order to define the state transition rules connecting these events and states, a predicate V(e, f) is needed. (A predicate is a condition that is either true or false.)

V(e, f): Elevator e is Visiting (stopped at) floor f (7.5)

Now the formal transition rules can be stated. If elevator button (e, f) is off [current state] and elevator button (e, f) is pressed [event] and elevator e is not visiting floor f [predicate], then the button is turned on. In the format of (7.2) this becomes

$$\text{EBOFF(e, f) and EBP(e, f) and not V(e, f)} \Rightarrow \text{EBON(e, f)} \quad (7.6)$$

If the elevator is currently visiting floor f, nothing happens. In Kampen's formalism, events that do not trigger a transition may indeed occur, but if they do they are ignored.

Conversely, if the elevator arrives at floor f and the light is on, then it is turned off. This is expressed as

$$\text{EBON(e, f) and EAF(e, f)} \Rightarrow \text{EBOFF(e, f)} \quad (7.7)$$

Now the floor buttons are considered. FB(d, f) denotes the button on floor f that requests an elevator traveling in direction d. The FSM for floor button FB(d, f) is shown in Figure 7.11. More precisely, the states are:

FBON(d, f):	Floor Button (d, f) ON	(7.8)
FBOFF(d, f):	Floor Button (d, f) OFF	

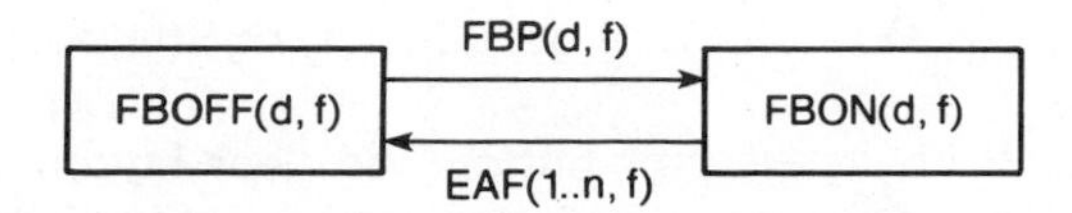

Figure 7.11 FSM for floor button [Kampen, 1987]. (© 1987 IEEE.)

If the button is on and an elevator arrives at floor f traveling in the correct direction d, then the light is turned off. Conversely, if the light is off and the button is pressed, then the light comes on. Again there are two events involved, namely

$$\begin{array}{ll} \text{FBP(d, f):} & \underline{\text{F}}\text{loor } \underline{\text{B}}\text{utton (d, f) } \underline{\text{P}}\text{ressed} \\ \text{EAF(1..n, f):} & \underline{\text{E}}\text{levator 1 or } \ldots \text{ or n } \underline{\text{A}}\text{rrives at } \underline{\text{F}}\text{loor f} \end{array} \tag{7.9}$$

Note the use of 1..n to denote disjunction. Throughout this section an expression such as P(a, 1..n, b) denotes

$$\text{P(a, 1, b) or P(a, 2, b) or } \ldots \text{ or P(a, n, b)} \tag{7.10}$$

In order to define the state transition rules connecting these events and states, a predicate is again needed. In this case it is S(d, e, f), which is defined as follows:

S(d, e, f): Elevator e is visiting floor f and its direction of motion is either up (d = U), down (d = D), or no requests are pending (d = N) (7.11)

This predicate is actually a state. In fact, the formalism allows both events and states to be treated as predicates.

Using S(d, e, f), the formal transition rules are then:

$$\begin{array}{r} \text{FBOFF(d, f) and FBP(d, f) and not S(d, 1..n, f)} \Rightarrow \text{FBON(d, f)} \\ \text{FBON(d, f) and EAF(1..n, f) and S(d, 1..n, f)} \Rightarrow \text{FBOFF(d, f)} \\ \text{d = U or D} \end{array} \tag{7.12}$$

That is to say, if the floor button at floor f for motion in direction d is off, and the button is pushed, and none of the elevators are currently visiting floor f about to move in direction d, then the floor button is turned on. Conversely, if the light is on and at least one of the elevators arrives at floor f, and the elevator is about to move in direction d, then the light is turned off. The notation 1..n in S(d, 1..n, f) and EAF(1..n, f) was defined in (7.10). The predicate V(e, f) can be defined in terms of S(d, e, f) as follows:

$$\text{V(e, f) = S(U, e, f) or S(D, e, f) or S(N, e, f)} \tag{7.13}$$

The states of the elevator button and floor button were straightforward to define. Turning now to the elevators, complications arise. The state of an elevator essentially consists of a number of component substates. Kampen

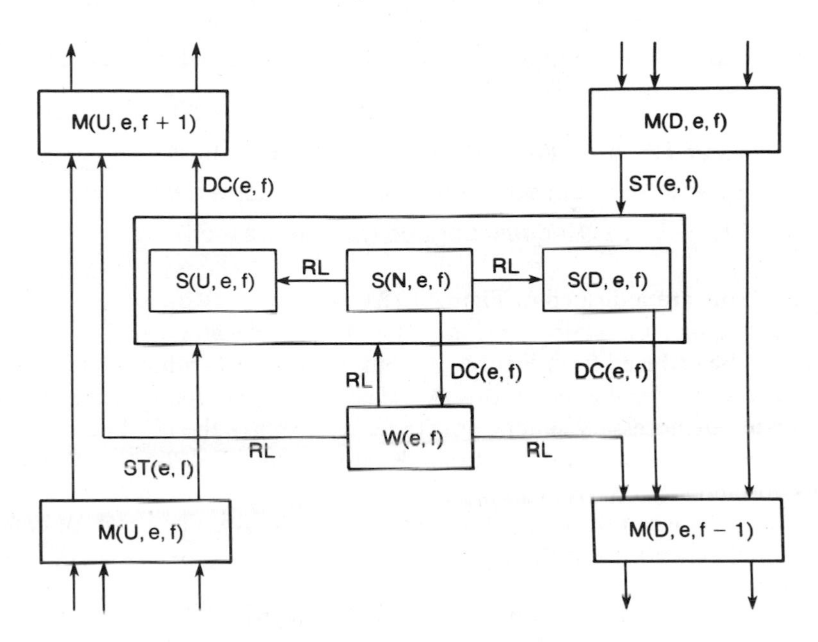

Figure 7.12 FSM representation of elevator problem [Kampen, 1987]. (© 1987 IEEE.)

identifies several, such as the elevator slowing and stopping, the door opening, the door open with a timer running, or the door closing after a timeout [Kampen, 1987]. He makes the reasonable assumption that the elevator controller (the mechanism that directs the motion of the elevator) initiates a state such as S(d, e, f) and that the controller then moves the elevator through the substates. Three elevator states can now be defined, one of which, namely S(d, e, f), was defined in (7.11), but is included here for completeness.

M(d, e, f):	Elevator e is <u>M</u>oving in direction d (floor f is next)	
S(d, e, f):	Elevator e is <u>S</u>topped (d-bound) at floor f	(7.14)
W(e, f):	Elevator e is <u>W</u>aiting at floor f (door closed)	

These states are shown in Figure 7.12. Note that the three stopped states S(U, e, f), S(N, e, f), and S(D, e, f) have been grouped into one larger state in order to simplify the diagram.

The events that can trigger state transitions are: DC(e, f), the closing of the door of elevator e at floor f; ST(e, f) which occurs when the sensor on the elevator is triggered as it nears floor f and the elevator controller must decide whether or not to stop the elevator at that floor; and RL which occurs whenever an elevator button or a floor button is pressed and enters its **ON** state.

DC(e, f):	Door Closed for elevator e, at floor f	
ST(e, f):	Sensor Triggered as elevator e nears floor f	(7.15)
RL:	Request Logged (button pressed)	

These events are indicated in Figure 7.12.

Finally, the state transition rules for an elevator can be presented. They can be deduced from Figure 7.12, but in some cases additional predicates are necessary. To be more precise, Figure 7.12 is nondeterministic; the predicates are necessary, among other reasons, to make the FSM deterministic. The interested reader should consult [Kampen, 1987] for the complete set of rules; for brevity's sake, the only rules presented here are those which state what happens when the door closes. The elevator moves up, down, or enters a wait state, depending on the current state.

$$\begin{aligned} &S(U, e, f) \text{ and } DC(e, f) \Rightarrow M(U, e, f + 1) \\ &S(D, e, f) \text{ and } DC(e, f) \Rightarrow M(D, e, f - 1) \\ &S(N, e, f) \text{ and } DC(e, f) \Rightarrow W(e, f) \end{aligned} \qquad (7.16)$$

The first rule states that if elevator e is in state S(U, e, f), that is, stopped at floor f, about to go up, and the doors close, then the elevator will move up towards the next floor. The second and third rules correspond to the cases of the elevator about to go down or with no requests pending.

The format of these rules reflects the power of finite state machines for specifying complex products. Instead of having to list a complex set of preconditions which have to hold for the product to do something and then having to list all the conditions that hold after the product has done it, the specifications take the simple form

current state and **event** and **predicate** ⇒ **next state**

This type of specification is easy to write down, easy to validate, and easy to convert into a design and into code. In fact, it is straightforward to construct an automatic tool that will translate an FSM specification directly into source code. Maintenance is then achieved by replay. That is to say, if new states and/or events are needed, the specifications are modified, and a new version of the product generated directly from the new specifications.

The finite state machine approach is more precise than the graphical method of Gane and Sarsen presented in Section 7.2, but is almost as easy to understand. Like Gane and Sarsen's method, timing considerations are not handled in Kampen's formalism. The next formal method, however, can handle timing issues.

7.5 Petri Nets

A major difficulty with specifying real-time systems is coping with timing. This difficulty can manifest itself in many different ways such as synchronization problems, race conditions, and deadlock. While it is true that timing problems can arise as a consequence of a poor design or a faulty implementation, such designs and implementations are often the natural consequence of poor specifications. That is to say, if the specifications are not properly drawn up, then there is a very real risk that the design and implementation will be inadequate. One powerful method for specifying systems with potential timing problems is Petri nets. A further advantage of this method is that it can be used for the design as well.

Petri nets were invented by Carl Adam Petri [Petri, 1962]. Originally of interest only to automata theorists, Petri nets have found wide applicability in computer science, being used in such fields as performance evaluation, operating systems, and software engineering. But before the use of Petri nets for specifications can be demonstrated, a brief introduction to Petri nets is given for those readers who may be unfamiliar with them.

A Petri net consists of four parts: a set of places P, a set of transitions T, an input function I, and an output function O. Consider the Petri net shown in Figure 7.13.

The set of places P is $\{p_1, p_2, p_3, p_4\}$.
The set of transitions T is $\{t_1, t_2\}$.
The input functions for the two transitions, represented by the arrows from

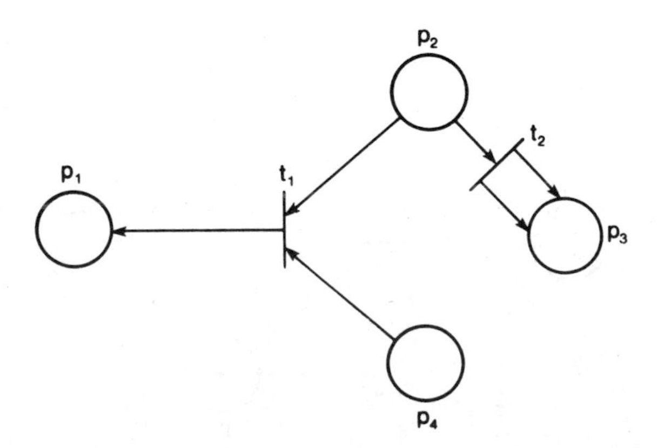

Figure 7.13 Petri net.

places to transitions, are:

$$I(t_1) = \{p_2, p_4\}$$

$$I(t_2) = \{p_2\}$$

The output functions for the two transitions, represented by the arrows from transitions to places, are:

$$O(t_1) = \{p_1\}$$

$$O(t_2) = \{p_3, p_3\}$$

Note the duplication of p_3; there are two arrows from t_2 to p_3.

More formally [Peterson, 1981], a Petri net structure is a 4-tuple $C = (P, T, I, O)$.

$P = \{p_1, p_2, \ldots, p_n\}$ is a finite set of *places*, $n \geq 0$.
$T = \{t_1, t_2, \ldots, t_m\}$ is a finite set of *transitions*, $m \geq 0$, with P and T disjoint.
$I : T \rightarrow P^\infty$ is the *input* function, a mapping from transitions to bags of places.
$O : T \rightarrow P^\infty$ is the *output* function, a mapping from transitions to bags of places. (A *bag*, or *multiset*, is a generalization of a set which allows for multiple instances of an element.)

A *marking* of a Petri net is an assignment of tokens to that Petri net. In Figure 7.14 there are four tokens, one in p_1, two in p_2, none in p_3, and one in p_4. The marking can be represented by the vector $(1, 2, 0, 1)$. Transition t_1 is enabled (ready to fire), because there are tokens in p_2 and in p_4; in general, a

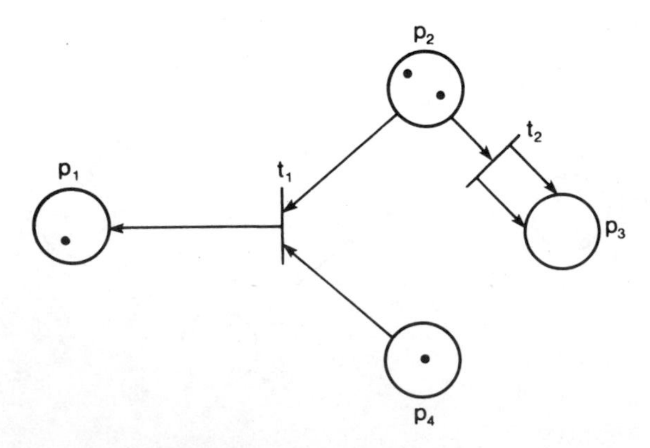

Figure 7.14 Marked Petri net.

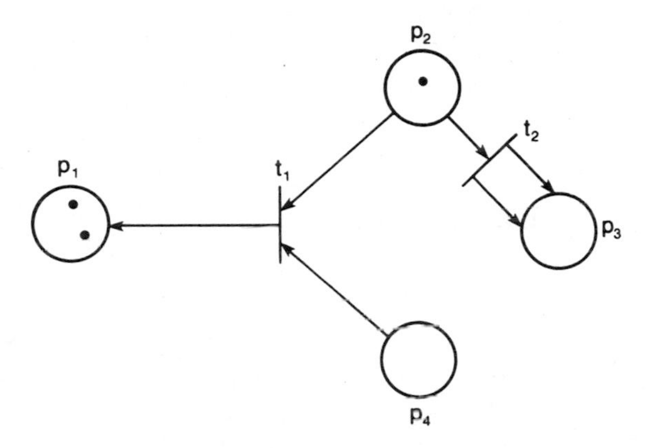

Figure 7.15 Petri net after firing transition t_1.

transition is enabled if each of its input places has as many tokens in it as there are arcs from the place to that transition. If t_1 fires, one token is removed from p_2 and one from p_4, and one new token is placed in p_1. The number of tokens is not conserved—two tokens are removed, but only one new one is placed in p_1. Transition t_2 is also enabled, because there are tokens in p_2. If t_2 fires, one token is removed from p_2, and two new tokens are placed in p_3.

Petri nets are nondeterministic, that is to say, if more than one transition is able to fire, then any one of them may be fired. Consider Figure 7.14 with marking $(1, 2, 0, 1)$. Both t_1 and t_2 are enabled; suppose that t_1 fires. The resulting marking $(2, 1, 0, 0)$ is shown in Figure 7.15. Now only t_2 is enabled. It fires, the enabling token is removed from p_2, and two new tokens are placed in p_3. The marking is now $(2, 0, 2, 0)$ as shown in Figure 7.16.

More formally [Peterson, 1981], a marking M of a Petri net $C = (P, T, I, O)$ is a function from the set of places P to the set of nonnegative integers

$$M : P \rightarrow \{0, 1, 2, \ldots\}$$

A marked Petri net is then a 5-tuple (P, T, I, O, M).

An important extension to a Petri net is an inhibitor arc. Referring to Figure 7.17, the inhibitor arc is marked by a small circle rather than an arrowhead. Transition t_1 is enabled because there is a token in p_3 but no token in p_2. In general, a transition is enabled if there is at least one token on each of its (normal) input arcs and no tokens on any of its inhibitor input arcs. This extension will be used in a Petri net specification of the elevator problem presented in Section 7.4.1 [Guha, Lang, and Bassiouni, 1987].

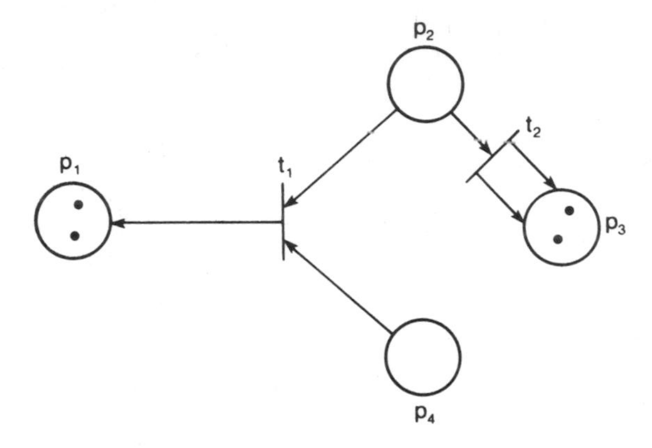

Figure 7.16 Petri net after firing transition t_2.

7.5.1 Petri Net Case Study

Recall that an n elevator system is to be installed in a building with m floors. In this Petri net specification each floor in the building will be represented by a place F_f, $1 \leq f \leq m$, in the Petri net; an elevator is represented by a token. A token in F_f denotes that an elevator is at floor f.

The first constraint is:

> C_1: Each elevator has a set of buttons, one for each floor. These illuminate when pressed and cause the elevator to visit the corresponding floor. The illumination is canceled when the corresponding floor is visited by the elevator.

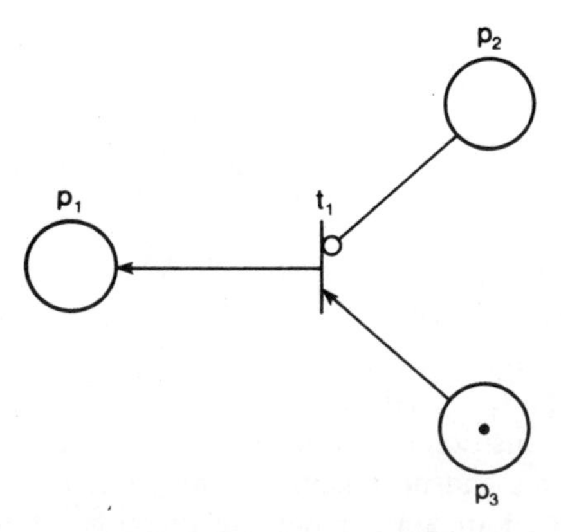

Figure 7.17 Petri net with inhibitor arc.

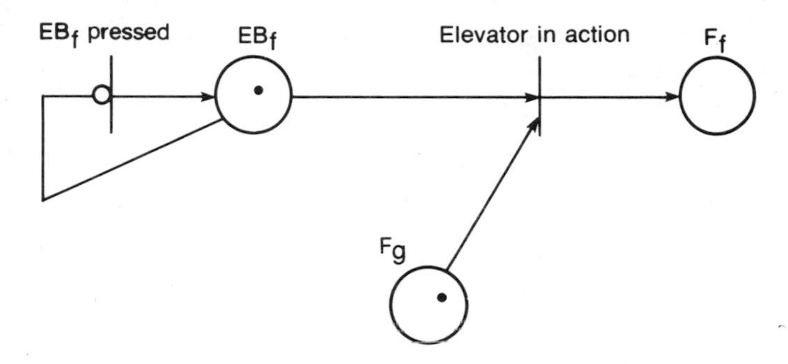

Figure 7.18 Petri net representation of elevator button [Guha, Lang, and Bassiouni, 1987]. (© 1987 IEEE.)

To incorporate this into the specifications, further places are needed. The elevator button for floor f is represented in the Petri net by place EB_f, $1 \leq f \leq m$. More precisely, since there are n elevators the place should be denoted $EB_{f,e}$ with $1 \leq f \leq m$, $1 \leq e \leq n$. But for the sake of simplicity of notation, the subscript e representing the elevator will be suppressed. A token in EB_f denotes that the elevator button for floor f is illuminated. Since the button must be illuminated the first time the button is pressed and subsequent button presses must be ignored, this is specified using a Petri net as shown in Figure 7.18. First, suppose that button EB_f is not illuminated. There is no token in place and hence, because of the presence of the inhibitor arc, transition EB_f pressed is enabled. The transition fires, and a new token is placed in EB_f as in the figure. Now, no matter how many times the button is pressed, the combination of the inhibitor arc and the presence of the token means that transition EB_f pressed cannot be enabled. There can therefore never be more than one token in place EB_f. Suppose that the elevator is to travel from floor g to floor f. Since the elevator is at floor g a token is in place F_g, as shown in Figure 7.18. Transition Elevator in action is enabled and then fires. The tokens in EB_f and F_g are removed, thereby turning off the light in button EB_f, and a new token appears in F_f; the firing of this transition brings the elevator from floor g to floor f. This motion from floor g to floor f cannot take place instantaneously. To handle this and similar issues, such as the fact that it is physically impossible for a button to illuminate at the very instant it is pressed, timing must be added to the Petri net model. That is to say, in classical Petri net theory transitions are instantaneous. In practical situations such as the elevator problem, timed Petri nets [Merlin, 1974] are needed in order to be able to associate a nonzero time with a transition.

The second constraint is:

C_2: Each floor, except the first floor and top floor, has two buttons, one to request an up-elevator and one to request a down-elevator. These buttons

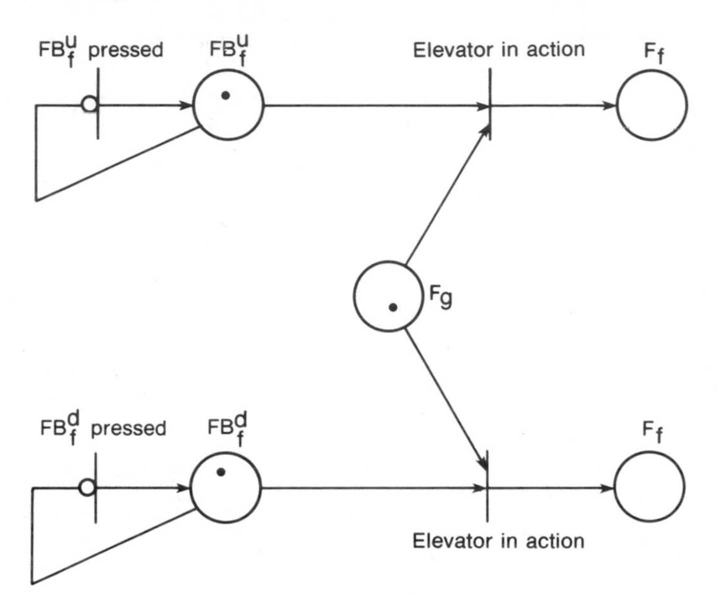

Figure 7.19 Petri net representation of floor buttons [Guha, Lang, and Bassiouni, 1987]. (© 1987 IEEE.)

illuminate when pressed. The illumination is canceled when an elevator visits the floor and moves in the desired direction.

The floor buttons are represented by places FB_f^u and FB_f^d representing the buttons for requesting up- and down-elevators, respectively. More precisely, floor 1 has a button FB_1^u, floor m has a button FB_m^d, and the intermediate floors each have two buttons, FB_f^u and FB_f^d, $1 < f < m$. The situation when an elevator reaches floor f from floor g with one or both buttons illuminated is shown in Figure 7.19. If both the buttons are illuminated, only one is turned off. To ensure that the correct light is turned off requires a more complex Petri net model; see, for example, [Ghezzi and Mandrioli, 1987].

The third constraint is:

C_3: When an elevator has no requests to service, it should remain at its current floor with its doors closed.

This is easily achieved, because if there are no requests, no Elevator in action transition is enabled.

Not only can Petri nets be used to represent the specifications, they can be used for the design as well [Guha, Lang, and Bassiouni, 1987]. But even at this stage of the development of the product, it is clear that Petri nets possess the expressive power necessary for specifying timing aspects of real-time systems.

7.6 Executable Specifications

A worthwhile approach to the problem of specifying a software product is to use executable specifications, that is to say, to write the specifications in a language that can be interpreted on a simulator. One example of an executable language is SREM, mentioned in Section 7.3. An executable specification is a formal model of the product which, when executed by an interpreter, simulates the behavior of the product. There are some similarities between rapid prototyping and executable specifications, but they are by no means the same thing. To see the difference, consider the following specifications [Zave, 1984].

> The product must sample the temperature of each of 10 machines once every second. The software must cause an alarm to be sounded if the temperature of a machine rises above 250°F or rises by more than 5°F over the last five readings. The alarm must continue to sound until the temperature is within permissible limits. The sensor values and the alarm activator flag are accessible as special-purpose hardware registers.

Since a prototype is an implementation, when constructing a rapid prototype of the product a decision has to be made as to how the product is to be implemented. For example, the product could be structured as a single module which samples the temperature of each of the 10 machines in turn and sounds the alarm if necessary (Figure 7.20). Alternatively, the product could be implemented as a main module calling each of 10 monitor modules in turn. If a monitor module sends back a flag indicating an unacceptable temperature rise, then the main module calls the alarm module. This implementation is depicted in Figure 7.21.

Both of the preceding implementations run on a uniprocessor. An alternative implementation consists of 10 independent monitoring processes each sampling the temperature of one particular machine with each module running on its own processor. All 10 modules can communicate with an alarm module when necessary, and the alarm module also runs on its own processor (Figure 7.22).

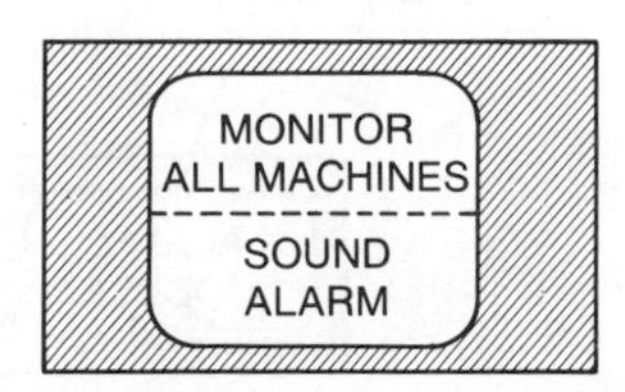

Figure 7.20 Monitor implemented as single module running on uniprocessor.

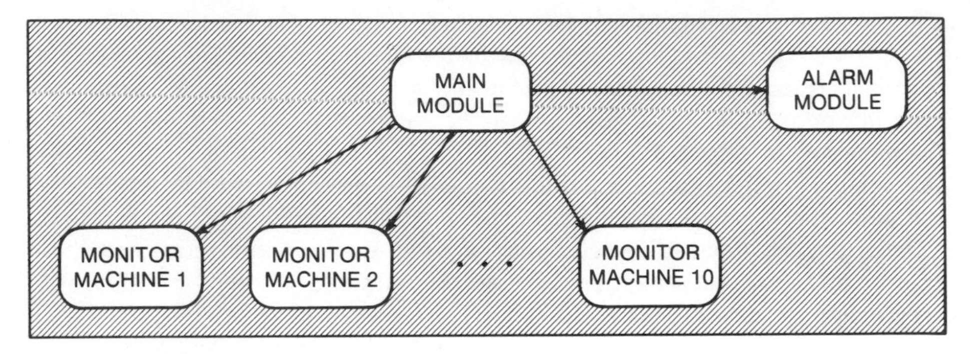

Figure 7.21 Monitor implemented as 12 modules running on uniprocessor.

When building a prototype of the monitoring system, a decision has to be taken as to how the product will be implemented. The implementation may correspond to one of those depicted in Figures 7.20 through 7.22, or it may be a different implementation. But whatever implementation is chosen for the prototype, a decision has been made that goes beyond what is stated in the specifications. This is not the case with executable specifications. When run on an interpreter, the behavior is what is encapsulated in the specifications; no design or implementation details are involved.

To take another example, the preceding specifications require that a list of the five most recent temperature readings be maintained for each machine. This list can be implemented as a linear list, as a linked list, or as some other data structure. If the information needed from the prototype depends critically on how the list is implemented, then executable specifications cannot provide the information. But if the purpose of the prototype is to determine if the product does what the user really needs, then executable specifications can be a powerful tool.

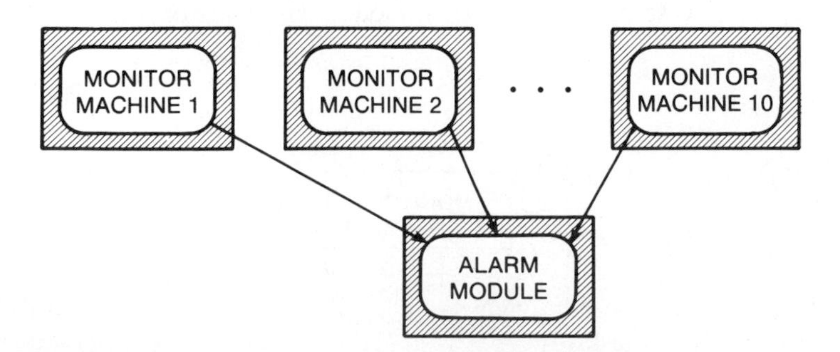

Figure 7.22 Monitor implemented as 11 modules running on 11 processors.

In many respects an executable specification goes further than a rapid prototype. In theory, a specification written in any executable specification language can be checked for internal consistency. After all, such a specification language has a syntax and semantics. As shown in Section 6.3.1, code can be proved correct by performing certain mathematical operations on the code. This can be done because code is expressed in a formal language, unlike the same set of instructions expressed in a natural language or graphically. By using a formal representation of specifications it is possible, in theory, to achieve similar results. In practice, however, some formal languages are easier to check for internal consistency than others. This applies especially to PAISLey, an executable specification language that was designed to make such consistency checking easy [Zave and Schell, 1986].

The idea of executable specifications seems to offer much to software developers. In order to evaluate its practicality, the executable specification language PAISLey is now examined.

7.6.1 PAISLey

PAISLey [Zave and Schell, 1986] is an executable language for describing specifications. It was designed for specifying real-time and distributed products. In PAISLey, a system is described as a set of asynchronous processes. Processes can represent data objects, buffers, system functions, and so on. Each process has a state and can go through an infinite sequence of discrete state changes.

A process is defined in terms of its successor mapping, a mapping which takes the process from its current state to the next state, as shown in Figure 7.23. Each step of the process consists of one evaluation of the successor mapping. The current state of the process is the argument given to the successor mapping. The successor mapping then produces a value, the next state. At the end of the process step, the current state is replaced by the next state. For example, consider a process representing the location of a packet in a packet-switching network. The successor mapping takes as input the current node at which the packet is located and returns the next node. In PAISLey, this is declared as follows:

```
return-next-node:
  NODE → NODE;                                   (7.17)
```

Lowercase letters are used for mapping names, uppercase letters for sets such as NODE.

In addition to the preceding *declaration*, a *definition* of return-next-node would be supplied which would specify how return-next-node carries out its task. In order to understand more about this approach, a PAISLey case study is given.

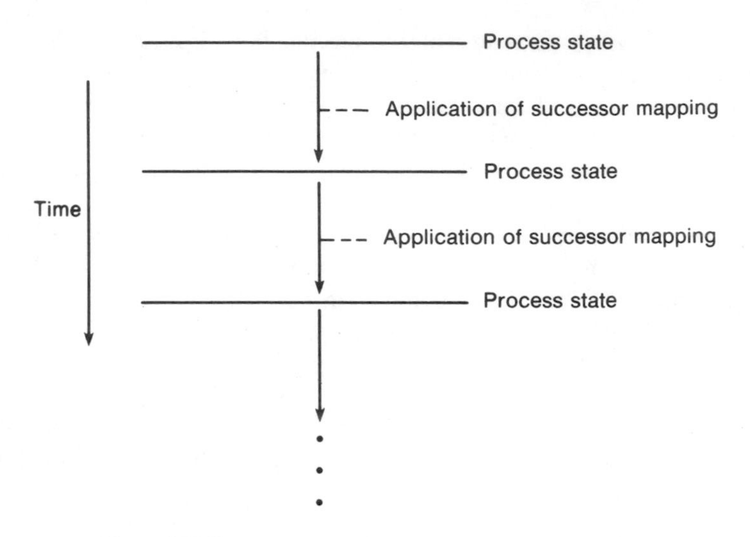

Figure 7.23 Process states and successor mapping in PAISLey.

7.6.2 PAISLey Case Study

The case study examined here is the temperature-monitoring product described in Section 7.6 which is to cause an alarm to be sounded if the temperature rises above predetermined limits [Zave, 1984]. One mapping in the PAISLey specification is update-last-five-temperatures. This mapping takes two inputs, a 5-tuple containing the last five temperatures to date and the next reading, and returns the latest five temperatures, that is to say, it updates the temperature history. Thus the domain of update-last-five-temperatures is the Cartesian product of LAST-FIVE-TEMPERATURES and TEMPERATURE. This is expressed in PAISLey as follows:

```
update-last-five-temperatures:
   LAST-FIVE-TEMPERATURES × TEMPERATURE                (7.18)
    → LAST-FIVE-TEMPERATURES;
```

The alarm is sounded if the current temperature of a machine is above 250°F or if the temperature rises by more than 5°F over the last five readings. Furthermore, the alarm must continue to sound until neither of these conditions holds. Since both the current temperature and the history of the last five temperatures are needed, the domain of check-alarm-m, which causes the alarm to be sounded as long as the temperature of machine m is unacceptable, is the Cartesian product of LAST-FIVE-TEMPERATURES and TEMPERATURE. The range is empty, because check-alarm-m causes the alarm to be sounded as a side effect; it does not need to return a value. The declaration of check-alarm-m is then

```
check-alarm-m:
  LAST-FIVE-TEMPERATURES × TEMPERATURE → FILLER;          (7.19)
```

FILLER denotes that this mapping returns a null value.

The projection mapping proj is used to select a portion of the output of a mapping to be returned as the value of the mapping. Thus

```
proj [(1, (first-parameter, second-parameter)
      )                                                   (7.20)
     ];
```

returns first-parameter, while

```
proj [(2, (first-parameter, second-parameter)
      )                                                   (7.21)
     ];
```

returns second-parameter. In each case, the parameter not selected is simply discarded. With regard to the notation of (7.20) and (7.21), parentheses are used to denote tuples (ordered sets of elements), while arguments of mappings are in square brackets.

Returning to the temperature-monitoring example, the next mapping to be considered is check-current-m-temperature. This mapping takes two inputs, a 5-tuple containing the last five temperatures and the current temperature of machine-m. It invokes update-last-five-temperatures to update the 5-tuple and invokes check-alarm-m to determine whether the latest reading should cause the alarm to be sounded. It returns the updated value of the 5-tuple. The declaration of check-current-m-temperature is thus

```
check-current-m-temperature:
  LAST-FIVE-TEMPERATURES × TEMPERATURE                    (7.22)
   → LAST-FIVE-TEMPERATURES;
```

The mapping is defined as follows:

```
check-current-m-temperature [(hist, temp)] =
  proj [(1, (update-last-five-temperatures [(hist, temp)],
            (check-alarm-m [(hist, temp)])                (7.23)
        )
       ];
```

What this means is that the two arguments of check-current-m-temperature, namely hist, the 5-tuple containing the temperature history, and temp, the current temperature, are passed to the two mappings it calls. The values of these arguments remain fixed until the current process step has been completed. Mapping update-last-five-temperatures takes the two arguments and returns the new value of the 5-tuple hist, while check-alarm-m uses the values of hist and temp to determine whether the alarm should be sounded; in PAISLey these two operations are carried out in parallel. Now mapping proj comes into play and ensures that the result of check-alarm-m is discarded, so that what is returned as the value of check-current-m-temperature is the new 5-tuple containing the latest five temperatures.

Mapping obtain-current-m-temperature simply returns the current temperature of machine-m. Its declaration is

```
obtain-current-m-temperature:
  → TEMPERATURE;                                              (7.24)
```

In terms of the specifications, machine-m must be monitored every second. This process is called process-machine-m. Recall that a mapping function takes a process from its current state to the next state. The state of process-machine-m is the list of the latest five temperatures. The domain of the corresponding successor mapping function machine-m-step is the current state, that is, the current temperature and the previous four temperature readings. The range is the next state, that is, the latest five readings, namely the newest reading, the previous current reading, and the three readings before that. Thus the declaration for machine-m-step is

```
machine-m-step:
  LAST-FIVE-TEMPERATURES → LAST-FIVE-TEMPERATURES;            (7.25)
```

The successor mapping is defined as follows:

```
machine-m-step [hist] = check-current-m-temperature
                        [(hist, obtain-current-m-temperature)];   (7.26)
```

In other words, the 5-tuple temperature history hist is passed to check-current-m-temperature, together with the current temperature reading, which is returned by obtain-current-m-temperature; check-current-m-temperature then behaves as specified in (7.23) and returns the new value of hist. At the end of the step, the old state is replaced by this new value.

The previous specification is incomplete; for example, no definition has been given for update-last-five-temperatures or check-alarm-m. As will be explained, even incomplete PAISLey specifications can be executed on an interpreter. This and other aspects of PAISLey will now be examined.

7.6.3 Strengths and Weaknesses of PAISLey

In [Zave and Schell, 1986] a description is given of the major strengths of PAISLey. The first is that PAISLey supports full parallelism; any parallelism that might be inherent in the problem can be expressed in PAISLey. For example, as remarked previously, the two mapping invocations in (7.23), namely **update-last-five-temperatures [(hist, temp)]** and **check-alarm-m [(hist, temp)]**, are evaluated in parallel. While this is being done, the values of **hist** and **temp** cannot change. Only at the end of each step of **check-current-m-temperature** is the value of **hist** updated.

PAISLey tolerates incompleteness. If a specification consists of any set of syntactically correct declarations and definitions, then it can be executed. As programming languages go, this property seems to be unique to PAISLey. The closest that any other language comes is Ada, which was designed to permit development by stepwise refinement of a product. Thus a top down subtree of Ada modules can certainly be compiled, but it cannot be executed until everything used has been completely defined. In contrast, any PAISLey specification can be executed. The user may specify **default**, **random**, or **interactive** mode. If the user chooses **default** mode, then PAISLey selects the appropriate default value, for example, 0 if the range is INTEGER. In **random** mode the interpreter chooses an arbitrary element of the range as the value of the mapping, while in **interactive** mode the interpreter prompts the user for the value. This ability to handle only partially defined specifications is a powerful tool supporting stepwise refinement of processes.

Another powerful aspect of PAISLey is its ability to handle timing constraints. For every mapping the evaluation time of that mapping may be specified. More precisely, any or all of an upper time bound, a lower time bound, or a distribution may be specified for the evaluation time of each mapping. The interpreter will then simulate the product, taking all timing constraints into account. Before execution the interpreter will report a timing *inconsistency*. An example of such an inconsistency would be if mapping M_{12} consists of mapping M_1 followed by mapping M_2, and mappings M_1 and M_2 both have a lower time constraint of 10 seconds, and therefore it will take at least 20 seconds to evaluate them sequentially, but M_{12} has an upper time constraint of 5 seconds. If a timing fault occurs during simulation this, too, will be reported by the interpreter to the user.

At first sight it might seem that PAISLey would be difficult to use. But, as described in the following discussion, experience with PAISLey has shown that, with training, it can indeed be used in practice for drawing up executable specifications whose consistency can then be easily checked.

7.6.4 PAISLey in Practice

PAISLey has been used in a number of projects, including the development of an experimental numerical problem solver [Zave and Cole,

1983] and for the user interface to its own environment [Zave, 1986]. In those projects the PAISLey expertise was supplied by a PAISLey expert. More significant is the experiment in technology transfer conducted in 1985 in which PAISLey specifications were written by experts in the application area, rather than by PAISLey experts [Berliner and Zave, 1987].

At AT&T Bell Laboratories, the need arose to construct a software product for the SL communications system, an undersea transmission system which uses optical fiber technology. PAISLey, itself a product of Bell Laboratories, was put forward as a good way of expressing the requirements for the system, because PAISLey provides powerful tools for validating specifications, that is, for making sure that they do what the user needs and for showing that they are consistent. It was decided that parts of one subsystem of the complete SL project would be specified in PAISLey, and that the specifications would be written by members of the SL team. The first step was therefore to provide training for SL team members. This was done by Pamela Zave, the author of PAISLey. Some members of the team received in all considerably less than a month's training, but two staffers worked together with Zave over a period of months, not all full-time, until they were able to write PAISLey specifications on their own.

The experiment was a success in that the two members of the SL team were indeed able to write PAISLey specifications that passed a formal review. But there were some difficulties. Of the eight team members involved, two were successfully trained as specification writers as mentioned previously, but the other six were much less successfully trained as specification readers. The training for the specification readers was not well designed, and the would-be readers spent too little time learning to read PAISLey. A contributory factor might have been that the readers had no formal training in computer science.

But there is another way of looking at the SL project. It could be that the project succeeded because of the environment and the individual. With regard to the environment, the experiment was conducted at Bell Laboratories, the "home" of PAISLey. Management at Bell Laboratories was committed to the project in a way that is unlikely to be repeated at another organization, one for which PAISLey would be just another specifications tool.

The individual involved was Pamela Zave, the author of PAISLey, Again, it is unlikely that any other person would have the same motivation to have the experiment succeed. Furthermore, the results of experiments in computer science are sometimes ascribed to the participants, rather than the method being investigated. For example, the overwhelming success of the chief programmer team approach at *The New York Times* (Section 10.8.1) has never been repeated, and one possible explanation is that the success of the project was due to the chief programmer (F. Terry Baker), rather than to the chief programmer team approach. In the same way, it is possible that Zave's enthusiasm and drive were the primary reason that PAISLey worked so well.

There is no doubt that using PAISLey to represent specifications is harder than using, say, structured systems analysis. Using a functional approach with domains and ranges and Cartesian products requires that the computer personnel using PAISLey have a background both in computer science and in the relevant mathematics. Equally important is a commitment to invest the time needed to be trained in PAISLey. Once that commitment has been made, PAISLey is a powerful tool that provides the consistency checking and timing checking that can prove invaluable when specifying real-time and distributed systems.

7.7 Other Formal Methods

Many other formal methods have been proposed. These methods are extremely varied. For example, Anna [Luckham and von Henke, 1985] is a formal specification language for Ada. Some of them are knowledge-based such as Refine [Smith, Kotik, and Westfold, 1985] and Gist [Balzer, 1985]. Gist was designed so that users can describe processes in a way that is as close as possible to the way that we think about processes. This was to be achieved by formalizing the constructs used in natural languages. In practice, Gist specifications are as hard to read as most other formal specifications, so much so that a paraphraser from Gist to English had to be written. Gist is the specification language underlying the automated life cycle described in Section 16.2.

Vienna definition method (VDM) [Bjørner, 1987] is a technique based on denotational semantics [Gordon, 1979]. VDM can be applied not just to the specifications, but also to the design and implementation. VDM has been successfully used in a number of projects, most spectacularly by the Dansk Datamatik Center development of the DDC Ada Compiler System [Oest, 1986].

A different way of looking at specifications is to view them in terms of sequences of events, where an event is either a simple action or a communication that transfers data into or out of the system. For example, in the elevator problem, one event consists of pushing the elevator button for floor f on elevator e and its resulting illumination. Another event is elevator e leaving floor f in the downward direction and the canceling of the illumination of the corresponding floor button. The language Communicating Sequential Processes (CSP) invented by Hoare is based on the idea of describing the behavior of a system in terms of such events [Hoare, 1985]. In CSP, a process is described in terms of the sequences of events that the process will engage in with its environment. Processes interact with each other by sending messages to one another. CSP allows processes to be combined in a wide variety of ways, such as sequentially, in parallel, or interleaved nondeterministically.

The power of CSP lies in the fact that CSP specifications are executable [Delisle and Schwartz, 1987] and thus have all the advantages of executable specification languages described in Section 7.6. But CSP goes further than that. It provides a framework for going from specifications to design to implementation by a sequence of steps that preserve validity. In other words, if the specifications are correct and if the transformations are correctly performed, then the design and implementation will be correct as well. Going from design to implementation is particularly straightforward if the implementation language is Ada.

But CSP also has its disadvantages. In particular, it is not an easy language to learn. An attempt was made to include a CSP specification for the elevator problem in this book [Schwartz and Delisle, 1987]. But the quantity of essential preliminary material and the level of detail of explanation needed to describe each CSP statement adequately were simply too great to permit inclusion in a book as general as this one. The relationship between the power of a specification language and its degree of difficulty of use will be expanded in the next section.

7.8 Comparison of Specification Methods

The main lesson of this chapter is that every development organization has to decide what type of specification language is appropriate for the product about to be developed. An informal method is easy to learn, but does not have the power of a semiformal or formal method. Conversely, each formal method supports a variety of features which may include executability, consistency checking, or transformability to design and implementation through a series of correctness preserving steps. But while it is generally true that the more formal the method the greater its power, it is also generally true that formal methods can be difficult to learn and use. In other words, there is a trade-off to be made between ease of use and the power of a specification language.

In some circumstances, the choice of specification language type is easy. For example, if the vast majority of the members of the development team do not have training in computer science, then it is virtually impossible to use anything other than an informal or semiformal specification method. Conversely, where a mission-critical real-time system is being built in a research laboratory, the power of a formal specification method will almost certainly be required.

An additional complicating factor is that many of the newer formal methods are virtually untested under practical conditions. There is a considerable risk involved in using such a method. Large sums of money will be needed to pay for training of the relevant members of the development team. Then more money will be spent while the team adjusts from using the language in the classroom to using it on the actual project. It may well happen that the language's supporting software tools do not work properly, as happened with SREM [Scheffer, Stone, and Rzepka, 1985], with the resulting additional expense and time slippage. But if everything works, and if the software project management plan takes into account the additional time and money needed when a new technology is used in practice, the possible gains may be huge.

What specification method should be used for a specific project? It will depend on the project, on the development team, on the management team, and on myriad other factors. As with so many other aspects of software engineering, trade-offs have to made. Unfortunately, there is no simple rule for deciding which specification method to use.

Chapter Review

Specifications can be expressed informally (Section 7.1), semiformally (Sections 7.2 and 7.3), or formally (Sections 7.4 through 7.7). One semiformal method, namely Gane and Sarsen's structured systems analysis, is described in some detail (Section 7.2). Formal methods that are described include finite state machines (Section 7.4), Petri nets (Section 7.5), and PAISLey, an executable specifications language (Section 7.6). Case studies of each of these formal methods are presented (Sections 7.4.1, 7.5.1, and 7.6.2). The major theme of this chapter is that informal methods are easy to use, but imprecise; this is demonstrated by means of a case study (Section 7.1.1). Conversely, formal methods are powerful, but require a nontrivial investment in training time.

For Further Reading

The classical texts on specific semiformal methods are the books by DeMarco, and Gane and Sarsen [DeMarco, 1978; Gane and Sarsen, 1979]. These ideas have been updated in [Martin and McClure, 1985] and in [Yourdon, 1989]. Data-oriented design, described in Section 9.5, is a design method that is integrated with a different class of semiformal specification methods; details are to be found in [Orr, 1981], [Warnier, 1981], and [Jackson, 1983]. SADT is described in [Ross, 1985], and PSL/PSA is described in [Teichroew and Hershey, 1977]. Two sources of information on SREM are [Alford, 1985] and [Scheffer, Stone, and Rzepka, 1985].

A good collection of papers on formal methods can be found in the special section on specification and verification in the March 1984 issue of *IEEE Transactions on Software Engineering*. An interesting paper on the application of formal specifications in industry is [Hayes, 1985], while a comparison of formal and informal specification methods can be found in [Gehani, 1982].

An early reference to the finite state machine approach is [Naur, 1964], where it is unfortunately referred to as the Turing machine approach. The finite state machine approach is described in [Ferrentino and Mills, 1977] and [Linger, 1980]. A FSM model for real-time systems is given in [Chandrasekharan, Dasarathy, and Kishimoto, 1985]. A graphical tool for FSMs is described in [Jacob, 1985], and the use of FSMs in specifying the interaction between humans and computers is found in [Wasserman, 1985].

[Peterson, 1981] is an excellent introduction to Petri nets and their applications. The use of Petri nets in prototyping is described in [Bruno and Marchetto, 1986]. A paper discussing automatic implementation of Petri nets is [Nelson, Haibt, and Sheridan, 1983]. Timed Petri nets are described in [Merlin, 1974] and [Coolahan and Roussopoulos, 1983].

With regard to PAISLey, the reader should consult [Zave, 1988] for full details about the specification language.

An interesting and wide-ranging collection of papers on specification techniques can be found in [Gehani and McGettrick, 1986]. Specification languages are described in the April 1985 issue of *IEEE Computer*. The proceedings of the International Workshops on Software Specification and Design are a preeminent source for research ideas regarding specifications; the proceedings of the third and subsequent workshops are published by IEEE Computer Society Press.

Problems

7.1 Consider the following recipe for grilled pockwester.

Ingredients: 1 large onion
1 can frozen orange juice
Freshly squeezed juice of 1 lemon
$\frac{1}{2}$ cup bread crumbs
Flour
Milk
3 medium-sized shallots
2 medium-sized eggplants
1 fresh pockwester
$\frac{1}{2}$ cup Pouilly Fuissé
1 garlic
Parmesan cheese
4 free-range eggs

The night before, take one lemon, squeeze it, strain the juice, and freeze it. Take one large onion and three shallots, dice them, and grill them in a skillet. When clouds of black smoke start to come off, add 2 cups of fresh orange juice. Stir vigorously. Slice the lemon into paper-thin slices and add to the mixture. In the meantime, coat the mushrooms in flour, dip them in milk, and then shake them up in a paper bag with the bread crumbs. In a saucepan heat $\frac{1}{2}$ cup of Pouilly Fuissé. When it reaches 170° add the sugar and continue to heat. When the sugar has caramelized add the mushrooms. Blend the mixture for 10 minutes or until all lumps have been removed. Add the eggs. Now take the pockwester and kill it by sprinkling it with frobs. Skin the pockwester, break it into bite-sized chunks, and add it to the mixture. Bring to the boil and simmer, uncovered. The eggs should previously have been vigorously stirred with a wire whisk for 5 minutes. When the pockwester is soft to the touch, place it on a serving platter, sprinkle with Parmesan cheese, and broil for not more than 4 minutes.

Determine the ambiguities, omissions, and contradictions in the preceding specification.

7.2 Consider the problem of determining whether a bank statement is correct. The data needed include the balance at the beginning of the month, the number, date, and amount of each check, the date and amount of each deposit, and the balance at the end of the month. Write a precise English specification of the problem.

7.3 Draw a data flow diagram of the specification you drew up for Problem 7.2. Ensure that your DFD simply reflects the flow of data and that no assumptions regarding computerization have been made.

7.4 Consider an automated library circulation system. Every book has a bar code affixed to it, and every borrower has a card bearing a bar code. When a borrower wishes to check out a book, the librarian scans the bar code on the book and on the borrower's card and enters C at the computer terminal. Similarly, when a book is returned, it is again scanned, and the librarian enters R. Librarians can add books (+) to the library collection or remove them (−). Borrowers can go to a terminal and determine all the books in the library by a particular author (the borrower enters A = followed by the author's name), all the books with a specific title (T = followed by the title), or all the books in a particular subject area (S = followed by the subject area). Finally, if a borrower wants a book that is currently checked out by another borrower, the librarian can place a hold on the book so that when it is returned it will be held for the borrower who requested it (H followed by the number of the book). Write down precise specifications for the library circulation system.

7.5 Draw a data flow diagram showing the operation of the library circulation system of Problem 7.4.

7.6 Complete the specifications for the library circulation system of Problem 7.4 using Gane and Sarsen's method. Where data have not been specified (e.g., the number of books checked in/out each day) make your own assumptions, but clearly indicate any assumptions you have made.

7.7 A fixed-point binary number consists of an optional sign, followed by one or more bits, followed by a binary point, and one or more bits. Examples of fixed-point binary numbers include 1011.0100, −0.000001, and +1101101.0. More formally, this can be expressed as:

```
⟨fixed point binary⟩::={⟨sign⟩}⟨bitstring⟩⟨binary point⟩⟨bitstring⟩
⟨sign⟩              ::=+|−
⟨bitstring⟩         ::=⟨bit⟩{⟨bitstring⟩}
⟨binary point⟩      ::=.
⟨bit⟩               ::=0|1
```

(The notation {...} denotes an optional item and a|b denotes a or b.) Specify a finite state machine which will take as input a string of characters and determine whether or not that string constitutes a valid fixed-point binary number.

7.8 Use the finite state machine approach to specify the library circulation system of Problem 7.4.

7.9 Show how your solution to Problem 7.8 can be used to design and implement a menu-driven product for the library circulation system.

7.10 Use Petri nets to specify the circulation of a single book through the library of Problem 7.4. Include operations H, C, and R in your specifications.

7.11 (Term Project) Using a method specified by your instructor, draw up a specifications document for the Plain Vanilla Ice Cream Corporation product described in the Appendix.

7.12 (Readings in Software Engineering) Your instructor will distribute copies of [Meyer, 1985]. Critique his formal specifications from the viewpoint of ease in understanding. Do you think that an automatic tool can be constructed that could produce an English paraphrase of this type of formal specification?

References

[Alford, 1985] M. Alford, "SREM at the Age of Eight; The Distributed Computing Design System," *IEEE Computer* **18** (April 1985), pp. 36–46.

[Balzer, 1985] R. Balzer, "A 15 Year Perspective on Automatic Programming," *IEEE Transactions on Software Engineering* **SE-11** (November 1985), pp. 1257–1268.

[Berliner and Zave, 1987] E. F. BERLINER AND P. ZAVE, "An Experiment in Technology Transfer: PAISLey Specification of Requirements for an Undersea Lightwave Cable System," *Proceedings of the Ninth International Conference on Software Engineering*, Monterey, CA, March 1987, pp. 42–50.

[Bjørner, 1987] D. BJØRNER, "On the Use of Formal Methods in Software Development," *Proceedings of the Ninth International Conference on Software Engineering*, Monterey, CA, March 1987, pp. 17–29.

[Brady, 1977] J. M. BRADY, *The Theory of Computer Science*, Chapman and Hall, London, UK, 1977.

[Bruno and Marchetto, 1986] G. BRUNO AND G. MARCHETTO, "Process-Translatable Petri Nets for the Rapid Prototyping of Process Control Systems," *IEEE Transactions on Software Engineering* **SE-12** (February 1986), pp. 346–357.

[Chandrasekharan, Dasarathy, and Kishimoto, 1985] M. CHANDRASEKHARAN, B. DASARATHY, AND Z. KISHIMOTO, "Requirements-Based Testing of Real-Time Systems: Modeling for Testability," *IEEE Computer* **18** (April 1985), pp. 71–80.

[Coolahan and Roussopoulos, 1983] J. E. COOLAHAN, JR., AND N. ROUSSOPOULOS, "Timing Requirements for Time-Driven Systems Using Augmented Petri Nets," *IEEE Transactions on Software Engineering* **SE-9** (September 1983), pp. 603–616.

[Dart, Ellison, Feiler, and Habermann, 1987] S. A. DART, R. J. ELLISON, P. H. FEILER, AND A. N. HABERMANN, "Software Development Environments," *IEEE Computer* **20** (November 1987), pp. 18–28.

[Delisle and Schwartz, 1987] N. DELISLE AND M. SCHWARTZ, "A Programming Environment for CSP," *Proceedings of the Second ACM SIGSOFT / SIGPLAN Software Engineering Symposium on Practical Software Development Environments*, *ACM SIGPLAN Notices* **22** (January 1987), pp. 34–41.

[DeMarco, 1978] T. DEMARCO, *Structured Analysis and System Specification*, Yourdon Press, New York, NY, 1978.

[Ferrentino and Mills, 1977] A. B. FERRENTINO AND H. D. MILLS, "State Machines and Their Semantics in Software Engineering," *Proceedings of the First International Computer Software and Applications Conference, COMPSAC '77*, Chicago, IL, 1977, pp. 242–251.

[Gane and Sarsen, 1979] C. GANE AND T. SARSEN, *Structured Systems Analysis: Tools and Techniques*, Prentice-Hall, Englewood Cliffs, NJ, 1979.

[Gehani, 1982] N. GEHANI, "Specifications: Formal and Informal—A Case Study," *Software—Practice and Experience* **12** (May 1982), pp. 433–444.

[Gehani and McGettrick, 1986] N. GEHANI AND A. MCGETTRICK (Editors), *Software Specification Techniques*, Addison-Wesley, Reading, MA, 1986.

[Ghezzi and Mandrioli, 1987] C. GHEZZI AND D. MANDRIOLI, "On Eclecticism in Specifications: A Case Study Centered around Petri Nets," *Proceedings of the Fourth International Workshop on Software Specification and Design*, Monterey, CA, April 1987, pp. 216–224.

[Goodenough and Gerhart, 1975] J. B. GOODENOUGH AND S. L. GERHART, "Toward a Theory of Test Data Selection," *Proceedings of the Third International Conference on Reliable Software*, Los Angeles, CA, 1975, pp. 493–510. Also published in: *IEEE Transactions on Software Engineering* **SE-1** (June 1975), pp. 156–173. Revised version: J. B. Goodenough and S. L. Gerhart, "Toward a Theory of Test Data Selection: Data Selection Criteria," in: *Current Trends in Programming Methodology, Volume 2*, R. T. Yeh (Editor), Prentice-Hall, Englewood Cliffs, NJ, 1977, pp. 44–79.

[Gordon, 1979] M. J. C. GORDON, *The Denotational Description of Programming Languages, An Introduction*, Springer-Verlag, New York, NY, 1979.

[Guha, Lang, and Bassiouni, 1987] R. K. GUHA, S. D. LANG, AND M. BASSIOUNI, "Software Specification and Design Using Petri Nets," *Proceedings of the Fourth International Workshop on Software Specification and Design*, Monterey, CA, April 1987, pp. 225–230.

[Hayes, 1985] I. J. HAYES, "Applying Formal Specification to Software Development in Industry," *IEEE Transactions on Software Engineering* **SE-11** (February 1985), pp. 169–178.

[Hoare, 1985] C. A. R. HOARE, *Communicating Sequential Processes*, Prentice-Hall International, Englewood Cliffs, NJ, 1985.

[IWSSD, 1986] Call for Papers, Fourth International Workshop on Software Specification and Design, *ACM SIGSOFT Software Engineering Notes* **11** (April 1986), pp. 94–96.

[Jackson, 1983] M. A. JACKSON, *System Development*, Prentice-Hall, Englewood Cliffs, NJ, 1983.

[Jacob, 1985] R. J. K. JACOB, "A State Transition Diagram Language for Visual Programming," *IEEE Computer* **18** (August 1985), pp. 51–59.

[Kampen, 1987] G. R. KAMPEN, "An Eclectic Approach to Specification," *Proceedings of the Fourth International Workshop on Software Specification and Design*, Monterey, CA, April 1987, pp. 178–182.

[Leavenworth, 1970] B. LEAVENWORTH, Review #19420, *Computing Reviews* **11** (July 1970), pp. 396–397.

[Linger, 1980] R. C. LINGER, "The Management of Software Engineering. Part III. Software Design Practices," *IBM Systems Journal* **19** (No. 4, 1980), pp. 432–450.

[London, 1971] R. L. London, "Software Reliability through Proving Programs Correct," in: *Proceedings of the IEEE International Symposium on Fault-Tolerant Computing*, March 1971.

[Luckham and von Henke, 1985] D. C. Luckham and F. W. von Henke, "An Overview of Anna, a Specification Language for Ada," *IEEE Software* **2** (March 1985), pp. 9–22.

[Martin and McClure, 1985] J. P. Martin and C. McClure, *Diagramming Techniques for Analysts and Programmers*, Prentice-Hall, Englewood Cliffs, NJ, 1985.

[Merlin, 1974] P. Merlin, "A Study of the Recoverability of Computing Systems," Ph.D. Dissertation, University of California, Irvine, CA, 1974.

[Meyer, 1985] B. Meyer, "On Formalism in Specifications," *IEEE Software* **2** (January 1985), pp. 6–26.

[Naur, 1964] P. Naur, "The Design of the GIER ALGOL Compiler," in: *Annual Review in Automatic Programming, Volume 4*, Pergamon Press, Oxford, UK, 1964, pp. 49–85.

[Naur, 1969] P. Naur, "Programming by Action Clusters," *BIT* **9** (No. 3, 1969), pp. 250–258.

[Nelson, Haibt, and Sheridan, 1983] R. A. Nelson, L. M. Haibt, and P. B. Sheridan, "Casting Petri Nets into Diagrams," *IEEE Transactions on Software Engineering* **SE-9** (September 1983), pp. 590–602.

[Oest, 1986] O. N. Oest, "VDM from Research to Practice," *Proceedings of IFIP Congress, Information Processing '86*, 1986, pp. 527–533.

[Orr, 1981] K. Orr, *Structured Requirements Definition*, Ken Orr and Associates, Inc., Topeka, KS, 1981.

[Peterson, 1981] J. L. Peterson, *Petri Net Theory and the Modeling of Systems*, Prentice-Hall, Englewood Cliffs, NJ, 1981.

[Petri, 1962] C. A. Petri, "Kommunikation mit Automaten," Ph.D. Dissertation, University of Bonn, West Germany, 1962. (In German).

[Pollack, Hicks, and Harrison, 1971] S. L. Pollack, H. T. Hicks, Jr., and W. J. Harrison, *Decision Tables: Theory and Practice*, Wiley-Interscience, New York, NY, 1971.

[Ross, 1985] D. T. Ross, "Applications and Extensions of SADT," *IEEE Computer* **18** (April 1985), pp. 25–34.

[Scheffer, Stone, and Rzepka, 1985] P. A. Scheffer, A. H. Stone, III, and W. E. Rzepka, "A Case Study of SREM," *IEEE Computer* **18** (April 1985), pp. 47–54.

[Schwartz and Delisle, 1987] M. D. Schwartz and N. M. Delisle, "Specifying a Lift Control System with CSP," *Proceedings of the Fourth International Workshop on Software Specification and Design*, Monterey, CA, April 1987, pp. 21–27.

[Smith, Kotik, and Westfold, 1985] D. R. SMITH, G. B. KOTIK, AND S. J. WESTFOLD, "Research on Knowledge-Based Software Environments at the Kestrel Institute," *IEEE Transactions on Software Engineering* **SE-11** (November 1985), pp. 1278–1295.

[Teichroew and Hershey, 1977] D. TEICHROEW AND E. A. HERSHEY, III, "PSL/PSA: A Computer-Aided Technique for Structured Documentation and Analysis of Information Processing Systems," *IEEE Transactions on Software Engineering* **SE-3** (January 1977), pp. 41–48.

[Warnier, 1981] J. D. WARNIER, *Logical Construction of Systems*, Van Nostrand Reinhold, New York, NY, 1981.

[Wasserman, 1985] A. I. WASSERMAN, "Extending State Transition Diagrams for the Specification of Human–Computer Interaction," *IEEE Transactions on Software Engineering* **SE-11** (August 1985), pp. 699–713.

[Yourdon, 1989] E. YOURDON, *Modern Structured Analysis*, Yourdon Press, Englewood Cliffs, NJ, 1989.

[Zave, 1984] P. ZAVE, "The Operational versus the Conventional Approach to Software Development," *Communications of the ACM* **27** (February 1984), pp. 104–118.

[Zave, 1986] P. ZAVE, "Case Study: The PAISLey Approach to its Own Software Tools," *Computer Languages* **11** (January 1986), pp. 15–28.

[Zave, 1988] P. ZAVE, "PAISLey User Documentation, Volumes 1, 2, and 3," AT&T Bell Laboratories, 1988.

[Zave and Cole, 1983] P. ZAVE AND G. E. COLE, JR., "A Quantitative Evaluation of the Feasibility of, and Suitable Hardware Architectures for, an Adaptive, Parallel Finite-Element System," *ACM Transactions on Mathematical Software* **9** (September 1983), pp. 271–292.

[Zave and Schell, 1986] P. ZAVE AND W. SCHELL, "Salient Features of an Executable Specification Language and its Environment," *IEEE Transactions on Software Engineering* **SE-12** (February 1986), pp. 312–325.

Chapter 8 Modularity: and Beyond

The purpose of this chapter is to introduce the theoretical concepts regarding modularity. Reference has been made several times to "modules" even though no definition of a module, formal or informal, has been given. In this chapter the concept of a module is examined. This is not done for the sake of completeness, but rather because the concept of a module, and more specifically, what constitutes a good module, is central to good software design.

8.1 What Is a Module?

When a large product consists of a single monolithic block of code, maintenance is a nightmare. Even for the author of such a monstrosity, attempting to debug the code is extremely difficult; for another programmer to try to understand it is virtually impossible. The solution is to break up the product into smaller pieces, called modules. But what is a module? And furthermore, is the way that the product is broken into modules important in itself, or is the aim of the exercise simply to break up a large product into smaller pieces of code?

One of the earlier attempts to deal with these issues is the work of Stevens, Myers, and Constantine. They defined a module as, "A set of one or more contiguous program statements having a name by which other parts of the system can invoke it, and preferably having its own distinct set of variable names" [Stevens, Myers, and Constantine, 1974]. In other words, a module consists of a single block of code that can be called in the way that a procedure is called. This definition seems to be very broad. It includes procedures of all kinds, whether internal or separately compiled. It includes

COBOL paragraphs and sections, even though they cannot have their own variables, because the definition states that the property of possessing a distinct set of variable names is merely "preferable." But broad as it is, the definition does not go far enough. For example, an assembler macro is not invoked and therefore, by the preceding definition, is not a module. In C, a file of declarations that is included in a product is similarly not invoked; neither is an Ada package nor an Ada generic. In short, the definition given previously is too restrictive.

A broader definition was given by Yourdon and Constantine, namely that, "A module is a lexically contiguous sequence of program statements, bounded by boundary elements, having an aggregate identifier" [Yourdon and Constantine, 1979]. Examples of such boundary elements referred to are **begin**...**end** pairs in a block-structured language like Pascal or Ada, or {...} pairs in C. This definition not only includes all the cases excluded by the previous definition, but is broad enough to be used throughout this book. The reason why it is important to have a broad definition of a module is that most computer users have an intuitive idea of what constitutes a module, and to use a definition that excludes what many people would consider to be a perfectly acceptable module would be counterproductive.

To obtain an insight into the importance of modularization, consider the following somewhat fanciful example. A certain highly incompetent computer architect has still not discovered that both NAND gates and NOR gates are complete, that is, every circuit can be built with only NAND gates or with only NOR gates. He therefore decides to build an ALU, shifter, and 16 registers using AND, OR, and NOT gates. The resulting computer is shown in Figure 8.1. The three components are connected together in a simple fashion.

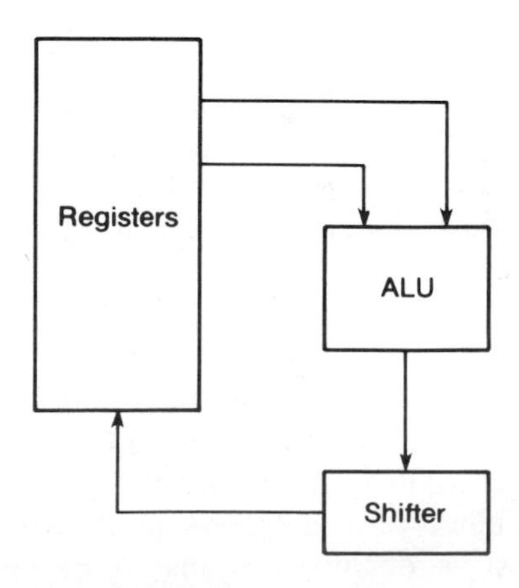

Figure 8.1 Design of computer.

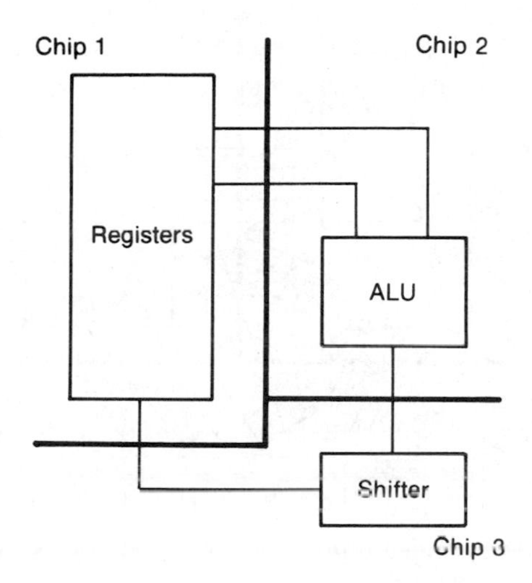

Figure 8.2 Computer of Figure 8.1 fabricated on three chips.

Now our architect friend decides that the circuit should be fabricated on three silicon chips, so he designs the three chips shown in Figure 8.2. One chip has all the gates of the ALU, a second comprises the shifter, and the third is for the registers. At this point he vaguely recalls that some guy in a bar told him that it is best to build chips in such a way that the chip has only one kind of gate. So, starting again with the circuits of Figure 8.1, he redesigns his chips. On chip 1 he puts all the AND gates, on chip 2 all the OR gates, and all the NOT gates go onto chip 3. The resulting "work of art" is shown schematically in Figure 8.3.

Figures 8.2 and 8.3 arc functionally equivalent, that is to say, they do exactly the same thing. But the two designs have markedly different properties. First, it is an understatement that Figure 8.3 is considerably harder to *understand* than Figure 8.2. If the circuits comprising the chips of Figure 8.2 are shown to almost anyone with a knowledge of digital logic, it will be immediately apparent that the chips comprise an ALU, a shifter, and a set of registers. But show Figure 8.3 even to a leading hardware expert and it is unlikely that he or she will be able to understand what the various AND, OR, and NOT gates are doing.

Second, *corrective maintenance* of the circuits shown in Figure 8.3 is difficult. Should there be a design fault in the computer—and anyone capable of coming up with Figure 8.3 is undoubtedly going to make lots and lots of design faults—it will be difficult to determine where the fault is located. On the other hand, if there is a fault in the design of the computer represented by Figure 8.2, the fault can be localized by determining whether the fault appears

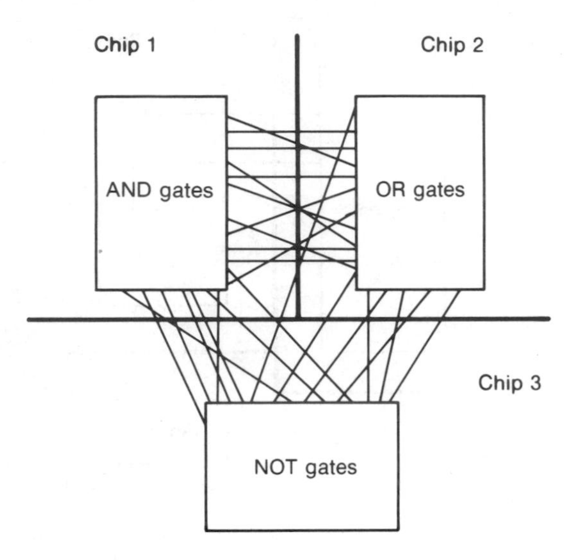

Figure 8.3 Computer of Figure 8.1 fabricated on three other chips.

to be in the way the ALU works, the way the shifter works, or the way the registers work. But if there is something wrong with the design of Figure 8.3, there is no way of determining on which chip the design fault has been made. Similarly, if the computer of Figure 8.2 breaks down, it is relatively easy to determine which chip to replace, while it is probably best to replace all three chips of Figure 8.3 if that computer should break down.

Third, the computer of Figure 8.3 is difficult to *extend* or *enhance*. If a new type of ALU is needed or if faster registers are required, it is back to the drawing board with Figure 8.3. But in the computer of Figure 8.2, all that needs to be done is to replace the appropriate chip.

Perhaps worst of all, the chips of Figure 8.3 cannot be *reused* in any new product. There is no way that those three specific combinations of AND, OR, and NOT gates can be utilized for any product other than the one for which they were designed. But the three chips of Figure 8.2 can, in all probability, be reused in other products that require an ALU, a shifter, or registers.

The point here is that software products have to be designed to look like Figure 8.2, where there is a maximal relationship within each chip and a minimal relationship between chips, and not like Figure 8.3. A module can be likened to a chip in that it performs a function or series of functions and is connected to other modules. The functionality of the product as a whole is fixed; what has to be determined is how to break up the product into modules. Composite/structured design [Stevens, Myers, and Constantine, 1974] provides a rationale for breaking up a product into modules in such a way as to reduce the cost of maintenance which, as pointed out in Chapter 1, is the major component of the total software budget. The maintenance effort, whether corrective, perfective, or adaptive, is reduced when there is maximal interaction within each module and minimal interaction between modules. Or in other words, the aim of composite/structured design (C/SD) is to ensure that the modular decomposition of the product resembles Figure 8.2, rather than Figure 8.3.

Myers quantified the idea of module *cohesion*, the degree of interaction within a module, and module *coupling*, the degree of interaction between modules [Myers, 1978b]. To be more precise, Myers used the term "strength" rather than "cohesion." The reason why "cohesion" is preferable is that modules can have high strength or low strength, and there is something inherently contradictory in the expression "low strength"—if something is not strong, it is weak. In order to prevent such terminological inexactitude detracting from C/SD, the term "cohesion" is now used. Stevens, Myers, and Constantine used the term "binding" in place of "cohesion" [Stevens, Myers, and Constantine, 1974]. Unfortunately, "binding" is also used in other contexts in computer science, such as the binding of values to variables. But "cohesion" does not have these overtones and is therefore preferable.

It is necessary at this point to distinguish between the function of a module, the logic of a module, and the context of a module. The *function* of a module is what it does. For example, the function of module M is to compute the square root of its argument. The *logic* of a module is how the module performs its function; in the case of module M the specific method of computation is, say, Newton's method [Young and Gregory, 1972]. The *context* of a module is the specific usage of that module. For example, module M is used to compute the square root of a double precision integer. A key point in C/SD is that the name assigned to a module is its function, and not its logic or its context. Thus, in C/SD, module M should be named COMPUTE SQUARE ROOT; its logic and its context are irrelevant from the viewpoint of its name.

8.2 Cohesion

Myers defined seven categories or levels of cohesion [Myers, 1978b]. In the light of modern theoretical computer science, the first two levels are considered to be equally good. The resulting ranking is shown in Figure 8.4. This is not a linear scale of any sort. It is merely a relative ranking, a way of determining which types of cohesion are high (good), and which are low (bad).

In order to understand what constitutes a module with high cohesion, it is necessary to start at the other end and consider the lower cohesion levels.

7.	Functional cohesion Informational cohesion	(good)
5.	Communicational cohesion	
4.	Procedural cohesion	
3.	Temporal cohesion	
2.	Logical cohesion	
1.	Coincidental cohesion	(bad)

Figure 8.4 Levels of cohesion.

8.2.1 Coincidental Cohesion

A module has coincidental cohesion if it performs multiple, completely unrelated functions, or if its function cannot be defined, that is, the module must be described in terms of its logic, rather than its function. An example of a module with coincidental cohesion is a module named PRINT NEXT LINE, REVERSE THE STRING OF CHARACTERS COMPRISING THE SECOND PARAMETER, ADD 7 TO THE FIFTH PARAMETER, CONVERT THE FOURTH PARAMETER TO FLOATING POINT.

An obvious question is: How can such modules possibly arise in practice? The most common cause is as a consequence of silly rules such as "every module will consist of between 35 and 50 executable statements." If a software organization insists that modules must neither be too big nor too small, then two undesirable things will happen. First, two or more otherwise ideal smaller modules have to be lumped together to create a larger module of coincidental cohesion. Second, pieces that have to be hacked off well-designed modules that management considers too large will be combined, again resulting in modules with coincidental cohesion. Another cause is when a product is too large to fit into the available storage space and overlay phases have to be created. What happens in practice is that pieces are chopped off here and there and consolidated to form modules of coincidental cohesion.

Why is coincidental cohesion so bad? Modules with coincidental cohesion suffer from two serious drawbacks. First, such modules degrade the maintainability of the product, both corrective maintenance and enhancement. In fact, it has been shown that, from the viewpoint of trying to understand a product, modularization with coincidental cohesion is worse than no modularization at all [Shneiderman and Mayer, 1975]. Second, they are not reusable. It is extremely unlikely that the preceding module with coincidental cohesion could be reused in any other product.

Lack of reusability is an important drawback. The cost of building software is so great that it is essential to try to reuse modules wherever possible. Designing, coding, documenting, and above all, testing a module is a time-consuming, and hence costly, process. If an existing well-designed, thoroughly tested, and properly documented module can be used in another product, then management should insist that the existing module be reused. But there is no way that a module with coincidental cohesion can be reused, and the money devoted to developing it can never be recouped by reusing it in another product.

It is generally easy to solve the problem of having a module of coincidental cohesion—since it performs multiple functions, break it up into separate modules each performing one function.

8.2.2 Logical Cohesion

A module has logical cohesion when it performs a series of related functions, one of which is selected by the calling module. The following are all examples of modules with logical cohesion:

Example 1: Module NEW_OPERATION which is invoked as follows:

```
FUNCTION_CODE := 7;
NEW_OPERATION (FUNCTION_CODE, DUMMY_1, DUMMY_2,
                    DUMMY_3);
-- DUMMY_1, DUMMY_2, and DUMMY_3 are dummy variables,
-- not used if FUNCTION_CODE is equal to 7
```

In this example, NEW_OPERATION is called with four parameters, but as stated in the Ada comment lines, three of them are not needed if FUNCTION_CODE is equal to 7. This degrades readability, with the usual implications for maintenance, both corrective and enhancement.

Example 2: A module performing all input and output.

Example 3: A module performing editing of insertions and deletions and modifications of master file records.

Example 4: A module in an early version of OS/VS2 that performed 13 different functions; its interface contained 21 pieces of data [Myers, 1978b].

There are two problems with modules of logical cohesion. First, the interface is difficult to understand, Example 1 being a case in point, and comprehensibility of the module as a whole may suffer as a result. Second, the code for more than one function may be intertwined, leading to severe maintenance problems. For instance, a module that performs all input and output may be structured as shown in Figure 8.5. If a new tape unit is installed, it may be necessary to modify the sections numbered 1, 2, 3, 4, 6, 9, and 10. These changes may adversely impact other forms of input/output

MODULE PERFORMING ALL INPUT AND OUTPUT
1. Code for all input and output
2. Code for input only
3. Code for output only
4. Code for disk and tape input/output
5. Code for disk input/output
6. Code for tape input/output
7. Code for disk input
8. Code for disk output
9. Code for tape input
10. Code for tape output
⋮ ⋮ ⋮
37. Code for keyboard input

Figure 8.5 Module that performs all input and output.

such as line-printer output, because the line printer will be affected by changes to sections 1 and 3. This intertwined property is characteristic of modules of logical cohesion. A further consequence of this intertwining is that it is difficult to reuse such a module in other products.

8.2.3 Temporal Cohesion

A module has temporal cohesion when it performs a series of functions related in time. An example of a module with temporal cohesion is one named OPEN OLD_MASTER_FILE, NEW_MASTER_FILE, TRANSACTION_FILE, AND PRINT_FILE, INITIALIZE SALES_DISTRICT_TABLE, READ FIRST TRANSACTION_RECORD, READ FIRST OLD_MASTER_FILE_RECORD. In the "bad old days" before C/SD, such a module would be called PERFORM INITIALIZATION.

The functions of this module are weakly related to one another, but more strongly related to functions in other modules. Consider, for example, the SALES_DISTRICT_TABLE. It is initialized in this module, but procedures such as UPDATE SALES_DISTRICT_TABLE and PRINT SALES_DISTRICT_TABLE are located in other modules. Thus, if the structure of the SALES_DISTRICT_TABLE is changed, perhaps because the organization is expanding into areas of the country where it has previously not done business, a number of modules will have to be changed. Not only is there more chance of a regression fault (a fault caused by a change being made to an apparently unrelated part of the product), but if the number of affected modules is large, there is a good chance that one or two modules will be overlooked. It is much better to have all the operations on the SALES_DISTRICT_TABLE in one module; this is shown in Section 8.2.6. In addition, a module with temporal cohesion is unlikely to be reusable in a different product.

8.2.4 Procedural Cohesion

A module has procedural cohesion if it performs a series of functions related by the procedure to be followed by the product. An example of a module with procedural cohesion is READ PART_NUMBER FROM DATABASE AND UPDATE REPAIR_RECORD ON MAINTENANCE FILE.

This is clearly better than temporal cohesion—at least the functions performed have something to do with one another other than the fact that they are performed at the same time. But even so, the functions are still weakly connected, and hence the module is again unlikely to be reusable in another product.

8.2.5 Communicational Cohesion

A module has communicational cohesion if it performs a series of functions related by the procedure to be followed by the product, but in addition all the functions operate on the same data. Two examples of modules with communicational cohesion are UPDATE RECORD IN DATABASE AND WRITE *IT* TO THE AUDIT TRAIL and CALCULATE NEW TRAJECTORY AND SEND *IT* TO THE PRINTER. This is better than procedural cohesion because the functions of the module are more closely connected, but it still has the same drawback as coincidental, logical, temporal, and procedural cohesion, namely lack of reusability.

In passing, it is interesting to note that Berry uses the term "flowchart cohesion" to refer to temporal, procedural, and communicational cohesion because the operations performed by such modules are adjacent in the product flowchart [Berry, 1978]. They are adjacent in the case of temporal cohesion because they are performed at the same time. For procedural cohesion, they are adjacent because the algorithm requires them to be performed in series. Communicational cohesion means that the functions are performed on the same data, and it is natural that these operations should be adjacent in the flowchart.

8.2.6 Informational Cohesion

A module has informational cohesion if it performs a number of functions, each with its own entry point, with independent code for each function, all performed on the same data structure. An example is given in Figure 8.6. This does not violate the tenets of structured programming (Section 10.3); each piece of code has exactly one entry point and one exit, and the code pieces themselves are totally independent. Furthermore, a major difference between logical cohesion and informational cohesion is that the various functions of a module with logical cohesion are intertwined, while in a module with informational cohesion each function is completely independent.

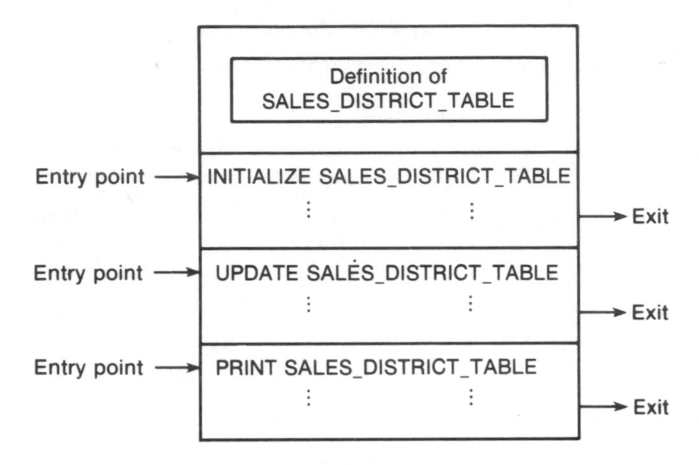

Figure 8.6 Module with informational cohesion.

A module with informational cohesion is essentially an implementation of an abstract data type, as explained in Section 8.5. All the advantages of using an abstract data type are gained when a module of informational cohesion is used. This is discussed further after abstract data types have been introduced.

8.2.7 Functional Cohesion

When a module performs exactly one function or achieves a single goal, then it has functional cohesion. Examples of such modules are GET TEMPERATURE OF FURNACE, COMPUTE ORBITAL OF ELECTRON, WRITE TO FLOPPY DISK, and CALCULATE SALES COMMISSION.

A module with functional cohesion is generally reusable in other contexts. After all, the one function that it performs will often need to be performed in other products, and it makes good economic sense to reuse it. A properly designed, thoroughly tested, and well-documented module of functional cohesion is a valuable asset to any software organization and should be reused as often as possible.

Maintenance is easier with a module of functional cohesion. First, functional cohesion leads to fault isolation. If it is clear that the temperature of the furnace is not being read correctly, then the fault is almost certainly in module GET TEMPERATURE OF FURNACE. Similarly, if the orbital of an electron is incorrectly computed, then the first place to look is COMPUTE ORBITAL OF ELECTRON.

Once the fault has been localized to a single module, the next step is to make the required changes. Since a module of functional cohesion performs one and only one function, such a module will generally be easier to understand than a module with lower cohesion. This ease in understanding also simplifies the maintenance task. Finally, when a change is made, the chance of that change impacting other modules is slight, especially if the coupling between modules is low (Section 8.3).

Functional cohesion is also valuable when a product has to be extended. For example, suppose that a personal computer has an internal floppy disk drive, but that the manufacturer now wishes to market a more powerful model of the computer with an internal hard disk. Reading through the list of modules, the maintenance programmer finds a module named WRITE TO FLOPPY DISK. The obvious thing to do is to throw away that module and replace it with a new one entitled WRITE TO HARD DISK.

In passing, it should be pointed out that the three "modules" of Figure 8.2 have functional cohesion, and the arguments made in Section 8.1 for favoring the design of Figure 8.2 over that of Figure 8.3 are precisely those made in the preceding discussion for favoring functional cohesion.

8.2.8 Cohesion Example

For further insight into cohesion, consider the example shown in Figure 8.7. One module in particular merits comment. The reader may be

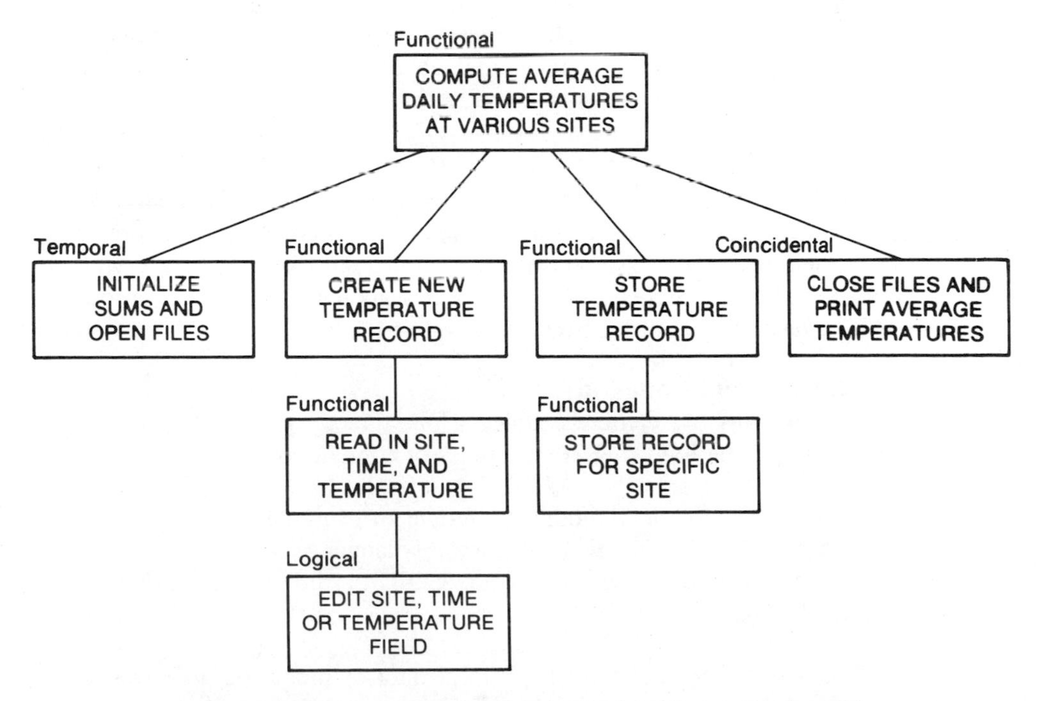

Figure 8.7 Module interconnection diagram showing cohesion of each module.

5. Data coupling (good)
4. Stamp coupling
3. Control coupling
2. Common coupling
1. Content coupling (bad)

Figure 8.8 Levels of coupling.

somewhat surprised that the module INITIALIZE SUMS AND OPEN FILES has been labeled as having temporal cohesion, but the corresponding module CLOSE FILES AND PRINT AVERAGE TEMPERATURES has coincidental cohesion. Why is there this apparent contradiction? INITIALIZE SUMS AND OPEN FILES is a module that performs two functions related in time in that both have to be done before any calculations can be performed, and therefore it has temporal cohesion. While the two functions of CLOSE FILES AND PRINT AVERAGE TEMPERATURES are indeed performed at the end of the calculation, there is another factor involved. Printing the average temperature is related to the problem, while closing files has nothing to do with the problem itself. The rule when two or more different levels of cohesion could be assigned to a module is to assign the lowest possible level. Thus, in the example, since CLOSE FILES AND PRINT AVERAGE TEMPERATURES could have either temporal or coincidental cohesion, the lower of the two cohesions, namely coincidental, is assigned to that module.

8.3 Coupling

Recall that cohesion is the degree of interaction within a module. Coupling is the degree of interaction between two modules. As before, a number of levels can be distinguished, as shown in Figure 8.8. As with cohesion, in order to highlight what constitutes good coupling, the various levels will be described in order from the worst to the best.

8.3.1 Content Coupling

Two modules are content-coupled if one directly references the contents of the other. The following are examples of content coupling:

Example 1: Module A modifies a statement of module B.

This practice is not restricted to assembly language programming. The **alter** verb, now mercifully removed from COBOL, did precisely that; it modified another statement.

Example 2: Module A refers to local data of module B in terms of some numerical displacement within B.

Example 3: Module A branches into a local label of module B.

Suppose that module A and module B are content-coupled. One of the many dangers is that almost any change to B, even recompiling B with a new compiler or assembler, requires a change to A. While content coupling is easy to implement when programming in assembly language, surprisingly enough it is also possible to implement content coupling in Ada through the use of overlays implemented via address clauses [Hammons and Dobbs, 1985]. To be more precise, paragraph 13.5 of the Ada Reference Manual [ANSI/MIL-STD-1815A, 1983] states that the use of address clauses to achieve overlays is erroneous. But it also states in paragraph 1.6(c) that compilers are not required to detect erroneous usages either during compilation or at run time. Thus, while content coupling implemented this way is forbidden in Ada, there is a good chance that it will not be detected either by the compiler or at run time.

8.3.2 Common Coupling

Two modules are common-coupled if they have access to global data. The situation is depicted in Figure 8.9. Instead of communicating with one another by passing parameters, modules CC_A and CC_B can access and change the value of GLOBAL_VARIABLE. The most common situation in which this arises is when CC_A and CC_B both have access to the same database and can both read and write the same record. For common coupling it is necessary that both modules can read *and* write to the database; if the database access mode is read-only, then it is not common coupling. But there are other ways of implementing common coupling, including use of the **common** statement in FORTRAN, the (nonstandard) **common** statement in COBOL, and the **global** statement in COBOL-80.

This form of coupling is undesirable for a number of reasons. First, it contradicts the spirit of structured programming (Section 10.3) in that the resulting code is virtually unreadable. Consider the code fragment shown in Figure 8.10. If GLOBAL_VARIABLE is a global variable, then its value may

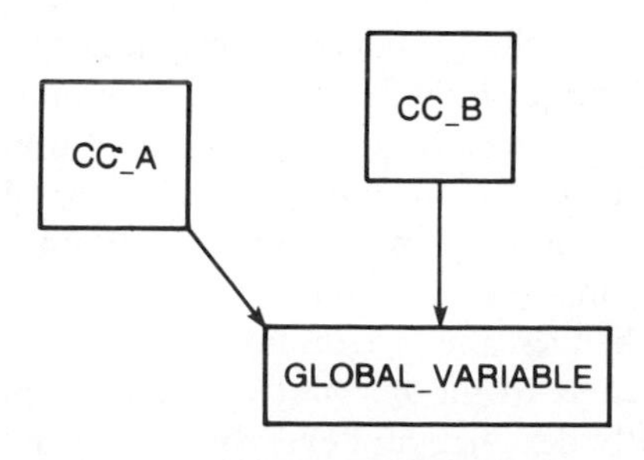

Figure 8.9 Common coupling.

```
while GLOBAL_VARIABLE = 0 loop
   if PARAMETER_XYZ > 25 then
      PROCEDURE_3;
   else
      PROCEDURE_4;
   end if;
end loop;
```

Figure 8.10 Code fragment reflecting common coupling.

be changed by PROCEDURE_3, PROCEDURE_4, or any module called by them. Determining under what conditions the loop terminates is then a nontrivial question; if a run-time error occurs it may be difficult to reconstruct what happened, because any one of a number of modules could have changed the value of GLOBAL_VARIABLE.

A second difficulty is that modules can have side effects, thus affecting their readability. To see this, consider the call EDIT THIS TRANSACTION (RECORD_7). If there is common coupling, this call could change not just the value of RECORD_7, but also any global variable that can be accessed by the module. In short, the entire module must be read to find out precisely what it does.

Third, if in the course of maintenance a change is made in one module to the declaration of a global variable, then every module that can access that global variable has to be changed. Furthermore, all these changes have to be made consistently.

A fourth problem is that modules that are common-coupled are difficult to use in future products. The identical list of global variables has to be supplied each time the module is used, thus virtually precluding the reuse of such a module.

The fifth problem is potentially the most dangerous. As a consequence of common coupling, a module may be exposed to more data than are strictly necessary. This defeats any attempts to control data access and may ultimately lead to computer crime. For many types of computer crime, some form of collusion is needed. A single dishonest programmer in general does not have access to all the data and/or modules needed to commit a crime. For example, a programmer writing that part of a payroll product that prints the checks has to have access to employee records, but in a well-designed product such access will be exclusively in read-only mode, thus precluding the possibility of the programmer making unauthorized changes to his or her monthly salary. In order to make such changes, the programmer has to find some other dishonest employee who has access to the relevant records in update mode. But if the product has been badly designed and every module can access the database in update mode, then an unscrupulous programmer can, without the collusion of anyone else, make unauthorized changes to any record in the database. In other words, the presence of global variables allows anyone to alter such a record unilaterally.

While the previous arguments are hopefully strong enough to dissuade all but the most daring of readers from using common coupling, there are situations where the use of common coupling might seem to be preferable to the alternatives. Consider, for example, a product that performs computer-aided design of petroleum storage tanks [Schach and Stevens-Guille, 1979]. A tank is specified by a large number of descriptors such as height, diameter, maximum wind speed to which the tank will be subjected, and insulation thickness. The descriptors have to be initialized but do not change in value thereafter, and most of the modules in the product need to have access to the values of the descriptors. Suppose that there are 55 tank descriptors. If all these descriptors are passed as parameters to every module, then the interface to each module will consist of at least 55 parameters, and the potential for faults is huge. Even if a language like Ada, which requires strict type checking of parameters, is used, it is still possible to interchange two parameters of the same type, a fault that would not be detected by a type-checker. One solution is to put all the tank descriptors in a database and to design the product in such a way that one module initializes the values of all the descriptors, while the other modules access the database exclusively in read-only mode. However, if the database solution is impractical, perhaps because the specified implementation language cannot be interfaced with the available database management system, the best that can be done is to use common coupling, but in a controlled way. That is to say, the product should be designed so that the 55 descriptors are initialized by one module, but none of the other modules changes the value of a descriptor. This programming style has to be enforced by management, unlike the database solution where enforcement is imposed by the software. Thus, in situations where there is no good alternative to the use of common coupling, close supervision by management can reduce some of the possible risks.

8.3.3 Control Coupling

Two modules are control-coupled if one passes an element of control to the other module, that is, one module explicitly controls the logic of the other. For example, control is passed when a function code is passed to a module of logical cohesion (Section 8.2.2). Another example of control coupling is when a control switch is passed as an argument.

Note that if module A calls module B and B passes back a flag to A that says, "I am unable to complete my task," then B is passing *data*. But if the flag means, "I am unable to complete my task; write error message PQR123," then A and B are control-coupled. In other words, if B passes information back to A and A then decides what action to take as a consequence of receiving that information, then B is passing back data. But if B passes back not only information but also informs module A as to what action A must take, then control coupling is present.

The major difficulty that arises as a consequence of control coupling is that the two modules are not independent; module B, the called module, has to be aware of the internal structure and logic of module A. As a result, the possibility of reusability is reduced. In addition, control coupling is generally associated with modules of logical cohesion. Thus, when there is control coupling, the difficulties associated with logical cohesion will usually also be present.

8.3.4 Stamp Coupling

In some programming languages only simple variables such as PART_NUMBER, SATELLITE_ALTITUDE, or DEGREE_OF_MULTIPROGRAMMING can be passed as parameters. But many languages also support the passing of data structures such as records or arrays as parameters. In such languages valid parameters would include PART_RECORD, SATELLITE_COORDINATES, or SEGMENT_TABLE. Two modules are stamp-coupled if a data structure is passed as a parameter, but the called module operates on some but not all of the individual components of that data structure.

Consider, for example, the call CALCULATE WITHHOLDING (EMPLOYEE_RECORD). It is not clear, without reading the entire CALCULATE WITHHOLDING module, which fields of the employee record it accesses or changes. Passing it the employee's salary is obviously essential for computing the withholding, but it is difficult to see how, say, the employee's home telephone number is needed for this purpose. Instead, module CALCULATE WITHHOLDING should be passed only those fields it actually needs for computing the withholding. Not only is the resulting module, and particularly its interface, easier to understand, but it is likely to be reusable in a variety of other products that compute withholding.

Perhaps even more important, since the call CALCULATE WITHHOLDING (EMPLOYEE_RECORD) passes more data than are strictly necessary, the problems of uncontrolled data access, and conceivably computer crime, once again can arise. This issue was discussed in Section 8.3.2.

There is nothing at all wrong with passing a data structure as a parameter, provided that all the components of the data structure are accessed and/or changed, and not just some of them. For example, calls like INVERT MATRIX (ORIGINAL_MATRIX, INVERTED_MATRIX) or PRINT INVENTORY_RECORD (WAREHOUSE_RECORD) pass a data structure as a parameter, but they operate on all the components of that data structure. Stamp coupling is present when a data structure is passed as a parameter, but only some of the components are used by the called module.

A subtle form of stamp coupling can occur when a pointer to a record is passed as a parameter. Consider the call CHECK ALTITUDE (POINTER_TO_POSITION_RECORD). At first sight, what is being passed is a simple variable. But the called module has access to all of the fields in the POSITION_RECORD pointed to by POINTER_TO_POSITION_RECORD. Because of the potential problems, it is a good idea to examine the coupling closely whenever a pointer is passed as a parameter.

8.3.5 Data Coupling

Two modules are data-coupled if all parameters are homogeneous data items. That is to say, every parameter is either a simple parameter or a data structure all of whose elements are used by the called module. Examples include DISPLAY TIME OF ARRIVAL (FLIGHT_NUMBER), COMPUTE PRODUCT (FIRST_NUMBER, SECOND_NUMBER), and DETERMINE JOB WITH HIGHEST PRIORITY (JOB_QUEUE).

Data coupling should be the goal of every design. To put it in a negative way, if a product exhibits data coupling exclusively, then the difficulties of content, common, control, and stamp coupling will not be present. From a more positive viewpoint, if two modules are data-coupled, then maintenance is easier, because a change to one module is less likely to cause a regression fault in the other.

To clarify certain aspects of coupling, an example is now given.

8.3.6 Coupling Example

Consider the example shown in Figure 8.11. The numbers on the arcs represent interfaces that are defined in greater detail in Figure 8.12. Thus, for example, when module P calls module Q (interface 1) it passes one parameter,

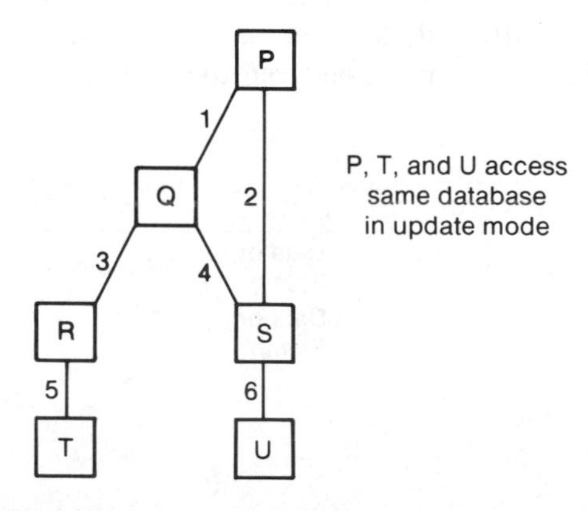

Figure 8.11 Module interconnection diagram for coupling example.

Interface number	In	Out
1	Aircraft type	Status flag
2	—	List of aircraft parts
3	Function code	—
4	—	List of aircraft parts
5	Part number	Part manufacturer
6	Part number	Part name

Figure 8.12 Interface description for Figure 8.11.

namely the type of the aircraft. When Q returns control to P, it passes a status flag back. Using the information in Figures 8.11 and 8.12, the coupling between every pair of modules can be deduced. The results are shown in Figure 8.13.

Some of the entries in Figure 8.13 are obvious. For instance, the data coupling between P and Q (interface 1 in Figure 8.11), between R and T (interface 5), and between S and U (interface 6) is a direct consequence of the fact that a simple variable is passed in each direction. The coupling between P and S (interface 2) would be data coupling if all the elements of the list of parts passed from S to P are used and/or updated, but it is stamp coupling if P operates on only certain elements of the list. The coupling between Q and S (interface 4) is similar. Since it is not clear from Figures 8.11 and 8.12 as to precisely what the various modules do, there is no way of determining whether the coupling is data or stamp. The coupling between Q and R (interface 3) is control coupling, because a function code is passed from Q to R.

Perhaps somewhat surprising are the three entries marked common coupling in Figure 8.13. The three modules in Figure 8.11 that are furthest apart, namely P and T, P and U, and T and U, at first appear not to be coupled in any way. After all, there is no interface of any kind connecting them, so the very idea of coupling between them, let alone common coupling,

	Q	R	S	T	U
P	Data	—	Data or Stamp	Common	Common
Q		Control	Data or Stamp	—	—
R			—	Data	—
S				—	Data
T					Common

Figure 8.13 Coupling between pairs of modules of Figure 8.11.

requires some explanation. The answer lies in the annotation on the right-hand side of Figure 8.11, namely that P, T, and U all access the same database in update mode. The result is that there are a number of global variables that can be changed by all three modules, and hence they are pairwise common-coupled.

Hopefully, the reader has been convinced that a design in which modules have high cohesion and low coupling is a good design. The obvious question is then: How can such a design be achieved? Since this chapter is devoted to theoretical concepts surrounding design, the answer to the question, namely data flow analysis, is presented in Section 9.3. In the meantime, those qualities that distinguish a good design are further examined and refined.

8.4 Data Encapsulation

Consider the problem of designing an operating system for a large mainframe computer. It has been decided that any job submitted to the computer will be classified as high priority, medium priority, or low priority. Scheduling, such as deciding which job to load next into memory when space becomes available, deciding which of the jobs in memory gets the next time slice and how long that time slice should be, or deciding which of the jobs that require disk access has priority, must take into account the class of each job; the higher the priority of a job, the sooner it should be assigned the resources of the computer. One way of achieving this is to maintain separate job queues for each job class. The job queues have to be initialized, and facilities must exist for adding a job to a job queue when the job requires memory, CPU time, or disk access, as well as removing a job from a queue when the operating system decides to allocate the required resource to that job.

To simplify matters, consider the restricted problem of batch jobs queuing up for memory access. There are three queues for incoming batch jobs, one for each job type. When a job is submitted by a user, the job is added to the appropriate queue, and when the operating system decides that a job is ready to be run, it is removed from its queue and memory is allocated to it.

There are a number of different ways to build this portion of the product. One possible design is shown in Figure 8.14, which depicts modules for manipulating one of the three job queues. Here, the procedure of module M_1 is responsible for the initialization of the job queue, and addition and deletion of jobs are done by the procedures of modules M_2 and M_3, respectively. Module M_123 calls all three procedures in order to manipulate the job queue. Note that in order to concentrate on data encapsulation, issues such as underflow (trying to remove a job from an empty queue) and overflow (trying to add a job to a full queue) have been suppressed here, as well as in the remainder of this chapter.

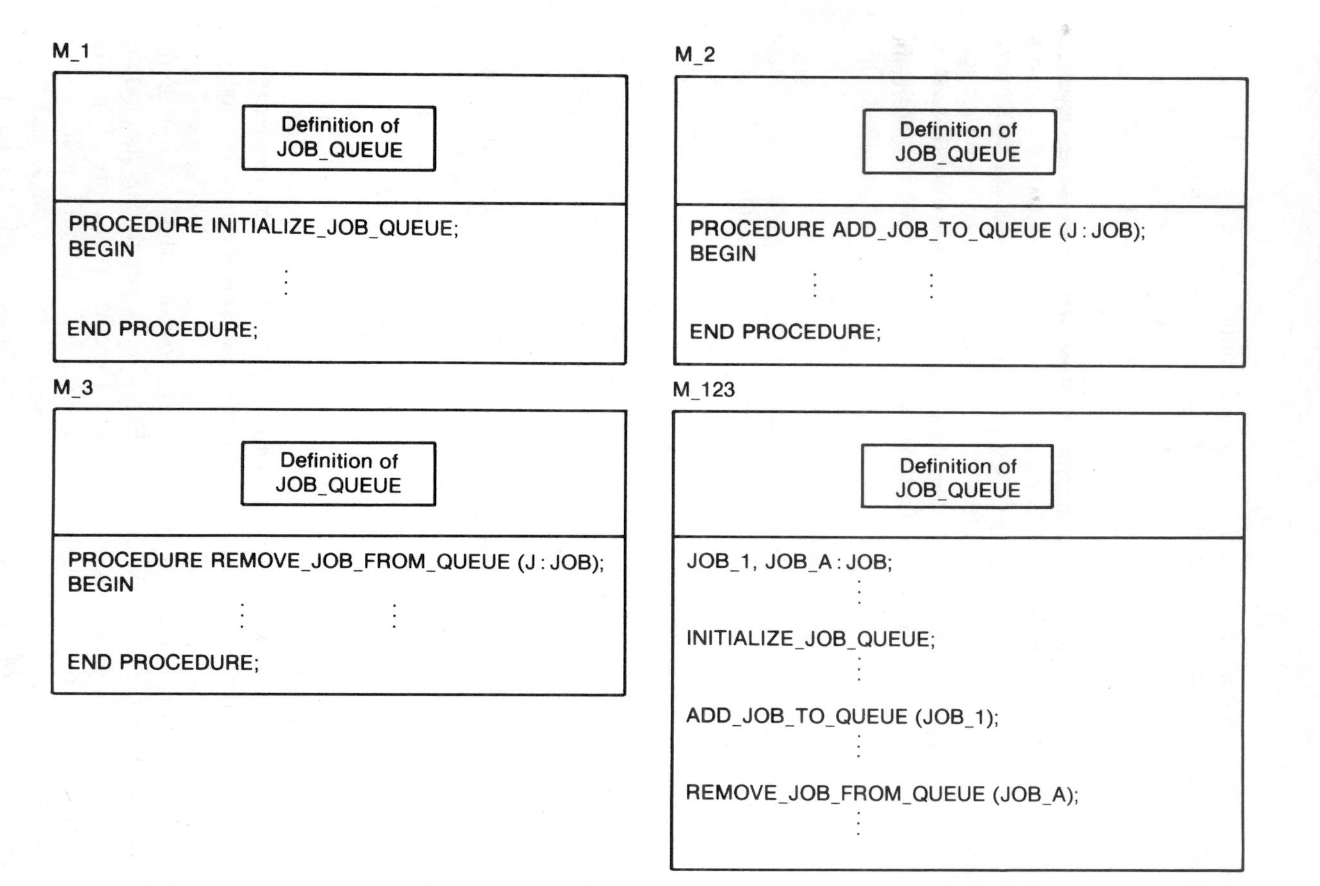

Figure 8.14 One possible design of job queue portion of operating system.

The modules of the design of Figure 8.14 have low cohesion, because operations on the job queue are spread all over the product. If a decision is made to change the way JOB_QUEUE is implemented, for example, as a linear list of records instead of as a linked list, then modules M_1, M_2, and M_3 have to be drastically revised, and M_123 also has to be changed.

Now suppose that the design of Figure 8.15 is chosen instead. The module on the right-hand side of the figure has informational cohesion (Section 8.2.6) in that it performs a number of operations, each with its own entry point and exit, with independent code for each function, all performed on the same data structure. Module M_ENCAPSULATION in Figure 8.15 is an implementation of *data encapsulation*, that is to say, a data structure, in this case the job queue, together with the operations to be performed on that data structure.

An obvious question to ask at this point is: What is the advantage of designing a product using data encapsulation? This will be answered in two ways, namely from the viewpoint of development and from the viewpoint of maintenance.

8.4.1 Data Encapsulation and Product Development

Data encapsulation is an example of *abstraction*. Returning to the job queue example, a data structure, namely the job queue, has been defined, together with three operations, namely initialize the job queue, add a job to the queue, and delete a job from the queue. The developer is able to conceptualize the problem at a higher level, the level of jobs and job queues, rather than at the lower level of records or arrays.

The basic theoretical concept behind abstraction is, once again, stepwise refinement. First, a design for the product is produced in terms of high-level concepts such as jobs, job queues, and the operations that are performed on job queues. At this stage it is entirely irrelevant how the job queue is implemented. Once a complete design has been obtained, the second step is to design the lower-level components in terms of which the data structure and operations on the data structure will be implemented. The data structure, that is, the job queue, will be implemented in terms of records or arrays; the three operations, namely initialize the job queue, add a job to the queue, and remove a job from the queue, will be implemented as procedures. The key point is that while this lower level is being designed, the designer totally ignores what use will be made of the jobs, job queue, and operations. Thus during the first step the existence of the lower level is assumed, even though at this stage no thought at all has been given to the lower level; during the second step, namely the design of the lower level, the existence of the higher level is ignored. At the higher level the concern is with the behavior of the data structure, namely the job queue; at the lower level the implementation of that behavior is the primary concern. Of course, in a larger product there

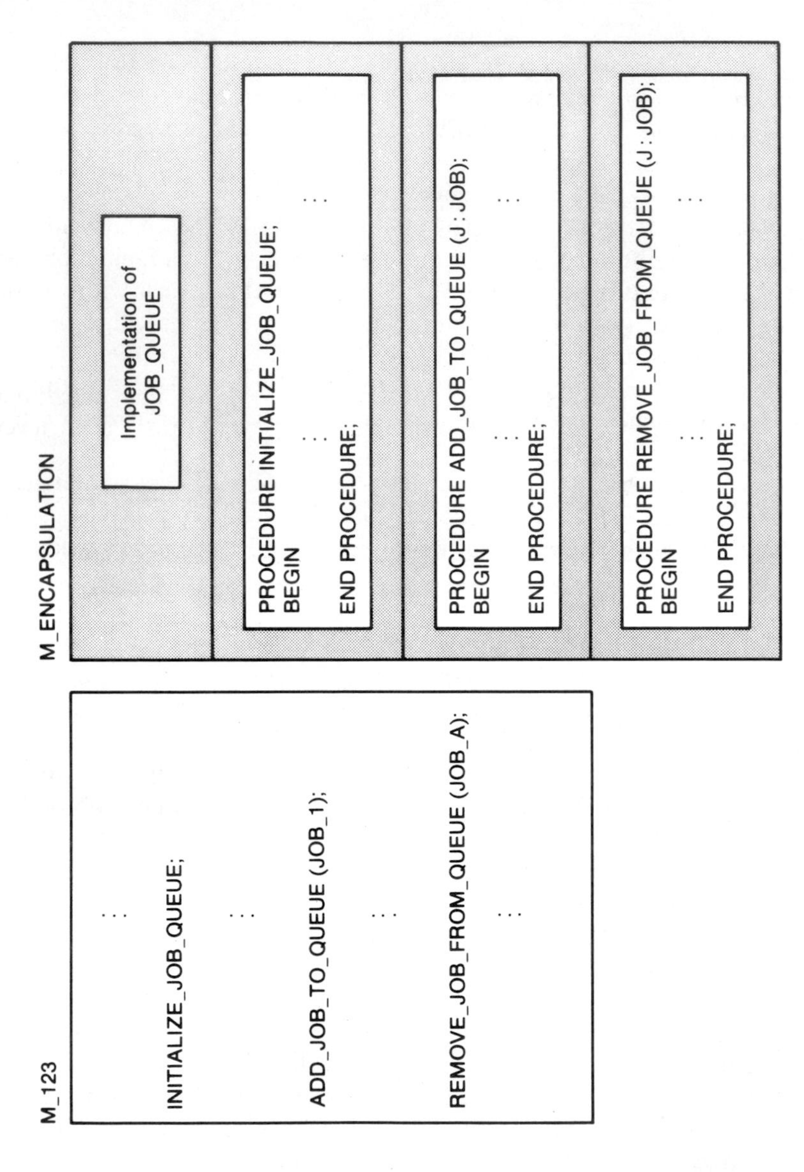

Figure 8.15 Design of job queue portion of operating system using data encapsulation.

will be many levels of abstraction. Examples of such multilevel products are now given.

The concept of levels of abstraction was described by Dijkstra in a different context in terms of layers of an operating system [Dijkstra, 1968b]. The highest level, level 5, is the computer operator. Level 4 is the user program level. Each of the higher levels of abstraction is then implemented using some of the functions of the levels below it. For example, level 1, the memory segment controller, utilizes some of the functions of level 0, where clock interrupt handling and processor allocation are implemented. If the implementation of a lower-level function is changed, then provided that the interface between that function and the rest of the machine is not altered, that is, the number and types of the parameters are not changed, the higher levels should not even be aware that the change was made.

The same concept can be applied to computer organization. A computer can be thought of as consisting of levels of abstract machines [Tanenbaum, 1984]. The highest level, level 5, is the problem-oriented language level (Figure 8.16). Level 4 is the assembly language level. Level 3 is the operating system level, and level 2 is the machine code level. Suppose a product is written in C (level 5). It is compiled down to machine code (level 2). However, some compilers translate from C (level 5) to assembler (level 4), and from assembler to machine code. Level 1 is the microprogramming level; machine code (level 2) is interpreted by the microprogram of level 1. The microcode is directly executed by the hardware of level 0, the digital logic level. This level is implemented in terms of silicon chips. Again there is independence of the various levels. Provided that the functionality and interfaces are not changed, it is possible to replace the digital logic level by faster hardware without altering any other aspect of the computer. The microprogram can be rewritten to achieve further speed improvement, and, provided again that the functionality and interfaces are not changed, this should not require any sort of change to products at level 5 or level 4, to the operating system level (level 3), or to the machine code level, level 2. By designing and implementing a computer in this way, each level of abstraction is independent. This allows a user working at level L to think conceptually at level L without having to descend to a lower level. Also, changes made at one level should not impact other levels in any way.

Level 5: Problem-oriented language level
Level 4: Assembly language level
Level 3: Operating system level
Level 2: Machine code level
Level 1: Microprogramming level
Level 0: Digital logic level

Figure 8.16 Levels of abstraction of computer.

The description of abstraction in this section has glossed over the fact that different types of abstraction exist. Consider Figure 8.15. In that figure there are two types of abstraction. Data encapsulation, that is, a data structure, together with the operations to be performed on that data structure, is an example of *data abstraction*; the procedures are an example of *procedural abstraction*. Abstraction, as stated previously, is simply a means of achieving stepwise refinement by suppressing unnecessary details and accentuating relevant details. In the case of data abstraction, data encapsulation allows the designer to think at the level of the data structure and the operations performed on it, and only later be concerned with the details of how that data structure and operations are to be implemented. Turning now to procedural abstraction, consider the result of defining a procedure, INITIALIZE_JOB_QUEUE, say. The effect is to extend the language by supplying the developer with another operation, one that is not part of the language as originally defined. The developer can now use INITIALIZE_JOB_QUEUE in the same way as predefined procedures like **sqrt** or **tan**.

The implications of procedural abstraction for design are as powerful as those of data abstraction. The designer can conceptualize the product in terms of high-level procedures. At the next level, these procedures can be defined in terms of lower-level procedures, until, finally, the lowest level is reached. At this level the procedures are expressed in terms of the predefined constructs of the programming language in which the product is to be coded. At each level the designer is concerned only with expressing the product in terms of procedures appropriate to that level. The designer can ignore the level below, which will be handled at the next level of abstraction, that is, the next refinement step. The designer can also ignore the level above, a level that is irrelevant from the viewpoint of designing the current level.

8.4.2 Data Encapsulation and Product Maintenance

Approaching data encapsulation from the viewpoint of maintenance, a basic issue is to identify what aspects of the product are likely to change and to design the product in such a way as to minimize the effects of change. Data structures as such are unlikely to change; if a product includes, say, job queues, then it is likely that future versions will incorporate them. At the same time, the way that job queues are implemented may well change, and data encapsulation provides a means of coping with that change.

An Ada implementation of M_ENCAPSULATION (Figure 8.15) is in two parts. Figure 8.17 shows the package specification, that is to say, information about the package that is visible to other modules. Figure 8.18 is the package body. It incorporates the implementation of the data structure JOB_QUEUE as an array of up to 25 job numbers, as well as the implementation of the three procedures. Information in the package body is inaccessible to other modules. It should be pointed out that the code examples in this

```
package M_ENCAPSULATION is

   procedure INITIALIZE_JOB_QUEUE;
   procedure ADD_JOB_TO_QUEUE (JOB_NUMBER : in INTEGER);
   procedure REMOVE_JOB_FROM_QUEUE (JOB_NUMBER : out INTEGER);

end M_ENCAPSULATION;
```

Figure 8.17 Package specification of Ada implementation of M_ENCAPSULATION.

```
package body M_ENCAPSULATION is

   type QUEUE_ARRAY is array (1 .. 25) of INTEGER;

   JOB_QUEUE : QUEUE_ARRAY;
        -- the job queue contains up to 25 job numbers

   QUEUE_LENGTH : INTEGER range 0 .. 25;
        -- number of jobs currently in the job queue

   procedure INITIALIZE_JOB_QUEUE is
   begin
      QUEUE_LENGTH := 0;
   end INITIALIZE_JOB_QUEUE;

   procedure ADD_JOB_TO_QUEUE (JOB_NUMBER : in INTEGER) is
   begin
        -- add job to end of job queue
      QUEUE_LENGTH := QUEUE_LENGTH + 1;
      JOB_QUEUE(QUEUE_LENGTH) := JOB_NUMBER;
   end ADD_JOB_TO_QUEUE;

   procedure REMOVE_JOB_FROM_QUEUE (JOB_NUMBER : out INTEGER) is
   begin
        -- remove job from head of queue and move up remaining jobs
      JOB_NUMBER := JOB_QUEUE(1);
      QUEUE_LENGTH := QUEUE_LENGTH - 1;
      for K in 1 .. QUEUE_LENGTH loop
          JOB_QUEUE(K) := JOB_QUEUE(K + 1);
      end loop;
   end REMOVE_JOB_FROM_QUEUE;

end M_ENCAPSULATION;
```

Figure 8.18 Package body of Ada implementation of M_ENCAPSULATION.

```
with M_ENCAPSULATION; use M_ENCAPSULATION;
procedure M_123 is
  JOB_1, JOB_A : INTEGER;
begin
    -- various statements
  INITIALIZE_JOB_QUEUE;
    -- more statements
  ADD_JOB_TO_QUEUE(JOB_1);
    -- still more statements
  REMOVE_JOB_FROM_QUEUE(JOB_A);
    -- further statements
end M_123;
```

Figure 8.19 Ada implementation of M_123.

chapter were deliberately written in such a way as to highlight data abstraction issues, at the cost of what is considered good programming practice. For example, the number 25 in the definition of **type** QUEUE_ARRAY should be coded as a **constant**, and parameter JOB_NUMBER should be a **subtype** rather than an INTEGER.

Figure 8.19 shows how module M_123 may be implemented. The **with** clause in Figure 8.19 is essentially the mechanism that enables M_123 to make use of those portions of M_ENCAPSULATION that are accessible to it. The **use** clause allows an abbreviated form of names declared in M_ENCAPSULATION to be used in M_123. For example, the name INITIALIZE_JOB_QUEUE can be used in M_123; there is no need to write it out in full as M_ENCAPSULATION.INITIALIZE_JOB_QUEUE.

Now suppose that the implementation of JOB_QUEUE is to be changed to a two-way linked list of job records. Each job record will have three components, the job number as before, a pointer to the job record in front of it in the linked list, and a pointer to the job record behind. Because this formulation is recursive in that the definition of a job record includes a pointer to a job record, such a job record must be defined in Ada in two stages. First, an incomplete type declaration of JOB_RECORD is made, followed by a type declaration of LINK, a pointer (access type) to a JOB_RECORD.

```
type JOB_RECORD;
            -- incomplete type declaration
type LINK is access JOB_RECORD;
            -- LINK is a pointer to a JOB_RECORD
```

Now JOB_RECORD can be defined.

```
type JOB_RECORD is
   record
      JOB_NO : INTEGER;
            -- number of the job is stored in this component
      IN_FRONT, IN_REAR : LINK;
            -- pointers to the job records in front and behind
   end record;
```

The key point is: What changes have to be made as a result of this modification? In fact, only the package body of M_ENCAPSULATION has to be changed; the changed package body is shown in Figure 8.20. The package specification need not be changed in any way. After all, the implementation of JOB_QUEUE is in the body of the package, not the specification. The source code of module M_123 (Figure 8.19) need not be changed since that module merely makes use of package M_ENCAPSULATION.

The use of packages or their equivalent in other programming languages such as modules in Modula-2 [Wirth, 1985] is a powerful tool that supports the implementation of data abstraction in such a way that product maintenance is simplified.

8.5 Abstract Data Types

The difficulty with both implementations of M_ENCAPSULATION proposed for the job queue problem (Figures 8.18 and 8.20) is that they apply only to one queue. Rather than specifying a data *structure* together with the relevant operations, it would be more useful to have a data *type*, together with the operations to be performed on instantiations of that data type. Such a construct is called an *abstract data type*. An Ada implementation of the job queue abstract data type is shown in Figures 8.21 and 8.22; the use of this package is shown in Figure 8.23. (Notation J_Q.QUEUE_LENGTH denotes the QUEUE_LENGTH component of record J_Q, and similarly for J_Q.JOB_QUEUE.)

Abstract data types are a widely applicable design tool. For example, suppose that a product is to be written in which a large number of operations have to be performed on rational numbers, that is, numbers that can be represented in the form N / D, where N and D are integers, $D \neq 0$. Rational numbers can be represented in a number of different ways such as two elements of a one-dimensional array of integers or as two components of a record. To implement rational numbers in terms of an abstract data type, a suitable representation for the data structure is chosen. In Ada it could be

```
package body M_ENCAPSULATION is

  type JOB_RECORD;                    -- incomplete type declaration
  type LINK is access JOB_RECORD;     -- LINK is a pointer to a JOB_RECORD

  type JOB_RECORD is
     record
       JOB_NO : INTEGER;              -- number of the job is stored in this component
       IN_FRONT, IN_REAR : LINK;      -- pointers to the job records in front and behind
     end record;

  FRONT_OF_QUEUE, REAR_OF_QUEUE : LINK;

  procedure INITIALIZE_JOB_QUEUE is
  begin
     FRONT_OF_QUEUE := null;
     REAR_OF_QUEUE := null;
  end INITIALIZE_JOB_QUEUE;

  procedure ADD_JOB_TO_QUEUE (JOB_NUMBER : in INTEGER) is
  begin
          -- add job to rear of job queue
     REAR_OF_QUEUE := new JOB_RECORD'(JOB_NO => JOB_NUMBER,
                                      IN_FRONT => REAR_OF_QUEUE,
                                      IN_REAR => null);
          -- create a new job record, place JOB_NUMBER in the JOB_NO field, and
          -- set IN_FRONT equal to the current value of REAR_OF_QUEUE,
          -- thereby linking the new record to the rear of the queue.
          -- REAR_OF_QUEUE now points to this new record.
     if FRONT_OF_QUEUE = null then
       FRONT_OF_QUEUE := REAR_OF_QUEUE;
          -- special case of first record
     end if;
  end ADD_JOB_TO_QUEUE;

  procedure REMOVE_JOB_FROM_QUEUE (JOB_NUMBER : out INTEGER) is
  begin
          -- set JOB_NUMBER to the contents of the JOB_NO field of the record
          -- at the front of the queue, update FRONT_OF_QUEUE pointer,
          -- update the IN_FRONT field of what is now the record at the head of the queue
     JOB_NUMBER := FRONT_OF_QUEUE.JOB_NO;
     FRONT_OF_QUEUE := FRONT_OF_QUEUE.IN_REAR;
     if FRONT_OF_QUEUE /= null then
       FRONT_OF_QUEUE.IN_FRONT := null;
     else
       REAR_OF_QUEUE := null;
          -- special case of last record
     end if;
  end REMOVE_JOB_FROM_QUEUE;

end M_ENCAPSULATION;
```

Figure 8.20 Implementation of package body of M_ENCAPSULATION using two-way linked list.

```
package JOB_QUEUE_ADT is

 type QUEUE_ARRAY is array (1 .. 25) of INTEGER;

 type JOB_QUEUE_TYPE is
    record
      JOB_QUEUE : QUEUE_ARRAY;
      QUEUE_LENGTH : INTEGER range 0 .. 25;
        -- moved from package body of Figure 8.18 to specification here
    end record;

 procedure INITIALIZE_JOB_QUEUE (J_Q : in out JOB_QUEUE_TYPE);
 procedure ADD_JOB_TO_QUEUE (JOB_NUMBER : in INTEGER;
                                                     J_Q : in out JOB_QUEUE_TYPE);
 procedure REMOVE_JOB_FROM_QUEUE (JOB_NUMBER : out INTEGER;
                                                            J_Q : in out JOB_QUEUE_TYPE);
end JOB_QUEUE_ADT;
```

Figure 8.21 Package specification of job queue implemented as abstract data type.

```
package body JOB_QUEUE_ADT is

 procedure INITIALIZE_JOB_QUEUE (J_Q : in out JOB_QUEUE_TYPE) is
 begin
    J_Q.QUEUE_LENGTH := 0;
 end INITIALIZE_JOB_QUEUE;

 procedure ADD_JOB_TO_QUEUE (JOB_NUMBER : in INTEGER;
                                                     J_Q : in out JOB_QUEUE_TYPE) is
 begin
          -- add JOB_NUMBER to end of job queue
    J_Q.QUEUE_LENGTH := J_Q.QUEUE_LENGTH + 1;
    J_Q.JOB_QUEUE(J_Q.QUEUE_LENGTH) := JOB_NUMBER;
 end ADD_JOB_TO_QUEUE;

 procedure REMOVE_JOB_FROM_QUEUE (JOB_NUMBER : out INTEGER;
                                                            J_Q : in out JOB_QUEUE_TYPE) is
 begin
          -- remove job from head of queue, and move up remaining jobs
    JOB_NUMBER := J_Q.JOB_QUEUE(1);
    J_Q.QUEUE_LENGTH := J_Q.QUEUE_LENGTH - 1;
    for K in 1 .. J_Q.QUEUE_LENGTH loop
       J_Q.JOB_QUEUE(K) := J_Q.JOB_QUEUE(K + 1);
    end loop;
 end REMOVE_JOB_FROM_QUEUE;

end JOB_QUEUE_ADT;
```

Figure 8.22 Package body of job queue implemented as abstract data type.

```
with JOB_QUEUE_ADT; use JOB_QUEUE_ADT;
procedure M_123 is
  JOB_1, JOB_A : INTEGER;
  J_Q_3 : JOB_QUEUE_TYPE;
begin
    -- various statements
  INITIALIZE_JOB_QUEUE(J_Q_3);
    -- more statements
  ADD_JOB_TO_QUEUE(JOB_1, J_Q_3);
    -- still more statements
  REMOVE_JOB_FROM_QUEUE(JOB_A, J_Q_3);
    -- further statements
end M_123;
```

Figure 8.23 Module making use of abstract data type of Figures 8.21 and 8.22.

defined as

```
type RATIONAL_NUMBER is
  record
    N : INTEGER;
    D : INTEGER;
  end record;
```

Then the various operations that are performed with rational numbers, such as constructing a rational number from two integers, adding two rational numbers, and multiplying two rational numbers, together with the data type RATIONAL_NUMBER, constitute an abstract data type.

Abstract data types support both data abstraction and procedural abstraction (Section 8.4.1). In addition, even if a product is drastically modified, it is unlikely that the abstract data types will be changed; at worst, additional procedures may have to be added to an abstract data type. Thus, both from the product development and the product maintenance viewpoints, abstract data types are an attractive tool for software producers.

8.6 Information Hiding

In addition to data abstraction and procedural abstraction, there is in fact a third type of abstraction, namely iteration abstraction [Liskov and Guttag, 1986]. Iteration abstraction allows a programmer to specify at a higher level that a loop is to be used, and then to describe at a lower level the exact elements over which the iteration is to be performed, and the order in which the elements are to be processed. These three types of abstraction, namely data abstraction, procedural abstraction, and iteration abstraction, are, in turn, instances of a more general design concept put forward by Parnas, namely *information hiding* [Parnas, 1971, 1972a, 1972b]. Parnas' ideas are directed towards future maintenance. Before a product is designed, a list should be made of those design decisions that are likely to change in the future. Modules should then be designed in such a way that those design decisions are hidden from other modules. Thus, if a change has to be made in the future, such a change is localized to one specific module. Since the implementation of the original design decision is not visible to other modules, changing the design clearly cannot affect any other module.

To see how these ideas can be used in practice, first consider the data encapsulation implementation of Figures 8.17 and 8.18. All that is visible to a user of the package are the names of the procedures that operate on a job queue. There is true information hiding. Modules that use this package have no information at all about the way that a job queue is implemented.

The most important difference between the data encapsulation implementation of Figures 8.17 and 8.18 and the abstract data type implementation of Figures 8.21 and 8.22 is that the definitions

```
JOB_QUEUE : QUEUE_ARRAY;
QUEUE_LENGTH : INTEGER range 0..25;
```

in the package *body* of Figure 8.18 now appear in the definition of the record type JOB_QUEUE_TYPE in the package *specification* of Figure 8.21. In other words, in Ada there is apparently a price to pay for the increase in abstraction, namely a decrease in information hiding; but see the following discussion. Full details of JOB_QUEUE are now freely accessible. For example, it is perfectly legal in Ada for a module to tamper with a job queue by declaring a variable J_Q of type JOB_QUEUE_TYPE and then to use the assignment statement

```
J_Q.JOB_QUEUE(7) := -5678;
```

Recall that record J_Q has two components, namely array JOB_QUEUE and integer QUEUE_LENGTH. The above statement causes JOB_QUEUE(7) to be set equal to −5678. Thus it is possible to alter the contents of a job queue without using any of the three procedures of the package. In addition to the implications this might have with regard to lowering cohesion and increasing coupling, unless management is aware of the fact that what was previously a secure data structure is now no longer hidden, there could be consequences with regard to computer fraud.

Fortunately, there is a way out. The designers of Ada provided for information hiding even within a package specification. This is shown in Figure 8.24 (the package body of Figure 8.22 is not changed). Now the only information visible to other modules is the fact that JOB_QUEUE_TYPE is a

```
package JOB_QUEUE_ADT is
 type QUEUE_ARRAY is array (1 .. 25) OF INTEGER;

 type JOB_QUEUE_TYPE is private;                                                -- new
        -- all that a module using this package knows is that JOB_QUEUE_TYPE is a type

 procedure INITIALIZE_JOB_QUEUE (J_Q : in out JOB_QUEUE_TYPE);
 procedure ADD_JOB_TO_QUEUE (JOB_NUMBER : in INTEGER;
                             J_Q : in out JOB_QUEUE_TYPE);
 procedure REMOVE_JOB_FROM_QUEUE (JOB_NUMBER : out INTEGER;
                                  J_Q : in out JOB_QUEUE_TYPE);

 private                                                                        -- new
    type JOB_QUEUE_TYPE is
      record
        JOB_QUEUE : QUEUE_ARRAY;
        QUEUE_LENGTH : INTEGER range 0 .. 25;
      end record;
        -- this definition of JOB_QUEUE_TYPE, even though in the specification,
        -- is not visible to other modules
end JOB_QUEUE_ADT;
```

Figure 8.24 Package specification achieving information hiding via private type (package body appears in Figure 8.22).

type and that there are three operations, with specified interfaces, that can operate on job queues. But the exact way that job queues are implemented is **private**, that is to say, invisible to the outside. This is diagrammatically depicted in Figure 8.25. Thus packages with private types enable Ada users to implement abstract data types with full information hiding. In other words, packages with private types enable Ada users to implement abstract data types without sacrificing any of the information hiding achievable through data encapsulation.

8.7 Objects

Object-oriented design is currently a major topic in computer science. There is an annual conference on object-oriented programming systems, languages, and applications (abbreviated OOPSLA) attended by over a thousand participants. The term "object-oriented" is starting to be used as widely, and as inaccurately, as the buzzword "structured" was used a few years ago.

An incomplete definition of object-oriented design is that the product is designed in terms of abstract data types, and the variables ("objects") are instantiations of abstract data types. But defining an object as an instantiation of an abstract data type is too simplistic. Something more is needed, namely *inheritance*, a concept first introduced in Simula 67 [Dahl, Myrhaug, and

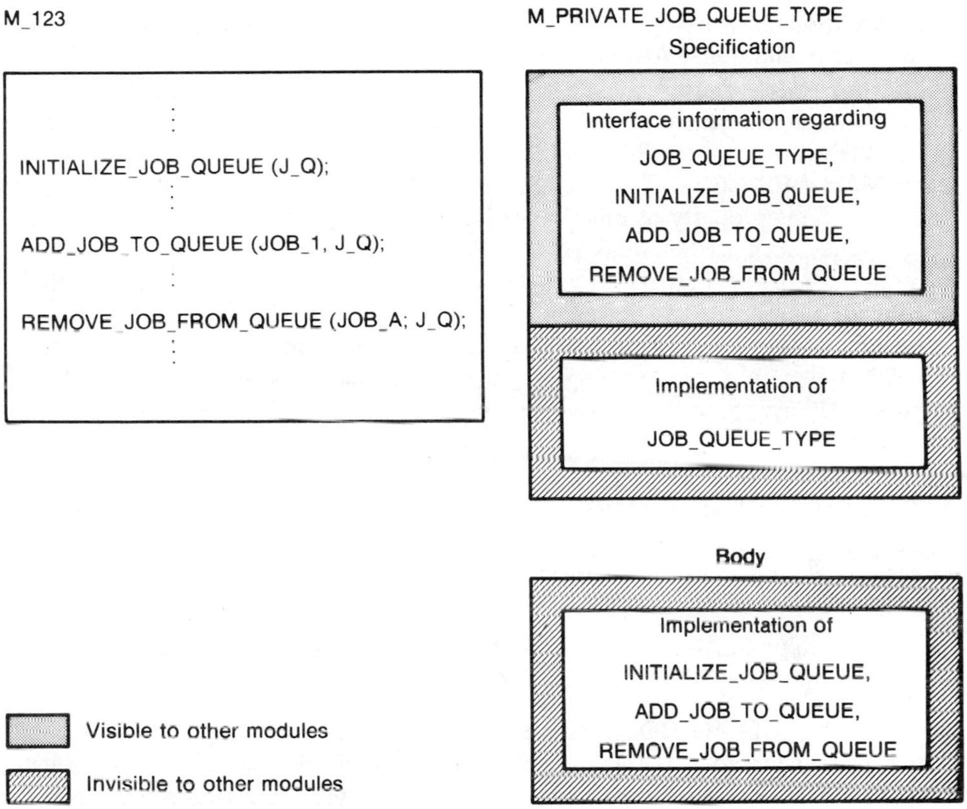

M_123

INITIALIZE_JOB_QUEUE (J_Q);

ADD_JOB_TO_QUEUE (JOB_1, J_Q);

REMOVE_JOB_FROM_QUEUE (JOB_A; J_Q);

M_PRIVATE_JOB_QUEUE_TYPE

Specification

Interface information regarding JOB_QUEUE_TYPE, INITIALIZE_JOB_QUEUE, ADD_JOB_TO_QUEUE, REMOVE_JOB_FROM_QUEUE

Implementation of JOB_QUEUE_TYPE

Body

Implementation of INITIALIZE_JOB_QUEUE, ADD_JOB_TO_QUEUE, REMOVE_JOB_FROM_QUEUE

Visible to other modules

Invisible to other modules

Figure 8.25 Diagrammatic representation of abstract data type with information hiding via private type (Figure 8.24 with Figure 8.22).

Nygaard, 1973]. Inheritance is supported by most object-oriented programming languages, the most important of which is Smalltalk [Goldberg, 1984]. The basic idea behind inheritance is that new data types can be defined as extensions of previously defined types, rather than having to be defined from scratch [Meyer, 1986].

In such object-oriented languages, a *class* can be defined. A class is an abstract data type that supports inheritance. To see how classes are used, consider the following example. HUMAN_BEINGS can be defined to be a class, and JOE can then be defined to be an *object*, an instance of that class. All HUMAN_BEINGS have certain attributes such as age, height, and sex, and values can be assigned to those attributes when describing the object JOE. Now suppose that PARENTS is defined to be a *subclass* of HUMAN_BEINGS. This means that PARENTS have all the attributes of HUMAN_BEINGS, but in addition may have attributes of their own such as spouse's name and number of children. PARENTS *inherit* all the attributes of HUMAN_BEINGS, because class PARENTS is a subclass of class HUMAN_BEINGS. If FRED is an object and an instance of the class of PARENTS, then FRED has all the attributes of PARENTS, but also inherits all the attributes of HUMAN_BEINGS.

This property of inheritance is an essential feature of object-oriented languages such as Smalltalk, C++ [Stroustrup, 1986], Loops [Bobrow and Stefik, 1983], and Flavors [Moon, 1986]. However, neither inheritance nor the concept of class (every object is an instance of some class) is supported by languages such as C. Thus object-oriented design cannot be directly implemented even in a popular language such as C. Ada supports a restricted form of inheritance, namely in subtypes and derived types. Thus, given the type definition

```
type COLOR is (ORANGE, GREEN, RED, WHITE, BLUE, PURPLE);
```

a derived subtype FLAG_COLOR can be defined as follows:

```
type FLAG_COLOR is new COLOR range RED .. BLUE;
```

Items of type FLAG_COLOR have all the properties of items of type COLOR, except that the range is constrained to just RED, WHITE, and BLUE. In other words, Ada does not support objects in all their glory, but in a more limited way.

The advantages of using objects are precisely those of using abstract data types, including data abstraction and procedural abstraction. But, in addition, the inheritance aspects of objects as expressed in attributes, where supported by the implementation language, provide a further instance of data abstraction, leading to easier and less fault-prone product development. With regard to maintenance, objects, like abstract data types, appear to be features of products that will not change from version to version. Thus it is hoped that products designed in terms of objects will be easier to maintain than many present-day products that were designed before object-oriented design, described in Section 9.8, became an important design technique.

8.8 Languages for Implementing Data Abstraction

All the code examples in this chapter have so far been in Ada, and the incautious reader may conceivably have come to the wrong conclusion that Ada is the only language in which data abstraction can be achieved. The key point is that the major concepts of data abstraction were formulated in the 1970s. Thus languages like FORTRAN, BASIC, and COBOL predate the

```
      SUBROUTINE ENCAP (JOBNUM)
C
      INTEGER JOBNUM, QUELEN, JOBQUE(25), K
C
C
C Code equivalent to procedure INITIALIZE_JOB_QUEUE
C
      ENTRY INITAL
         QUELEN = 0
      RETURN
C
C Code equivalent to procedure ADD_JOB_TO_QUEUE
C
      ENTRY ADDJOB
         QUELEN = QUELEN + 1
         JOBQUE(QUELEN) = JOBNUM
      RETURN
C
C Code equivalent to procedure REMOVE_JOB_FROM_QUEUE
C
      ENTRY REMJOB
         JOBNUM = JOBQUE(1)
         QUELEN = QUELEN - 1
         DO 40 K = 1, QUELEN
40          JOBQUE(K) = JOBQUE(K + 1)
      RETURN
      END
```

Figure 8.26 FORTRAN implementation of data encapsulation (equivalent to Figures 8.17 and 8.18).

ideas described in this chapter. Since their design goals, unlike those of, say, Ada, did not include support for data abstraction, it is perhaps surprising that some, but not all, of the concepts described previously can be achieved in some, but not all, of the languages designed in the 1960s. For example, Figure 8.26 shows how FORTRAN can be used to achieve data encapsulation. The figure shows a FORTRAN implementation equivalent to Figures 8.17 and 8.18. However, since FORTRAN does not support type definitions, there is no way that abstract data types, and hence objects, can be directly implemented in FORTRAN.

Even though there are many other modern programming languages that support data abstraction, such as Smalltalk and CLU [Liskov, Snyder, Atkinson, and Schaffert, 1977], the examples in this chapter have been written in Ada. It was felt that readers would have less difficulty reading Ada than any of the other newer languages. In addition, a number of readers have already used Ada, or at least have previously encountered Ada; fewer readers are familiar with Smalltalk or CLU. But that is not to say that Ada is the best language for data abstraction, or even that Ada is the best language for examples in software engineering textbooks. After all, even its strongest protagonists grudgingly admit that Ada is somewhat baroque. Nevertheless, of all the languages that could have been used to illustrate the points made in this chapter, Ada is the least undesirable, and for this reason it was chosen.

The purpose of this chapter is to introduce the theoretical concepts regarding modularity. These concepts will now be used in a practical way in the design methods described in the next chapter.

Chapter Review

The chapter begins with a description of a module (Section 8.1). In the next two sections an analysis is given of what constitutes a well-designed module in terms of module cohesion and module coupling (Sections 8.2 and 8.3). Specifically, a module should have high cohesion and low coupling. Various types of abstraction are presented in Sections 8.4 through 8.7, including data abstraction and procedural abstraction. In data encapsulation (Section 8.4), a module comprises a data *structure*, together with the operations performed on that data structure. An abstract data type (Section 8.5) is a data *type*, together with the operations performed on instantiations of that type. Information hiding (Section 8.6) consists of designing a module so as to localize to a single module a design decision that may be changed in the future. The progression of increasing abstraction culminates in a description of an object, namely an abstract data type that supports inheritance (Section 8.7).

Objects (section 8.7)

⇑

Information hiding (Section 8.6)

⇑

Abstract data types (Section 8.5)

⇑

Data encapsulation (Section 8.4)

⇑

Modules with high cohesion and low coupling (Sections 8.2 and 8.3)

⇑

Modules (Section 8.1)

Figure 8.27 Major concepts of Chapter 8.

Figure 8.27 is an overview of the major concepts of Chapter 8. In some sense the diagram can be considered as a hierarchy, with each element as a special case of the element above it. Figure 8.27 explains the title of this chapter, Modularity: and Beyond.

For Further Reading

Many of the ideas in this chapter were originally put forward by Parnas [Parnas, 1971, 1972a, 1972b].

The primary source on cohesion and coupling is the work of Stevens, Myers, and Constantine [Stevens, Myers, and Constantine, 1974; Myers, 1975, 1978b; Yourdon and Constantine, 1979]. The ideas of composite/structured design have subsequently been extended to Ada [Hammonds and Dobbs, 1985].

The use of abstract data types in software development was put forward by Liskov and Zilles [Liskov and Zilles, 1974]. In addition to presenting probably the first paper on abstract data types, they described a precursor to the programming language CLU [Liskov, Snyder, Atkinson, and Schaffert, 1977]. In CLU, a Pascal-like language, an abstract data type is implemented in terms of a *cluster*, a mechanism for achieving information hiding of the type depicted in Figure 8.15. In many ways the CLU cluster is the precursor of the Ada package. Another important early paper is [Guttag, 1977]. Later papers on abstraction include [Feldman, 1981], in which a practical approach to data abstraction is presented, and [Shaw, 1984], a discussion of how abstraction is achieved in modern programming languages. A useful book on abstraction is [Liskov and Guttag, 1986]. [Berry, 1985] discusses information hiding in Ada. The application of information hiding to the U.S. Navy A-7 project appears in [Parnas, Clements, and Weiss, 1985]. In [Neumann, 1986], a description is given of the use of abstraction for designing safer software for controlling life-critical systems in areas such as air-traffic control or medical care. An account of experiences with abstraction-based software development techniques in a university environment is found in [Berzins, Gray, and Naumann, 1986].

Introductory material on objects can be found in [Booch, 1986], [Cox, 1986], and [Stefik and Bobrow, 1986]. [Halbert and O'Brien, 1987] is a tutorial on inheritance, while [Buzzard and Mudge, 1985] is an analysis of Ada as an object-oriented language. [Perez, 1988] is a description of how inheritance may be simulated in Ada. The use of objects in commercial data processing is described in [van Hoeve and Engmann, 1987]. Ways in which object-oriented programming promotes reusability are put forward in [Meyer, 1987]. The proceedings of the conferences on object-oriented programming systems, languages, and applications (OOPSLA), sponsored by the ACM and published as special issues of *ACM SIGPLAN Notices*, have a wide selection of research papers. [Shriver and Wegner, 1987] contains many worthwhile papers on various aspects of object-oriented software development.

Problems

8.1 Choose any programming language with which you are familiar. Consider the two definitions of modularity given in Section 8.1. Determine which of the two definitions includes what you intuitively understand to constitute a module in the language you have chosen.

8.2 Determine the cohesion of the following modules:

INITIALIZE MESSAGE BUFFER AND READ FIRST RECORD
EDIT NAME AND ADDRESS RECORD
EDIT NAME RECORD AND ADDRESS RECORD
TEST REASONABLENESS OF ACID CONCENTRATION
MEASURE ALLOY TEMPERATURE AND SOUND ALARM
IF NECESSARY

8.3 You are a software engineer involved in product development. Your manager asks you to investigate ways of ensuring that from now on modules designed by your group will be as reusable as possible. What do you tell him?

8.4 Your manager now asks you to determine how existing modules can be reused. Your first suggestion is to break up each module of coincidental cohesion into separate modules with functional cohesion. Your manager correctly points out that the separate modules have not been tested, nor have they been documented. What do you say now?

8.5 What is the influence of coupling on reusability?

8.6 Carefully distinguish between data encapsulation and abstract data types.

8.7 Carefully distinguish between abstraction and information hiding.

8.8 It has been suggested that Ada supports implementation of abstract data types, but only by giving up information hiding. Discuss this claim.

8.9 It has also been suggested that Ada is the best language for object-oriented programming. Discuss this claim.

8.10 Your instructor will distribute a software product. Analyze the modules from the viewpoints of information hiding, levels of abstraction, coupling, and cohesion.

8.11 (Term Project) Determine the cohesion and coupling of the modules you designed for Problem 5.7.

8.12 (Readings in Software Engineering) Your instructor will distribute copies of [Meyer, 1987]. To what extent do you agree with Meyer that the object-oriented approach promotes reusability? What experimental evidence does he bring to support his claims?

References

[ANSI/MIL-STD-1815A, 1983] "Reference Manual for the Ada Programming Language," ANSI/MIL-STD-1815A, American National Standards Institute, Inc., United States Department of Defense, 1983.

[Berry, 1978] D. M. BERRY, Personal Communication, 1978.

[Berry, 1985] D. M. BERRY, "On the Application of Ada and its Tools to the Information Hiding Decomposition Methodology for the Design of Software Systems," in: *Methodologies for Computer System Design*, W. K. Giloi and B. D. Shriver (Editors), Elsevier North-Holland, Amsterdam, The Netherlands, 1985, pp. 308–321.

[Berzins, Gray, and Naumann, 1986] V. BERZINS, M. GRAY, AND D. NAUMANN, "Abstraction-Based Software Developments," *Communications of the ACM* **29** (May 1986), pp. 402–415.

[Bobrow and Stefik, 1983] D. G. BOBROW AND M. STEFIK, *The Loops Manual*, Intelligent Systems Laboratory, Xerox Corporation, Palo Alto, CA, 1983.

[Booch, 1986] G. BOOCH, "Object-Oriented Development," *IEEE Transactions on Software Engineering* **SE-12** (February 1986), pp. 211–221.

[Buzzard and Mudge, 1985] G. D. BUZZARD AND T. N. MUDGE, "Object-Based Computing and the Ada Programming Language," *IEEE Computer* **18** (March 1985), pp. 11–19.

[Cox, 1986] B. J. COX, *Object-Oriented Programming: An Evolutionary Approach*, Addison-Wesley, Reading, MA, 1986.

[Dahl, Myrhaug, and Nygaard, 1973] O.-J. DAHL, B. MYRHAUG, AND K. NYGAARD, *SIMULA begin*, Auerbach, Philadelphia, PA, 1973.

[Dijkstra, 1968b] E. W. DIJKSTRA, "The Structure of the 'THE' Multiprogramming System," *Communications of the ACM* **11** (May 1968), pp. 341–346.

[Feldman, 1981] M. B. FELDMAN, "Data Abstraction, Structured Programming, and the Practicing Programmer," *Software—Practice and Experience* **11** (July 1981), pp. 697–710.

[Goldberg, 1984] A. GOLDBERG, *Smalltalk-80: The Interactive Programming Environment*, Addison-Wesley, Reading, MA, 1984.

[Guttag, 1977] J. GUTTAG, "Abstract Data Types and the Development of Data Structures," *Communications of the ACM* **20** (June 1977), pp. 396–404.

[Halbert and O'Brien, 1987] D. C. HALBERT AND P. D. O'BRIEN, "Using Types and Inheritance in Object-Oriented Programming," *IEEE Software* **4** (September 1987), pp. 71–79.

[Hammons and Dobbs, 1985] C. HAMMONS AND P. DOBBS, "Coupling, Cohesion and Package Unity in Ada," *ACM SIGAda Ada Letters* **IV** (No. 6, 1985), pp. 49–59.

[Liskov and Guttag, 1986] B. Liskov and J. Guttag, *Abstraction and Specification in Program Development*, The MIT Press, Cambridge, MA, 1986.

[Liskov and Zilles, 1974] B. Liskov and S. Zilles, "Programming with Abstract Data Types," *ACM SIGPLAN Notices* **9** (April 1974), pp. 50–59.

[Liskov, Snyder, Atkinson, and Schaffert, 1977] B. Liskov, A. Snyder, R. Atkinson, and C. Schaffert, "Abstraction Mechanisms in CLU," *Communications of the ACM* **20** (August 1977), pp. 564–576.

[Meyer, 1986] B. Meyer, "Genericity versus Inheritance," *Proceedings of the Conference on Object-Oriented Programming Systems, Languages and Applications, ACM SIGPLAN Notices* **21** (November 1986), pp. 391–405.

[Meyer, 1987] B. Meyer, "Reusability: The Case for Object-Oriented Design," *IEEE Software* **4** (March 1987), pp. 50–64.

[Moon, 1986] D. A. Moon, "Object-Oriented Programming with Flavors," *Proceedings of the Conference on Object-Oriented Programming Systems, Languages and Applications, ACM SIGPLAN Notices* **21** (November 1986), pp. 1–8.

[Myers, 1975] G. J. Myers, *Reliable Software Through Composite Design*, Petrocelli/Charter, New York, NY, 1975.

[Myers, 1978b] G. J. Myers, *Composite/Structural Design*, Van Nostrand Reinhold, New York, NY, 1978.

[Neumann, 1986] P. G. Neumann, "On Hierarchical Design of Computer Systems for Critical Applications," *IEEE Transactions on Software Engineering* **SE-12** (September 1986), pp. 905–920.

[Parnas, 1971] D. L. Parnas, "Information Distribution Aspects of Design Methodology," *Proceedings of the IFIP Congress*, Ljubljana, Yugoslavia, 1971, pp. 339–344.

[Parnas, 1972a] D. L. Parnas, "A Technique for Software Module Specification with Examples," *Communications of the ACM* **15** (May 1972), pp. 330–336.

[Parnas, 1972b] D. L. Parnas, "On the Criteria to be Used in Decomposing Systems into Modules," *Communications of the ACM* **15** (December 1972), pp. 1053–1058.

[Parnas, Clements, and Weiss, 1985] D. L. Parnas, P. C. Clements, and D. M. Weiss, "The Modular Structure of Complex Systems," *IEEE Transactions on Software Engineering* **SE-11** (March 1985), pp. 259–266.

[Perez, 1988] E. P. Perez, "Simulating Inheritance with Ada," *ACM SIGAda Ada Letters* **VIII** (No. 5, 1988), pp. 37–46.

[Schach and Stevens-Guille, 1979] S. R. Schach and P. D. Stevens-Guille,

"Two Aspects of Computer-Aided Design," *Transactions of the Royal Society of South Africa* **44** (Part 1, 1979), pp. 123–126.

[Shaw, 1984] M. SHAW, "Abstraction Techniques in Modern Programming Languages," *IEEE Software* **1** (October 1984), pp. 10–26.

[Shneiderman and Mayer, 1975] B. SHNEIDERMAN AND R. MAYER, "Towards a Cognitive Model of Programmer Behavior," Technical Report TR-37, Indiana University, Bloomington, IN, 1975.

[Shriver and Wegner, 1987] B. SHRIVER AND P. WEGNER (Editors), *Research Directions in Object-Oriented Programming*, The MIT Press, Cambridge, MA, 1987.

[Stefik and Bobrow, 1986] M. STEFIK AND D. G. BOBROW, "Object-Oriented Programming: Themes and Variations," *The AI Magazine* **6** (No. 4, 1986), pp. 40–62.

[Stevens, Myers, and Constantine, 1974] W. P. STEVENS, G. J. MYERS, AND L. L. CONSTANTINE, "Structured Design," *IBM Systems Journal* **13** (No. 2, 1974), pp. 115–139.

[Stroustrup, 1986] B. STROUSTRUP, *The C++ Programming Language*, Addison-Wesley, Reading, MA, 1986.

[Tanenbaum, 1984] A. S. TANENBAUM, *Structured Computer Organization*, Second Edition, Prentice-Hall, Englewood Cliffs, NJ, 1984.

[van Hoeve and Engmann, 1987] F. VAN HOEVE AND R. ENGMANN, "An Object-Oriented Approach to Application Generation," *Software—Practice and Experience* **17** (September 1987), pp. 623–645.

[Wirth, 1985] N. WIRTH, *Programming in Modula-2*, Third Corrected Edition, Springer-Verlag, Berlin, West Germany, 1985.

[Young and Gregory, 1972] D. M. YOUNG AND R. T. GREGORY, *A Survey of Numerical Mathematics, Volume I*, Addison-Wesley, Reading, MA, 1972.

[Yourdon and Constantine, 1979] E. YOURDON AND L. L. CONSTANTINE, *Structured Design: Fundamentals of a Discipline of Computer Program and Systems Design*, Prentice-Hall, Englewood Cliffs, NJ, 1979.

Chapter 9 Design Methods

Over the past 30 or so years, literally hundreds of design methods have been put forward. Some are variations of existing methods, others are radically different from anything that has previously been proposed. A few design methods have been used by tens of thousands of software engineers, many have been used only by their authors. Some design strategies, particularly those developed by academics, have a firm theoretical basis. Others, including many drawn up by academics, are more pragmatic in nature; they were put forward because their authors found that they worked well in practice. Most design methods are manual, but automation is increasingly becoming an important aspect of design, if only to assist in management of documentation. This aspect is described in greater detail in Chapter 12.

Notwithstanding this plethora of design methods, there is a certain underlying pattern. Two aspects of a product are its various processes and the data on which the processes operate. The two basic ways of designing a product are process-oriented design and data-oriented design. In *process-oriented design*, the emphasis is on the processes. An example is composite/structured design (C/SD), where the objective is to design modules with high cohesion and low coupling, as described in Sections 8.2 and 8.3. That is to say, in C/SD each process, or group of closely related processes, becomes a module. In *data-oriented design*, the data come first. Thus in Jackson's method (Section 9.6) the structure of the data is determined first. Then the procedures are designed in such a way as to conform to the structure of the data.

In addition to pure process-oriented design and pure data-oriented design, hybrid methods are becoming increasingly important. For example, in object-oriented design the objective is to identify the objects and to build the product around those objects. One aspect of an object is that it is a data structure, together with the processes that operate on that data structure. Object-oriented design is therefore both data-oriented and process-oriented.

In this chapter examples of all three approaches to design will be given. But first, some general remarks regarding the design process must be made.

9.1 Design and Abstraction

The software design phase consists of three activities: architectural design, detailed design, and design testing. The input to the design process is the specifications, a description of *what* the product is to do. The output is the design document, a description of *how* the product is to achieve this.

During *architectural design*, also known as *general design*, a modular decomposition of the product is developed. That is to say, the specifications are carefully analyzed, and a module structure is produced which has the desired functionality. The output from this activity is a description of the modules, and how they are to be interconnected. From the viewpoint of abstraction, what is going on during the architectural design phase is that the existence of certain modules is assumed; the design is then developed in terms of those modules.

The next activity is *detailed design*, also known as *modular design*. Here each of the modules is designed in detail. For example, specific algorithms are selected, and data structures are chosen. Again from the viewpoint of abstraction, during this activity the fact that the modules are to be interconnected to form a complete product is ignored.

This two-stage process is typical of abstraction as described in Chapter 8. First, the top level, the overall product, is designed in terms of modules that do not yet exist. Then each module in turn is designed independently of the fact that it is to be a component of the product as a whole.

It was stated previously that the design phase comprises three activities and that the third activity is testing. The word "activity" was used, rather than, say, "stage" or "step" because testing is an integral part of design, just as it is an integral part of the entire software development and maintenance process; it is not something that is performed only after the architectural design and detailed design have been completed. In Chapter 6 a comprehensive description was given of testing during the design phase. To avoid repetition, in this chapter the word "testing" is simply mentioned from time to time, as a gentle reminder to the reader that testing is an activity that must be continuously conducted in parallel with all software production activities, including design.

A variety of design methods are now presented, first process-oriented methods and then data-oriented methods.

9.2 Process-Oriented Design

In Sections 8.2 and 8.3 a case was made for decomposing a product into modules of high cohesion and low coupling. Being a theoretical chapter, no indication was given there regarding practical techniques for achieving this. A description is now given of two methods, namely data flow analysis (Section 9.3) and transaction analysis (Section 9.4). In theory, data flow analysis can be applied whenever the specifications can be represented by a data flow diagram (DFD). Since, in theory at least, every product can be represented by a DFD, this would mean that data flow analysis is universally applicable. In practice, however, there are a number of situations where there are more appropriate design techniques, specifically for designing products where the flow of data is secondary to other considerations. Examples where other design techniques would be indicated include rule-based systems (expert systems), databases, and transaction-processing products; transaction analysis (Section 9.4) is a good way of decomposing transaction-processing products into modules.

9.3 Data Flow Analysis

Data flow analysis (DFA) is a design method for achieving modules of high cohesion. It can be used in conjunction with most specification methods. Here DFA will be presented in conjunction with structured systems analysis (Section 7.2). The input to the method is a data flow diagram (DFD). A key point is that once the DFD has been completed, the software designer has precise and complete information regarding the input to and output from every process.

Consider the flow of data in the product represented by the DFD of Figure 9.1. The product is somehow transforming input into output. At some point in that figure the input ceases to be input and becomes some sort of internal data. Then, at some further point, these internal data take on the quality of output. This is shown in more detail in Figure 9.2. The point at which the input loses the quality of being input and simply becomes internal data operated on by the product is termed the point of highest abstraction of input. The point of highest abstraction of output is similarly the first point in the flow of data at which the output can be identified as such, rather than as some sort of internal data.

Using the points of highest abstraction of input and output, the product is decomposed into three modules, namely the INPUT MODULE, TRANSFORM MODULE, and OUTPUT MODULE. Now each module is

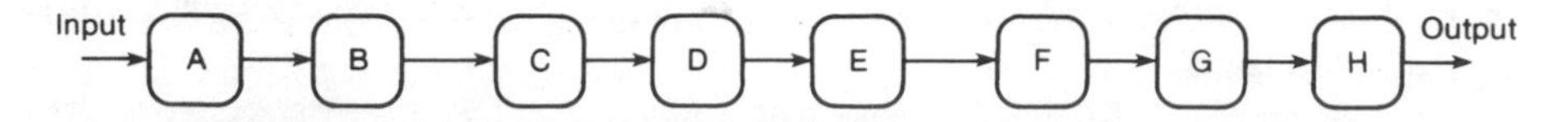

Figure 9.1 Data flow diagram showing flow of data and processes of product.

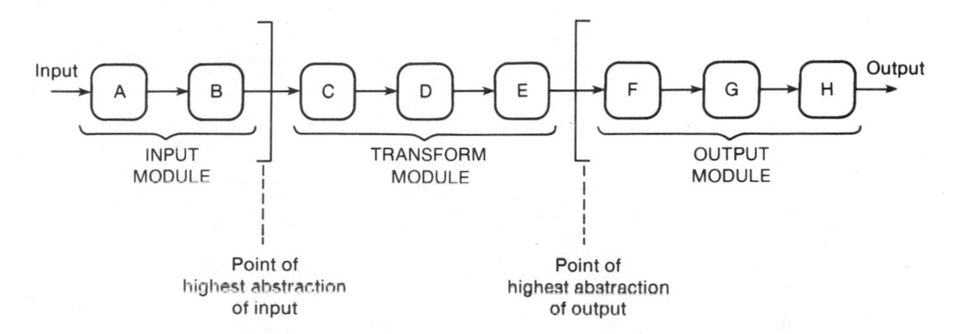

Figure 9.2 Points of highest abstraction of input and output.

taken in turn, its points of highest abstraction found, and the module is again decomposed. This procedure is continued stepwise until each module performs a single function, that is, the design consists of modules of high cohesion. As with so many other software engineering techniques, stepwise refinement is the basic underlying principle.

In fairness, it should be pointed out that minor modifications might have to be made to the decomposition in order to achieve the lowest possible coupling. Data flow analysis is a way of achieving high cohesion. The aim of composite/structured design is high cohesion but also low coupling. In order to achieve the latter it is sometimes necessary to make minor modifications to the design. For example, since DFA does not take coupling into account, control coupling may inadvertently arise in a design constructed using DFA. In such a case, all that is needed is to modify the two modules involved so that data, and not control, are passed between them.

9.3.1 Data Flow Analysis Case Study

Consider the problem of designing a product which will take as input a file name and return the number of words in that file, similarly to the UNIX *wc* command.

Figure 9.3 depicts the data flow diagram (DFD). There are five processes. READ FILE NAME reads the name of the file which is then validated by VALIDATE FILE NAME. The validated name is then passed to COUNT NUMBER OF WORDS which does precisely that. The word count is then passed on to FORMAT WORD COUNT. The formatted word count is finally passed to DISPLAY WORD COUNT for output.

Examining the data flow, the initial input is File name. When this becomes Validated file name it is still a file name, and therefore has not lost its quality of being input data. But consider process COUNT NUMBER OF WORDS. Its input is Validated file name and its output is Word count. The

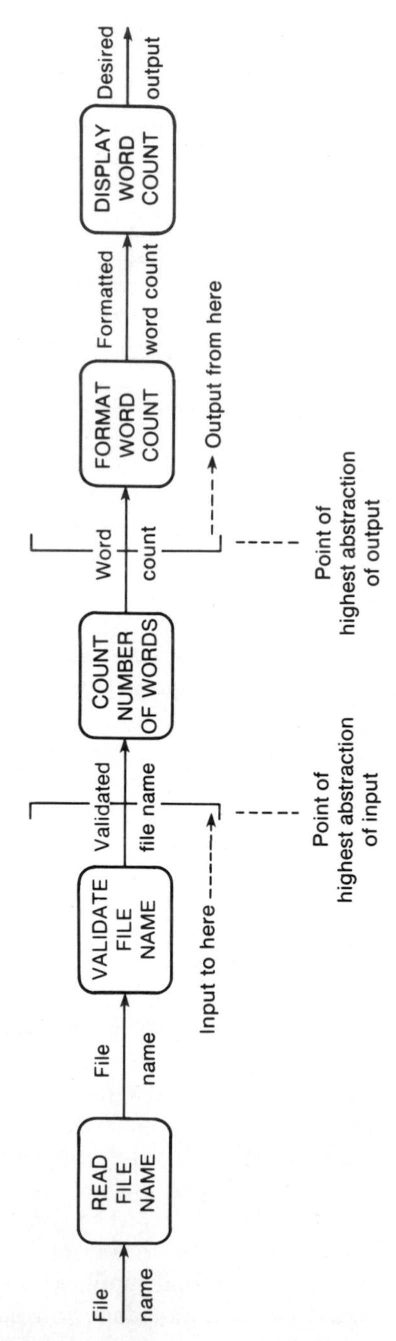

Figure 9.3 Data flow diagram: first refinement.

output from this process is totally different in quality to the input to the product as a whole. It is clear that the point of highest abstraction of input is as indicated on Figure 9.3. Similarly, even though the output from COUNT NUMBER OF WORDS undergoes some sort of formatting, it is essentially output from the time it emerges from process COUNT NUMBER OF WORDS. The point of highest abstraction of output is therefore as shown in the figure.

The result of decomposing the product using these two points of highest abstraction is shown in the structure chart of Figure 9.4. Figure 9.4 also reveals that the data flow diagram of Figure 9.3 is somewhat too simplistic. The DFD does not show the logical flow corresponding to what happens if the file specified by the user does not exist. READ AND VALIDATE FILE NAME must return a Status flag to PERFORM WORD COUNT. If the name is invalid, then it is ignored by PERFORM WORD COUNT and an error message of some sort is printed. But if the name is valid it is passed on to COUNT NUMBER OF WORDS.

In Figure 9.4 there are two modules of communicational cohesion, namely READ AND VALIDATE FILE NAME and FORMAT AND DISPLAY WORD COUNT. These must be broken down further. The final result is shown in Figure 9.5. All eight modules have functional cohesion, and there is either no coupling or data coupling between them.

Now that the architectural design has been completed, the next step is the detailed design. Here data structures are chosen and algorithms selected. The detailed design of each module is then handed to the programmers for implementation. Just as with virtually every other phase of software production, time constraints usually require that the implementation be done by a team, rather than having a single programmer responsible for coding all the

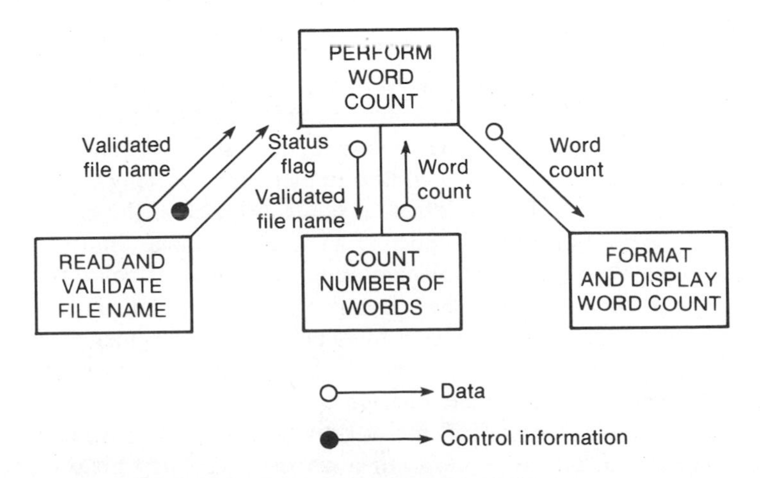

Figure 9.4 Structure chart: first refinement.

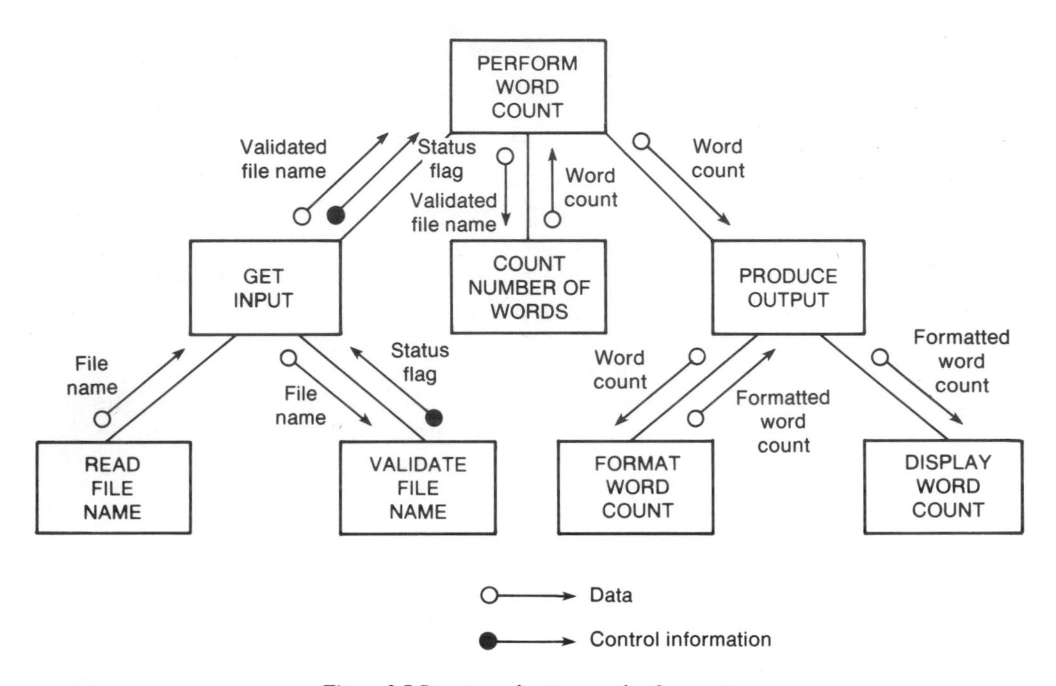

Figure 9.5 Structure chart: second refinement.

modules. For this reason, the detailed design of each module must be presented in such a way that it can be understood independently of any other module. The detailed design of four of the eight modules of the case study appears in Figure 9.6; the other four modules will be presented in a different format.

While the design of Figure 9.6 is language independent, if management has decided on an implementation language before the detailed design is started, then the use of pseudocode for representing the detailed design is an attractive alternative. Pseudocode essentially consists of comments connected by the control statements of the chosen implementation language. Figure 9.7 shows a detailed design for the remaining four modules of the product written in an Ada-like pseudocode. Pseudocode has the advantage that it is generally clear and concise, and the implementation step usually consists merely of translating the comments into the relevant programming language. The disadvantage is that there is sometimes a tendency for the designer to go into too much detail, and to produce a complete code implementation of a module rather than a pseudocode detailed design.

The detailed design is handed over to the implementation team for coding. The product then proceeds through the remaining phases of the life cycle.

Module name	READ_FILE_NAME
Module type	procedure
Input parameters	none
Output parameters	File_name : string
Error messages	none
Files accessed	none
Files changed	none
Modules called	none
Narrative	The product is invoked by the user by means of the command string WORDCOUNT File_name Using an operating system call, this module accesses the contents of the command string that was input by the user, and extracts File_name.

Module name	VALIDATE_FILE_NAME
Module type	procedure
Input parameters	File_name : string
Output parameters	Status_flag : Boolean
Error messages	none
Files accessed	none
Files changed	none
Modules called	none
Narrative	This module makes an operating system call to determine whether file File_name exists. Status_flag is set to TRUE if the file exists, and FALSE otherwise.

Module name	COUNT_NUMBER_OF_WORDS
Module type	procedure
Input parameters	Validated_file_name : string
Output parameters	Word_count : integer
Error messages	none
Files accessed	none
Files changed	none
Modules called	none
Narrative	This module determines whether File_name is a text file, that is, a file divided into lines of characters. If not, Word_count is set equal to −1, and control is returned to the calling module. Otherwise, Word_count is set equal to the number of words in the text file.

Figure 9.6 Detailed design of four modules of case study.

Module name	PRODUCE_OUTPUT
Module type	procedure
Input parameters	Word_count : integer
Output parameters	none
Error messages	none
Files accessed	none
Files changed	none
Modules called	FORMAT_WORD_COUNT Parameters: Word_count : integer Formatted_word_count : string DISPLAY_WORD_COUNT Parameters: Formatted_word_count : string
Narrative	This module takes the integer Word_count passed to it by the calling module, and calls FORMAT_WORD_COUNT to have that integer formatted according to the specifications. Then it calls DISPLAY_WORD_COUNT to have the line printed.

Figure 9.6 Detailed design of four modules of case study (continued).

```
procedure PERFORM_WORD_COUNT is
  VALIDATED_FILE_NAME : STRING;
  STATUS_FLAG : BOOLEAN;
  WORD_COUNT : INTEGER;
begin
  GET_INPUT (VALIDATED_FILE_NAME, STATUS_FLAG);
  if STATUS_FLAG is FALSE then
    Print "error 1: file does not exist";
  else
    COUNT_NUMBER_OF_WORDS (VALIDATED_FILE_NAME, WORD_COUNT);
    if WORD_COUNT is equal to -1 then
      Print "error 2: file is not a text file";
    else
      PROCESS_OUTPUT (WORD_COUNT);
    end if;
  end if;
end PERFORM_WORD_COUNT;
```

Figure 9.7 Pseudocode representation of detailed design of four modules of case study.

```
procedure GET_INPUT (VALIDATED_FILE_NAME : out STRING;
                     STATUS_FLAG : out BOOLEAN) is
  FILE_NAME : STRING;
begin
  READ_FILE_NAME (FILE_NAME);
  VALIDATE_FILE_NAME (FILE_NAME, STATUS_FLAG);
  if STATUS_FLAG is TRUE then
    VALIDATED_FILE_NAME := FILE_NAME;
  end if;
end GET_INPUT;
```

```
procedure DISPLAY_WORD_COUNT (FORMATTED_WORD_COUNT : in STRING) is
begin
  Print string FORMATTED_WORD_COUNT, left justified
end DISPLAY_WORD_COUNT;
```

```
procedure FORMAT_WORD_COUNT (WORD_COUNT : in INTEGER;
                             FORMATTED_WORD_COUNT : out STRING) is
  WORD_STRING : STRING;
begin
  WORD_STRING := WORD_COUNT'IMAGE;
    -- For any value X, X'IMAGE is the sequence of characters representing the value in
    -- display form
  FORMATTED_WORD_COUNT := "File contains " & WORD_STRING & " words";
    -- Operator & is the concatenation operator in Ada
end FORMAT_WORD_COUNT;
```

Figure 9.7 Pseudocode representation of detailed design of four modules of case study (continued).

9.3.2 Extensions

The reader may well feel that this example is somewhat artificial in that the data flow diagram (Figure 9.3) has but one input stream and one output stream. To see what happens in more complex situations, consider Figure 9.8. Now there are four input streams and five output streams, a situation that corresponds more closely to reality.

When there are multiple input and output streams, the way to proceed is to find the point of highest abstraction of input for each input stream and the point of highest abstraction of output for each output stream. Use these points to decompose the given data flow diagram into modules with fewer input/output streams than the original. Continue in this way until each resulting module has high cohesion. Finally, determine the coupling between each pair of modules and make any adjustments that may be needed.

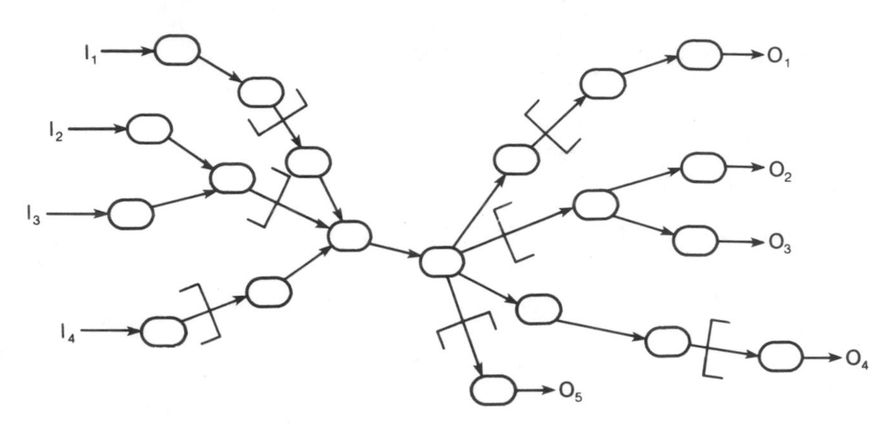

Figure 9.8 Data flow diagram with multiple input and output streams.

9.4 Transaction Analysis

Data flow analysis is inappropriate for the transaction-processing type of product in which a number of related functions, similar in outline but differing in detail, must be performed. A typical example is the software controlling an automated teller machine (ATM). The customer inserts a magnetized card into a slot, keys in a password, and then may perform functions such as deposit to a checking, savings, or credit card account, withdraw from an account, or determine the balance in an account. This type of product is depicted in Figure 9.9. A good way to design such a product is to break it up into two pieces, the analyzer, and the dispatcher. The analyzer determines the transaction type and passes this information on to the dispatcher, which then performs the transaction.

A poor design of this type is shown in Figure 9.10. There are two modules of logical cohesion, namely EDIT ANY TRANSACTION and UPDATE ANY FILE. In addition, there is control coupling. At the same time, it seems a waste of effort to have five very similar edit modules and five very similar update modules. The solution is to use "almost reusable code." Construct a basic edit module and then instantiate it five times. Each version will be slightly different, but the differences will be small enough to make this approach worthwhile. Similarly, a basic update module can be instantiated five times and slightly modified to cater for the five different types of update. This results in a design with high cohesion and low coupling.

9.5 Data-Oriented Design

The basic principle behind data structure design is to design the product according to the structure of the data on which it is to operate. There are a number of methods of this type, including those of Warnier

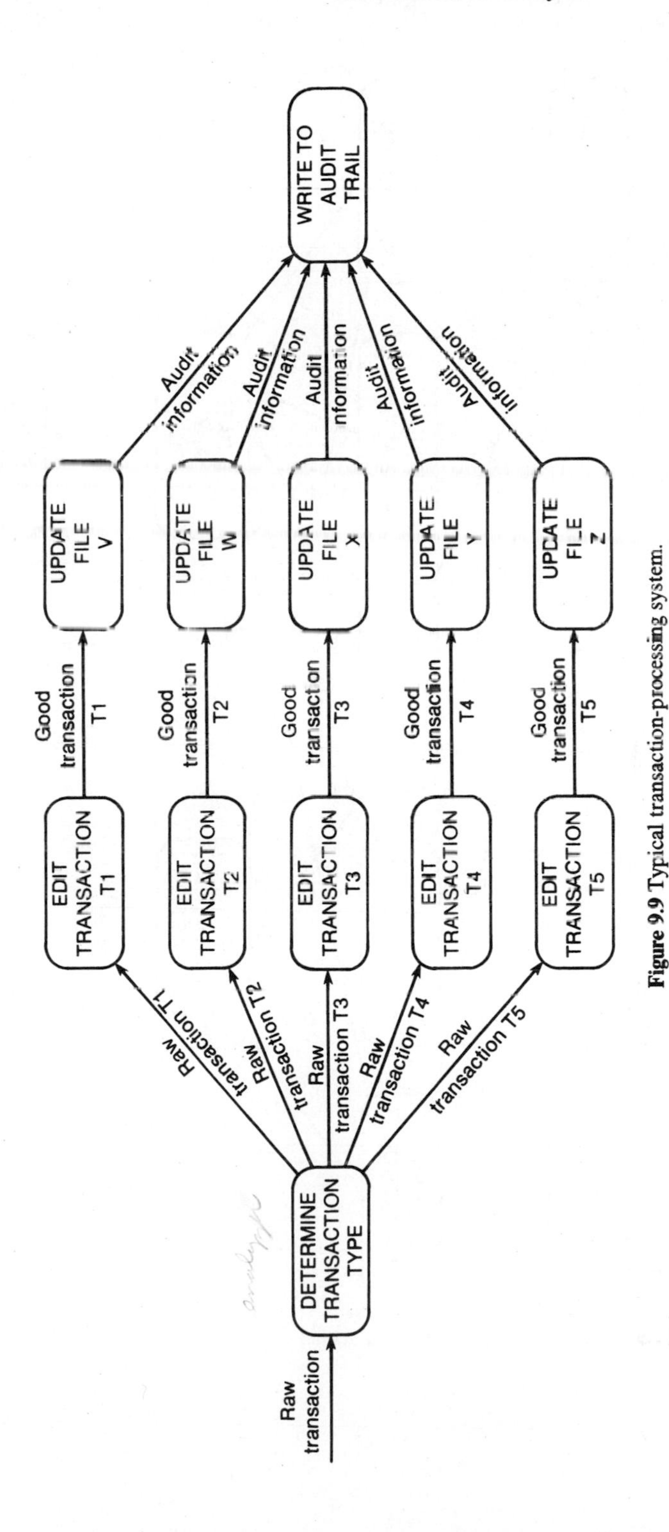

Figure 9.9 Typical transaction-processing system.

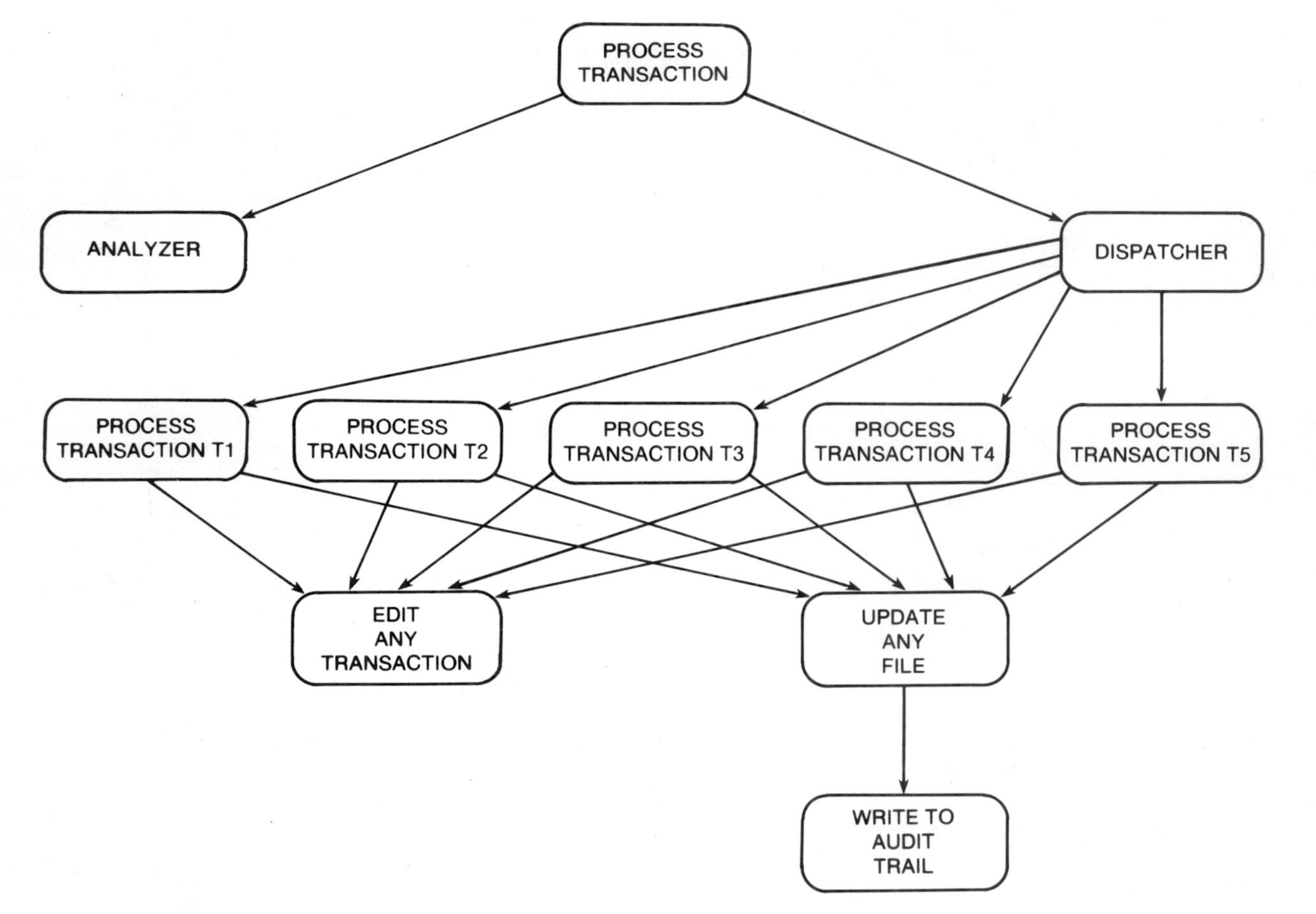

Figure 9.10 Bad design for transaction-processing system.

[Warnier, 1976, 1981] and Orr [Orr, 1981]. Jackson, too, has developed a systematic approach that addresses a large portion of the life cycle, from specifications to design and implementation [Jackson, 1975, 1983]. The three methods share many similarities. For reasons of space, Jackson's method is presented in detail, followed by a brief description of the other two methods.

9.6 Jackson System Development

9.6.1 Overview of Jackson System Development

Before describing Jackson's method [Jackson, 1983], some preliminary concepts are needed. Jackson is not concerned with functional decomposition, that is to say, Jackson system development (JSD) does not begin with some sort of functional requirement in which it is stated what the product must do and in which the inputs and the outputs of the product are laid down. Instead, the basis of JSD is real-world modeling. The developer builds a *model* of the real world, or rather, of the subset of the real world that is relevant to the product to be built, and then implements that model on the computer. The model of the real world is described in terms of two concepts, namely *entities* and the *actions* performed by or on them. Jackson does not use the term "entity" in quite the same way as it is used in database design; for Jackson, an entity must have a real-world action performed by it or performed on it. For example, if a product is to be built to control a bank of elevators, entities include BUTTON and ELEVATOR. With regard to actions, a passenger must PRESS a button to summon an elevator or to request an elevator to travel to a specific floor. An elevator will ARRIVE at a floor or LEAVE a floor. Actions must take place in the real world that is being modeled; aspects specific to the product that have no real-world counterparts such as "display menu" or "print list of unmatched tokens" are not actions in JSD.

Jackson has developed his own notation for describing software [Jackson, 1975]. There are two types of notation, graphical and textual. Figure 9.11 shows the three different control structures for actions using both notations. Figure 9.11(a) shows sequencing; action A-SEQ consists of action A-1 followed by action A-2. The small circles in Figure 9.11(b) denote selection; action A-SEL consists of action A-3, or of action A-4, or of action A-5. The asterisk in Figure 9.11(c) stands for iteration 0 or more times; action A-ITR consists of 0 or more repetitions of action A-6. Note that the symbols are placed one level lower than might be expected. Thus, while action A-ITR consists of 0 or more repetitions of A-6, the asterisk is placed in A-6, rather than in A-ITR.

In JSD, a product is built out of processes. Processes can be connected in only two ways, with a data stream connection or with a state vector connection. Figure 9.12(a) shows a data stream connection. Process DS-0

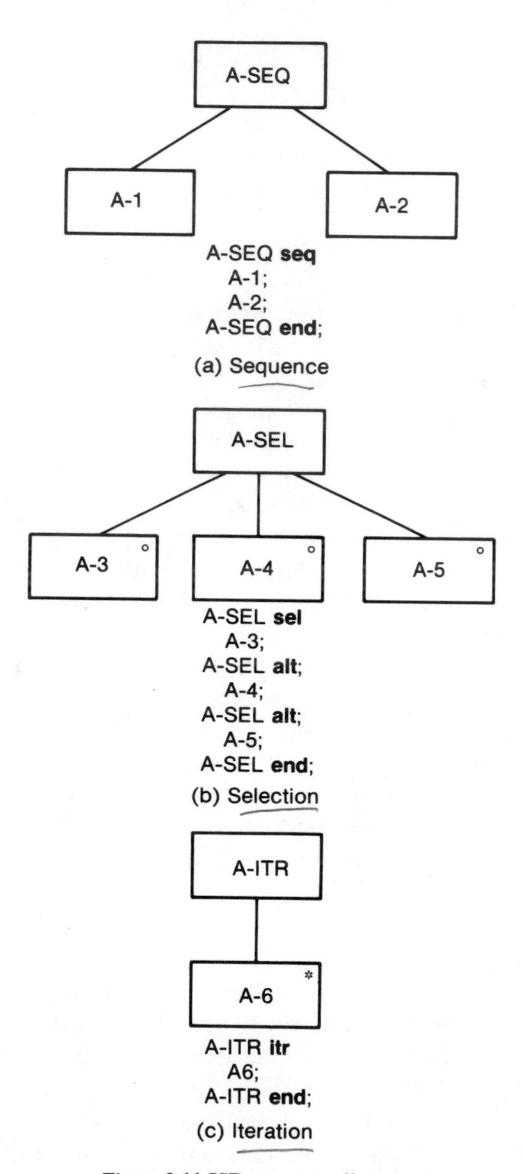

Figure 9.11 JSD structure diagram.

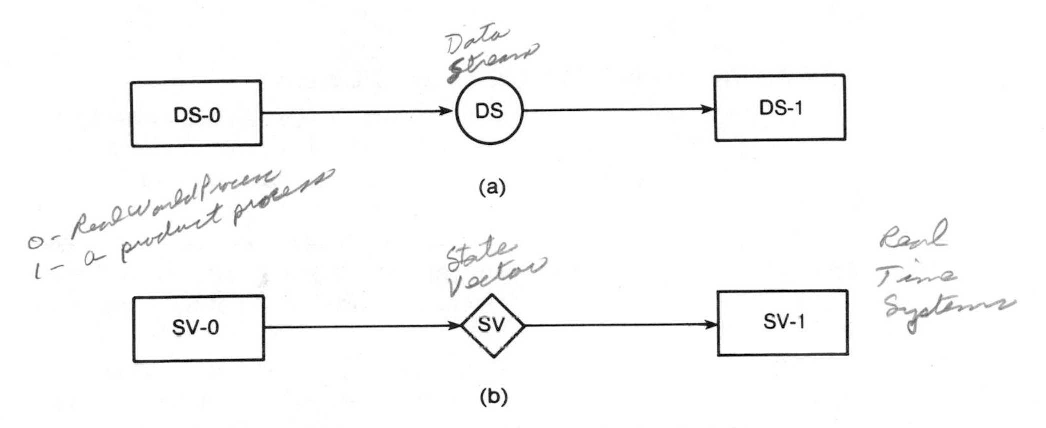

Figure 9.12 JSD structure diagram depicting two ways of interconnecting JSD processes.

sends a stream of data to process DS-1. An example of this type of situation is where process DS-0 writes a stream of records, and process DS-1 reads those records. With such a data stream connection, the elements written by DS-0 are read by DS-1 in the order in which they are sent. In contrast, in Figure 9.12(b) process SV-1 inspects the state vector of process SV-0. This type of connection frequently occurs in real-time systems, where process SV-1 continually monitors process SV-0. Process SV-1 inspects the state of SV-0 whenever it wishes; with this sort of connection there is no sequence of data written from SV-0 to SV-1.

The JSD convention is that the suffix 0 represents a real-world process, and the suffix 1 represents a product process. Figure 9.12 is an example of a JSD structure diagram.

Jackson's method consists of six steps.

1. *Entity action step*. The real-world area of interest is delimited. The entities and actions of the proposed product are listed.
2. *Entity structure step*. The actions performed by or on the entities are displayed using Jackson's notation. The time ordering of the actions is taken into account.
3. *Initial model step*. The entities and actions are represented by a process model. Correspondences are established between the model and the real world.
4. *Function step*. Functions are specified that will result in the required outputs from the product.
5. *Product timing step*. Process scheduling aspects are examined.
6. *Implementation step*. Decisions are taken as to how the processes will be mapped to the available processors (hardware).

9.6.2 Why JSD Is Presented in this Chapter

Looking through the six steps of JSD, steps 1 through 5 are essentially specification steps, and the product is implemented at step 6. What has happened to design? Working through the case study presented in Section 9.6.3, the reader will appreciate that steps 1 through 5 are not pure specification steps. Instead, each step also has a design component, and the proportion of design component increases with increasing step number. Thus steps 4 and 5 have a large design component, even though the product is not fully specified until the end of step 5.

At first sight it would seem that because JSD covers both specification and design, the specification aspects should be presented in Chapter 7 (Specification Methods), and the design aspects in this chapter (Design Methods). But because the specification and design components are so closely intermingled, this is not possible. JSD has to be presented as a whole. Since design concepts have not been introduced by Chapter 7, Chapter 9 is the correct place for JSD.

But JSD does not really fit in anywhere. The method is unique. To highlight some of the features of JSD, a case study is now presented.

9.6.3 JSD Case Study

The JSD method will be applied to a modified version of the so-called elevator problem that has already been used for a number of case studies in this book.

> An n elevator product is to be installed in a building with m floors. The problem concerns the logic to move elevators between floors according to the following constraints:
>
> C_1: Each elevator has a set of buttons, one for each floor. These illuminate when pressed and cause the elevator to visit the corresponding floor. The illumination is canceled when the corresponding floor is visited by the elevator.
> C_2: Each floor, except the first floor and top floor, has two buttons, one to request an up-elevator and one to request a down-elevator. These buttons illuminate when pressed. The illumination is canceled when an elevator visits the floor and moves in the desired direction.
> C_3: When an elevator has no requests to service, it should remain at its current floor with its doors closed.

For the sake of simplicity, the discussion in this section is restricted to one elevator. It is amusing to note that in the treatment of the elevator problem in [Jackson, 1983] on which this case study is based, the problem is stated in terms of two elevators, but quite soon the problem is simplified to the case of only one elevator. The general case of n elevators is therefore "obvious" and "left as an exercise to the reader."

Step 1. Entity action step: A candidate list of entities is obtained by reading through the description and listing the nouns. This gives the following list: ELEVATOR, BUILDING, FLOOR, BUTTON, ILLUMINATION, REQUEST, and DOOR. From this list, BUILDING, FLOOR, and DOOR can be deleted, because they lie outside the model boundary. The model is concerned with the control of the elevator—the physical surroundings in which the elevator operates must be excluded. REQUEST corresponds to the pressing of a button, as does ILLUMINATION, so these two may also be deleted. This leaves only two entities, namely ELEVATOR and BUTTON. Note that USER, which is not a noun that occurs in the description, is not an entity. If it were, its actions would include ENTER-ELEVATOR, LEAVE-ELEVATOR, and PRESS-BUTTON. Without some sort of photoelectric cell there is no way to detect that a user has entered or left an elevator, and without a surveillance system there is no way to determine which user pressed a particular button. Undetectable entities and events have to be excluded from a JSD model. Selecting candidate entities from the nouns in the description sometimes has the useful side effect of automatically excluding such undetectable entities.

Turning now to the actions, a candidate list of actions is similarly drawn up from the verbs in the description. This gives the following list: ILLUMINATE, PRESS, VISIT, CANCEL, REQUEST, MOVE-UP, MOVE-DOWN, SERVICE, and REMAIN. Verbs VISIT and REMAIN relate to FLOOR, which has already been excluded as a entity because it lies outside the model boundaries. Verbs ILLUMINATE, CANCEL, REQUEST, and SERVICE are incorporated in the functionality of BUTTON. This leaves the entity BUTTON with action PRESS, and the entity ELEVATOR with actions MOVE-UP and MOVE-DOWN. But the problem of undetectability again arises. In most elevators, all that can be detected is that the elevator has arrived at a floor or has left a floor; there is no way of measuring its direction of motion. Making the necessary changes, this leaves the following entities and actions:

Entity: BUTTON
    Action: PRESS (the button is depressed and released once)
Entity: ELEVATOR
    Action: ARRIVE(f) (the elevator arrives at floor f, from above or from below)
    Action: LEAVE(f) (the elevator leaves floor f, about to move up or move down)

Having identified the entities and their actions, step 1 is complete.

Step 2. Entity structure step: Using the notation of Figure 9.11, the actions performed by or on the entities are shown in Figures 9.13(a) and (b). With regard to BUTTON, each BUTTON, whether in an elevator or at a floor, can be pressed 0 or more times. The ELEVATOR must DEPART from floor 1, and then alternately arrives at and departs from floor f. Note that there are constraints on the value of f. Such constraints do not appear in a JSD diagram, but are added in the form of a note. In this instance, the constraints are:

1. $1 \le f \le m$.
2. For the first instance of FLOOR(f), f = 1 or 2.
3. For any two successive instances of FLOOR(f), the values of f must either be identical or must differ by 1.

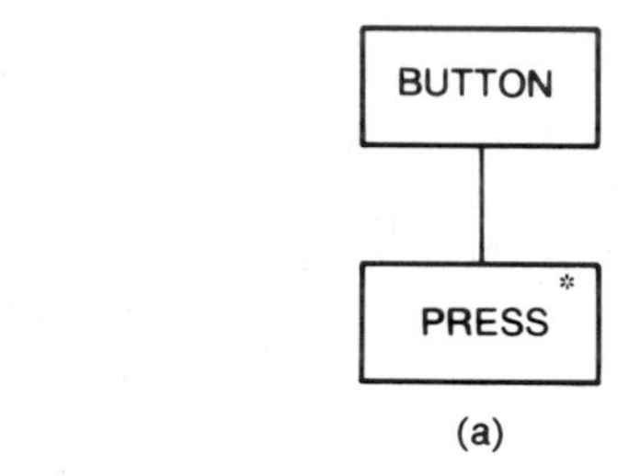

(a)

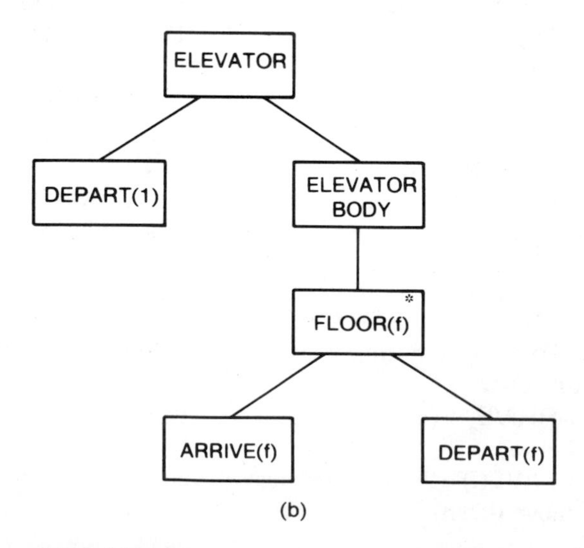

(b)

Figure 9.13 JSD structure diagrams for elevator problem.

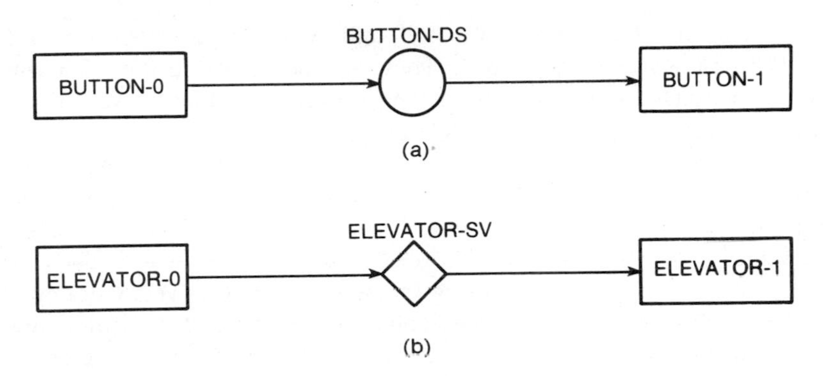

Figure 9.14 JSD structure diagram showing connection between real-world processes and model processes

The point is that when an elevator is installed in a building it is at the first floor. When put into service it leaves the first floor, and thereafter cannot skip floors. If it is currently at, say, the sixth floor, then it can travel next to the fifth floor or to the seventh floor, and it can also return to the sixth floor, but it cannot travel next to, say, the third floor. It does not have to stop at every floor, but it certainly must pass through every floor.

Step 3. Initial model step: At this step the entities and actions are represented by a process model. Figure 9.14(a) shows the data stream mapping between the real-world button process BUTTON-0 and the model button process BUTTON-1. The latter must read the value of BUTTON-DS, that is to say, register the pulses sent each time the button is pressed. This could be represented graphically, but for many purposes a textual version of the corresponding JSD structure diagram is more useful. This is given in Figure 9.15. The process consists of an initial read of the button data stream, and

```
BUTTON-1 seq
    read BUTTON-DS;
    BUTTON-BODY itr
        PRESS;
        read BUTTON-DS;
    BUTTON-BODY end;
BUTTON-1 end;
```

Figure 9.15 Structure text for BUTTON-1.

then a repetition of button pressing, followed by reading the data stream. Note the ***read*** BUTTON-DS commands. These are the ways that the model button process inspects the real-world button process. In other words, the ***read*** connects the real world to the model.

The state vector mapping between the real-world elevator process and the model elevator process is shown in Figure 9.14(b). The textual version of the model elevator process is shown in Figure 9.16. In this case, the ***getsv*** (get state vector) commands sense a change of state in the real world. The elevator process consists of an initial sensing of the elevator state vector followed by departure from the first floor. Then comes a repetition of arriving at floor f, and departing again, detected by sensing the elevator state vector at the appropriate times.

The presentation in this section is somewhat oversimplified. For instance, an elevator does not continuously move from floor to floor. Instead, after halting at a floor it waits for passengers to get in or get out, and only then does the elevator move up or down. To implement this, a set of state vectors ATF(1), ATF(2), . . . , ATF(m) is needed to denote that the elevator is at floor 1, 2, . . . , m, respectively. A further state vector, MOVING, denotes that the elevator is in motion between floors. The structure diagram is shown in Figure 9.17, and the corresponding structure text in Figure 9.18.

Step 4. Function step: In this step, the developer specifies the functions of the product. A JSD function is specified as follows: When a certain set

```
ELEVATOR-1 seq
   getsv ELEVATOR-SV;
   DEPART(1);
   ELEVATOR-BODY itr
      FLOOR(f) seq
         getsv ELEVATOR-SV;
         ARRIVE(f);
         getsv ELEVATOR-SV;
         DEPART(f);
      FLOOR(f) end;
   ELEVATOR-BODY end;
ELEVATOR-1 end;
```

Figure 9.16 Structure text for ELEVATOR-1.

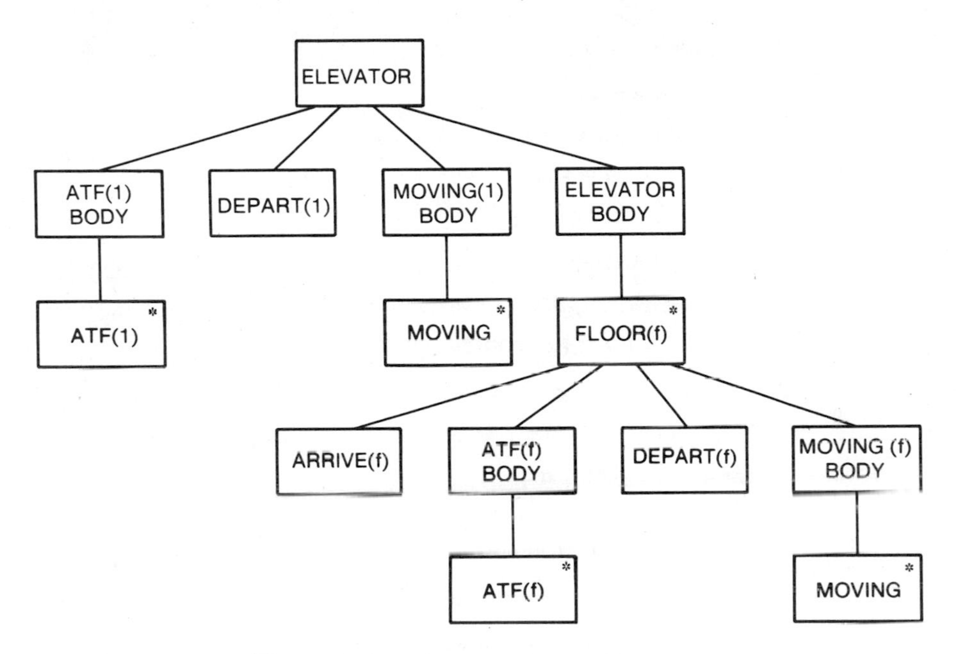

Figure 9.17 Refined structure diagram for elevator.

of conditions occurs in the real world, the following outputs should be produced by the product. Jackson distinguishes a number of different types of functions. The most elementary type is an *embedded function*, which simply produces a report when a specified set of circumstances holds. This is achieved by attaching the relevant operation to the structure diagram at the appropriate place. In the elevator problem, for example, a light is illuminated whenever a button is pressed. The structure diagram is shown in Figure 9.19(a). Similarly, the elevator process ELEVATOR-1 must issue commands to the elevator motor as shown in Figure 9.19(b).

If there is to be a display at each floor showing the current position of the elevator, two additional sets of commands must be added. LIGHT-ON(f) causes the light corresponding to floor f in each display to be illuminated, $1 \leq f \leq m$, and similarly for LIGHT-OFF(f). The structure text of Figure 9.18 must be modified so that LIGHT-ON(f) is inserted immediately after ARRIVE(f), and LIGHT-OFF(f) must appear immediately before DEPART(f). Details of this, and of inserting LIGHT-ON(1) and LIGHT-OFF(1), are left as an exercise to the reader (Problem 9.7).

Step 5. Product timing step: During this step, timing constraints on the product are specified. In the case of the elevator, suppose that sensors are

```
ELEVATOR-1 seq
   getsv ELEVATOR-SV;
   ATF(1)-BODY itr while (ATF(1))
      getsv ELEVATOR-SV;
   ATF(1)-BODY end;
   DEPART(1);
   MOVING(1)-BODY itr while (MOVING)
      getsv ELEVATOR-SV;
   MOVING(1)-BODY end;
   ELEVATOR-BODY itr
      FLOOR(f) seq
         getsv ELEVATOR-SV;
         ARRIVE(f);
         ATF(f)-BODY itr while (ATF(f))
            getsv ELEVATOR-SV;
         ATF(f)-BODY end;
         getsv ELEVATOR-SV;
         DEPART(f);
         MOVING(f)-BODY itr while (MOVING)
            getsv ELEVATOR-SV;
         MOVING(f)-BODY end;
      FLOOR(f) end;
   ELEVATOR-BODY end;
ELEVATOR-1 end;
```

Figure 9.18 Refined structure text for elevator.

placed at each floor in such a way that when the elevator comes within 6 inches of floor f the state changes from MOVING to ATF(f). Suppose further that the elevator travels at a speed of 2 feet per second. At that speed, it would cover the 6 inches in one-quarter of a second. Within that time, process ELEVATOR-1 (Figure 9.19) must detect that the elevator has arrived, and must send a stop command to the elevator motor. Also within that time, the motor must receive the stop command and take the appropriate action to

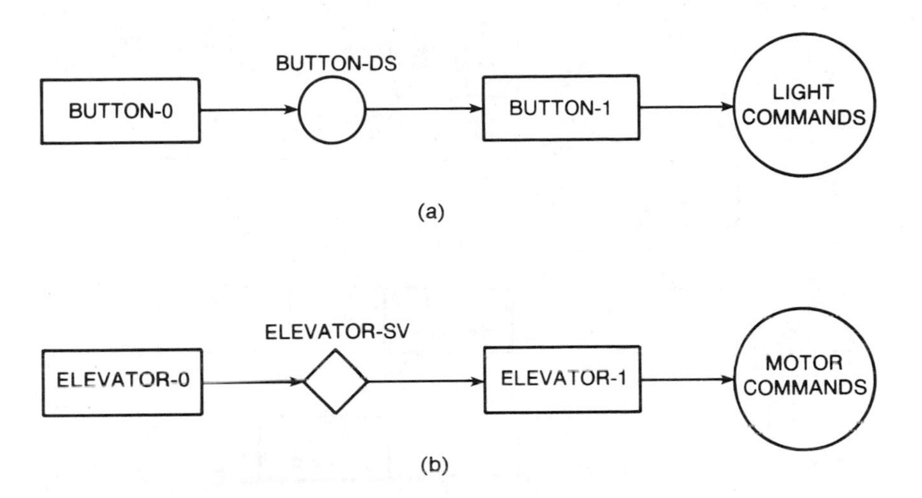

Figure 9.19 JSD structure diagram with embedded functions added.

ensure that the elevator stops at the correct place. During step 6, the implementation step, the developer checks that these timing constraints are satisfied.

It was remarked that the specifications are not complete until the end of step 5. As an example of this, the requirement that buttons light up when pressed was not incorporated until step 4. Furthermore, timing constraints were not even mentioned until step 5. But while the specifications were being produced, the design was being developed in parallel with the specifications.

Step 6. Implementation step: A strength of Jackson's method is that, once the detailed design is complete, implementation is virtually automatic.

Throughout this book, a consistent effort has been made to avoid drowning the reader in a level of detail. The elevator problem is entirely appropriate for illustrating the first five steps of Jackson's method. However, the implementation of a complete solution to the elevator problem is lengthy, and as a result, key aspects of the implementation step are lost. The reader who wishes to see a complete implementation is urged to consult [Jackson, 1983]; in this book, the implementation step will be illustrated by means of the following simpler problem.

> A file contains temperature records and pressure records. A temperature record consists of the letter T followed by an integer denoting a temperature in degrees Fahrenheit. A pressure record consists of the letter P followed by an integer denoting a pressure in pounds per square inch. The first record contains the letter F, while the last record contains the letter L. Determine the lowest temperature and the highest pressure.

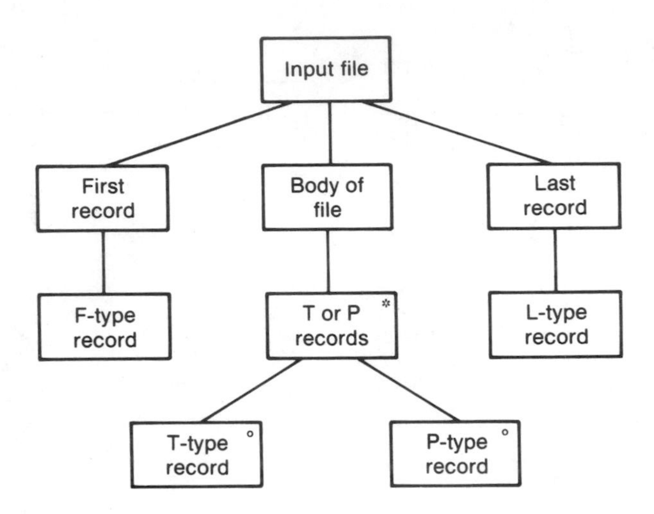

Figure 9.20 Structure of file containing temperature and pressure records.

The structure of the input file is shown in Figure 9.20. The first record is of type F. It is followed by 0 or more records of type T or type P. Finally, the last record is of type L. The structure text of a product to process this file is shown in Figure 9.21. Note the almost one-to-one correspondence between data in Figure 9.20 and procedures in Figure 9.21. Except for the addition of the prefix PROCESS-, the names of the procedures are identical to those of the data that they process. It is a fundamental tenet of Jackson's method that

```
PROCESS-INPUT-FILE seq
   PROCESS-FIRST-RECORD;
   PROCESS-BODY-OF-FILE itr
      PROCESS-T-OR-P-RECORD sel
         PROCESS-T-TYPE-RECORD;
      PROCESS-T-OR-P-RECORD alt
         PROCESS-P-TYPE-RECORD;
      PROCESS-T-OR-P-RECORD end;
   PROCESS-BODY-OF-FILE end;
   PROCESS-LAST-RECORD;
PROCESS-INPUT-FILE end;
```

Figure 9.21 Structure text for temperatures and pressures problem.

the structure of a module must reflect the structure of the data on which that module operates.

Implementation of this structure text is now straightforward. The **procedure division** of a COBOL implementation is shown in Figure 9.22. For the sake of brevity, details of **ERROR-ROUTINE** have been omitted. (The unexpected **read INPUT-RECORD** sentence in paragraph **PROCESS-LAST-RECORD** is necessary; in COBOL, the end-of-file condition is set to true only after an attempt has been made to read a record beyond the last record.)

Notice the close correspondence between the structure text and the code. In Chapter 16 the possibility of automatic programming is put forward.

```
procedure division.
PROCESS-INPUT-FILE.
    perform PROCESS-PRELIMINARY.
    perform PROCESS-FIRST-RECORD.
    perform PROCESS-BODY-OF-FILE until ((not TYPE-T) or (not TYPE-P)).
    perform PROCESS-LAST-RECORD.
    perform PROCESS-TERMINATE.
PROCESS-PRELIMINARY.
    open input INPUT-RECORD-FILE.
    move 10000 to MINIMUM-TEMPERATURE.
    move -10000 to MAXIMUM-PRESSURE.
    read INPUT-RECORD at end move "YES" to FILE-EOF.
PROCESS-FIRST-RECORD.
    if (not TYPE-F) perform ERROR-ROUTINE.
    read INPUT-RECORD at end move "YES" to FILE-EOF.
PROCESS-BODY-OF-FILE.
    if TYPE-T perform PROCESS-T-TYPE-RECORD
    else if TYPE-P perform PROCESS-P-TYPE-RECORD
    else perform ERROR-ROUTINE.
PROCESS-LAST-RECORD.
    if (not TYPE-L) perform ERROR-ROUTINE.
    read INPUT-RECORD at end move "YES" to FILE-EOF.
PROCESS-TERMINATE.
    display MINIMUM-TEMPERATURE, MAXIMUM-PRESSURE.
    close INPUT-RECORD-FILE.
    stop run.
PROCESS-T-TYPE-RECORD.
    if TEMPERATURE is less than MINIMUM-TEMPERATURE
        move TEMPERATURE to MINIMUM-TEMPERATURE.
    read INPUT-RECORD at end move "YES" to FILE-EOF.
PROCESS-P-TYPE-RECORD.
    if PRESSURE is greater than MAXIMUM-PRESSURE
        move PRESSURE to MAXIMUM-PRESSURE.
    read INPUT-RECORD at end move "YES" to FILE-EOF.
```

Figure 9.22 Portion of COBOL implementation of temperatures and pressures problem.

When a design has been produced using JSD, the process of translating the design into a programming language is in its broadest sense automatic.

9.6.4 Analysis of JSD

Notwithstanding its many strengths as a product development method, there are some difficulties associated with JSD. Consider, for example, paragraph PROCESS-TERMINATE in Figure 9.22. It consists of three sentences, namely

```
display MINIMUM-TEMPERATURE, MAXIMUM-PRESSURE.
close INPUT-RECORD-FILE.
stop run.
```

Even the reader who is totally unfamiliar with COBOL will quickly realize that the paragraph is a module of coincidental cohesion. It is not that JSD is incompatible with composite/structured design, but rather that the aims of JSD bear no relation to those of C/SD. As a result, it frequently happens that the module structure resulting from a JSD design is unacceptable from the viewpoint of C/SD.

The major advantages of C/SD are that the resulting product should be easily and safely maintainable, and that many of the component modules should be reusable. In [Jackson, 1983], there is no mention of product maintenance or of reusability. Does this mean that JSD is unacceptable as a development method? Not at all. There are two reasons why Jackson's method is important. First, JSD, as well as the two related methods, LCS [Warnier, 1981] and DSSD [Orr, 1981], has had a number of successes. These successes have not been restricted just to development, but have extended to maintenance. This means that while the techniques of C/SD are *sufficient* for product maintenance, they are by no means *necessary*. Second, many of Jackson's ideas have proved to be important within the broader context of software engineering. Even if JSD itself never achieves universal acceptance, many of the ideas underlying JSD have already taken hold.

The related methods of Warnier and of Orr have been mentioned a number of times. For the sake of completeness, a brief description of each is now given.

9.7 Methods of Jackson, Warnier, and Orr

Logical construction of systems (LCS) [Warnier, 1981] is an extension of Warnier's logical construction of programs (LCP) [Warnier, 1976]. Independently of Jackson, Warnier put forward the idea that the structure of a process should be the same as that of the data on which the process operates. In order to implement this idea he also needed a notation. Figure 9.23 depicts the Warnier diagram corresponding to Figure 9.20, the file of temperature and pressure records represented in Jackson's notation. The symbol $\oplus$ denotes selection, corresponding to **sel** ... **alt** in Jackson's notation.

Warnier diagrams, like Jackson's, are hierarchical, but they differ in that the number of times an item is repeated may be explicitly stated. For example, if an item occurs between 1 and n times, then the notation used is ITEM (1, n). If an item may or may not occur, then the notation is ITEM (0, 1).

In Jackson's method, the structure of a module reflects the structure of the data processed by that module. It may happen that the module has to process two or more types of data which are incompatible. For example, suppose that the input to a module is an array and the output is the transpose of the same array. In this case there is a *structure clash*. Jackson has developed a powerful technique for dealing with this type of situation, namely *program inversion* [Jackson, 1975]. In contrast, in Warnier's method the structure of a module is the structure of the input data; as a result, structure clashes do not occur.

As shown in Section 9.6.3, the implementation step of JSD is virtually a straightforward translation of the structure text into the chosen programming language. LCP similarly supports the development of the code from the Warnier diagram; this technique is termed *detailed organization*. For more information regarding this and other aspects of Warnier's techniques, the reader should consult [Warnier, 1976] and [Warnier, 1981].

JSP [Jackson, 1975] is a method for detailed design and implementation. Jackson extended his method to the complete software development process; the resulting method is termed JSD. In the same way, Orr's data structured systems development (DSSD) can be viewed as an extension of Warnier's LCP from the realm of just detailed design and implementation to all phases of software development. The product is described using Warnier's notation, an example of which was given in Figure 9.23. In fact, such diagrams are frequently termed Warnier–Orr diagrams to reflect the role played by Orr in extending Warnier's technique. For full details of Orr's method the reader should refer to [Orr, 1981] or [Hansen, 1983].

```
           ( First record { F-type record
           (
           (                                  ( T-type record
Input file { Body of file { T or P records {        ⊕
           (                                  ( P-type record
           (
           ( Last record { L-type record
```

Figure 9.23 Warnier diagram corresponding to Figure 9.20.

9.8 Object-Oriented Design

The aim of object-oriented design (OOD) is to determine the objects in a product and then to design the product in terms of those objects. As pointed out in Section 8.5, a design which uses an *abstract data type* is superior to one which merely uses a *data structure* together with the operations performed on that data structure. In the same way, a design in terms of *classes* is superior to one which uses only *objects*; recall that an object is an instance of a class. Thus a secondary aim of OOD is to identify object classes where appropriate and then to use them to design the product.

As was mentioned in Section 8.7, objects are supported by very few programming languages because the concept of inheritance, a necessary prerequisite for an object-oriented language, is found in only a few programming languages such as Simula 67 [Dahl, Myrhaug, and Nygaard, 1973], Smalltalk [Goldberg, 1984], C++ [Stroustrup, 1986], Loops [Bobrow and Stefik, 1983], and Flavors [Moon, 1986]. At the same time, popular languages such as C, COBOL, FORTRAN, and Pascal do not support objects as such, and even Ada has only a limited form of inheritance. Since the majority of the readers of this book are probably unlikely to develop software in one of these somewhat specialized languages, one possible approach would be to suggest to the few readers who may use those languages that they should consult a book such as [Cox, 1986] and omit any treatment of OOD in this book.

That would not be a good idea. While it is true that OOD as such is not supported by the majority of popular languages, a large subset of OOD can be used. As explained in Section 8.7, a class is an abstract data type with inheritance, an object is an instance of a class. Since inheritance is not generally supported, the solution is to utilize those aspects of OOD that can be achieved in the programming language used in the project, that is to say, to use *abstract data type design*. Abstract data types can be implemented in virtually any language that supports **type** statements. In Ada, for example, a good way of implementing an abstract data type is by means of a package with a private type (Section 8.6). Even in those languages which do not support type statements as such, and hence cannot support abstract data types, it may still be possible to implement data encapsulation. For example, in Figure 8.26 it is shown how data encapsulation may be achieved in FORTRAN. The widespread use of parallel variants of FORTRAN in supercomputers means that FORTRAN is still an important programming language. A FORTRAN product written for, say, the Cray Y-MP, is no different, from the viewpoint of maintenance, to any other software product. Every possible way of easing the maintenance problem is to be encouraged, including the use of OOD. Thus, rather than omitting OOD from this book on the grounds that full OOD is not practical to achieve in the majority of programming languages, the approach taken here will be to focus on abstract data type design. In this way, the major ideas of OOD can be used.

In fact, this approach has been taken by others, the majority of whom fail to point out that what they are describing is not object-oriented design (OOD), but rather abstract data type design (ADTD). Because of the widespread misuse of the term OOD in this way and because of the fact that ADTD is rarely identified as such, from now on the incorrect term OOD will be used, with the hope that the reader will appreciate that what is really meant is ADTD, or "OOD without inheritance."

OOD consists of four steps [Abbott, 1983; Booch, 1983; Yau and Tsai, 1986].

1. Define the problem as concisely as possible.
2. Develop an informal strategy, a general sequence of steps for satisfying the specifications subject to the constraints.
3. Formalize the strategy:
   a. Identify the objects and their attributes.
   b. Identify operations to be applied to the objects.
   c. If possible, identify object classes.
4. Proceed to detailed design, and implementation.

Steps 1 and 2 are performed during the specifications phase. OOD itself is a technique applicable only to the architectural design phase. Any reasonable specification technique may be used in conjunction with OOD.

Steps 2 and 3 may be applied stepwise until the architectural design is satisfactory. For large products, this stepwise approach is all but mandatory. This is not unexpected; stepwise refinement is used throughout software engineering in order to reduce the number of chunks of information that must be dealt with at any one time, as described in Chapter 5.

To show how OOD is used, a case study is presented. As before, the elevator problem will be handled. By using the same example, the reader is put in the position of being able to compare methods, without having to worry about the ramifications of the problem itself.

9.8.1 Object-Oriented Design Case Study

Step 1. Concise problem definition: The first step in OOD is to define the product as briefly and concisely as possible, preferably in a single sentence. One possible way of doing this is:

> Buttons in elevators and on the floors are to be used to control the motion of n elevators in a building with m floors.

Step 2. Informal strategy: In order to come up with an informal strategy for solving the problem, the constraints must be determined. The three constraints of the elevator problem appear on page 272. Now the informal strategy can be expressed, preferably in a single paragraph. One possible paragraph is:

- Buttons in elevators and on the floors control the movement of n elevators in a building with m floors. Buttons illuminate when pressed to request the elevator to stop at a specific floor; the illumination is canceled when the request has been satisfied. When an elevator has no requests to service, it should remain at its current floor with its doors closed.

As mentioned previously, these first two steps are carried out during the specifications phase. Now the object-oriented design itself is developed.

Step 3. Formalize the strategy: The heart of OOD is identifying the objects and the operations to be performed on those objects. The way this is usually done in OOD is by identifying the nouns and the verbs in the informal strategy, excluding those that lie outside the problem boundary, and then using what is left as the basis for the objects and the operations, respectively.

At this point the reader may be somewhat confused, and understandably so. After all, the technique of identifying nouns and verbs has just been used in the JSD case study of Section 9.6.3, and now the same technique is being recommended for object-oriented design. The reason for this duplication is that in both JSD and OOD the designer has to identify the data of the product, as well as the operations performed on or by that data. In the case of JSD, the data are called *entities* and the operations are called *actions*. For OOD, they are termed *objects* and *operations*, respectively. The similarity is therefore not unexpected. At the same time, it is important to realize that the similarities begin and end with the identification of nouns and verbs—JSD and OOD are markedly different from this point on.

Returning to OOD, the informal strategy is now reproduced, but this time with the nouns identified by being printed in uppercase and in a different typeface.

BUTTONS in ELEVATORS and on the FLOORS control the MOVEMENT of n ELEVATORS in a BUILDING with m FLOORS. BUTTONS illuminate when pressed to request the ELEVATOR to stop at a specific FLOOR; the ILLUMINATION is canceled when the REQUEST has been satisfied. When an ELEVATOR has no REQUESTS to service, it should remain at its current FLOOR with its DOORS closed.

There are eight different nouns, namely **BUTTON, ELEVATOR, FLOOR, MOVEMENT, BUILDING, ILLUMINATION, REQUEST**, and **DOOR**. Three of these nouns, namely **FLOOR, BUILDING**, and **DOOR**, lie outside the problem boundary and may therefore be ignored. Three of the remaining nouns, namely **MOVEMENT, ILLUMINATION**, and **REQUEST**, are abstract nouns. That is to say, they "identify ideas or quantities that have no physical existence" [World Book Encyclopedia, 1982]. It is a useful rule of thumb that abstract nouns rarely end up corresponding to objects. Instead, they frequently are attributes of objects. For example, **ILLUMINATION** will turn out to be an attribute of **BUTTON**. Other abstract nouns are simply extraneous to the problem description; **REQUEST** is an example of this. This leaves two candidate objects, namely **ELEVATOR** and **BUTTON**.

Following the rule of thumb of deleting abstract nouns may occasionally result in an object being deleted. If this happens, the situation will be corrected at the next stepwise refinement of the design. OOD is an iterative process that makes use of stepwise refinement, like so many other of the techniques of software engineering.

Now the verbs in the informal strategy are identified and related to the candidate objects, namely **ELEVATOR** and **BUTTON**. The verbs are *control*, *illuminate*, *press*, *request*, *stop*, *cancel*, *satisfy*, *service*, and *remain*. The result of associating the relevant verbs with the two candidate objects and discarding the others is shown in Figure 9.24.

This first refinement is clearly unsatisfactory. Operations *stop* and *remain* are associated with **ELEVATOR**, but there does not seem to be any way of making the elevator go! This has to be fixed. Rereading the informal strategy, it appears that the abstract noun **MOVEMENT** that previously was correctly rejected as an object should in fact be considered an operation, namely *move*. Also, the participle "closed" has been ignored, while it will be necessary to open and close the doors of elevator. Finally, attributes of both objects and operations have also been ignored. This is corrected in Figure 9.25, which shows the second refinement.

Object	Operation
ELEVATOR	*request* *stop* *remain*
BUTTON	*press* *illuminate* *cancel*

Figure 9.24 Object-operations table: first refinement.

Object	Attribute	Operation	Attribute
ELEVATOR		*request* *move* *stop*	 direction
BUTTON	illuminated	*press* *cancel*	initial

Figure 9.25 Object-operations table: second refinement.

Two objects have been identified, namely ELEVATOR and BUTTON. The operations performed on ELEVATOR are *request*, *move*, and *stop*. Operation *move* has an attribute direction; it can take on the values up or down. Two operations are performed on a BUTTON, namely *press* and *cancel*. A BUTTON may be illuminated; when it is pressed, it is important whether or not this is the initial press.

But still there are problems, because there is some degree of redundancy in the second refinement. The operation *press* a BUTTON includes two different types of operation, namely pressing an elevator button and pressing a floor button. This latter operation is identical to the operation *request* an ELEVATOR. The solution is to remove operation *request* from object ELEVATOR. Furthermore, operation *stop* is also superfluous in that it can be achieved simply by not invoking operation *move*. Finally, it is simpler to have the two operations *move one floor up* and *move one floor down*, rather than operation *move* with attribute direction. Making these changes yields the third refinement, depicted in Figure 9.26.

Having extracted the objects, the architectural design is developed around them. The first refinement is shown in Figure 9.27. Process CONTROLLER uses the objects BUTTON and ELEVATOR in the sense that it invokes the procedures of each object to operate on the data structure of that

Object	Attribute	Operation	Attribute
ELEVATOR		*move one floor up* *move one floor down* *open door* *close door*	
BUTTON	illuminated	*press* *cancel*	initial

Figure 9.26 Object-operations table: third refinement.

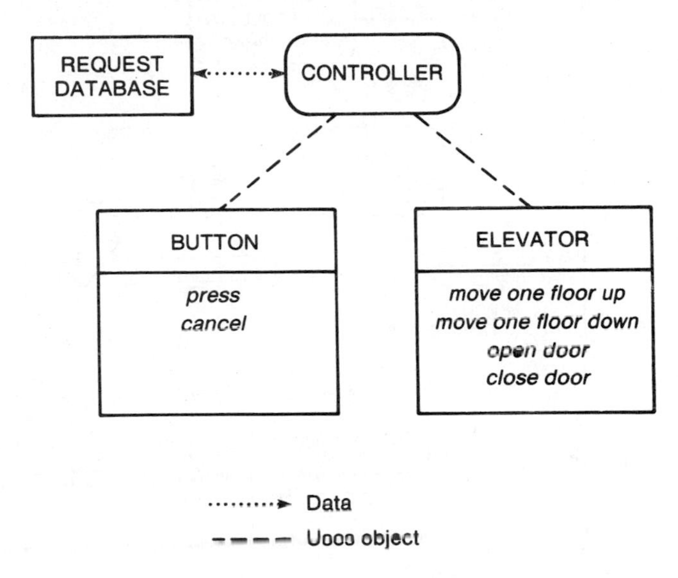

Figure 9.27 Architectural design: first refinement.

object. Now process **CONTROLLER** is expanded further, resulting in the second and final refinement shown in Figure 9.28. Since, as mentioned previously, OOD is a technique specifically for the architectural design, completing the architectural design formally concludes the OOD process.

Step 4. Detailed design and implementation: A detailed design is now developed centered around these two objects. Any suitable technique may be used, such as stepwise refinement as described in Chapter 5. The detailed design of one module is shown in Figure 9.29. Once the detailed design is complete, the product is implemented.

9.9 Detailed Design

Two techniques for detailed design have already been presented. In Chapter 5 a description of stepwise refinement was given. It was then applied to detailed design using flowcharts. Then, step 6 of JSD is detailed design plus implementation; this was illustrated at the end of Section 9.6.3 by means of an example, namely the file of temperature and pressure data.

In addition to stepwise refinement and Jackson's method, formal techniques can be used to advantage in detailed design. In Chapter 6 it was suggested that implementing a product and then proving it correct could be

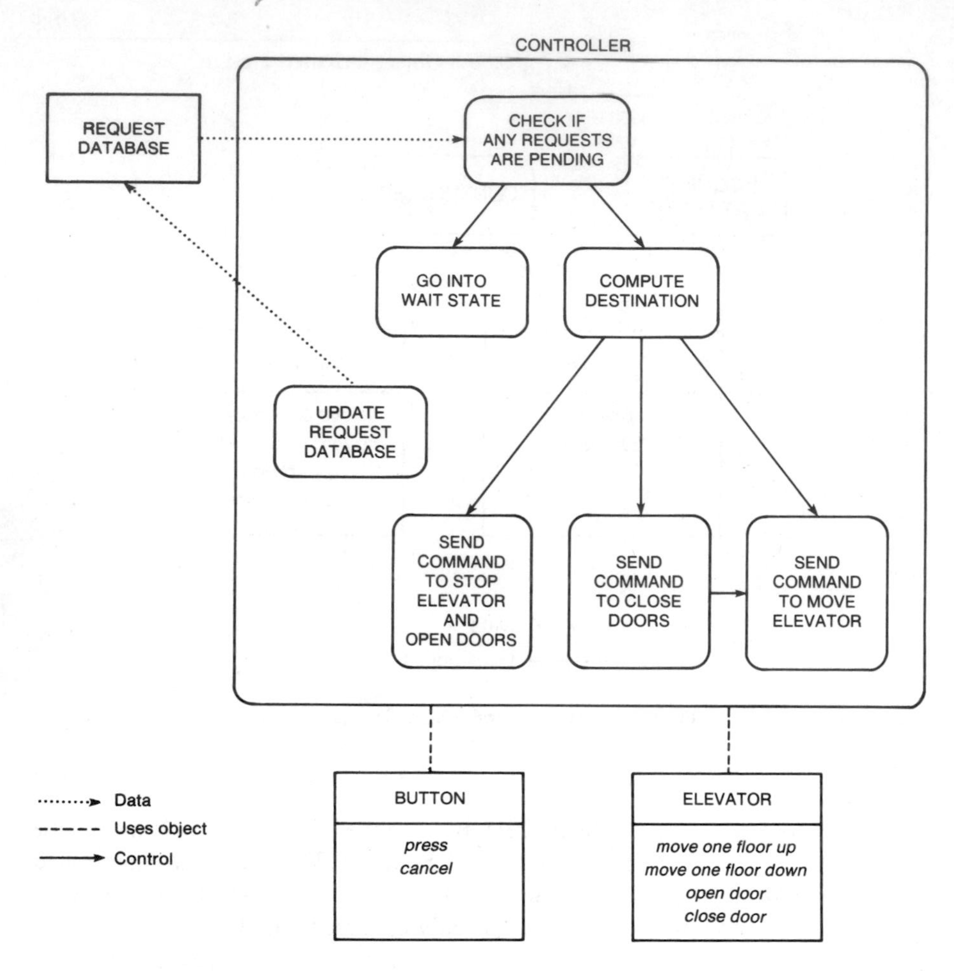

Figure 9.28 Architectural design: second refinement.

```
procedure CHECK_IF_ANY_REQUESTS_ARE_PENDING;
begin
  if there are one or more requests in REQUEST_DATABASE then
    if DIRECTION_OF_MOTION is UP then
      if there is a request for CURRENT_FLOOR then
        SEND_COMMAND_TO_STOP_ELEVATOR_AND_OPEN_DOORS;
      else
        SEND_COMMAND_TO_MOVE_ONE_FLOOR_UP;
      end if;
    elsif DIRECTION_OF_MOTION is DOWN then
      (similar to UP case)
    else elevator is currently at rest, so send elevator in direction of request;
    end if;
  else there are no requests, so
    GO_INTO_WAIT_STATE;
  end if;
end CHECK_IF_ANY_REQUESTS_ARE_PENDING;
```

Figure 9.29 Detailed design of module CHECK_IF_ANY_REQUESTS_ARE_PENDING.

counterproductive. But developing the proof and the detailed design in parallel, and carefully testing the code as well, is quite a different matter. Formal techniques at the detailed design phase can greatly assist in three ways. First, the state of the art in correctness proving is such that, while it cannot be applied to a product as a whole, it can be applied to module-sized pieces of a product. Second, developing a proof together with the design should lead to a design with fewer faults than if correctness proving were not used. Third, if the same programmer is responsible for both the detailed design and the implementation, as is often the case, then that programmer will feel confident that the detailed design is correct. This positive attitude towards the design should lead to fewer faults in the code.

9.10 Comparison of Process-, Data-, and Object-Oriented Design

Examples of each of the three basic approaches to design have so far been presented. First, a process-oriented method, namely data flow analysis, was described. This was followed by Jackson system development, a data-oriented approach. The third approach was object-oriented design, an example of a hybrid technique. The obvious question is: Which is best?

In one sense, the question is unanswerable, because of the duality of data and processes [Goldberg, 1986]. That is to say, it is all very well to base a design on the data or to base it on the processes, but data and processes are two sides of the same coin. The processes operate on the data, while the data are operated on by the processes. Goldberg gives the analogy of the duality of nouns and verbs, a particularly relevant analogy in the light of the role played by nouns and verbs in JSD and OOD. Furthermore, comparison of Figures 9.20 and 9.21 reflects Jackson's assertion regarding the equivalence of data and processes: The structure of the process must in every way reflect the structure of the data operated on by that process. In short, although process-oriented design and data-oriented design appear to be diametric opposites, fundamentally they are equivalent. And object-oriented design, combining as it does process- and data-oriented design, is then fundamentally no different than the other two.

While the three approaches are indistinguishable from a theoretical viewpoint, it is certainly possible to contrast specific instances of each approach. In other words, it is meaningful to compare data flow analysis (DFA) and object-oriented design (OOD) with Jackson system development (JSD).

Data flow analysis is a top-down, decompositional technique. That is to say, the product as a whole is broken down into smaller parts until each part is a module of functional cohesion. Object-oriented design is similarly a top-down, decompositional technique; starting with the product as a whole, the objects are extracted. However, Cameron, the co-inventor of JSD with

Jackson, suggests that any decompositional technique is critically dependent on the initial decomposition [Cameron, 1986]. As Cameron puts it, "... the first decomposition is a commitment to a system structure; the commitment has to be made from a position of ignorance." In other words, at the beginning of the decomposition process a decision has to be taken at a stage when the designer as yet knows little about the product. But that decision will critically affect the final product, because by decomposing the product in that way at the beginning, the designer may be precluding the possibility of decomposing the product in a different way at a later stage. Cameron suggests that techniques based on decomposition work well only when the designer is working on a product that he or she really understands. In contrast, JSD is a bottom-up, compositional technique. The designer starts with nothing, and incrementally adds pieces until the design is complete.

Thus there is a fundamental difference between JSD on the one hand, and OOD and DFA on the other, in that OOD and DFA seem to be appropriate when the designer is familiar with the application area. But when the application area is new to the software developer, and this applies particularly when the product is research-oriented, then JSD would appear to be a better choice.

A second way of comparing design methods is to consider maintenance issues. As mentioned previously, JSD is not really maintenance-oriented, while the aim of DFA is to arrive at modules with high cohesion, with all the attendant advantages with regard to maintenance. Similarly, OOD is frequently recommended from the viewpoint of maintainability of the resulting design. While DFA has been in use for more than a decade, object-oriented design is somewhat newer, and there is not as yet specific evidence to back up the claim that OOD results in products that are easier to maintain than when other design methods are used.

A third way of comparing methods is from the viewpoint of reusable code. Again, DFA and OOD seem to have the edge over JSD. Nothing in Jackson's method indicates that using it will result in reusable modules. On the contrary, the fact that JSD, as shown in Section 9.6.4, can lead to modules of coincidental cohesion implies that JSD may in fact be counterproductive from the reusability viewpoint.

To complicate the issue of choosing between design methods still further, as stated at the beginning of this chapter there are in fact many hundreds of different design methods from which to choose, not just the three that have been presented as representatives of the three main approaches to design. When faced with having to decide between competing design techniques, one way of reducing the risk is for the software manager to choose a technique that has previously been used with some success in applications similar to the one to be designed. This approach will preclude the use of previously untried techniques, or at least, techniques previously untried in that application area. But while this may mean that some new, outstanding method may be overlooked until it has become more widely accepted, the manager can be confident that there is a good chance that the chosen technique will result in a high-quality product.

9.11 Design of Real-Time Systems

9.11.1 Difficulties Associated with Real-Time Systems

A real-time system is characterized by the fact that its inputs come from the real world, and that the software has no control over the timing of those inputs. Furthermore, each input must be processed before the next input arrives. An example of such a system is a computer-controlled nuclear reactor. Inputs such as the temperature of the core and the level of the water in the reactor chamber are continually being sent to the computer which must be able to read the value of each input and perform the necessary processing before the next input arrives. Another example is a computer-controlled intensive care unit. There are two types of patient data, namely routine information such as heart rate, temperature, and blood pressure of each patient and emergency information when the system deduces that the condition of one or more patients has become critical. When such emergencies occur, the software must be able to process both the routine inputs as well as the emergency-related inputs from one or more patients.

A characteristic of many real-time systems is that they are implemented on distributed hardware. For example, software controlling a war plane may be implemented on five computers, one to handle navigation, another the weapons system, a third for electronic countermeasures, a fourth to control the flight hardware such as wing flaps and engines, and the fifth to propose tactics in combat. Because hardware is not totally reliable, there may be additional computers that can automatically take the place of a malfunctioning unit. Not only does the design of such a system have major communications implications, but timing issues arise as a consequence of the distributed nature of the system. For example, it can happen under combat conditions that the tactical computer might suggest that the pilot should climb, while the weapons computer recommends that the pilot go into a dive so that a particular weapon may be launched under optimal conditions. But the human pilot decides to move the stick to the right, thereby sending a signal to the flight hardware computer to make the necessary adjustments so that the plane banks in the indicated direction. All this information must be carefully managed in such a way that the actual motion of the plane takes precedence in every way over suggested maneuvers. Furthermore, the actual motion must be relayed to the tactical and weapons computers so that new suggestions can be formulated in the light of real, rather than suggested, conditions.

A further difficulty of real-time systems is the problem of synchronization. Suppose that a real-time system is to be implemented on distributed hardware. For example, there may be two computers C_1 and C_2 connected to one another, as well as to a file server. Processes A and B, running on C_1 and C_2, respectively, both require exclusive use of records R_1 and R_2. One possible scenario is shown in Figure 9.30(a). Process A requests exclusive use of each record in turn, the requests are granted, and process A then relinquishes control of both records. Then process B goes through the same steps. But unless care is taken in the design process, the scenario shown in Figure 9.30(b) could also occur. Process A requests from the file server exclusive use of record R_1, and this request is granted. Now process B, running concurrently on C_2, requests and is granted exclusive use of R_2. When this request has been granted by the file server, process A requests exclusive use of R_2. But exclusive use of R_2 has already been granted to process B. As a result, process A enters a wait state until the request can be granted. Now process B requests R_1. But this request cannot be granted either, so process B running on C_2 also enters a wait state. The system as a whole is thus in a state of *deadlock*, or so-called *deadly embrace*. Of course, deadlock does not occur only in real-time systems implemented on distributed hardware. But it is particularly troublesome in real-time systems, where, as stated previously, there is no control over the order or timing of the inputs, and the situation can be complicated by the distributed nature of the hardware. In addition to deadlock, other synchronization problems are possible, including race conditions; for details, the reader may refer to [Deitel, 1983] or other operating systems textbooks.

From these examples it is clear that the major difficulty with regard to the design of real-time systems is determining that the timing constraints are met by the design. That is to say, the design method should provide a mechanism for checking that, when implemented, the design will be able to read and process incoming data fast enough. Furthermore, it should be possible to show that synchronization issues in the design have been correctly addressed. In brief, what characterizes real-time systems is timing; the difficulties that are encountered with real-time systems are essentially difficulties that arise as a direct consequence of time being a major component of the real-world system being controlled by the real-time software.

9.11.2 Real-Time Design Methods

Since the beginning of the computer age, advances in hardware technology have outstripped, in almost every respect, advances in software technology. Thus, while the hardware exists to handle every aspect of the real-time systems described previously, software design technology has lagged considerably behind. In some areas of real-time software engineering major progress has been made. For instance, many of the specification methods of

Time	Process A running on computer C_1	Process B running on Computer C_2
↓	Request exclusive use of R_1 (Request granted by file server)	
	Request exclusive use of R_2 (Request granted by file server)	
	...	
	...	
	Relinquish R_2	
	Relinquish R_1	
	...	
		Request exclusive use of R_2 (Request granted by file server)
		Request exclusive use of R_1 (Request granted by file server)
		...
		...
		Relinquish R_1
		Relinquish R_2

(a) No Deadlock

Time	Process A running on computer C_1	Process B running on computer C_2
↓	Request exclusive use of R_1 (Request granted by file server)	
	...	Request exclusive use of R_2 (Request granted by file server)
	...	
	Request exclusive use of R_2 (Process A enters wait state)	...
		...
		Request exclusive use of R_1 (Process B enters wait state)

(b) Deadlock

Figure 9.30 Possible timing problems with distributed real-time systems.

Chapter 7 can be used to specify real-time systems. But software design has not reached the same level of sophistication. Great strides are indeed being made, but the state of the art is not yet comparable to what has been achieved with regard to specification methods. Because almost any design method for real-time systems is preferable to no method at all, a number of real-time design methods are being used in practice. But there is still a long way to go before it will be possible to design real-time systems such as those described previously and to be certain that, before the system has been implemented, every real-time constraint will be met and that synchronization problems cannot arise.

Most real-time design methods fall into one of two categories: extensions of nonreal-time methods to the real-time domain and Ada-specific methods. For example, structured development for real-time systems (SDRTS) [Ward and Mellor, 1985] is essentially an extension of structured systems analysis (Section 7.2), data flow analysis (Section 9.3), and transaction analysis (Section 9.4) to real-time software. The development method includes a component for real-time design. Another extension of the same group of methods to the real-time domain is design approach for real-time systems (DARTS) [Gomaa, 1986]. Software cost reduction (SCR) [Kmielcik et al., 1984] is a real-time design method based on the concept of information hiding [Parnas, 1971, 1972a, 1972b].

Turning now to Ada-specific real-time methods, process abstraction method for embedded large applications (PAMELA) was developed by Cherry as a graphics-based method for describing an Ada design [Cherry, 1986]. The method was developed around the Ada tasking mechanism, which is used in Ada for interprocess communication and synchronization, and the method is therefore of limited use outside of the Ada context.

Object-oriented design described in Section 9.8 is not, in itself, an Ada-specific method. But as popularized by Booch [Booch, 1983], OOD has acquired an Ada-oriented flavor. The strength of OOD lies in its support for abstract data types; it is not specifically a real-time design method. Thus timing and synchronization are not explicitly handled. Nevertheless, because of its Ada orientation and because Ada has been mandated by the Department of Defense for embedded real-time products (Chapter 14), OOD is being used as a real-time design method for real-time Ada products. The fact that timing is not explicitly handled should not come as too much of a surprise. After all, Ada does not support any mechanism for ensuring that code will be executed by a given real-time deadline. In fact, the Ada programmer has very little control over process scheduling.

Kelly has provided an interesting comparison of four real-time design methods [Kelly, 1987]. The four, namely SDRTS, SCR, OOD, and PAMELA, were described by Kelly as methods "of current interest in real-time systems development." Nevertheless, at the time he wrote (1987), limited experience had been gained in the use of any of the four methods. This follows partly as consequence of the newness of the methods, but also because none of them fully support all the features needed for a real-time design method. As mentioned previously, the state of the art in real-time design methods is not where we would like it to be.

Chapter Review

The design phase consists of architectural design, followed by detailed design (Section 9.1). Three approaches to design are described, namely process-oriented design (Section 9.2), data-oriented design (Section 9.5), and hybrid design. Examples of each approach are given, and case studies are presented to illustrate the methods. Two instances of process-oriented design are described, namely data flow analysis (Section 9.3) and transaction analysis (Section 9.4). The methods of Jackson, Warnier, and Orr (Section 9.7) are all data-oriented; Jackson's method is described in some detail in Section 9.6. Object-oriented design, a hybrid method, is presented in Section 9.8. Methods for detailed design are put forward in Section 9.9. As a consequence of the duality of data and processes, the three approaches are theoretically equivalent; a comparison of specific instances of the three approaches is given in Section 9.10. Finally, design aspects of real-time systems are described in Section 9.11. The lack of an all-encompassing real-time design method is pointed out.

For Further Reading

A useful overall source on design techniques is [Freeman and Wasserman, 1983]. The tutorial covers a wide variety of methods, including those described in this chapter. [Yau and Tsai, 1986] is a survey of design techniques. The authors cite over 80 references to a broad variety of approaches to design. A wide variety of design methods can also be found in the February 1986 issue of *IEEE Transactions on Software Engineering*, which is devoted exclusively to software design methods.

Data flow analysis and transaction analysis are described in books such as [Gane and Sarsen, 1979] and [Yourdon and Constantine, 1979]. Related methods are to be found in [Martin and McClure, 1985]. Jackson's method is described in [Jackson, 1975] and [Jackson, 1983]; another excellent source is the work of Cameron, Jackson's associate, [Cameron, 1986, 1988]. For readers interested in Warnier's work, the original sources are [Warnier,

1976] and [Warnier, 1981]. The approach of Orr can be found in [Orr, 1981] and [Hansen, 1983]. Turning now to object-oriented design, information can be obtained from [Abbott, 1983], [Booch, 1986], [Cox, 1986], [Stefik and Bobrow, 1986], and [Peterson, 1988]. Formal design methods are described in [Hoare, 1987].

With regard to real-time design, specific methods are to be found in [Kmielcik et al., 1984], [Ward and Mellor, 1985], [Gomaa, 1986], and [Faulk and Parnas, 1988]. A comparison of four real-time design methods is to be found in [Kelly, 1987]. Ada-specific design methods are described in [Buhr, 1984], [Cherry, 1986], and [Nielsen and Shumate, 1987].

Problems

9.1 Starting with the DFD you drew for Problem 7.3, use data flow analysis to design a product for determining whether a bank statement is correct.

9.2 Consider the problem of designing the software for an automatic teller machine (ATM). The user puts his or her card into a slot, keys in a four-digit PIN (personal identification number), and can then perform the following operations on up to four different bank accounts: deposit any amount, withdraw up to $200 in units of $20 (the account may not be overdrawn), determine account balance, or transfer funds between two accounts. Use transaction analysis to design the software for the ATM. At this stage, omit error-handling capabilities.

9.3 Now take your design for Problem 9.2 and add to it modules to perform error handling. Carefully examine the resulting design and determine the cohesion and coupling of the modules. Be on the lookout for situations such as that depicted in Figure 9.10.

9.4 Starting with your data flow diagram for the automated library circulation system (Problem 7.4), design the circulation system using data flow analysis.

9.5 Repeat Problem 9.4 using transaction analysis. Which of the two methods did you find to be more appropriate?

9.6 A file contains expense account records. The records are of three types: accommodation expenses, travel expenses, and food expenses. Each record includes the date the expense was incurred and the name of the employee who incurred it. The records are sorted by name and by date within name. Use JSD notation to specify the structure of the file.

9.7 Modify Figure 9.18 by inserting LIGHT-ON and LIGHT-OFF commands at the appropriate places.

9.8 Starting with the specifications for the automated library circulation system (Problem 7.4), design the library system using Jackson's method.

9.9 Design the ATM software (Problem 9.2) using Jackson's method.

9.10 Repeat Problem 9.6 using Warnier's notation.

9.11 Repeat Problem 9.8 using object-oriented design.

9.12 Design the ATM software (Problem 9.2) using object-oriented design.

9.13 (Term Project) Starting with your specifications of Problem 7.11, design the Plain Vanilla Ice Cream Corporation product. Use the design method specified by your instructor.

9.14 (Readings in Software Engineering) Your instructor will distribute copies of [Kelly, 1987]. Determine criteria for deciding when to use each of the four methods described by Kelly.

References

[Abbott, 1983] R. J. ABBOTT, "Program Design by Informal English Descriptions," *Communications of the ACM* **26** (November 1983), pp. 882–894.

[Bobrow and Stefik, 1983] D. G. BOBROW AND M. STEFIK, *The Loops Manual*, Intelligent Systems Laboratory, Xerox Corporation, Palo Alto, CA, 1983.

[Booch, 1981] E. G. BOOCH, "Describing Software Designs in Ada," *ACM SIGPLAN Notices* **16** (September 1981), pp. 42–47.

[Booch, 1983] G. BOOCH, *Software Engineering with Ada*, Second Edition, Benjamin/Cummings, Menlo Park, CA, 1983.

[Booch, 1986] G. BOOCH, "Object-Oriented Development," *IEEE Transactions on Software Engineering* **SE-12** (February 1986), pp. 211–221.

[Buhr, 1984] R. J. A. BUHR, *System Design in Ada*, Prentice-Hall, Englewood Cliffs, NJ, 1984.

[Cameron, 1986] J. R. CAMERON, "An Overview of JSD," *IEEE Transactions on Software Engineering* **SE-12** (February 1986), pp. 222–240.

[Cameron, 1988] J. CAMERON, *JSP & JSD: The Jackson Approach to Software Development*, Second Edition, IEEE Computer Society Press, Washington, DC, 1988.

[Cherry, 1986] G. CHERRY, *The PAMELA Designer's Handbook*, Thought Tools, Reston, VA, 1986.

[Cox, 1986] B. J. COX, *Object-Oriented Programming: An Evolutionary Approach*, Addison-Wesley, Reading, MA, 1986.

[Dahl, Myrhaug, and Nygaard, 1973] O.-J. DAHL, B. MYRHAUG, AND K. NYGAARD, *SIMULA begin*, Auerbach, Philadelphia, PA, 1973.

[Deitel, 1983] H. M. DEITEL, *An Introduction to Operating Systems*, Addison-Wesley, Reading, MA, 1983.

[Faulk and Parnas, 1988] S. R. FAULK AND D. L. PARNAS, "On Synchronization in Hard-Real-Time Systems," *Communications of the ACM* **31** (March 1988), pp. 274–287.

[Freeman and Wasserman, 1983] P. FREEMAN AND A. I. WASSERMAN, *Tutorial: Software Design Techniques*, Fourth Edition, IEEE Computer Society Press, Washington, DC, 1983.

[Gane and Sarsen, 1979] C. GANE AND T. SARSEN, *Structured Systems Analysis: Tools and Techniques*, Prentice-Hall, Englewood Cliffs, NJ, 1979.

[Goldberg, 1984] A. GOLDBERG, *Smalltalk-80: The Interactive Programming Environment*, Addison-Wesley, Reading, MA, 1984.

[Goldberg, 1986] R. GOLDBERG, "Software Engineering: An Emerging Discipline," *IBM Systems Journal* **25** (No. 3/4, 1986), pp. 334–353.

[Gomaa, 1986] H. GOMAA, "Software Development of Real-Time Systems," *Communications of the ACM* **29** (July 1986), pp. 657–668.

[Hansen, 1983] K. HANSEN, *Data Structured Program Design*, Ken Orr and Associates, Topeka, KS, 1983.

[Hoare, 1987] C. A. R. HOARE, "An Overview of Some Formal Methods for Program Design," *IEEE Computer* **20** (September 1987), pp. 85–91.

[Jackson, 1975] M. A. JACKSON, *Principles of Program Design*, Academic Press, New York, NY, 1975.

[Jackson, 1983] M. A. JACKSON, *System Development*, Prentice-Hall, Englewood Cliffs, NJ, 1983.

[Kelly, 1987] J. C. KELLY, "A Comparison of Four Design Methods for Real-Time Systems," *Proceedings of the Ninth International Conference on Software Engineering*, Monterey, CA, March 1987, pp. 238–252.

[Kmielcik et al., 1984] J. KMIELCIK ET AL., "SCR Methodology User's Manual," Grumman Aerospace Corporation, Report SRSR-A6-84-002, 1984.

[Martin and McClure, 1985] J. P. MARTIN AND C. MCCLURE, *Diagramming Techniques for Analysts and Programmers*, Prentice-Hall, Englewood Cliffs, NJ, 1985.

[Moon, 1986] D. A. MOON, "Object-Oriented Programming with Flavors," *Proceedings of the Conference on Object-Oriented Programming Systems, Languages and Applications, ACM SIGPLAN Notices* **21** (November 1986), pp. 1–8.

[Nielsen and Shumate, 1987] K. W. NIELSEN AND K. SHUMATE, "Designing Large Real-Time Systems in Ada," *Communications of the ACM* **30** (August 1987), pp. 695–715.

[Orr, 1981] K. ORR, *Structured Requirements Definition*, Ken Orr and Associates, Inc., Topeka, KS, 1981.

[Parnas, 1971] D. L. PARNAS, "Information Distribution Aspects of Design Methodology," *Proceedings of the IFIP Congress*, Ljubljana, Yugoslavia, 1971, pp. 339–344.

[Parnas, 1972a] D. L. PARNAS, "A Technique for Software Module Specification with Examples," *Communications of the ACM* **15** (May 1972), pp. 330–336.

[Parnas, 1972b] D. L. PARNAS, "On the Criteria to be Used in Decomposing Systems into Modules," *Communications of the ACM* **15** (December 1972), pp. 1053–1058.

[Peterson, 1988] G. E. PETERSON (Editor), *Tutorial: Object-Oriented Computing, Volume 2: Implementations*, IEEE Computer Society Press, Washington, DC, 1988.

[Stefik and Bobrow, 1986] M. STEFIK AND D. G. BOBROW, "Object-Oriented Programming: Themes and Variations," *The AI Magazine* **6** (No. 4, 1986), pp. 40–62.

[Stroustrup, 1986] B. STROUSTRUP, *The C++ Programming Language*, Addison-Wesley, Reading, MA, 1986.

[Ward and Mellor, 1985] P. T. WARD AND S. J. MELLOR, *Structured Development for Real-Time Systems, Volumes 1, 2 and 3*, Yourdon Press, New York, NY, 1985.

[Warnier, 1976] J. D. WARNIER, *Logical Construction of Programs*, Van Nostrand Reinhold, New York, NY, 1976.

[Warnier, 1981] J. D. WARNIER, *Logical Construction of Systems*, Van Nostrand Reinhold, New York, NY, 1981.

[World Book Encyclopedia, 1982] *World Book Encyclopedia*, World Book-Childcraft International, Inc., Chicago, IL, 1982, Volume N, p. 430.

[Yau and Tsai, 1986] S. S. YAU AND J. J.-P. TSAI, "A Survey of Software Design Techniques," *IEEE Transactions on Software Engineering* **SE-12** (June 1986), pp. 713–721.

[Yourdon and Constantine, 1979] E. YOURDON AND L. L. CONSTANTINE, *Structured Design: Fundamentals of a Discipline of Computer Program and Systems Design*, Prentice-Hall, Englewood Cliffs, NJ, 1979.

Chapter 10 Implementation

Implementation is the process of translating the detailed design into code. When this is done by a single individual, the process is relatively well understood. But most real-life products today are too large to be completed by one programmer within the given time constraints. Instead, the product is implemented by a team, all working at the same time on different components of the product. This is termed *programming-in-the-many*. While the objective of the programmers is to implement the design, this can be marred by poor cooperation and communication between the members of the development team. The problems associated with programming-in-the-many are examined in this chapter.

10.1 Choice of Programming Language

In most cases, the issue of choice of programming language simply does not arise. Suppose the client wants a product to be written in, say, BASIC. It may be the case that, in the opinion of the development team, BASIC is entirely unsuitable for the product. Such an opinion is irrelevant. Management has only two choices: implement the product in BASIC or turn down the job.

If the product has to be implemented on a specific computer, and the only language available on that computer is, say, assembler, then again there is no choice. If no other language is available, either because no compiler has yet been written for any high-level language on that computer or because management is not prepared to pay for, say, a C compiler for the stipulated computer, then it is again clear that the issue of choice of programming language is not relevant.

A more interesting question is: A contract specifies that the product is to be implemented in "the most suitable" programming language. What language should be chosen? In order to answer that question, consider the following scenario. QQQ Corporation has been writing COBOL products for over 25 years. The software staff of QQQ number over 200 employees, all of whom, from the most junior programmer to the vice-president for software, have COBOL expertise. Why on earth should the most suitable programming language be anything but COBOL? The introduction of a new language, Ada for example, would mean having to hire new programmers, or, at the very least, existing staff would have to be intensively retrained. Having invested all that money and effort in Ada training, management might well decide that future products should also be written in Ada. But all the existing COBOL products would still have to be maintained. There would then be two classes of programmers, COBOL maintenance programmers and Ada programmers writing the new applications. Quite undeservedly, maintenance is almost always considered to be an inferior activity to developing new applications, so there would be distinct unhappiness among the ranks of the COBOL programmers. This unhappiness would be compounded by the fact that Ada programmers are usually paid more than COBOL programmers because Ada programmers are in short supply. Although QQQ has excellent development tools for COBOL, an Ada compiler would have to be purchased, as well as appropriate Ada support tools and, in all probability, an Ada software development environment as described in Section 12.2.1. Additional hardware may have to be purchased or leased to run this amount of new software. Perhaps most serious of all, what QQQ has built up is many hundreds of person-years of expertise with COBOL, the kind of expertise that can be gained only through hands-on experience, such as what to do when a certain cryptic error message appears on the screen, or how to handle the quirks of the compiler. In brief, it would seem that "the most suitable" programming language could only be COBOL—any other choice would be financially suicidal, either from the viewpoint of the costs involved or as a consequence of plummeting staff morale leading to poor quality code.

And yet, the most suitable programming language for QQQ Corporation's latest project may indeed be some language other than COBOL. Notwithstanding its popularity, COBOL is really only suitable for one class of software products—essentially, business data-processing applications. The fact that the vast majority of software worldwide falls into this class is one of the two reasons for the preeminence of COBOL as a programming language; the other reason is given in Chapter 14. But if QQQ Corporation has software needs outside this class, then COBOL rapidly loses its attractiveness. For example, if QQQ wishes to construct a knowledge-based product using artificial intelligence (AI) techniques, then an AI language such as Lisp is a prerequisite; COBOL is totally unsuitable for AI applications. If there is large-scale communications software to be built, perhaps because QQQ requires satellite links to hundreds of branch offices all over the United States and worldwide, then a language such as C would prove to be more suitable than COBOL. If QQQ is to go into the business of writing systems software such as operating systems, compilers, and linkers, then COBOL is very definitely unsuitable. And if QQQ Corporation has decided to go into defense contracting, the Department of Defense directive mandating the use of Ada for all embedded, real-time systems totally precludes the use of COBOL in this area, in the unlikely event of a misguided QQQ manager actually thinking that COBOL can be used for real-time embedded software.

The issue of which programming language to use can almost always be decided by cost–benefit analysis. That is to say, management must compute a dollar cost of an implementation in COBOL as well as the dollar benefits, present and future, of using COBOL. This computation must then be repeated for every language under consideration. The language with the largest expected gain, that is, the difference between estimated benefits and estimated costs, is then the appropriate implementation language.

For example, in the case of QQQ Corporation, management may decide that the COBOL expertise that the corporation has acquired over the years is worth $450,000, while the cost of trying to write a communications package in COBOL would be $400,000. Thus the benefits of using COBOL exceed the cost by $50,000; the expected gain is therefore $50,000. On the other hand, the cost of training existing programmers in C is estimated at $300,000, the lowering of morale among COBOL programmers appears to cost $75,000, but the benefits of writing the communications package in C are estimated at $500,000. This latter figure is high because management realizes that, as a consequence of the wide availability of C compilers, the resulting portable code can be implemented on any number of different computers within the communications network. Thus management estimates that the benefits of using C will exceed the costs by $125,000. The expected gain is therefore $125,000. Thus C is more attractive than COBOL for the project; the value $125,000 minus $50,000, or $75,000, can be attached to its attractiveness over COBOL. The reader who is gripped by a sense of growing amazement as to the source of the dollar amounts in this paragraph—for instance, how on earth can management claim that the cost of lowering morale among COBOL programmers is $75,000—should read the explanation of cost–benefit analysis in Section 4.2.3.

Another way of deciding which programming language to select is to use risk analysis. For each language under consideration a list is made of the potential risks and ways to try to resolve them. The language for which the overall risk is the smallest is then selected.

What about implementation in a fourth-generation language? This issue is addressed in the next section.

10.2 Fourth-Generation Languages

The first computers had neither interpreters nor compilers. They were programmed in binary, either hard-wired with plug boards or by setting switches. Such a binary machine code was a *first-generation language*. The *second-generation languages* were assemblers, developed in the late 1940s and early 1950s. Instead of having to program in binary, instructions could be expressed in symbolic notation such as

```
MOV     $17, NEXT
```

In general, each assembler instruction is translated into one machine code instruction. Thus, although assembler is easier to write than machine code and easier for maintenance programmers to comprehend, programs are still the same length.

The idea behind a *third-generation language* (or high-level language) such as FORTRAN, ALGOL 60, or COBOL is that one statement of a high-level language is compiled to 5 or 10 machine code instructions. High-level language code is thus considerably shorter than the equivalent assembler code. It is also simpler to understand than assembler. Thus high-level language code is easier to maintain than assembler code. The fact that the high-level language code may not be quite as efficient as the equivalent assembler code is generally a small price to pay for ease in maintenance.

This concept was then taken further in the late 1970s. A major objective in the design of a *fourth-generation language* (4GL) is that each 4GL statement should be equivalent to 30, or even 50, machine code instructions. Products written in 4GL would then be short, and hence quicker to develop, and easier to maintain.

It is difficult to program in machine code. It is somewhat easier to program in assembler, and easier still to use a high-level language. A second major design objective of a 4GL is ease in programming. In particular, many 4GLs are *nonprocedural*. The meaning of the term "nonprocedural" can be explained by the following. Some years ago the author hailed a cab outside Grand Central Station in New York City, and said to the driver, "Please take me to Lincoln Center." This was a nonprocedural request because the desired result was expressed, but it was left to the driver to decide how to achieve that result. It turned out that the driver was an immigrant from Central Europe who had been less than two days in America, and who knew virtually nothing about the geography of New York City or the English language. As a result, the nonprocedural request was replaced by a procedural request of the form,

"Straight, straight. Take a right on Broadway at the next light. I said right. Right, yes, here, right! Now straight. Slow down, please. I said, please slow down. For heaven's sake, slow down!" and so on. In other words, a sequence of instructions was issued, and the driver carried them out one after the other.

A third-generation language is procedural. Every step has to be specified. In contrast, a 4GL is frequently nonprocedural. For example, consider the command

```
for every SURVEYOR
   if RATING is EXCELLENT
      add 6500 to SALARY
```

It is up to the compiler of the 4GL to translate this nonprocedural instruction into a sequence of machine code instructions that can be executed procedurally.

There is price to pay for the ease of use gained through nonprocedurality. The compiler or interpreter of such a 4GL is generally large and slow, and it is critical that it should be as fault-free as possible. Many 4GLs are more suited to management information systems than to data-processing applications. In a management information system, the objective is to extract information and present it to the manager in the desired format. In contrast, in data processing large volumes of data have to be handled and computations performed, and many 4GLs are simply too slow for this purpose. This is one example of trade-offs in the design of a 4GL, the ease of use of a nonprocedural language as opposed to the computing power of a procedural language.

It is too soon to know all the answers regarding 4GLs. Some organizations that previously have used COBOL have reported a tenfold increase in productivity through use of a 4GL. Other organizations have tried a 4GL, and have been bitterly disappointed by the results. The reason for this inconsistency is that no one 4GL is appropriate for all situations. On the contrary, it is important to select the correct 4GL for the specific product. For example, Playtex used IBM's Application Development Facility (ADF) and reported an 80 to 1 productivity increase over COBOL. Notwithstanding this impressive result, Playtex has subsequently used COBOL for products that were deemed by management to be less well suited to ADF [Martin, 1985].

The attitudes of 43 organizations to 4GLs is reported in [Guimaraes, 1985]. It was found that use of a 4GL reduced user frustration because the data-processing department was able to respond more quickly when a user needed information to be extracted from the organization's database. But there were also a number of problems. 4GLs proved, on average, to be slow and inefficient, with long response times. One product consumed 60% of the CPU cycles on an IBM 4331, while supporting at most 12 concurrent users.

Overall, the 28 organizations that had been using a 4GL for over 3 years felt that the benefits outweighed the costs.

No one 4GL dominates the software market. Instead, there are literally hundreds of 4GLs, and dozens of them, such as FOCUS, NOMAD, RAMIS-II, SQL, and NATURAL, have a sizable user group. This widespread proliferation of 4GLs is further evidence of the fact that care has to be taken in selecting the correct 4GL. Of course, few organizations can support more than one 4GL. Once a 4GL has been chosen and used, for subsequent products the organization must either use that 4GL or fall back on the language that was used before the 4GL was introduced.

Notwithstanding the potential productivity gain, there is the potential danger of using a 4GL the wrong way. In many organizations there is currently a large backlog of products to be developed and a long list of maintenance tasks to be performed. A design objective of many 4GLs is *end user programming*, that is to say, programming by the person who will use the product. For example, before the advent of 4GLs the investment manager of an insurance company would ask the data-processing manager for a product that would display certain information regarding the bond portfolio. The investment manager would then wait a year or so for the data-processing group to find the time to develop the product. It was intended that a 4GL would be so simple to use that the investment manager, previously untrained in programming, would be able to write the desired product. End user programming was intended to help reduce the development backlog, leaving the professionals to maintain existing products.

In practice, end user programming can be dangerous. First, consider the situation when all product development is performed by computer professionals. Computer professionals are trained to mistrust computer output. After all, probably less than 1% of all output during product development is correct. On the other hand, the user is told to trust all computer output, because the product should not be delivered to the user until the computer professional has removed the faults. Now consider the situation when end user programming is encouraged. When a user who is inexperienced in programming writes his or her own code using a user-friendly, nonprocedural 4GL, the natural tendency is for that user to believe the output from the 4GL product that he or she wrote. After all, for years the user has been instructed to trust computer output. As a result, many business decisions have been taken on the basis of data generated by hopelessly incorrect end user code. In some cases the user friendliness of some 4GLs has led to financial catastrophes. Another source of problems is 4GL compiler faults; end users are rarely trained to test their products thoroughly to check that a correct product does not give wrong answers as a consequence of a fault in the compiler. Yet another danger lies in the fact that in some organizations, end users have been allowed to write 4GL products that update the organization's database. A fault in such a product may eventually result in the corruption of the entire database. The lesson is clear: Programming by inexperienced or inadequately trained end users can be exceedingly dangerous, if not fatal, to the financial health of a corporation.

The ultimate decision as to the choice of a 4GL will be taken by management. In taking such a decision, management should be guided by the many success stories resulting from the use of a 4GL. At the same time, management should also carefully analyze the failures caused by using an inappropriate 4GL or by poor management of the development process. For example, a common cause of failure is neglecting to train the development team thoroughly in all aspects of the 4GL, including relational database theory [Date, 1986] where appropriate. Management should carefully study both the successes and failures in the specific application area and learn from the mistakes that have been made in the past. Choosing the correct 4GL can spell the difference between a major success and dismal failure.

Having decided on the implementation language, the next issue is how software engineering principles can lead to better quality code.

10.3 Structured Programming

Structured programming has been credited with everything from preventing tooth decay to curing athlete's foot. When structured programming was first put forward, it was seen as the panacea of all ills. While structured programming has not fulfilled every aspect of its early promise, it has certainly led to improved code quality.

What is structured programming? Is it merely programming without **goto** statements, or is there something more? To help answer these questions, a historical perspective on structured programming must first be given.

10.3.1 History of Structured Programming

The development of structured programming took place between 1966 and 1974. Three major events can be highlighted during this period:

1. The mathematical theory of 1966.
2. Dijkstra's 1968 letter.
3. Knuth's landmark article of 1974.

Mathematical Theory

In 1966, two Italian computer scientists published an article that essentially showed that any software product that can be represented by a flowchart can be rewritten using only three types of control structures, namely concatenation, selection, and iteration [Böhm and Jacopini, 1966]. These three control structures are depicted in Figure 10.1.

The reaction of the computer science community to this paper was—so what? The fact that a product could be put together using only F **followed-by**

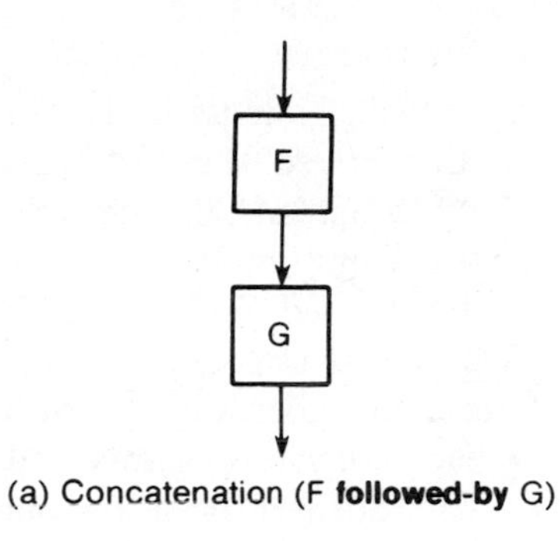

(a) Concatenation (F **followed-by** G)

(b) Selection (**if** P **then** F **else** G)

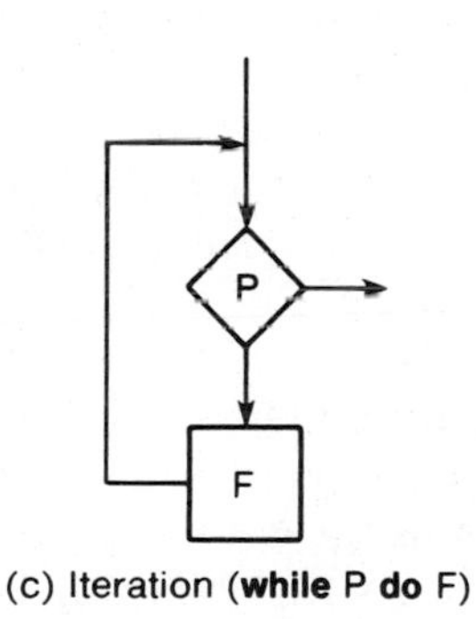

(c) Iteration (**while** P **do** F)

Figure 10.1 The three control structures of structured programming.

G (concatenation), **if** P **then** F **else** G (selection), and **while** P **do** F (iteration) was of interest to hardly anybody at the time. In 1966, the popular and important languages were COBOL II, FORTRAN IV, and assembler, and in all three of those languages the **goto** statement was the major control structure. The fact that, in principle, it was *possible* to program without the **goto** was just not relevant. Why on earth would anyone want to?

Dijkstra's Letter

In 1968, Dijkstra published a letter in the *Communications of the ACM* entitled "**goto** Statement Considered Harmful" [Dijkstra, 1968a]. The letter caused a major furor. Many disagreed that the **goto** is harmful in any way; letters of this type still appear from time to time. Others said—so what? Even if the **goto** was in fact to be avoided, nothing could be done about it. In the popular languages of 1968, it was impossible to program without the **goto**. After all, there is no alternative to using the **goto** in assembler. Because structured programming had not yet been invented, structured FORTRAN and structured COBOL had not yet been developed; there was thus no alternative to using the **goto** in both those languages as well. And since few people had read Böhm and Jacopini's article, and fewer still had appreciated its full significance, there was a third school of thought that averred that it was impossible to program without the **goto** statement.

Knuth's Article

The arguments ranged back and forth. It is surprising that no duels were fought over the issue, such were the passions that structured programming evoked. Finally, in 1974, Knuth published a 40-page article entitled "Structured Programming with **goto** Statements" [Knuth, 1974]. The title appears to be a contradiction in terms. But the point of the article is that structured programming is *not* **goto**-less programming. Instead, it is a way of making products more readable. In some cases, a module that contains a **goto** is easier to understand than its **goto**-less counterpart.

This article brought the major controversy to a stop, but the embers are still glowing and flare up occasionally. Nevertheless, the overwhelming preponderance of the software world is united in the view that **goto** statements are almost always harmful and should be avoided.

10.3.2 Why the goto Statement Is Considered Harmful

After placing structured programming in its historical perspective, the question still remains: Why is the **goto** considered harmful? There are three reasons. First, a **goto** decreases readability, and hence maintainability. Even worse is a conditional **goto** (**if — then goto —**), because the flow of control depends on the values of the variables at execution time. Depending on whether or not the conditional is satisfied, execution of the module can follow two very different paths, both of which must be considered.

Second, the presence of a **goto** can mar the validity of procedures. For example, consider the COBOL paragraph shown in Figure 10.2. The PRINT-PARAGRAPH requires the variable SWITCH-1 to be set to 15 before the rest of paragraph is executed. Provided that the paragraph is invoked via a call such as **perform** PRINT-PARAGRAPH **thru** END-PRINT-PARAGRAPH, then all will be well. But if the product contains **goto** DANGEROUS-ENTRY,

```
PRINT-PARAGRAPH.
      move 15 to SWITCH-1.
      ...
      ...
DANGEROUS-ENTRY.
      ...
      ...
END-PRINT-PARAGRAPH.
```

Figure 10.2 Print paragraph with initialization condition that can be bypassed by a **goto** statement.

then the initialization statement will be bypassed, with possibly dangerous results.

Third, a forward **goto**, that is, a jump to a point in the code ahead of the **goto** statement, is less harmful than a backwards **goto**. The reason is that a backwards **goto** is nothing more or less than a loop statement, and readability is considerably enhanced by coding a loop as such, rather than disguising it with a **goto**.

In the light of these observations, structured programming can now be defined.

> *Structured programming is the writing of code with* **goto** *statements kept to a minimum, preferably only when an error is detected, and always in the forward direction.*

To see the connection with error handling, consider the following analogy. It is customary to enter and leave a room by the door, not the window. But if the room is on fire, the demands of etiquette are irrelevant. If the safest and quickest way to exit from a burning room is via the window, then the window should be used. In the same way, the use of a **goto** is generally deleterious to readability, and hence to maintainability. But if an error condition arises, then the important thing is to exit from the current routine to an error-handling routine as quickly as possible before, say, the database is corrupted or the stack can overflow. Under error conditions the primary duty is survival, and this can often be best achieved by means of a forward **goto**.

Contrast the preceding definition with the classical definition

> *A product is structured if the code blocks are connected by concatenation, selection, and iteration only and every block has exactly one entry and one exit.*

Concatenation, selection, and iteration are shown in Figure 10.1.

This classical definition has been widened to include a fourth type of control statement, namely the **call**. But even with this, the definition is too narrow. Structured programming is not **goto**-less programming, but rather the writing of readable code. In most instances readable code is indeed **goto**-less code. But the objective of readability sometimes requires the use of a **goto** statement.

10.4 Good Programming Practice

Many recommendations on good coding style are language-specific. For example, suggestions regarding use of, say, COBOL 88-level entries or parentheses in Lisp, are of little interest to programmers implementing a product in, say, Pascal. The reader who is actively involved in implementation is urged to read one of the many books, such as those by Henry Ledgard, on good programming practice for the specific language in which the product is being implemented. In addition, some recommendations regarding language-independent good programming practice are now given.

Use of Consistent and Meaningful Variable Names

As stated in Chapter 1, about two-thirds of most software budgets are devoted to maintenance. This implies that the programmer developing a module is merely the first of many other programmers who will work on that module. It is counterproductive for a programmer to give names to variables that are meaningful to only that programmer; within the context of software engineering, the term "meaningful" means "meaningful from the viewpoint of future maintenance programmers."

This point is illustrated by events at a small software production organization in Johannesburg, South Africa, in the late 1970s. The organization consisted of two programming teams. Team A was made up of émigrés from Mozambique. They were of Portuguese extraction, and their mother tongue was Portuguese. Their code was well written. Variable names were meaningful, but unfortunately only to a speaker of Portuguese. Team B comprised Israeli immigrants whose mother tongue was Hebrew. Their code was equally well written, and the names they chose for their variables were equally meaningful—but only to a speaker of Hebrew. And one fine day, team A resigned en masse, together with their team leader. Team B was totally unable to maintain any of the excellent code that team A had written, because they could not speak a word of Portuguese. The variable names, meaningful as they were to Portuguese speakers, were incomprehensible to the Israelis whose linguistic abilities were restricted to Hebrew and English. The owner of the software organization was unable to hire enough Portuguese-speaking programmers to replace team A, and the company soon went into bankruptcy, under the weight of numerous lawsuits from disgruntled customers whose code was essentially nonmaintainable. The situation could so easily have been avoided. The head of the company should have insisted from the start that all variable names be in English, the language understood by every South African computer professional. Variable names would then have been meaningful to any maintenance programmer.

In addition to the use of meaningful variable names, it is equally essential that variable names be consistent. For example, a module includes the following four variables: AVERAGE_FREQ, FREQUENCY_MAXIMUM, MIN_FR, and FRQNCY_TOTL. A maintenance programmer who is trying to understand the code has to know if FREQ, FREQUENCY, FR, and FRQNCY all refer to the same thing. If yes, then the identical word should be used, preferably FREQUENCY, although FREQ or FRQNCY are marginally acceptable; FR is not. But if one or more variable names refer to a different quantity, then a totally different name, such as RATE, should be used. A second aspect of consistency is the ordering of the components of variable names. For example, if one variable is named FREQUENCY_MAXIMUM, then the name MINIMUM_FREQUENCY will be confusing; it should rather be FREQUENCY_MINIMUM. Thus, in order to make the code clear and unambiguous for future maintenance programmers, the four variables listed previously should be named FREQUENCY_AVERAGE, FREQUENCY_MAXIMUM, FREQUENCY_MINIMUM, and FREQUENCY_TOTAL, respectively. Alternatively, the FREQUENCY component can appear at the end of all four variables names, yielding the variable names AVERAGE_FREQUENCY, MAXIMUM_FREQUENCY, MINIMUM_FREQUENCY, and TOTAL_FREQUENCY. It clearly does not matter which of the two sets is chosen; what is important is that all the names be from the one set or from the other.

The Issue of Self-Documenting Code

When asked why their code contains no comments whatsoever, programmers often reply, "I write self-documenting code." The implication is that the variable names are so carefully chosen and the code so beautifully crafted that there is no need for comments of any sort. Self-documenting code does exist, but it is exceedingly rare. Instead, the usual scenario is that the programmer himself or herself appreciates every nuance of the code at the time the module is written. It is conceivable that the programmer uses the same style for every module, and that in 5 years' time, say, the code will still be crystal clear in every respect to the original programmer. Unfortunately, this is irrelevant. The important point is whether the module can be understood easily and unambiguously by all the other programmers who have to read the code, starting with the SQA group and including a number of different maintenance programmers. The problem becomes more acute in the light of the unfortunate practice of assigning maintenance tasks to inexperienced programmers and not supervising them closely. The undocumented code of the module may be only partially comprehensible to an experienced programmer. How much worse, then, is the situation when the maintenance programmer is inexperienced.

To see the sort of problems that can arise, consider the variable X_COORDINATE_OF_POSITION_OF_ROBOT_ARM. Such a variable name is undoubtedly self-documenting in every sense of the word, but few programmers are prepared to use a 37-character variable name, especially if that name is used frequently. Instead, a shorter name is used, X_COORD, for example. The reasoning behind this is that if the entire module deals with the movement of the arm of a robot, X_COORD can refer only to the x-coordinate of the position of the arm of the robot. While that argument holds water within the context of software development, it is not necessarily true for maintenance. The maintenance programmer may not have sufficient knowledge of the product as a whole to realize that, within this module, X_COORD can refer only to the arm of the robot. In addition, there may simply not be the necessary documentation available from which a maintenance programmer may learn what the module as a whole does, let alone what a variable such as X_COORD represents. The way to avoid this sort of problem is to insist that every variable name be explained at the beginning of the module, in the *prologue comments*. If this rule is followed, the maintenance programmer will quickly understand that variable X_COORD is used for the x-coordinates of the position of the robot arm.

Prologue comments are mandatory at the top of every single module. The minimum information that must be provided at the top of every module is: module name; a brief description of what the module does; programmer's name; date module was coded; date module was approved, and by whom; module parameters; list of variable names, preferably in alphabetical order, and their uses; files accessed by this module, if any; files changed by this module, if any; module input/output, if any; error-handling capabilities; name of file containing test data, to be used later for regression testing; list of modifications made, their dates, and who approved them; and known faults, if any.

Even if a module is clearly written, it is unreasonable to expect someone to have to read every line in order to understand what the module does and how it does it. Prologue comments make it easy for others to understand the key points. Only a member of the SQA group or a maintenance programmer who is modifying that specific module should be expected to have to read every line of a module.

In addition to prologue comments, comments are essential whenever the code is written in a nonobvious way or makes use of some subtle aspect of the language. For example, in languages such as FORTRAN, if an operand of type INTEGER is divided by another operand of type INTEGER, then the result has type INTEGER. Thus 7.0 / 2.0 = 3.5, but 7 / 2 = 3, because truncation from 3.5 to 3 is performed. Suppose that variable R is defined to be a REAL variable. Then the statement

```
R = 7 / 2
```

is computed as follows. First, 7 is divided by 2. Since both operands are of type INTEGER, the result is the integer 3. The result is then stored in REAL variable R. Thus, after the assignment statement has been executed, R contains the value 3.0.

A FORTRAN statement like

```
SPEED = DIST / TIME
```

can therefore give an unexpected result if both DIST and TIME are INTEGER variables, while SPEED is of type REAL. In such cases, it is incumbent on the original programmer to insert a comment in the code to assist the maintenance programmer in understanding the module. Of course, the preceding example is poor programming practice. A better way of coding such a confusing statement is to define an integer variable named ISPEED and then to use two assignments, namely

```
ISPEED = DIST / TIME
SPEED = ISPEED
```

In general, if code is written in such a way that an inline comment is necessary, it would be better to rewrite the code in a more straightforward way. While this practice should reduce the need for inline comments, prologue comments remain essential in every way.

Use of Parameters

There are very few genuine constants, that is, variables whose values will *never* change. For instance, satellite photographs have caused changes to be made in submarine navigation systems incorporating the latitude and longitude of Pearl Harbor, Hawaii, in order to reflect more accurate geographic data regarding the exact coordinates of Pearl Harbor. To take another example, sales tax is not a genuine constant; legislators tend to change the sales tax rate from time to time. Suppose that the sales tax rate is currently 6%. If the value 6.0 has been hard-coded in a number of modules of a product, then changing the product is a major exercise, with the likely outcome of one or two instances of the "constant" 6.0 being overlooked. A better solution is to initialize a variable SALES_TAX_RATE, say, to 6.0. Then, wherever the value of the sales tax rate is needed, the variable SALES_TAX_RATE should be used, and not the number 6.0. If the sales tax rate changes, then only the line containing the value of SALES_TAX_RATE need be altered. Better still, the value of the sales tax rate should be read in from a parameter file at the beginning of the run. All such apparent constants should be treated as parameters. If a value should change for whatever reason, this change can then be implemented quickly and effectively.

Code Layout for Increased Readability

It is relatively simple to make a module easy to read. For example, no more than one statement should appear on a line, even though many programming languages permit it. Indentation is perhaps the most important technique for increasing readability. Just imagine how difficult it would be to read the code examples in Chapter 8 if indentation had not been used to assist in understanding the code. Indentation can be used to connect corresponding **begin**...**end** pairs. It also shows which statements belong in a given block. In fact, correct indentation is too important to be left to human beings. Instead, as described in Section 10.10.1, tools should be used to ensure that indentation is correctly done.

Another useful aid is blank lines. Procedures should be separated by blank lines; in larger modules, it is often helpful to break up large blocks of code in this way. The blank space breaks up the code into smaller pieces that the reader can comprehend.

Nested **if** Statements

Consider the following example. A map consists of two squares, as shown in Figure 10.3. It is required to write code to determine whether a point on the Earth's surface lies in MAP SQUARE 1 or MAP SQUARE 2, or is not on the map. The solution of Figure 10.4 is so badly formatted that it is

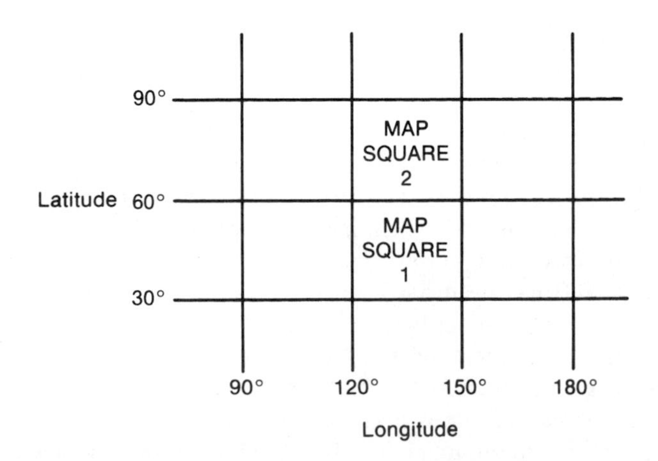

Figure 10.3 Coordinates for map.

```
if LATITUDE > 30 and LONGITUDE > 120 then if LATITUDE <= 60 and LONGITUDE
  <= 150 then MAP_SQUARE_NO := 1; elsif LATITUDE <= 90 and LONGITUDE <= 150
then MAP_SQUARE_NO := 2; else PUT ("Not on the map"); end if; else PUT ("Not on
the map"); end if;
```

Figure 10.4 Badly formatted nested **if** statements.

```
if LATITUDE > 30 and LONGITUDE > 120 then
   if LATITUDE <= 60 and LONGITUDE <= 150 then
      MAP_SQUARE_NO := 1;
   elsif LATITUDE <= 90 and LONGITUDE <= 150 then
      MAP_SQUARE_NO := 2;
   else
      PUT ("Not on the map");
   end if;
else
   PUT ("Not on the map");
end if;
```

Figure 10.5 Well-formatted, but badly constructed, nested **if** statements.

incomprehensible. A properly formatted version appears in Figure 10.5. However, the combination of **if-then-if** and **if-then-else-if** constructs is so complex that it is difficult to check whether the code fragment is correct.

When faced with complex code containing the **if-then-if** construct, one way to simplify it is to make use of the fact that the **if-then-if** combination

```
if A > 0 then
   if B > 0
```

is equivalent to the single condition

```
if A > 0 and B > 0
```

provided that expression B is defined even if condition A > 0 does not hold. This was used to construct the simpler version shown in Figure 10.6. In

```
if LONGITUDE > 120 and LONGITUDE <= 150 and LATITUDE > 30 and LATITUDE <= 60
   then MAP_SQUARE_NO := 1;
elsif LONGITUDE > 120 and LONGITUDE <= 150 and LATITUDE > 60 and LATITUDE <= 90
   then MAP_SQUARE_NO := 2;
else
   PUT ("Not on the map");
end if;
```

Figure 10.6 Acceptably nested **if** statements.

addition to the **if-then-if** construct, nesting **if** statements too deeply also leads to code that can be difficult to read. As a rule of thumb, **if** statements nested to a depth of greater than 3 should be avoided as poor programming practice.

10.5 Coding Standards

Coding standards can be both a blessing and a curse. It was pointed out in Section 8.2.1 that modules of coincidental strength generally arise as a consequence of rules such as, "every module will consist of between 35 and 50 executable statements." Instead of stating a rule in that dogmatic fashion, a better formulation is, "programmers should consult their managers before constructing a module with fewer than 35 or more than 50 executable statements." The point is that no coding standard can ever be applicable under all possible circumstances.

Coding standards imposed from above tend to be ignored. As mentioned previously, a useful rule of thumb is that **if** statements should not be nested to a depth of greater than 3. If programmers are shown examples of unreadable code resulting from nesting **if** statements too deeply, then it is likely that they will conform to such a regulation. But they are unlikely to adhere to a list of coding rules imposed on them without discussion or explanation. Furthermore, such standards are likely to lead to friction between programmers and their managers.

Unless a coding standard can be checked by machine, it is either going to waste a lot of the SQA group's time, or it will simply be ignored by the programmers and the SQA group alike. On the other hand, consider the following rules:

Nesting of **if** statements should not exceed a depth of 3, except with prior approval from the team leader.

Modules should consist of between 35 and 50 statements, except with prior approval from the team leader.

The use of **goto** statements should be avoided. However, with prior approval from the team leader, a forward **goto** may be used for error handling.

Such rules may be checked by machine, provided that some mechanism is set up for capturing the data relating to permission to deviate from the standard.

Some organizations have strict standards regarding names of procedures and variables. For example, there is a certain COBOL software organization that has laid down highly restrictive coding standards. For example, if a subprogram is named, say, SUB-23, then its sections are numbered A01-SUB-23 through A99-SUB-23, and paragraphs within, say, A34-SUB-23 are named B01-A34-SUB-23 through P99-A34-SUB-23. There are special naming conventions for error-handling paragraphs, paragraphs that perform input, paragraphs that perform output, and the EXIT-PROGRAM paragraph.

In addition, there are even more complex rules for naming variables. Such artificial names are meaningful only to someone who has been forced to learn the standard.

The aim of coding standards is to make maintenance easier. But if the effect of a standard is to make the life of software developers difficult, then such a standard should be modified, even in the middle of a project. Overly restrictive coding standards are counterproductive in that the quality of software production must inevitably suffer if programmers have to develop software within such a framework. On the other hand, standards such as those listed previously regarding **goto** statements, nesting of **if** statements, and module size, coupled with a mechanism for deviating from those standards, can lead to improved software quality, which is, after all, the goal of software engineering.

10.6 Team Organization

Most products are too large to be completed by a single software professional within the time constraints. As a result, the product must be assigned to a group of professionals organized as a *team*. For example, consider the specifications phase. In order to specify the target product within, say, three months, it may be necessary to assign the task to three specification specialists organized as a team under the direction of the specifications manager. Similarly, the design task may be shared between members of the design team.

Suppose now that a product has to be coded within 3 months, even though there is 1 person-year of coding involved (a person-year is the amount of work that can be done by one person in 1 year). The solution is simple: If one programmer can code the product in 1 year, four programmers can do it in 3 months.

This, of course, is nonsense. In practice, the four programmers will probably take nearly a year, and the quality of the resulting product may well be lower than if one programmer had coded the entire product unaided. The reason for this is that some tasks can be shared, others cannot. For instance, if one farmhand can pick a strawberry field in 10 days, then the same strawberry field can be picked by 10 farmhands in 1 day. On the other hand, one woman can produce a baby in 9 months, but nine woman cannot possibly produce that same baby in 1 month.

In other words, tasks like strawberry picking can be fully shared; others, like baby production, cannot be shared at all. Unlike baby production, it is possible to share coding tasks between members of a team by distributing the modules among the team members. But team programming is also unlike strawberry picking in that team members have to interact with one another in a meaningful and effective way. For example, suppose Jane and Ned have to code two modules, M_A and M_B, say. A number of things can go wrong. For instance, both Jane and Ned may code M_A and ignore M_B. Or, Jane may code M_A, and Ned may code M_B. But when M_A calls M_B it passes four parameters; Ned has coded M_B in such a way that it requires five parameters. Or, the order of the parameters in M_A and M_B may be different. Or, the order may be the same, but the data types may be slightly different. Such problems are usually caused by a decision taken during the design phase that is not propagated throughout the development organization. The issue has nothing whatsoever to do with the technical competency of the programmers involved. Team organization is a managerial issue. It is the responsibility of management to organize the programming team in such a way that the team is highly productive.

A different type of difficulty that arises from team development of software is shown in Figure 10.7. There are three channels of communication between the three computer professionals working on the project. Now suppose that the work is slipping, a deadline is rapidly approaching, but the task is not nearly complete. The obvious thing to do is to add a fourth professional to the team. But the first thing that must happen when the fourth professional joins is for the other three to explain in detail what has been accomplished to date and what is still incomplete. In other words, adding additional personnel to a team when a product is late has the effect of making the product even later [Brooks, 1975].

In a large organization, teams are used in every phase of software production, but especially in the implementation phase. The reason is that programmers can work independently on separate modules, so this phase appears to be a prime candidate for sharing the task among several computer professionals. In some smaller organizations, one individual may be responsible for both the specifications and the design, while the implementation is done by a team of two or three programmers. Because teams are thus most

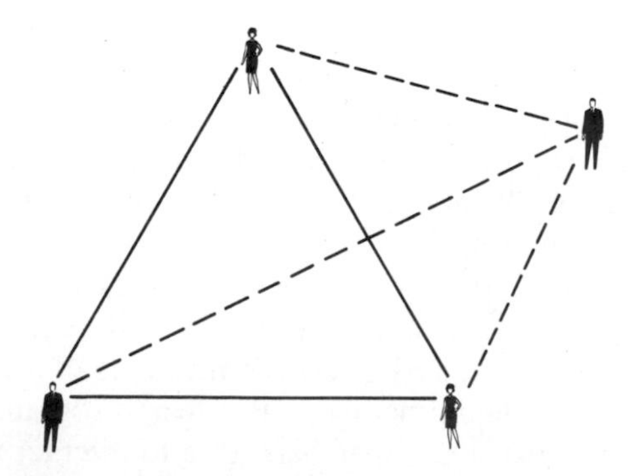

Figure 10.7 Communications paths between three computer professionals and what happens when fourth professional joins them.

heavily used during the implementation phase, the problems of team organization are most acutely felt during implementation. In this book, team organization is therefore presented within the context of implementation, even though the problems and their solution are also applicable to all the other phases.

There are two extreme approaches to programming team organization, namely democratic teams and chief programmer teams. The approach taken in this chapter is to describe each of the two approaches in turn, highlighting their strengths and weaknesses, and then to suggest other ways of approaching the problem of organizing a programming team that incorporate the best features of the two extremes.

10.7 Democratic Team Approach

The democratic team organization was first described by Weinberg in 1971 [Weinberg, 1971]. The basic concept underlying the democratic team is *egoless programming*. Weinberg points out that programmers can be highly attached to their code. Sometimes they even name their modules after themselves; they therefore see their modules as an extension of themselves. The difficulty with this is that if a programmer sees a module as an extension of his ego, that programmer is certainly not going to try to find all the faults in "his" code. And if there is a fault, it is termed a *bug*, like some insect that has crept unasked into the code, and could have been prevented if the code had only been guarded more zealously against invasion by the bug. Some years ago, when software was still input on punched cards, that attitude was amusingly lampooned by the marketing of an aerosol spray named Shoo-Bug. The instructions on the label solemnly explained that spraying one's card deck with Shoo-Bug would ensure that no bugs could possibly infest the code.

Weinberg's solution to the problem of programmers being too closely attached to their own code is egoless programming. The social environment must be restructured, as must programmers' values. A programmer must encourage the other members of the team to find faults in his or her code. The presence of a fault must not be considered something bad, but rather a normal and accepted event. When a module is taken to a fellow programmer for review, if any faults are found then the attitude of the reviewer should be appreciation at being asked for advice, rather than ridicule of the programmer for making coding mistakes. The team as a whole will develop an ethos, a group identity. Modules will not belong to any one individual, but rather to the team as a whole.

A group of up to 10 egoless programmers constitutes a *democratic team*. Weinberg warns that management may have difficulty working with such a team. After all, consider the managerial career path. When a manager is promoted to a higher level, his or her fellow managers at the previous level are not promoted and must strive to attain the higher level at the next round of promotions. In contrast, a democratic team is a group working for a common cause with no single leader, with no programmers trying to get promoted to the next level. What are important are the team identity and mutual respect.

Weinberg tells of a democratic team that developed an outstanding product. Management decided to give a cash award to the team's nominal manager; by definition, a democratic team has no leader. He refused to accept it personally, saying that it had to be shared equally among all members of the team. Management thought that he was angling for more money, and anyhow the team, and especially its nominal manager, seemed to have some very unhealthy ideas. Management forced the nominal manager to accept the money, which he then divided equally among the team. The entire team then resigned and joined another company as a team.

The advantages and disadvantages of democratic teams are now presented.

10.7.1 Analysis of the Democratic Team Approach

A major advantage of the democratic team approach is the positive attitude to the finding of faults. The more that are found, the happier are the members of a democratic team. This positive attitude leads to more rapid detection of faults, and hence to high-quality code. But there are some major problems. As pointed out previously, managers may have difficulty accepting egoless programming. But it is probably worse for programmers used to the more familiar attached approach to modules. A programmer with, say, 15 years of experience, is unlikely to be too keen about allowing his or her code to be subjected to appraisal by fellow programmers, especially beginners.

Weinberg feels that egoless teams spring up spontaneously and cannot be imposed from outside. Little experimental research has been done on democratic programming teams, but the experience of Weinberg is that democratic teams are enormously productive. Mantei has analyzed the democratic team organization using arguments based on theories of and experiments on group organization in general, rather than specifically on programming teams [Mantei, 1981]. She points out that decentralized groups work best when the problem is difficult and suggests that democratic teams should function well in a research environment. It has been the author's experience that a democratic team also works well in an industrial setting when there is a hard problem to solve. On a number of occasions he has been a member of democratic teams that have sprung up spontaneously among computer professionals with research experience. But once the task has been reduced to the implementation of a hard-won solution, the team must then be reorganized in a more hierarchical fashion, such as the chief programmer team approach described in the next section.

10.8 Classical Chief Programmer Team Approach

Consider the six-person team shown in Figure 10.8. There are 15 two-person communication channels. In fact, the total number of two-, three-, four-, five-, and six-person groups is 57. That is the major reason why a six-person team structured as in Figure 10.8 is unlikely to be able to do 6 person-months of work in 6 months; much of their time is wasted in conferences of two or more team members at a time.

Now consider the six-person team shown in Figure 10.9. Again there are six programmers, but now there are only five lines of communication. This is the basic concept behind what is now termed the *chief programmer* team. A related idea was put forward by Brooks who drew the analogy of a chief

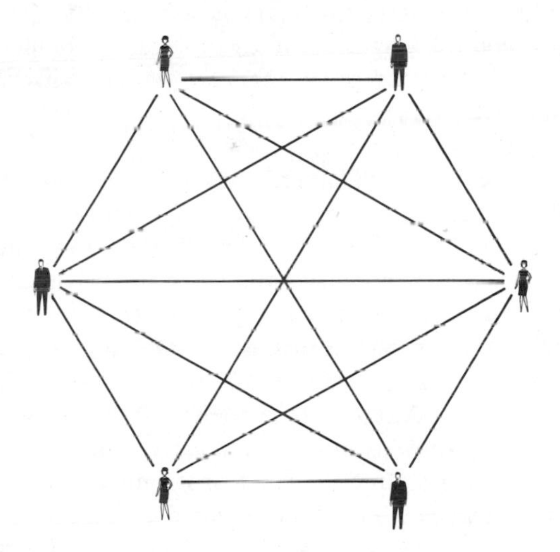

Figure 10.8 Communications paths between six computer professionals.

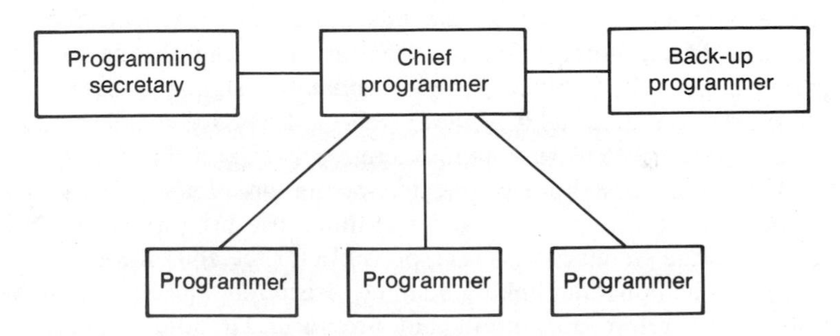

Figure 10.9 Structure of classical chief programmer team.

surgeon directing an operation [Brooks, 1975]. The surgeon is assisted by other surgeons, the anesthesiologist, and a variety of nurses. In addition, when necessary the team makes use of experts in other areas such as cardiologists or nephrologists. This analogy highlights two key aspects. The first is *specialization*; each member of the team carries out only those tasks for which he or she has been trained. The second aspect is *hierarchy*. The chief surgeon is in charge and directs the actions of all the other members of the team. At the same time, he or she is responsible for every aspect of the operation.

The chief programmer team concept was formalized by Mills. A classical chief programmer team, as described by Baker [Baker, 1972], is shown in Figure 10.9. It consists of the chief programmer, who is assisted by the back-up programmer, the programming secretary, and from one to three programmers. When necessary, the team is assisted by specialists in other areas, such as job control language (JCL) or legal or financial matters. The *chief programmer* is both a successful manager and a highly skilled programmer. He or she does the architectural design, then allocates the detailed design and the coding among the team members. Critical or complex sections of the code are done by the chief programmer, making use of his or her coding skills. As shown in the figure, there are no lines of communication between the programmers; all interfacing issues are handled by the chief programmer. Finally, the chief programmer reviews the work of the other team members, because the chief programmer is personally responsible for every line of code.

The *back-up programmer* is necessary only because the chief programmer is human and may therefore get ill, fall under a bus, or change jobs. Thus the back-up programmer must be as competent as the chief programmer in every respect, and must know as much about the project as the chief programmer. In addition, in order to free the chief programmer to concentrate on the architectural design, the back-up programmer does black-box test case planning and other tasks that are independent of the design process.

The word "secretary" has a number of meanings. On the one hand, a secretary assists a busy executive by answering the telephone, typing correspondence, and so on. But when we talk about the American Secretary of State or the British Foreign Secretary we are referring to one of the most senior members of the Cabinet. The *programming secretary* is not a part-time clerical assistant, but a highly skilled, well-paid, central member of a chief programmer team. The programming secretary is responsible for maintaining the product production library, the documentation of the project. This includes source code listings, JCL, and test data. The programmers hand their source code to the secretary who is responsible for the conversion to machine-readable form, compilation, linking, loading, execution, and running test cases. *Programmers* therefore do nothing but program. All other aspects of their work are handled by the programming secretary. Recall that what is being described here are Mills' and Baker's original ideas, dating back to 1971, when keypunches were still widely used. Coding is no longer done that way. Programmers have their own terminals or workstations in which they enter their code, edit it, test it, and so on.

10.8.1 *The New York Times* Project

The chief programmer team concept was first used in 1971 by IBM to automate the clipping file ("morgue") of *The New York Times*. The clipping file contains abstracts and full articles from *The New York Times* and other publications. This information bank is used by reporters and other editorial staff of the paper as a reference source.

The facts of the project are astounding. For example, 83,000 source lines of code (LOC) were written in 22 calendar months, representing 11 person-years. After the first year, only the file maintenance system consisting of 12,000 LOC had been written. Most of the code was written in the last 6 months. Only 21 faults were detected in the first 5 weeks of acceptance testing; only 25 further faults in the first year of operation. Principal programmers averaged one detected fault and 10,000 LOC per person-year. The file maintenance system, delivered 1 week after coding was completed, operated 20 months before a single fault was detected. Almost half the subprograms, usually 200 to 400 lines of PL/I, were correct at the first compilation [Baker, 1972].

Nevertheless, after this fantastic success, no comparable claims for the chief programmer team concept have been made. Yes, many successful projects have been carried out using chief programmer teams, but the figures reported, though good, are not as impressive as those obtained for *The New York Times* project. Why was *The New York Times* project such a success, and why have similar results not been obtained on other projects?

One possible explanation is that this was a prestige project for IBM. It was the first real trial for PL/I, a language developed by IBM. An organization known for its superb software experts, IBM set up a team comprising what can only be described as their crème de la crème from one division. Second, technical back-up was very strong. PL/I compiler writers were on hand to assist the programmers in every way they could with aspects of what was then a new language. Also, JCL experts assisted with the job control language. A third possible explanation is the chief programmer, F. Terry Baker. He is what is now called a *superprogrammer*, a programmer whose output is four or five times that of an average good programmer. In addition, Baker is a superb manager and leader, and it could be that his skills, enthusiasm, and personality were the reasons underlying the success of the project.

If the chief programmer is competent, then the chief programmer team organization works well. While it is true that the remarkable success of *The New York Times* project described previously has not been repeated, successful project after successful project has employed variants of the approach. The reason for the phrase "variants of the approach" is that the classical chief programmer team, as described by Baker, is impractical in many ways. This is discussed in the next section.

10.8.2 Impracticality of the Classical Chief Programmer Team Approach

Consider the chief programmer, a combination of a highly skilled programmer and successful manager. Such individuals are difficult to find; there is a shortage of highly skilled programmers, as well as a shortage of successful managers, and the job description of a chief programmer requires both abilities. In addition, it has been suggested that, in general, programmers and managers "are not made that way." That is to say, the qualities needed to be a highly skilled programmer are different from those needed to be a successful manager, so the chances of finding a chief programmer are small.

But if chief programmers are hard to find, back-up programmers are as rare as hen's teeth. After all, the back-up programmer is expected to be in every way as good as the chief programmer, but has to take a back seat, and a lower salary, while waiting for something to happen to the chief programmer. Neither top programmers nor top managers would be prepared to do that.

A programming secretary is also difficult to find. Software professionals are notorious for their aversion to paperwork, but the programming secretary is expected to do nothing but paperwork all day.

Thus chief programmer teams, at least as proposed by Baker, are impractical to implement. Democratic teams were also shown to be impractical, but for different reasons. Furthermore, neither technique seems to be able to handle products for which, say, 20 or even 120 programmers are needed for the implementation phase. What is needed is a way of organizing programming teams that makes use of the strengths of democratic teams and chief programmer teams and can be extended to the implementation of larger products.

10.9 Beyond Chief Programmer and Democratic Teams

Democratic teams have a major strength, namely a positive attitude to the finding of faults. A number of organizations use chief programmer teams in conjunction with code walkthroughs or inspections (Section 6.11.3). However, there is a potential pitfall. As pointed out previously, the chief programmer is personally responsible for every line of code and must therefore be present at the walkthrough or inspection. Furthermore, a chief programmer is also a manager. It was pointed out in Chapter 6 that neither walkthroughs nor inspections should be used for any sort of performance appraisal. Thus, since the chief programmer is also the manager who will be responsible for the primary evaluation of the programmers in his or her team, at the same time it

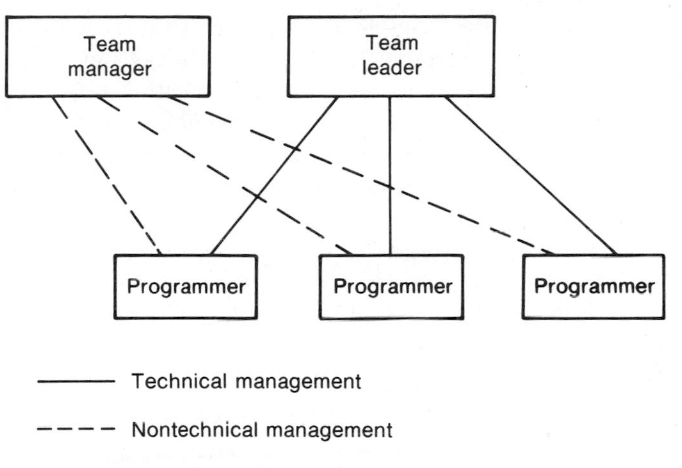

Figure 10.10 Modified programming team structure.

is inadvisable for the chief programmer to be present at either a walkthrough or an inspection.

The way out of this contradiction is to remove much of the managerial role from the chief programmer. After all, the difficulty of finding one individual who is both a highly skilled programmer and successful manager has been pointed out previously. Instead, the chief programmer should be replaced by two individuals: a team leader who is in charge of technical aspects of the team's activities and a team manager who is responsible for all nontechnical managerial decisions. The structure of the resulting team is shown in Figure 10.10. It is important to realize that this organizational structure does *not* violate the fundamental managerial principle that no employee should be responsible to more than one manager. The areas of responsibility are clearly delineated. The team leader is responsible only for technical management. Thus budgetary and legal issues are not handled by the team leader, nor are performance appraisals. On the other hand, the team leader has sole responsibility on technical issues. The team manager therefore has no right to promise that the product will be delivered within 4 weeks; promises of that sort have to be made by the team leader. The team leader naturally participates at walkthroughs and inspections; after all, he or she is personally responsible for every aspect of the code. At the same time, the team manager is not permitted at a walkthrough or inspection, because programmer performance appraisal is a function of the team manager. Instead, the team manager acquires knowledge of the technical skills of each programmer in the team during regularly scheduled team meetings.

It is important that areas which appear to be the responsibility of both the team manager and the team leader be clearly demarcated before implementation begins. For example, consider the issue of annual leave. The situation can arise that the team manager approves a leave application because leave is a nontechnical issue, only to find the application vetoed by the team leader because there is a deadline to be met in the near future. The solution to this and related issues is for higher management to draw up a policy regarding areas which both the team manager and the team leader consider to be their responsibility.

What about larger projects? This approach can be scaled up as shown in Figure 10.11, which shows the technical managerial organizational structure; the nontechnical side is similarly organized. Implementation of the product as a whole is under the direction of the project leader. The programmers report to their team leaders, and the team leaders report to the project leader. For even larger products, additional levels can be added to the hierarchy.

Another way of drawing on the best features of both democratic and chief programmer teams is to decentralize the decision-making process where appropriate. The channels of communication are then as shown in Figure 10.12. This scheme is useful for the sorts of problems for which the democratic approach is good, that is to say, in a research environment or whenever there is a hard problem that requires the synergistic effect of group interaction for its solution. Notwithstanding the decentralization, the arrows from level to level still point downwards; allowing programmers to dictate to the project leader can only lead to chaos.

Unfortunately, there is no one solution to the problem of programming team organization and, by extension, to the problem of organizing teams for all the other phases. The optimal way of organizing a team will depend on the product to be built, on previous experience with various team structures, and on the outlook of the heads of the organization. For example, if senior management are uncomfortable with decentralized decision making, then it will not be implemented. Unfortunately, not much research has been done on software development team organization, and many of the generally accepted principles are based on research on group dynamics in general, and not on software development teams. Until experimental results on team organization have been obtained within the software development industry, it will not be easy to determine the optimal team organization for a specific product.

10.10 Essential Coding Tools

In order to execute code written in a high-level language such as Ada, COBOL, or C, it is necessary to have a compiler to translate the product from that high-level language into machine code, a linker to link the various modules with the run-time routines, and a loader to load the executable version of the product into memory. Compilers, linkers, and loaders are considered to be *system software*, rather than *coding tools*. The term "coding

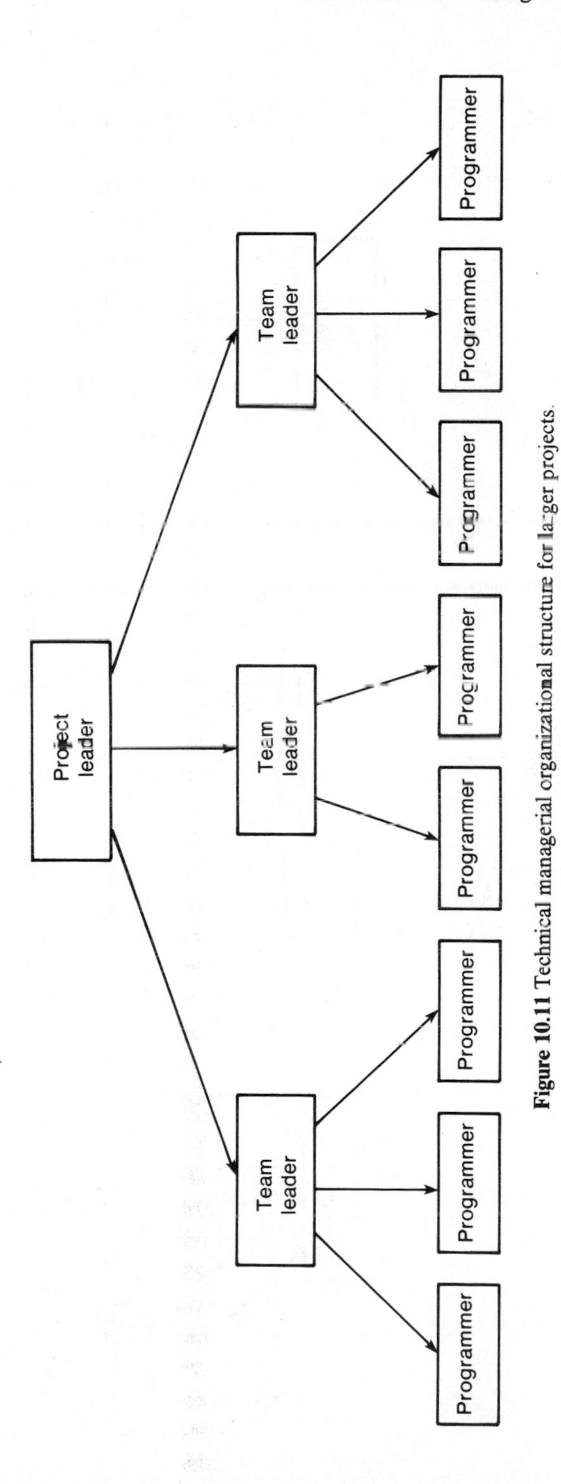

Figure 10.11 Technical managerial organizational structure for larger projects.

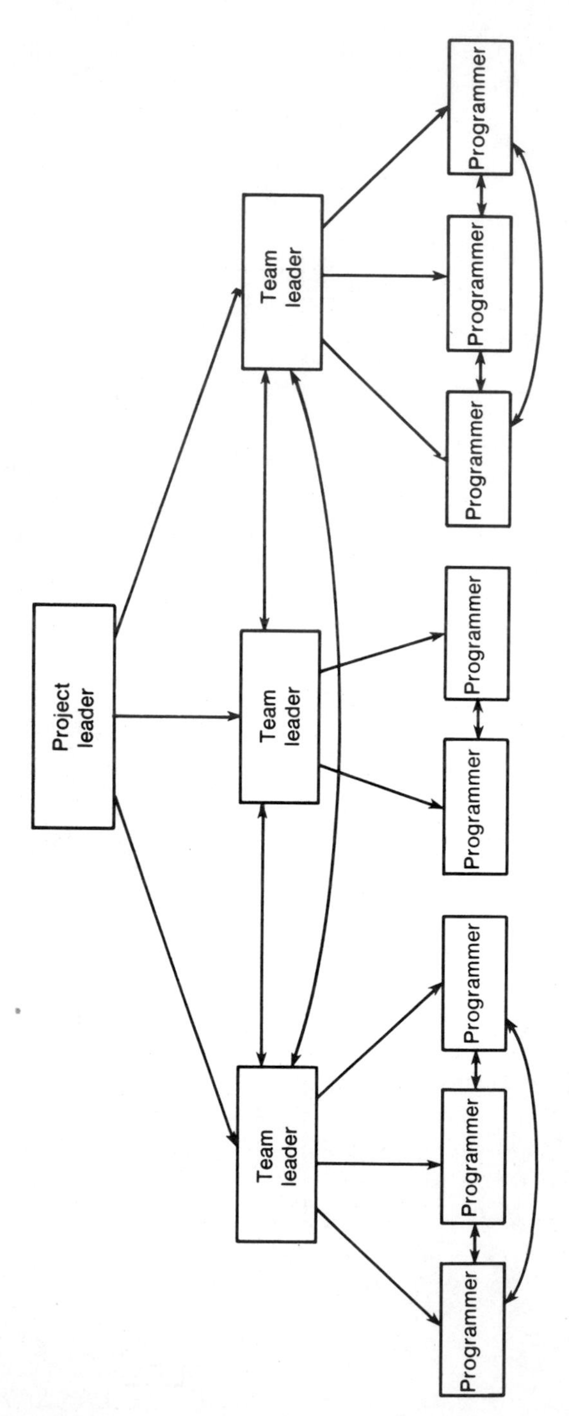

Figure 10.12 Decentralized decision-making version of team organization of Figure 10.11 showing communication channels.

tools" refers to software products such as text editors, debuggers, and pretty printers designed to simplify the programmer's task, to reduce the frustration that many programmers experience in their work, and to increase programmer productivity.

Before explaining the need for such tools, three definitions are required. *Programming-in-the-small* refers to software development at the level of the code of a single module, while *programming-in-the-large* is software development at the module level [DeRemer and Kron, 1976]. It includes aspects such as architectural design and integration. *Programming-in-the-many* refers to software production by a team. At times the team will be working at the module level, at times at the code level. Thus programming-in-the-many incorporates aspects of both programming-in-the-large and programming-in-the-small.

Now consider the conventional coding scenario for programming-in-the-small. Suppose a programmer has the task of writing a new module. The programmer sits at the terminal and invokes a text editor. The code is typed in line by line, and when the end of the module is reached, the programmer exits from the editor and invokes the compiler. The programmer has made two syntactic faults; one variable name is misspelled, and there is a missing **end**. So it is back to the text editor, and the changes are made. Then the text editor is exited, and the compiler is invoked a second time. Unfortunately, in the course of correcting the spelling mistake a semicolon is accidentally deleted, so the editor–compiler loop is invoked for the third time. The code is finally free of syntactic faults, and the linker is therefore invoked. At this point the programmer discovers that the linker cannot find a procedure called COMPUTE_AVERAGE which is called by the new module. After some thought and earnest inspection of directories of procedures already coded, the programmer realizes that the procedure is actually called COMPUTE_MEAN, so the programmer goes into the editor–compiler–linker loop again. The linking successfully accomplished, the programmer tries to execute the module—which immediately crashes. To find the fault, the programmer debugs the new module. Debugging corresponds to the editor–compiler–linker–execute loop and is repeated each time the programmer makes a change to the source code. Occasionally, another syntactic fault is made, necessitating a return to the editor–compiler loop. Finally, the module appears to work correctly. The programmer leans back, breathes a sigh of relief, and mutters, "There must be an easier way of doing this!"

There is.

10.10.1 Structure Editors

There are a number of tools currently available that will immeasurably improve on the familiar scenario for programming-in-the-small described previously. First, consider the problem of syntactic faults. What is needed is a *syntax-directed editor*, an editor that understands the syntax of the programming language being used. For example, a programmer is using a language in which variable names must be declared before use. Suppose that the programmer makes a typing mistake and misspells a variable name. For instance, the programmer types COUTNER instead of COUNTER, a previously declared variable. The syntax-directed editor immediately reacts. It prints the message

Variable COUTNER not declared

and gives the programmer the opportunity either to declare a new variable named COUTNER or to correct the spelling to COUNTER. Similarly, if the programmer's input is

```
if X > 3    B := A + 1;
```

then a syntax-directed editor will detect the absence of the **then** and position the cursor between the 3 and the B to enable the programmer to enter the missing **then**. Syntactic faults such as omitting an **end** or deleting a semicolon are quickly detected. Thus the use of a syntax-directed editor enables the programmer to obviate entirely the editor–compiler loop; the compiler cannot be called unless the code is free of syntactic faults.

In fact, an editor can go beyond syntax checking to checking the static semantics as well. For example, the statement

```
Y := X + 3;
```

is syntactically correct in Ada or Pascal. But if X is of type BOOLEAN while Y is an INTEGER, then a static semantic fault has been made. The term "static semantic" refers to a possible semantic fault that can be checked at compilation time. In contrast, if a variable K is defined to be an INTEGER in the range 1 .. 9, then GET(K) can lead to a semantic fault at run time if the value read in is less than 1 or greater than 9.

An editor that incorporates both syntactic and static semantic checking is termed a *structure editor*. A structure editor speeds up the implementation phase because time is not wasted on futile compilations. But it goes further than that. The user interface of an editor is generally very different to the user interface of a compiler. That is to say, output from a compiler is different in appearance to the way an editor displays the same module. If the programmer has to exit from the editor in order to invoke the compiler and then study the compiler output to decide what to edit next, then the programmer has to change gears mentally when switching from editor to compiler, and back. Since different thought patterns are needed for working with a compiler and with an editor, the programmer has continually to change his or her thinking mode. Mental energy is wasted on these adjustments, thus reducing programmer productivity while increasing the chance of a fault being introduced into the code. With a structure editor, the programmer works with only one piece of system software, namely the editor, and all his or her thinking is directed in terms of that editor. In addition, the programmer need concentrate on only two languages, namely the high-level language in which the module is being written and the language of the editing commands. In contrast, if a structure editor is not used, the programmer must mentally adjust to the job control language when invoking the compiler. For all these reasons, a structure editor should be considered as an essential software implementation tool.

Structure editors exist for a wide variety of languages, operating systems, and hardware. Because a structure editor has knowledge of the programming language, it is easy to incorporate a pretty printer (or formatter) into the editor, to ensure that the code always has a good visual appearance. For example, each **end** should be indented the same amount as its corresponding **begin**. An example of a syntax-directed editor that incorporates a formatter is the Macintosh Pascal editor [Apple, 1984]. Reserved words are automatically put in boldface so that they stand out, and indentation has been carefully thought out to aid readability.

10.10.2 Online Interface Checking

Now consider the problem of calling a procedure within the code, only to discover at linkage time either that the procedure does not exist or that it has been wrongly specified in some way. In the preceding scenario, the programmer tried to call procedure COMPUTE_AVERAGE, only to realize later that the actual name of the procedure is COMPUTE_MEAN. What is needed is for the structure editor to support *online interface checking*. That is to say, just as the structure editor has information regarding the name of every variable declared by the programmer, so it must also know the name of every procedure defined within the product. Interface checking is an important aspect of programming-in-the-large, that is, software development at the module level.

If the user enters a call such as

```
COMPUTE_AVERAGE (DATA_ARRAY, NUMBER_OF_VALUES, AVERAGE);
```

the editor immediately responds with a message such as

```
Procedure COMPUTE_AVERAGE not known
```

At this point, the programmer is given two choices, either to correct the name of the procedure or to declare a new procedure named COMPUTE_AVERAGE. If the second option is chosen, the programmer must also specify the parameters of the new procedure. A typical declaration might be:

```
procedure COMPUTE_AVERAGE (ARRAY_OF_DATA: in ARRAY_TYPE;
   NO_OF_ITEMS: in INTEGER; AVERAGE: out REAL);
```

That is to say, COMPUTE_AVERAGE takes as input two parameters: an array of data and the number of items whose average is to be determined. It returns the value of the average of those items.

Parameter types must be supplied when declaring a new procedure because the major reason for having online interface checking is precisely in order to be able to check full interface information, not just the names of procedures. A common fault is for procedure A to call procedure B passing, say, four parameters, while procedure B has been specified with five parameters. It is more difficult to detect the fault when the call correctly uses four parameters, but two of the parameters are transposed. For example, the declaration of procedure B might be

```
procedure B (R : in REAL; I : in INTEGER; S_1, S_2 : in STRING);
```

while the call is

```
B (INT_VAR, REAL_VAR, STRING_1, STRING_2);
```

The first two parameters have been transposed in the call statement. The presence of an online interface checker will lead to immediate detection of these and similar faults. In addition, if the editor has a help facility, the programmer can request online information as to the precise parameters of procedure B, before attempting to code the call to B. Better yet, the editor should generate a template for the call, showing the type of each of the parameters. The programmer merely has to replace each formal parameter by an actual parameter of the correct type.

A major advantage of this scheme is the programmer uses the editor to check that external references will be satisfied; there is no need to invoke a different tool, with a different user interface, to do the job. But an equally important advantage of online interface checking is that hard-to-detect faults caused by the calling of procedures with the wrong number of parameters, or with parameters of the wrong type, are immediately flagged for correction. As with the structure editor, online interface information is a necessity for the efficient production of high-quality software.

Online interface checking is particularly crucial when the software is produced by a team (programming-in-the-many). It is essential that online interface information regarding all modules be available to all programming team members at all times. Furthermore, if one programmer changes the interface of module M_5, perhaps by changing the type of one parameter from INTEGER to REAL or by adding an additional parameter, then every module that calls M_5 must automatically be disabled until the relevant call statements have been altered to reflect the new state of affairs. Online interface information is important when the product is developed by a single programmer. But it is vital when, as is usually the case, the product is being coded by a team, because of the possibility that one programmer may change the definition of a module interface thereby impacting the product as a whole.

There is still one remaining difficulty, namely that the programmer has to adjust from one language to another. Starting with the high-level programming language and the editing language, the programmer has to exit from the editor to invoke the compiler. There can be no syntactic or static semantic faults, but the compiler still has to be invoked to perform code generation. Then the linker has to be called. Again, the programmer can be sure that all external references will be satisfied as a consequence of the use of the online syntax checker, but the linker is still needed to link the product. So the user must use the job control language to invoke first the compiler and then the linker, and he or she must study the output of each to be sure that everything is satisfactory. As explained previously, having to adjust continually from the user interface of one component of system software to that of another reduces programmer efficiency and can lead to the needless introduction of faults into the code.

Fortunately, there is a way of handling this problem as well.

10.10.3 Operating System Front End

Operating systems can be cumbersome to use. Consider the simple task of listing the contents of a file EXPDATA, say, at a terminal. In some operating systems, this is achieved by a command such as

```
list EXPDATA
```

In UNIX, however, the correct syntax of the command is

```
cat EXPDATA
```

For new UNIX users, having to remember to use cat instead of list can be difficult. One of the many strengths of UNIX is that users are able to personalize the operating system ("shell") commands via aliasing. Thus, if the command

```
alias list cat
```

is given, from then on the command interpreter interprets list as an alias for

cat, and the UNIX programmer can quite happily type

list EXPDATA

knowing that the result will be that the contents of the file will be listed on the screen. In fact, the way in which UNIX allows users to personalize the operating system goes far beyond simple aliasing—for details, the reader should consult a UNIX text such as [Sobell, 1985].

What UNIX supports is a front end to the operating system. The user is able to set up and use his or her own operating system commands; the UNIX shell interprets the user's commands and causes the appropriate UNIX commands to be executed. For example, when the user types

list EXPDATA

the shell translates this into the UNIX command

cat EXPDATA

The UNIX command is then executed. The concept of a front end to an operating system is depicted in Figure 10.13.

Extending these ideas, the operating system front end should be incorporated within the editor. That is to say, it should not be necessary to have to exit from the editor in order to invoke the compiler or the linker. A programmer should be able to give operating system commands from within

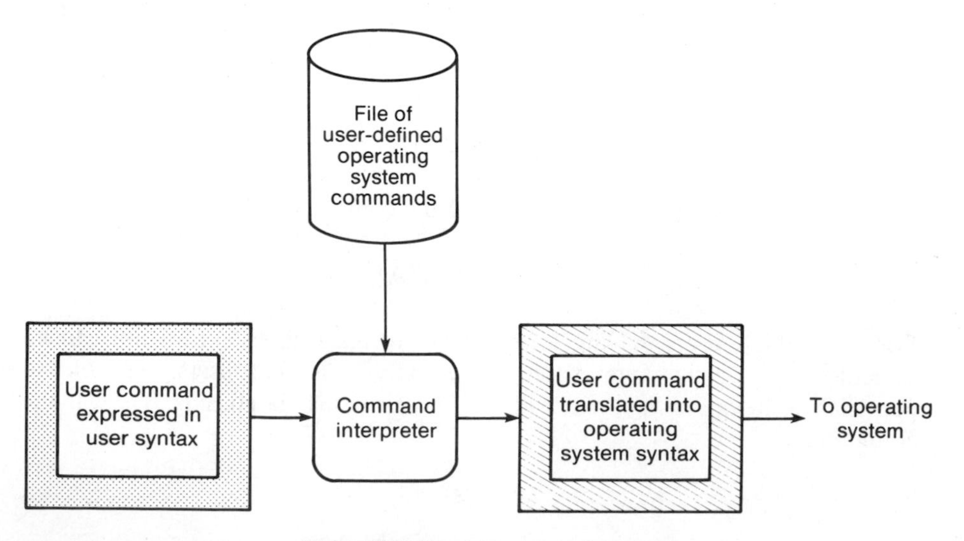

Figure 10.13 Operating system front end.

the editor. Suppose that the programmer has finished coding the module. Syntax checking, static semantic checking, and online interface checking ensure that the compiler and linker need be invoked solely to generate code and to link the module appropriately. The programmer should therefore be able to type a single command named GO or RUN or to use the mouse to choose the appropriate icon or menu selection. This will cause the editor to invoke the compiler, linker, loader, and any other system software needed to cause the module to be run. In UNIX, this can be achieved by using the make command or by invoking a shell script. But front ends can also be implemented in other operating systems.

10.10.4 Source Level Debugger

One of the most frustrating computing experiences is for a product to execute for a minute or so, then terminate abruptly, printing a message like

OVERFLOW AT 4B06

The programmer is working in a high-level language such as FORTRAN or Pascal, not a low-level language like assembler or machine code. The programmer is thinking at the level of the high-level language, not the low-level language. Up until now it has been possible for the programmer to stay within the editor, using only the programming language and the editing language, and thus having to cope with only one user interface, namely that of the editor. But when debugging support is of the OVERFLOW AT 4B06 variety, then the programmer is forced to examine machine code core dumps, assembler listings, linker listings, and a variety of similar low-level documentation, thereby destroying the whole advantage of programming in a high-level language.

In the event of a failure, the message shown in Figure 10.14 is a great improvement over the earlier terse error message. The programmer can immediately see that the module failed because of an attempt to divide by 0. Even more useful is for the operating system to enter edit mode and automatically display the line at which the failure was detected, namely line 56, together with the preceding and following four or five lines. The programmer can hopefully then see what caused the failure and make the necessary changes.

Another type of source level debugging is tracing. Before the advent of automatic tools, programmers had to insert print statements into the code

```
OVERFLOW ERROR
Module:   CYCLOTRON_ENERGY
Line 56:  NEW_VALUE := (OLD_VALUE + TEMP_VALUE) / TEMP_VALUE;
          OLD_VALUE = 3.9583        TEMP_VALUE = 0.0000
```

Figure 10.14 Output from source level debugger.

(a) Line 52: **procedure** COMPUTE_TAX_PENALTY entered

(b) Line 102: **if-then-else** block exited

(c) Line 36: **while** loop entered. Control expression (X + 2*Y) > 15 is TRUE

(d) Line 16: **for** loop entered. Loop counter IJ = 20

(e) Line 23: NEW_ROOT := **sqrt**(A*A − 4.0*B*C)
A = 17.8000 B = −13.9999 C = 15.2312 NEW_ROOT := 34.2021

Figure 10.15 Output from various debugging print statements.

by hand. Output from typical debugging print statements appears in Figure 10.15. Figure 10.15(a) results from a debugging print statement inserted into the code at the start of a procedure. It is often useful to know when an **if-then-else** block has been exited; this is shown in Figure 10.15(b). When loops are being debugged, the value of the control expression or loop counter can be helpful information. This is shown in Figures 10.15(c) and (d) for **while** loops and **for** loops, respectively. Finally, Figure 10.15(e) shows detailed trace output. This sort of debugging print statement is used when all else fails.

There are a number of difficulties associated with debugging print statements. The first is that it takes considerable effort to insert a sufficient number of appropriate debugging print statements to give the programmer adequate information. Second, debugging statements tend to generate hundreds, if not thousands, of lines of output. Third, this voluminous output is generated every time the module is executed. Fourth, Murphy's law tells us that it is inevitable that the moment a module appears to work correctly and the debugging print statements are removed or commented out, the module will fail again and all the debugging print statements will have to be reinserted.

One way of solving the latter two difficulties is to define a Boolean variable named PRINT_FLAG and then to insert debugging statements of the type

if PRINT_FLAG **then** PUT —

The value of PRINT_FLAG is set to TRUE just before pieces of code that the programmer feels may contain a fault, and set to FALSE immediately thereafter. If the programmer thinks that the code is correct, PRINT_FLAG is set to FALSE everywhere. If a new fault is detected, then PRINT_FLAG can be selectively set to TRUE again.

The difficulty of voluminous debugging output can be handled in two ways. One is to send only summary information to the terminal screen, with the bulk of the debugging data written to a file that can be examined selectively. The other solution is to define various levels of trace, utilizing a variable named DEBUGGING_LEVEL. For example, level 3 information could be that a procedure has been entered or exited, as in Figure 10.15(a). Level 2 information then relates to block entry or exit; Figures 10.15(b), (c), and (d) reflect block-level tracing. Level 1 information is a line-by-line trace displaying the line number, the statement itself, and the values of all the variables involved, as in Figure 10.15(e). The value of DEBUGGING_LEVEL can then be adjusted to obtain the relevant information. A typical level 2 debugging statement is

```
if (DEBUGGING_LEVEL <= 2) then
    PUT ("for loop block at line 381 entered");
```

This will be executed if the value of DEBUGGING_LEVEL is 2 or lower. It will not be executed if DEBUGGING_LEVEL is set to 3; in that case, the user will get level 3 information only.

But the major difficulty is inserting all these debugging statements by hand into the source code. A tool as essential as the previous three is a *source level debugger* that automatically causes various levels of trace output to be produced. The user merely specifies the level of tracing desired in each part of the module or in the module as a whole, and the tool causes the necessary output to be generated.

This sort of tool solves all the preceding difficulties, automatically providing the programmer with just the requested information at the requested level. Debugging information can be suppressed as easily as it can be turned on. For debugging purposes, an automatic source level debugger is a necessity.

Even better is an interactive source level debugger. Suppose that the value of variable ESCAPE_VELOCITY seems to be incorrect, and that procedure COMPUTE_TRAJECTORY seems to be faulty. The programmer can set breakpoints in the code. When a breakpoint is reached, execution stops and debugging mode is entered. The programmer now asks the debugger to trace variable ESCAPE_VELOCITY and procedure COMPUTE_TRAJECTORY. That is to say, every time the value of ESCAPE_VELOCITY is either used or changed, execution again halts. The programmer then has the option of entering further debugging commands, such as to request that the value of a specific variable be displayed. Alternatively, the programmer may choose to continue execution in debugging mode or to return to normal execution mode. The programmer can similarly interact with the debugger whenever procedure COMPUTE_TRAJECTORY is entered or exited. Such an interactive source level debugger offers almost every conceivable type of assistance to the programmer when the product fails.

10.10.5 Online Documentation

Programmers need online documentation. For example, online help information must be provided regarding the operating system, editor, programming language, and so on. Programming standards should also be available online, so that programmers can be certain that what they produce will be acceptable from the standards viewpoint. Depending on the project and on the organization, there can be instances when other information should also be made accessible online to programmers such as the design documentation or the user manual.

Programmers have to consult manuals of many kinds such as editor manuals and programming manuals. It is highly desirable that, wherever possible, such manuals be available online. Apart from the convenience of having everything needed at one's fingertips, it is generally quicker to query by computer than to try to find the appropriate manual and then plow through it to find the needed item. In addition, it is usually much easier to update a manual stored online than to try to find all hard-copy versions of a manual within an organization and make the necessary page changes. As a result, online documentation is likely to be more accurate than hard-copy versions of the same material, another reason for providing online documentation to programmers.

10.10.6 Stub Generator

Suppose that module M_1, currently being implemented, calls module M_2. When developing a product top-down (Section 6.14.2), M_2 has to be present as a stub, a dummy module which essentially does nothing. During the implementation of M_1, inserting the call to M_2 would cause an online interface checker to print a message pointing out that M_2 had not yet been defined. It would then allow the programmer to specify the interface of M_2, that is, the names and the types of the formal parameters.

When the time comes to link and execute M_1, it is useful if the editor is able to generate a stub for M_2 automatically, using the interface information provided by the programmer. An automatic stub generator is by no means hard to add to a structure editor that already has full online interface information. The advantage of such a generator is that it makes the programmer's task easier and speeds up the software development process. From the viewpoint of cost–benefit analysis (Section 4.2.3), the cost of adding stub generation capabilities to the structure editor is small, but the potential benefits can be large.

What has been described previously, namely a structure editor with online interface checking capabilities, operating system front end, source level debugger, online documentation, and stub generator, constitutes a simple programming environment.

This sort of environment is by no means new. All the preceding features were supported by the FLOW software development environment as far back as 1980 [Dooley and Schach, 1985]. And FLOW was by no means the first environment that provided some of these tools. For example, syntax-directed editing was described in [Hansen, 1971]. Thus what has been put forward as a minimal, but essential, toolkit does not require many years of research before a prototype can be tentatively produced. Quite the contrary, the technology has been in place for a decade or more, and it is therefore somewhat surprising that there are programmers who still implement code "the old-fashioned way."

Chapter Review

In this chapter various issues relating to the implementation of a product by a team are presented. These include choice of programming language (Section 10.1). The issue of fourth-generation languages is discussed in some detail in Section 10.2. Structured programming is the writing of readable code, and not necessarily **goto**-less code (Section 10.3). Good programming practice is described in Section 10.4, and the need for practical programming standards in Section 10.5. The issue of team organization (Section 10.6) is approached by first considering the democratic team (Section 10.7) and the classical chief programming team (Section 10.8), and then suggesting a team organization that makes use of the strengths of both approaches (Section 10.9). Finally, the need for a programming environment is stressed. Such an environment should incorporate a structure editor, online interface checker, operating system front end, source level debugger, online documentation, and a stub generator (Sections 10.10.1 through 10.10.6).

For Further Reading

A wide-ranging source of information on 4GLs is [Martin, 1985]. Two opposing views on 4GLs can be found in [Cobb, 1985] and [Grant, 1985]. [Horowitz, Kemper, and Narasimhan, 1985] is a paper on application generators, a class of 4GLs. The attitudes of 43 organizations to 4GLs are reported in [Guimaraes, 1985].

For a detailed perspective on structured programming, the three landmark articles mentioned in the text, namely [Böhm and Jacopini, 1966], [Dijkstra, 1968a], and [Knuth, 1974], should be read in their entirety. *ACM*

SIGPLAN Letters between 1968 and 1974 reflects the furor of the period regarding structured programming and especially the nature of the **goto** statement.

Excellent books on good programming practice include [Kernighan and Plauger, 1974], [Ledgard, 1975], and [Bentley, 1986]. With regard to programming team organization, the classical sources are [Weinberg, 1971], [Baker, 1972], and [Brooks, 1975]. Newer works on the subject include [Mantei, 1981], [Metzger, 1981], [Aron, 1983], [Licker, 1985], and [DeMarco and Lister, 1988].

Details of simple programming environments appear in [Teitelbaum and Reps, 1981] and [Dooley and Schach, 1985]. Information on how to design online help systems can be found in [Houghton, 1984]. For more information on environments, the reader should refer to the For Further Reading section of Chapter 12.

Problems

10.1 Your instructor has asked you to implement the product described in the Appendix using any programming language you like. Which language would you choose, and why? Of the various languages available to you, list their benefits and their costs. Do not attempt to attach dollar values to your answers.

10.2 Repeat Problem 10.1 for the elevator problem.

10.3 Repeat Problem 10.1 for the automatic teller machine (Problem 9.2).

10.4 Repeat Problem 10.1 for the automated library circulation system (Problem 7.4).

10.5 Show that if blocks are connected only by concatenation, selection, and iteration, and labels are forbidden, then every block will have exactly one entry and one exit.

10.6 Take a module that you have recently written and add prologue comments.

10.7 How do programming standards for a one-person software development company differ from those in organizations with 200 software professionals?

10.8 How would you organize a team to develop a payroll project? How would you do it for developing state-of-the-art military avionics software? Give reasons for your answers.

10.9 If your class is doing the software development project in the Appendix, how are the teams organized? Explain why they are organized that way.

10.10 You are the owner and sole employee of One-Person Software Company. You have bought the programming environment described in Chapter 10. List its six capabilities in order of importance to you, giving reasons.

10.11 You are now the vice-president for software technology and have 275 employees in your department. How do you rank the capabilities of the programming environment described in Chapter 10? Explain any differences between your answer to this problem and the previous one.

10.12 (Term Project) Implement the product you designed in Problem 9.13. Use the hardware and software specified by your instructor.

10.13 (Readings in Software Engineering) Your instructor will distribute copies of [Knuth, 1974]. Explain the title of the article.

References

[Apple, 1984] *Macintosh Pascal Users Guide*, Apple Computer, Inc., Cupertino, CA, 1984.

[Aron, 1983] J. D. ARON, *The Program Development Process. Part II. The Programming Team*, Addison-Wesley, Reading, MA, 1983.

[Baker, 1972] F. T. BAKER, "Chief Programmer Team Management of Production Programming," *IBM Systems Journal* **11** (No. 1, 1972), pp. 56–73.

[Bentley, 1986] J. BENTLEY, *Programming Pearls*, Addison-Wesley, Reading, MA, 1986.

[Böhm and Jacopini, 1966] C. BÖHM AND G. JACOPINI, "Flow Diagrams, Turing Machines, and Languages with Only Two Formation Rules," *Communications of the ACM* **9** (May 1966), pp. 366–371.

[Brooks, 1975] F. P. BROOKS, JR., *The Mythical Man-Month: Essays on Software Engineering*, Addison-Wesley, Reading, MA, 1975.

[Cobb, 1985] R. H. COBB, "In Praise of 4GLs," *Datamation* **31** (July 15, 1985), pp. 90–96.

[Date, 1986] C. J. DATE, *An Introduction to Database Systems*, Fourth Edition, Addison-Wesley, Reading, MA, 1986.

[DeMarco and Lister, 1988] T. DEMARCO AND T. LISTER, *Peopleware: Productive Projects and Teams*, Dorset House, New York, NY, 1988.

[DeRemer and Kron, 1976] F. DEREMER AND H. H. KRON, "Programming-in-the-Large Versus Programming-in-the-Small," *IEEE Transactions on Software Engineering* **SE-2** (June 1976), pp. 80–86.

[Dijkstra, 1968a] E. W. DIJKSTRA, "Go To Statement Considered Harmful," *Communications of the ACM* **11** (March 1968), pp. 147–148.

[Dooley and Schach, 1985] J. W. M. DOOLEY AND S. R. SCHACH, "FLOW: A Software Development Environment Using Diagrams," *Journal of Systems and Software* **5** (August 1985), pp. 203–219.

[Grant, 1985] F. J. GRANT, "The Downside of 4GLs," *Datamation* **31** (July 15, 1985), pp. 99–104.

[Guimaraes, 1985] T. GUIMARAES, "A Study of Application Program Development Techniques," *Communications of the ACM* **28** (May 1985), pp. 494–499.

[Hansen, 1971] W. HANSEN, "Creation of Hierarchic Text with a Computer Display," Ph.D. Thesis, Computer Science Department, Stanford University, Stanford, CA, 1971.

[Horowitz, Kemper, and Narasimhan, 1985] E. HOROWITZ, A. KEMPER, AND B. NARASIMHAN, "A Survey of Application Generators," *IEEE Software* **2** (January 1985), pp. 40–54.

[Houghton, 1984] R. C. HOUGHTON, JR., "Online Help Systems: A Conspectus," *Communications of the ACM* **27** (February 1984), pp. 126–133.

[Kernighan and Plauger, 1974] B. W. KERNIGHAN AND P. J. PLAUGER, *The Elements of Programming Style*, McGraw-Hill, New York, NY, 1974.

[Knuth, 1974] D. E. KNUTH, "Structured Programming with **go to** Statements," *ACM Computing Surveys* **6** (December 1974), pp. 261–301.

[Ledgard, 1975] H. LEDGARD, *Programming Proverbs*, Hayden Books, Rochelle Park, NJ, 1975.

[Licker, 1985] P. S. LICKER, *The Art of Managing Software Development People*, John Wiley and Sons, New York, NY, 1985.

[Mantei, 1981] M. MANTEI, "The Effect of Programming Team Structures on Programming Tasks," *Communications of the ACM* **24** (March 1981), pp. 106–113.

[Martin, 1985] J. MARTIN, *Fourth-Generation Languages, Volumes I, II, and III*, Prentice-Hall, Englewood Cliffs, NJ, 1985.

[Metzger, 1981] P. W. METZGER, *Managing a Programming Project*, Second Edition, Prentice-Hall, Englewood Cliffs, NJ, 1981.

[Sobell, 1985] M. G. SOBELL, *A Practical Guide to UNIX System V*, Benjamin/Cummings, Menlo Park, CA, 1985.

[Teitelbaum and Reps, 1981] T. TEITELBAUM AND T. REPS, "The Cornell Program Synthesizer: A Syntax-Directed Programming Environment," *Communications of the ACM* **24** (September 1981), pp. 563–573.

[Weinberg, 1971] G. M. WEINBERG, *The Psychology of Computer Programming*, Van Nostrand Reinhold, New York, NY, 1971.

Chapter 11 Maintenance

Once the product has passed its acceptance test, it is handed over to the client. Any changes after the client has accepted the product constitute maintenance. Since the product consists of more than just the source code, any changes to the documentation, manuals, or other components of the product also constitute maintenance. Some computer scientists prefer to use the term "evolution" rather than maintenance to indicate that products evolve in time.

In this chapter the various types of maintenance are considered. Then the problems of maintaining a product are described, and solutions to these problems put forward.

11.1 Why Maintenance Is Necessary

There are three main reasons for making changes to a product after it has been delivered to the client. The first reason is to correct any residual faults, whether specification faults, design faults, coding faults, documentation faults, or any other types of faults. This is termed *corrective maintenance*. Surprisingly enough, a study of 69 organizations showed that maintenance programmers spend only 17% of their time on corrective maintenance [Lientz, Swanson, and Tompkins, 1978]. Most of their time, namely 60%, was spent on the second type of maintenance, *perfective maintenance*. Here, changes are made to the code because the client thinks that these changes will improve the effectiveness of the product. For instance, the client may wish to have additional functionality added or might request that the product be modified so that it runs faster. Improving the maintainability of a product also constitutes perfective maintenance. The third reason for changing a product is *adaptive maintenance*. Here changes are made to the product in order to react to changes in the environment in which the product operates. For example, a product will almost certainly have to be modified if it has to be ported to a new compiler, operating system, and/or hardware. If there is a change to the Tax Code, then a product that prepares tax returns has to be modified accordingly. When the Post Office introduced nine-digit ZIP codes, products that had allowed for only five-digit ZIP codes had to be changed. Adaptive maintenance is thus not requested by a client; instead, it is externally imposed on the client. The study showed that 18% of software maintenance was adaptive in nature. The remaining 5% of maintenance time was devoted to other types of maintenance.

11.2 What Is Required of Maintenance Programmers

Within the life cycle of a product, more time is spent on maintenance than on any other phase. In fact, upwards of 50% of the total cost of a product accrues during the maintenance phase [Lientz, Swanson, and Tompkins, 1978; Zelkowitz, Shaw, and Gannon, 1979]. But many organizations, even today, assign the task of maintenance to beginners and to less competent programmers, leaving the "glamorous" job of product development to better programmers.

In fact, maintenance is the most difficult of all aspects of software production. A major reason is that maintenance incorporates aspects of all the other phases. Consider what happens when a fault report is handed to a maintenance programmer. A fault report is filed if, in the opinion of the user, the product is not working as specified in the user manual. There are a number of possibilities. First, there could be nothing wrong at all; the user has either misunderstood the user manual or is using the product incorrectly. Alternatively, if the fault does lie in the product, it might simply be that the user manual has been badly worded, and that there is nothing wrong with the software itself. Usually, however, the fault is in the software. But before making any changes, the maintenance programmer has to determine exactly where the fault lies. In order to do this, the maintenance programmer has at his or her disposal the fault report filed by the user, the source code—and often nothing else. Thus the maintenance programmer needs to have far above average debugging skills, because the fault could lie anywhere within the product. And the original cause of the fault might lie in the by now nonexistent specifications or design documents.

Now suppose that the maintenance programmer has located the fault. The problem is to fix it in such a way so as not to introduce inadvertently another fault elsewhere in the product, a so-called *regression fault*. If regression faults are to be minimized, detailed documentation for the product as a whole and for each individual module must be available. But software professionals are notorious for their dislike of paperwork of all kinds, especially documentation. It is quite common for there to be no documentation at all, or for the documentation to be incomplete and/or faulty. In these cases the maintenance programmer has to be able to deduce from the source code itself, the only valid form of documentation available, all the information needed to avoid introducing a regression fault.

Having determined the probable fault and having corrected it as well as he or she can, the next step is to test that the modification works correctly, and that no regression faults have in fact been introduced. In order to check the modification itself, the maintenance programmer must construct special test cases; checking for regression faults is done using the set of test data stored precisely for the purpose of performing regression testing. Then the test cases constructed for the purpose of checking the modification must be added to the set of stored test cases to be used for future regression testing of the modified product. Expertise in testing is therefore an additional prerequisite for maintenance. Finally, it is essential that the maintenance programmer document all changes made.

The preceding discussion relates to corrective maintenance. For that task, the maintenance programmer must first and foremost be a superb diagnostician, to be able to determine if there is a fault, and if so, what has to be done to fix it. But the major maintenance tasks are adaptive and perfective maintenance. To perform these, the maintenance programmer must go through the phases of requirements, specifications, design, and implementation, taking the existing product as starting point. For some types of changes, additional modules have to be designed and implemented. In other cases, changes to the design and implementation of existing modules are needed. Thus, while specifications are frequently produced by specifications experts, designs by design experts, and code by programming experts, the maintenance programmer has to be an expert in all three areas. Perfective and adaptive maintenance are adversely affected by a lack of adequate documentation just as corrective maintenance is. Furthermore, the ability to design suitable test cases and the ability to write good documentation are needed for perfective and adaptive maintenance, just as they are needed for corrective maintenance. Thus no form of maintenance is a task for a beginner or for anyone who is less than a top-ranked computer professional.

From the preceding discussion, it is clear that maintenance programmers have to possess almost every technical skill that a software professional could have. But what does he or she get in return? Maintenance is a thankless task in every way. Maintainers deal with dissatisfied users; if the user were happy with the product it would not need maintenance. Furthermore, the user's problems have frequently been caused by the individuals who developed the product, not the maintainer. The code itself may be badly written, adding to the frustrations of the maintainer. And finally, maintenance is looked down on by many software developers who consider development to be a glamorous job, while maintenance is thought to be drudge work fit only for junior programmers or incompetents.

Maintenance can be likened to after-sales service. The product has been delivered to the client. But now there is client dissatisfaction, either because the product does not work correctly, or because it does not do everything that the client currently wants, or because the circumstances for which the product was built have changed in some way. Unless the software organization provides the client with good maintenance service, the client will take his or her future product development business elsewhere. When the client and software group are part of the same organization and hence are inextricably tied from the viewpoint of future work, a dissatisfied client may use every means, fair or foul, to discredit the software group. This, in turn, leads to an erosion of confidence, from both outside and inside the software group, and to resignations and dismissals. It is important that every software organization providing maintenance services keeps its clients happy.

For product after product, maintenance is the most important phase of software production, the most difficult—and the most thankless.

How can this situation be changed? Managers must assign maintenance tasks to their best programmers. They must make it known that only the finest computer professionals merit maintenance assignments in their organization and pay them accordingly. If management faces up to the fact that maintenance is difficult and that good maintenance is critical for the success of the organization, attitudes towards maintenance will slowly change for the better.

Some of the problems that maintenance programmers face are now highlighted in a case study.

11.3 Maintenance Case Study

In countries with centralized economies, the government controls the distribution and marketing of agricultural products. In one such country, temperate fruits such as peaches, apples, and pears fell under the ambit of the Temperate Fruit Committee (TFC). One day, the chairman of TFC asked the government computer consultant assigned to his committee to computerize the operations of TFC. The chairman informed the consultant that there are exactly seven temperate fruits, namely apples, apricots, cherries, nectarines, peaches, pears, and plums. The database was to be designed for those seven fruits, no more and no less. After all, that was the way that the world was, and the consultant was not to waste time and money allowing for any sort of expansibility.

The product was duly delivered to TFC. About a year later, the chairman called in the maintenance programmer responsible for the product. "What do you know about kiwi fruit?" asked the chairman. "Nothing," replied the mystified programmer. "Well," said the chairman, "it seems that kiwi fruit is a temperate fruit that has just started to be grown in our country, and TFC is responsible for it. Please change the product accordingly."

The maintenance programmer discovered that the consultant had not carried out the chairman's original instructions to the letter. The idea of allowing for some sort of future expansion was too ingrained, so there were a number of unused fields in the relevant database records. By slightly rearranging certain items, the maintenance programmer was able to incorporate kiwi fruit, the eighth temperate fruit, into the product.

Another year went by, and the product functioned well. Then the maintenance programmer was again called to the chairman's office. The chairman was in a good mood. He expansively told the programmer that the government had reorganized the distribution and marketing of agricultural products. His committee was now responsible for all fruit produced in that country, not just temperate fruit, and so the product now had to be modified to incorporate the 26 additional kinds of fruit on the list he handed to the maintenance programmer. The programmer protested, pointing out that this change would take almost as long as rewriting the product from scratch. "Nonsense," replied the chairman. "You had no trouble adding kiwi fruit. Just do the same thing another 26 times!"

There are number of important lessons to be learnt from this. First, the problem with the product was caused by the developer, not the maintainer. The developer made the mistake of obeying the chairman's instruction regarding future expansibility of the product, but it was the maintenance programmer who suffered the consequences. In fact, unless he reads this book, the consultant who developed the product may never realize that his product was anything but a success. This is one of the more annoying aspects of maintenance, in that the maintainer is responsible for fixing other people's mistakes. The person who caused the problem either has other duties or has left the organization, but the maintenance programmer is left holding the baby. Second, the client frequently does not understand that maintenance can be difficult, or in some instances, all but impossible. The problem is exacerbated when the maintenance programmer has successfully carried out previous perfective and adaptive maintenance tasks, but now suddenly protests that a new assignment cannot be done, even though superficially it seems no different from what has been done before with little difficulty. Third, all software development activities must be carried out with an eye on future maintenance. If only the consultant had designed the product for an arbitrary number of different kinds of fruit, there would have been no difficulty in incorporating first the kiwi fruit and then the 26 other kinds of fruit. As has been stated many times, maintenance is almost always the most important phase of software production, and the one which consumes the most resources. During product development, it is essential that the development team never forget about the maintenance programmer who will be responsible for the product once it has been installed.

11.4 Software Versions

Whenever a product is maintained, there will be at least two versions of the product, the old version and the new version. Since a product is comprised of modules, there will also be two or more versions of each of the component modules that have been changed. Thus, in addition to the difficulties already listed, maintenance programmers have to cope with multiple versions of the software. It is helpful to distinguish between two types of versions, namely *revisions* and *variations*.

11.4.1 Revisions

If a fault is found in a module, then the module has to be fixed. After appropriate changes have been made there will be two versions of the module, the old one and the new one intended to replace it. The new version is termed a *revision*. The presence of multiple versions is apparently easy to solve—any incorrect versions should be thrown away, leaving just the correct one. But this is most unwise. Suppose that the previous version of the module was revision N, and that the new version is revision N + 1. First, there is no guarantee that revision N + 1 will be any more correct than revision N. Even though revision N + 1 may have been thoroughly tested by the SQA group both in isolation and linked to the rest of the product, there may be disastrous consequences when the new version of the product is run by the user on actual data. Revision N must be kept for a second reason. The product may have been distributed to a variety of sites, and not all of them may replace revision N by revision N + 1. If a fault report is now received from a site that is using revision N, then in order to analyze this new fault it is necessary to configure the product exactly in the way it is configured at the user's site, that is, incorporating revision N of the module. It is therefore necessary to retain a copy of every revision of each module.

Perfective maintenance is performed to widen the functionality of the product. In some instances new modules are written, in other cases existing modules are changed to incorporate this additional functionality. These new versions are also revisions of existing modules. So are modules that are changed when performing adaptive maintenance, that is, changes made to the product in response to changes in the environment in which the product operates. As with corrective maintenance, all previous versions must be retained.

11.4.2 Variations

Consider the following example. Most computers support more than one type of printer. Thus a personal computer may support a dot-matrix

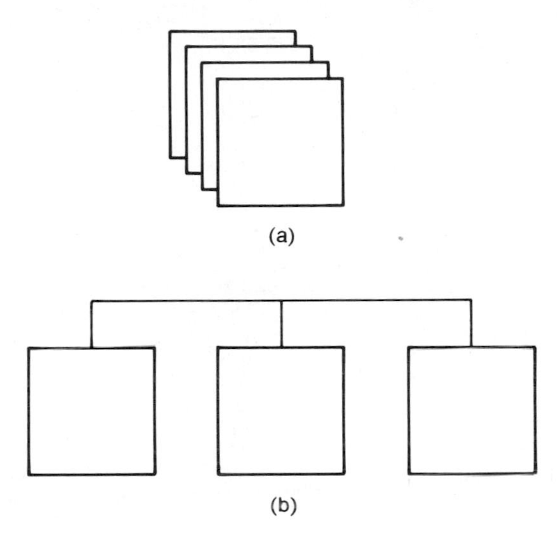

Figure 11.1 Schematic representation of multiple versions of modules, showing revisions (a) and variations (b).

printer or a laser printer. The operating system must therefore contain two *variations* of the printer driver, one for each type of printer. Unlike revisions, each of which is specifically written to replace its predecessor, variations are designed to coexist in parallel. Another situation where variations are needed is when a product is to be ported to a variety of operating systems and/or hardware. A different variation of virtually every module may have to be produced for each operating system/hardware combination.

Versions are schematically depicted in Figure 11.1, which shows both revisions and variations. To complicate matters further, in general there will be multiple revisions of each variation. In order for a software organization to avoid drowning in a morass of multiple versions, a configuration control tool is needed.

11.5 Configuration Control

Every module exists in three forms. First, there is the source code, nowadays generally written in a high-level language like Pascal, C, COBOL, or Ada. Then there is the object code, produced by compiling the source code. Finally, the object code for each module is combined with run-time routines to produce an executable load image. This is shown in Figure 11.2. The programmer has at his or her disposal a *library* of versions of modules. The specific version of each module from which a given version of the product is built is called the *configuration* of that version of the product.

Suppose that a maintenance programmer is given a fault report. One

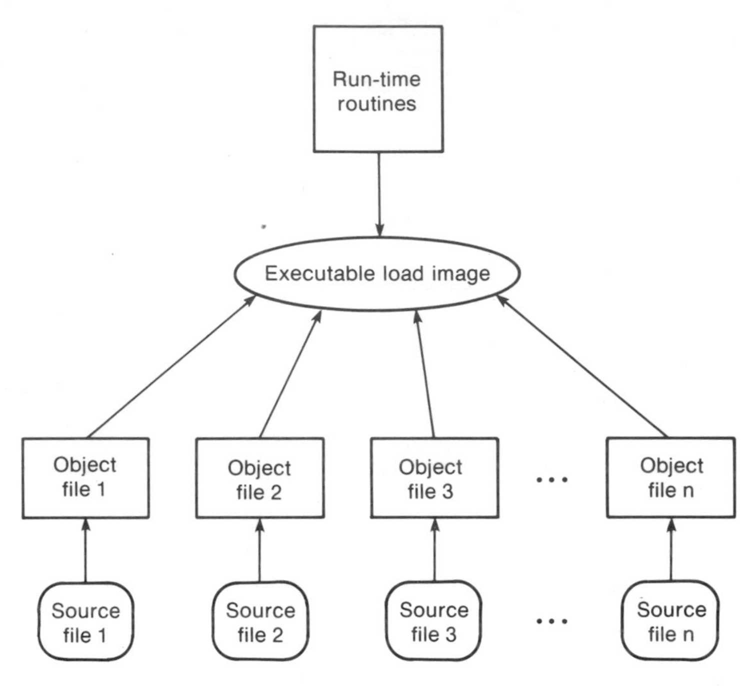

Figure 11.2 Components of executable load image.

of the first things to do is to attempt to recreate the failure. But how can the programmer determine which revisions of which variations went into the version of the product that crashed? Unless a configuration control tool is being used (Section 11.6.4), the only way to be certain is to look at the executable load image, in octal or hexadecimal format, and compare it to the object code, also in octal or hexadecimal. Then various versions of the source code have to be compiled and compared to the object code that went into the executable load image. While this can be done, it can take a long time, particularly if the product has dozens (if not hundreds) of modules, each with multiple versions. There are thus two problems that must be solved when dealing with multiple versions. First, it is necessary to be able to distinguish between versions so that the correct version of each module is compiled and linked into the product. Second, there is the inverse problem: Given a load image, determine which version of each of its components went into it.

11.5.1 Version Control

The first thing that is needed for configuration management is a version control tool. Many operating systems, particularly for mainframe computers, support version control. But many do not, so in that case a separate version control tool is needed. A common technique used in version control is for the name of each file to consist of two pieces, the file name itself and the revision number. Thus a module that acknowledges receipt of a message will have revisions ACKNOWLEDGE_MESSAGE/1, ACKNOWLEDGE_MESSAGE/2, and so on, as depicted in Figure 11.3(a). A maintenance programmer can then specify exactly which revision is needed for a given task.

With regard to multiple variations (slightly changed versions that fulfill the same role in different situations), one useful notation is to have a basic file name, followed by a variation name in parentheses [Babich, 1986]. Thus two printer drivers are given the names PRINTER_DRIVER (DOT_MATRIX) and PRINTER_DRIVER (LASER). Of course, there will be multiple revisions of each variation, such as PRINTER_DRIVER (LASER)/12, PRINTER_DRIVER (LASER)/13, and PRINTER_DRIVER (LASER)/14. This is depicted in Figure 11.3(b).

A version control tool is the first stage towards being able to manage multiple versions. Once it is in place, a detailed record (or *derivation*) of every version of the product must be kept. This comprises the name of each source code element, including the variation and revision, the versions of the various compilers and linker used, the name of the person who constructed the product, and, of course, the date and the time at which it was constructed.

Version control is a great help in managing multiple versions of modules, and the product as a whole. But more than just version control is needed, because of problems associated with maintaining multiple variations.

11.5.2 Maintaining Multiple Variations

Consider the two variations PRINTER_DRIVER (DOT_MATRIX) and PRINTER_DRIVER (LASER). Suppose that a fault is found in PRINTER_DRIVER (DOT_MATRIX) and suppose that the fault occurs in a part of the module that is common to both variations. Then not only PRINTER_DRIVER (DOT_MATRIX) has to be fixed, but also PRINTER_DRIVER (LASER). In general, if there are v variations of a module, all v of them have to be fixed. Not only that, but they have to be fixed in an identical way.

One way to solve this problem is to store just one variation, say PRINTER_DRIVER (DOT_MATRIX). Then any other variation is stored in terms of the list of changes that have to be made to go from the original to that variation. The list of differences is termed a *delta*. Thus what is stored is one variation and $v - 1$ deltas. Variation PRINTER_DRIVER (LASER) is retrieved by accessing PRINTER_DRIVER (DOT_MATRIX) and applying the delta. A change made just to PRINTER_DRIVER (LASER) is implemented by changing the appropriate delta. But any change made to PRINTER_DRIVER

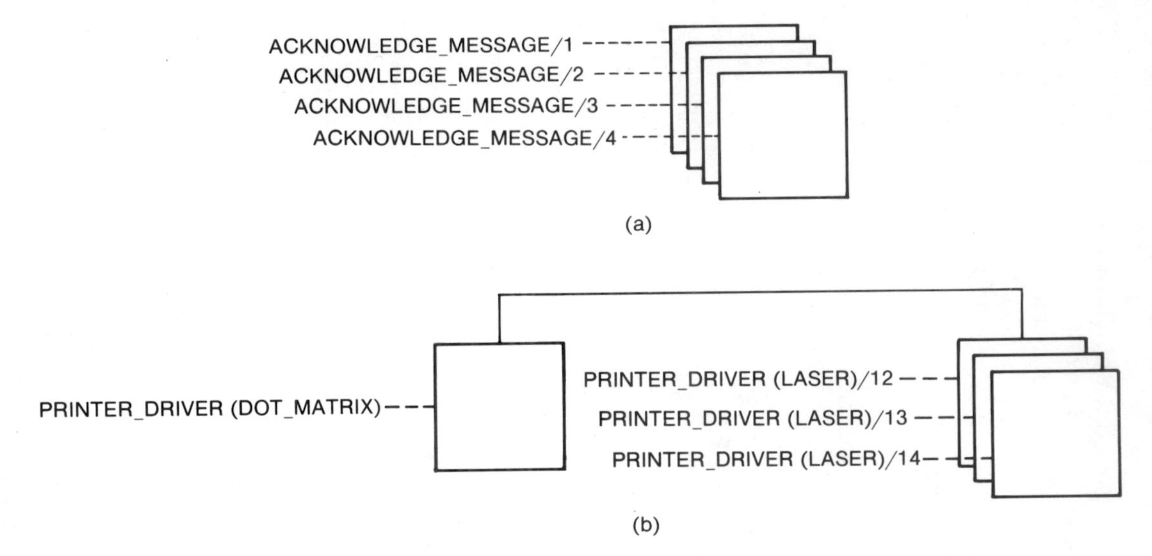

Figure 11.3 Multiple revisions and variations. (a) Four revisions of module ACKNOWLEDGE_MESSAGE. (b) Two variations of module PRINTER_DRIVER, with three revisions of variation PRINTER_DRIVER (LASER).

(DOT_MATRIX), the original variation, will automatically apply to all other variations.

Issues regarding management of maintenance are now considered.

11.6 Management of Maintenance

11.6.1 Fault Reports

The first thing that is needed when maintaining a product is to set up a mechanism for changing the product. With regard to corrective maintenance, that is, removing residual faults, if the product appears be functioning incorrectly, then a *fault report* should be filed by the user. This must include enough information to enable the maintenance programmer to recreate the problem, which will usually be some sort of software failure.

Ideally, every fault reported by a user should be fixed immediately. In practice, programming organizations are usually understaffed, and there is a backlog of work, both development and maintenance. If the fault is critical, such as if a payroll product crashes the day before payday, or overpays employees, or underpays them, immediate corrective action will have to be taken. Otherwise, the best that can be achieved is for each fault report to receive an immediate preliminary investigation.

The maintenance programmer should first consult the fault report file, containing all reported faults that have not yet been fixed, together with suggestions for working around them, that is, ways for the user to bypass the portion of the product that apparently is responsible for the failure until such time as the fault can be fixed. If the fault has previously been reported, any information in the fault report file should be given to the user. But if what the user has reported appears to be a new fault, then the maintenance programmer should study the problem, attempting to find the cause and a way to fix it. In addition, an attempt should be made to find a way to work around the problem, because it may take 6 or 9 months before someone can be assigned to make the necessary changes to the code. In the light of the serious shortage of programmers, and in particular, programmers who are good enough to perform maintenance, suggesting a way to live with the fault until it can be solved and then fixing it some months later is often the only way to deal with fault reports that cannot be considered to be emergencies.

The maintenance programmer's conclusions should then be filed in the fault report file, together with any supporting documentation such as listings, designs, and manuals used to arrive at those conclusions. The manager in charge of maintenance should consult the file regularly, prioritizing the various fixes. The file should also contain the client's requests for perfective and adaptive maintenance. Then the next modification to be made to the product will the one with the highest priority.

When copies of a product have been distributed to a variety of sites, copies of fault reports must be circulated to all users of the product, together with an estimate of when each fault can be fixed. Then if the same failure occurs at another site the user can consult the relevant fault report to determine if it is possible to work around the fault, and how long he or she will have to live with it until it can be fixed. It would of course be preferable to fix every fault immediately, and then distribute a new version of the product to all sites. But given the realities of software maintenance, distributing fault reports is probably the best that can be done under the circumstances of the current worldwide shortage of good programmers.

11.6.2 Authorizing Changes to the Product

Once a decision has been taken to perform corrective maintenance, a maintenance programmer will be assigned the task of determining the fault that caused the failure and then repairing it. After the code has been changed, the fix must be tested, as must the product as a whole (regression testing). Then the documentation must be updated to reflect the changes made. In particular, a detailed description of what was changed, why it was changed, by whom, and when, must be added to the prologue comments of any changed module. If necessary, design and/or specifications documents also have to be changed. A similar set of steps is followed when performing perfective or adaptive maintenance; the only real difference is that perfective and adaptive maintenance are initiated by a change in requirements rather than by a fault report.

At this point all that would seem to be needed would be to distribute the new version to the users. But what if the maintenance programmer has not tested the fix adequately? Before the product is distributed, it must be subjected to SQA performed by an independent group, that is, the members of the maintenance SQA group must not report to the same manager as the maintenance programmer. Reasons were given previously as to why maintenance is the most difficult of all software production activities. For those same reasons, maintenance is also the most fault-prone. Testing during the maintenance phase is both difficult and time-consuming, and the SQA group should not underestimate the implications of software maintenance with regard to testing. Once the new version has been approved by SQA, it can be distributed.

While the preceding procedure makes sense if only one programmer is working at maintaining the product, all sorts of difficulties can arise when more than one programmer at a time are simultaneously maintaining a product. For example, suppose two programmers are each assigned a fault report on a Monday morning. By coincidence, they both localize the fault they are to fix to different parts of the same module M_DUAL. Each programmer makes his or her own copy of the current version of the module, namely

M_DUAL/16, and they get to work. The first programmer fixes the first fault, has the changes approved, and replaces the module, now called M_DUAL/17. A day later the second programmer fixes the second fault, has the changes approved, and installs module M_DUAL/18. Unfortunately, revision 17 contains the changes of only the first programmer, while revision 18 contains those of only the second programmer. Thus all the changes of the first programmer have been overwritten.

While the idea of each programmer making individual copies of a module is far better than both working together on the same piece of software, it is clear that it is inadequate for maintenance by a team. What is needed is some mechanism that allows only one user at a time to change a module.

11.6.3 Baselines

The maintenance manager must set up a *baseline*, a configuration (set of versions) of all the modules in the product. When trying to find a fault, a maintenance programmer puts copies of any needed modules into his or her own *private workspace*. In this private workspace the programmer can change anything at all without impacting any other programmer in any way, because all changes are made to the programmer's private copy; the baseline version is untouched.

Once it has been decided which module has to be changed to fix the fault, the programmer *freezes* the current version of the module that he or she is going to alter. No other programmer may ever make changes to any frozen version. When the maintenance programmer has made changes and they have been tested, the new version of the module is then installed, thereby modifying the baseline. The previous version, now frozen, is retained because it may be needed in the future as explained previously, but it cannot be altered. Once a new version has been installed, any other maintenance programmer can in turn freeze it and make changes to it. The resulting module will, in turn, become the baseline version. A similar procedure is followed if more than one module has to be frozen simultaneously.

This scheme solves the problem with module M_DUAL. Both programmers make private copies of M_DUAL/16 and use those copies to analyze the respective faults that they have been assigned to fix. The first programmer decides what changes to make, freezes M_DUAL/16, and makes the changes to it to repair the first fault. After the changes have been tested, the resulting revision M_DUAL/17 becomes the baseline version. In the meantime, the second programmer has determined what the second fault is by experimenting with his or her private copy of M_DUAL/16. Changes cannot now be made to M_DUAL/16 because it has been frozen by the first programmer. Once M_DUAL/17 has been installed in the baseline by the first programmer, it is frozen by the second programmer, and the second set of changes is made. These changes are made on M_DUAL/17. The resulting module is now installed as M_DUAL/18, a version that will incorporate the changes of both programmers. Revisions M_DUAL/16 and M_DUAL/17 are retained for possible future reference, but they cannot be altered in any way in the future.

Management must ensure that procedures are carefully followed when the technique of baselines and private copies is used. Suppose that a programmer wishes to change a module M_TUE. The programmer freezes that module and makes copies of all other modules that will be needed to perform the required maintenance task; often this will be all the other modules in the product. The programmer makes the necessary changes to M_TUE, tests them, and the new revision of M_TUE is installed in the baseline. But when the modified product is delivered to the user, it immediately crashes. What went wrong is that the maintenance programmer tested the modified version of M_TUE using his or her private workspace copies, that is, the copies of the other modules that were in the baseline at the time that maintenance of M_TUE was started. In the meantime, certain other modules were updated by other maintenance programmers working on the same product. The lesson is clear: Before installing a module, it must be tested using the current baseline versions of all the other modules and not the programmer's private versions. This is a further reason for stipulating independent SQA—members of the SQA group simply do not have access to programmers' private workspaces.

Configuration control is clearly a necessity during maintenance. But is there a need for configuration control to be applied earlier in the life cycle?

11.6.4 Configuration Control during Product Development

While a module is being coded, versions are changing too rapidly for configuration control to be helpful. Nevertheless, once the module has been passed by SQA it should be subject to the same configuration control procedures as those of the maintenance phase. The reason for this is that once a module has been integrated into the product as a whole, any change to that module can impact the product in the same way as a change made during the maintenance phase. Thus configuration control is needed not only during maintenance, but also during the implementation and integration portions of the life cycle. Furthermore, management cannot monitor the development process adequately unless every module is subject to configuration control. When configuration control is properly applied, management is aware of the status of every module and can take early corrective action if it seems that project deadlines are slipping.

11.6.5 Maintenance Tools

It is unreasonable to expect maintenance programmers to keep track manually of the various revision numbers and to assign the next revision number each time a module is updated. Unless the operating system incorporates version control, a version control tool such as the UNIX tool ***sccs*** (source code control system) [Rochkind, 1975] or ***rcs*** (revision control system) [Tichy, 1985] is needed. It is equally unreasonable to expect manual control of the freezing technique described previously, or any other manual way of ensuring that revisions are updated appropriately. What is needed is a configuration control tool. A typical example of a commercial tool is CCC (change and configuration control).

Even if the software organization does not wish to purchase a complete configuration control tool, at the very least a *build* tool must be used in conjunction with a version control tool, that is, a tool that assists in selecting the correct version of each object code module to be linked to form a specific version of the product. At any time there will be multiple variations and revisions of each module in the product library. Any version control tool will assist users in distinguishing between versions of modules of source code. But keeping track of object code is harder, because some version control tools do not attach revision numbers to object versions.

To cope with this, some organizations automatically compile the latest version of each module every night, thereby ensuring that the object code in the baseline is always up to date. While this technique works, it can be extremely wasteful of computer time because a large number of unnecessary compilations will frequently be performed. The UNIX tool *make* can solve this problem [Feldman, 1979]. For each executable load image, the programmer sets up a Makefile specifying the hierarchy of source and object files that go into that particular configuration; such a hierarchy is shown in Figure 11.2. More complex dependencies, such as included files in C, can also be handled by *make*. When invoked by a programmer, the tool works as follows: UNIX, like virtually every other operating system, attaches a date and time stamp to each file. Suppose that the stamp on a source file is Friday, June 6 at 11:24 A.M., while the stamp on the corresponding object file is Friday, June 6 at 11:40 A.M. Then it is clear that the source file has not been changed since the object file was created by the compiler. But if the date and time stamp on the source file is later than that on the object file, then *make* calls the appropriate compiler or assembler to create a version of the object file that will correspond to the current version of the source file.

Next, the date and time stamp on the executable load image is compared to those on every object file in that configuration. If the executable load image was created later than all the object files, then there is no need to relink. But if an object file has a later stamp than that of the load image, then the load image does not incorporate the latest version of the corresponding object file. In this case *make* will call the linker and construct an updated load image.

Thus *make* checks whether the load image incorporates the current version of every module. If so, then nothing further is done, and no CPU time is wasted on needless compilations and/or linkage. But if not, then *make* calls the linker to create an up-to-date version of the product.

In addition, *make* simplifies the task of building an object file. There is no need for the user to specify each time what modules are to be used and how they are to be connected, because this information is already in the Makefile. Thus, to build a product comprising perhaps hundreds of modules, a single *make* command is all that is needed to ensure that the complete product is put together correctly.

Maintenance is difficult and frustrating. The very least that management can do is to provide the maintenance team with the tools that are needed for efficient and effective product maintenance.

11.6.6 Maintenance Techniques

Maintenance is not a one-time effort. A well-written product will go through a series of versions over its lifetime. As a result, it is necessary to plan for maintenance over the entire life cycle. During the design phase, for example, information hiding techniques (Section 8.6) should be employed; during implementation, variable names should be selected that will be meaningful to future maintenance programmers (Section 10.4). Documentation should be complete and correct and reflect the current version of every component module of the product.

During the maintenance phase, it is important that the maintainability that has been built into the product from the very beginning should not be compromised. In other words, just as software development personnel should be conscious at all times of the inevitable maintenance that is still to come, so software maintenance personnel should always be conscious of the equally inevitable further maintenance still to come. The principles leading to maintainability that have been described earlier in this book are equally applicable to the maintenance phase itself.

11.6.7 Problem of Repeated Maintenance

One of the more frustrating difficulties of product development is the so-called moving-target problem. What can happen is that as fast as the developer constructs the product, the client changes his or her requirements. Not only is this frustrating to the development team, but the frequent changes can result in the product itself being badly constructed. In addition, such changes add to the cost of the product. In theory, the way to cope with this is to start by constructing a prototype. Then it does not matter how often the client changes the requirements. Once the client is finally satisfied, the specifications are approved and the product itself is constructed. In practice, there is nothing to stop the client from changing the requirements the day after he or she has finally approved them. The main advantage to prototyping in this situation is that, by presenting the client with a working model, it *may* reduce the number and the frequency of the changes. But if the client is willing to pay the price, nothing can be done to prevent the requirements being changed every Monday and Thursday.

The problem is exacerbated during the maintenance phase. The more a completed product is changed, the more it will deviate from its original design, and the more difficult it will become to make further changes. Under repeated maintenance the documentation is likely to become even less reliable than usual, and the regression testing files may not be up to date. If still more maintenance has to be done, the product as a whole may first have to be completely rewritten using the current version as a prototype.

The problem of the moving target is clearly a management problem. In theory, if management is sufficiently firm with the client and explains the problem at the beginning of the project, then the specifications can be frozen from the time the prototype has been accepted until the product is delivered. Again, after each request for perfective maintenance, the specifications can be frozen for, say, 3 months or 1 year. In practice, it does not work that way. For example, if the client happens to be the president of the corporation and the development organization is the information systems division of that organization, then the president can indeed order changes every Monday and Thursday, and they will be implemented. The old proverb, "he who pays the piper calls the tune," is unfortunately only too relevant in such a situation. Perhaps the best that the vice-president for information systems can do is to try to explain to the president the effect on the product of repeated maintenance, and then simply have the complete product rewritten whenever further maintenance would be hazardous to the integrity of the product.

Trying to discourage additional maintenance by ensuring that the requested changes are implemented slowly only has the effect of the relevant personnel being replaced by others who are prepared to do the job faster. In short, if the person who requests the repeated changes has sufficient clout, then there is no solution to the problem of the moving target.

11.7 Maintenance Skills versus Development Skills

Earlier in this chapter much was said about the skills that are needed to be a maintenance programmer. For corrective maintenance, the ability to determine the cause of a failure of a large product was deemed essential. But this skill is not needed exclusively for product maintenance. It is used all the time during integration and during product testing. Furthermore, the ability to function effectively without adequate documentation was stressed as another necessity. Again, the documentation is rarely complete while integration and product testing are under way. It was also stressed that skills with regard to specification, design, implementation, and testing are essential for adaptive and perfective maintenance. These four activities are carried out during the development life cycle, and each requires specialized skills if it is to be performed correctly.

In other words, the skills that a maintenance programmer needs are in no way different to those needed by software professionals specializing in other phases of software production. The key point is that a maintenance programmer must not merely be skilled in a broad variety of areas, he or she must be *highly* skilled in *all* those areas. While the average software developer can specialize in one area of software development such as design or testing, the software maintainer must be a specialist in virtually every area of software production. After all, as was pointed out previously, maintenance is the same as development, only more so.

Chapter Review

Maintenance is almost always the most important, and most difficult, software activity (Section 11.2). This is illustrated by means of the case study of Section 11.3. The problem is compounded by the need to maintain multiple versions (Section 11.4). Version control tools are needed to cope with several revisions and variations of modules (Section 11.5). Management of maintenance requires the setting up of a suitable mechanism for determining what should be changed (Section 11.6.1) and authorizing such changes (Section 11.6.2). The changes must then be managed using baselines (Section 11.6.3), configuration control tools (Sections 11.6.4), and other maintenance tools (Section 11.6.5). Maintenance must be planned for at all times, including the maintenance phase itself (Section 11.6.6). The problem of repeated maintenance is discussed in Section 11.6.7. The skills that a maintenance programmer needs are the same as those of a developer; the difference is that a developer can specialize in one phase of the life cycle, while the maintainer must be an expert in all aspects of software production (Section 11.7).

For Further Reading

Sources of information on maintenance in general include [Lientz and Swanson, 1980], [Glass and Noiseux, 1981], [Babich, 1986], and [Parikh, 1988]. [Schneidewind, 1987] is an excellent overview of the state of the art.

Rochkind designed *sccs* [Rochkind, 1975], while *rcs* is described in [Tichy, 1985]. Detailed information on *make* can be found in [Feldman, 1979].

Ways of reducing the cost of maintenance are described in [Guimaraes, 1983]. Other papers on maintenance can be found in [Parikh and Zvegintzov, 1983], the March 1987 issue of *IEEE Transactions on Software Engineering*, and in the proceedings of the annual IEEE-sponsored Conference on Software Maintenance.

Problems

11.1 Why do you think that the mistake is made of considering software maintenance to be inferior to software development?

11.2 Consider the automated library circulation system (Problem 7.4). Describe why such a product is likely to have multiple variations of many of the modules.

11.3 Repeat Problem 11.2 for the automatic teller machine (ATM) of Problem 9.2.

11.4 Repeat Problem 11.2 for a product that checks whether a bank statement is correct (Problem 7.2).

11.5 You are the manager in charge of maintenance in a large software organization. What qualities do you look for when hiring new employees?

11.6 What are the implications of maintenance for a one-person software production organization?

11.7 Does a one-person software production organization need a version control tool, and if so, why? What about a configuration control tool?

11.8 You are the manager in charge of the software controlling the navigation system of a nuclear submarine. Three different user-reported faults have to be fixed, and you assign Alf, Brenda, and Clark to do the job. A day later you learn that in order to implement each of the three fixes, the same five modules must be changed. However, your configuration control tool is inoperative, so you will have to manage the changes yourself. How will you do it?

11.9 You have been asked to build a computerized fault report file. What sort of data would you store in the file? What sort of queries could be answered by your tool?

11.10 (Term Project) PVIC management have decided to change the objective from maximizing sales to minimizing shrinkage (a polite word for theft). The target shrinkage has been fixed at 2% of the previous month's sales. Modify the product accordingly.

11.11 (Readings in Software Engineering) Your instructor will distribute copies of [Schneidewind, 1987]. Which of the issues raised by Schneidewind do you consider to be the most important? Give reasons for your answer.

References

[Babich, 1986] W. A. Babich, *Software Configuration Management: Coordination for Team Productivity*, Addison-Wesley, Reading, MA, 1986.

[Feldman, 1979] S. I. Feldman, "Make—A Program for Maintaining Computer Programs," *Software—Practice and Experience* **9** (April 1979), pp. 225–265.

[Glass and Noiseux, 1981] R. L. GLASS AND R. A. NOISEUX, *Software Maintenance Guidebook*, Prentice-Hall, Englewood Cliffs, NJ, 1981.

[Guimaraes, 1983] T. GUIMARAES, "Managing Application Program Maintenance Expenditures," *Communications of the ACM* **26** (October 1983), pp. 739–746.

[Lientz and Swanson, 1980] B. P. LIENTZ AND E. B. SWANSON, *Software Maintenance Management: A Study of the Maintenance of Computer Applications Software in 487 Data Processing Organizations*, Addison-Wesley, Reading, MA, 1980.

[Lientz, Swanson, and Tompkins, 1978] B. P. LIENTZ, E. B. SWANSON, AND G. E. TOMPKINS, "Characteristics of Application Software Maintenance," *Communications of the ACM* **21** (June 1978), pp. 466–471.

[Parikh, 1988] G. PARIKH (Editor), *Techniques of Program and System Maintenance*, Second Edition, QED Information Sciences, Wellesley, MA, 1981.

[Parikh and Zvegintzov, 1983] G. PARIKH AND N. ZVEGINTZOV (Editors), *Tutorial: Software Maintenance*, IEEE Computer Society Press, Washington, DC, 1983.

[Rochkind, 1975] M. J. ROCHKIND, "The Source Code Control System," *IEEE Transactions on Software Engineering* **SE-13** (October 1975), pp. 255–265.

[Schneidewind, 1987] N. F. SCHNEIDEWIND, "The State of Software Maintenance," *IEEE Transactions on Software Engineering* **SE-13** (March 1987), pp. 303–310.

[Tichy, 1985] W. F. TICHY, "RCS—A System for Version Control," *Software —Practice and Experience* **15** (July 1985), pp. 637–654.

[Zelkowitz, Shaw, and Gannon, 1979] M. V. ZELKOWITZ, A. C. SHAW, AND J. D. GANNON, *Principles of Software Engineering and Design*, Prentice-Hall, Englewood Cliffs, NJ, 1979.

Part Four

Major Topics in Software Engineering

The six topics presented in Part Four were selected mainly because of their relevance to the current practice of software engineering. But, in addition, some attention was paid to the directions in which software engineering appears to be moving, so that the reader will be equipped to handle future trends in the subject.

Chapter 12 is entitled CASE (Computer-Aided Software Engineering). Stress is placed on the importance and scope of CASE. The many environments available today are categorized as language-centered, structure-oriented, toolkit, or method-based, and the four types are then compared. Finally, the ISTAR integrated project support environment (IPSE) is described, and an evaluation of IPSEs given.

The issues of portability and reusability are described in Chapter 13. Various barriers to portability, including hardware, operating system, numerical software, and compiler incompatibilities, are described; techniques for promoting portability of system software, application software, and even data, are suggested. Reusability of software refers to employing existing code, designs, and documentation in new products. Three case studies are presented to convince the reader that reusability is not merely a theoretical principle, but a practical technique that can be achieved in practice.

Ada and Software Engineering is the title of Chapter 14. This chapter has two main themes. First, the financial might of the U.S. Department of Defense (DoD) is so great that Ada is bound to sweep the world the same way that COBOL, the previous DoD language, did. Second, Ada embodies many of the software engineering principles described in this book, and the language therefore merits attention as a case study in software engineering.

Another major theme of this book is the importance of experiments performed to validate a particular technique or to show that one technique is superior to another. In Chapter 15, Experimentation in Software Engineering, the need for experimentation is stressed, as are the many difficulties associated with experiments for comparing methods for projects developed by a team. The chapter concludes with some suggestions for alternatives to controlled experimentation.

The final chapter, Chapter 16, is entitled Automatic Programming. It is concerned with the possibility that, in the future, software could be produced not by software engineers, but rather by an automatic tool. Intermediate steps towards this goal, namely domain-specific automatic programming and programming assistants, are presented. The chapter concludes with an assessment as to whether automatic programming can ever be achieved.

Chapter 12 CASE (Computer-Aided Software Engineering)

During the development of a software product, a number of very different operations and functions have to be carried out. Typical activities include estimating resource requirements, drawing data flow diagrams, coding, performing integration testing, and writing the user manual. Unfortunately, neither these activities nor the others in the software development life cycle can yet be fully automated and performed by a computer without human intervention.

But, computers can *assist* every step of the way. The title of this chapter, CASE, stands for computer-aided (or computer-assisted) software engineering. Computers can help by carrying out much of the drudge work associated with software development, including the organization of documentation of all kinds such as plans, specifications, designs, source code, data dictionaries, and management information. Documentation is essential for software development and maintenance, but the majority of individuals involved in software development are not fond of either creating or updating documentation.

But CASE is not restricted to assisting with documentation. It covers all aspects of computer support for software engineering. At the same time, it is important to remember that CASE stands for computer-*aided* software engineering, and not computer-*automated* software engineering—no computer can yet replace a human being with respect to development or maintenance of software. For the foreseeable future at least, the computer must remain a tool of the software professional.

In Section 10.10 it was shown how a set of tools, namely a structure editor with online interface checking capabilities, operating system front end, source level debugger, online documentation, and stub generator, could be combined to form a software development environment used by programmers to implement a product. But why should the computer be used in only the implementation phase? Surely the computer can help the development of software from the requirements phase through to maintenance? That is indeed the case, and it forms the subject of this chapter.

12.1 Scope of CASE

The simplest form of CASE is the software *tool*, a product that assists in just one aspect of the production of software. The tools mentioned in Section 10.10 such as the syntax-directed editor, source level debugger, and online interface checker are used during the implementation phase. In Chapter 11 version control tools were described. Tools are currently being used in all the other phases of the life cycle as well. For example, there are a variety of tools on the market, many of them for use with personal computers, that assist in the construction of graphical representations of software products, such as data flow diagrams (Figures 7.2 through 7.4) or structure charts (Figures 9.4 and 9.5). CASE tools that help the developer during the earlier life-cycle phases, namely the requirements, specifications, and design phases, are sometimes termed *upperCASE* tools, while those that assist with implementation and maintenance are termed *lowerCASE*.

Just as syntax-directed editors can be extended until they become complete programming *environments*, so additional features have been built onto graphical tools. An important example is the data dictionary, a computerized list of all data defined within the product. In a large product there will be tens, if not hundreds, of thousands of data items, and the computer is ideal for the task of storing information such as names and types of data items, and where each is defined. It is not necessary for the user to search manually through hundreds of DFDs like Figure 7.4. Instead, using a query language, the user can ask where Package_details is used and be informed that Package_details is passed from data store PACKAGE DATA to process VERIFY ORDER IS VALID. A specifications and design tool can integrate data flow diagrams, structure charts, and the data dictionary. The data can then be checked for consistency. For example, if the structure chart corresponding to Figure 7.4 omitted all reference to Package_details, then the computer will flag this. This ensures that the data flow diagrams, structure charts, and any other graphical specifications and design representations supported, such as entity relationship diagrams [Chen, 1983], transition diagrams (Figure 7.8), Nassi–Shneiderman charts [Nassi and Shneiderman, 1973], Petri nets (Figures 7.13 through 7.19), and module interconnection diagrams (Figure 6.15), are always consistent and are merely different views of the same product.

Another use of a data dictionary is to provide the data for report generators and screen generators. A report generator is used to generate the code needed for producing a report. A screen generator is used to assist the software developer in producing the code for a screen for capturing data. Suppose that a screen is being designed for entering the weekly sales at each branch of a chain of shoe stores. It has been decided that the branch number is to be entered on the screen three lines from the top. A branch number is a four-digit integer in the range 1000 to 4500 or 8000 to 8999. This information is given to the screen generator. The screen generator then automatically generates code to display the string BRANCH NUMBER ____ three lines from the top and to position the cursor at the first underline character. As the user enters each digit in turn it is displayed, and the cursor moves on to the next underline. The screen generator also generates code for checking that the user enters only digits, and that the resulting four-digit integer is in the specified range. If the data entered are invalid or if the user presses the ? key, help information is displayed to assist the user.

Use of such generators can result in prototypes being constructed very rapidly indeed. Furthermore, just as a syntax-directed editor can be broadened into a programming environment, so a graphical representation tool combined with a data dictionary, report generator, and screen generator together constitute a specifications and design environment that supports prototyping. An example of a commercial product that incorporates all these features is Excelerator. Excelerator supports a variety of different graphical methods, including data flow diagrams, entity relationship diagrams, module interconnection diagrams, and JSD structure diagrams. There is a data dictionary, a consistency checker, and both screen and report generators.

A necessary tool for programming-in-the-many is a version control tool. As described in Chapter 11, different versions of the same module will be produced in the course of maintaining a product. First, there will be *revisions*, newer versions of modules either to correct faults in previous versions or to incorporate additional functionality. Second, there may be parallel versions of the same module. For example, one version of a module may run under, say, UNIX, while another version of the same module runs under VAX/VMS. Apart from differences to allow for variances between the two operating systems, the two modules are functionally identical. These versions are termed *variants*. A version control tool is necessary to control all these versions of the hundreds of modules of a medium- or large-scale product. In addition, a build tool, such as the UNIX *make* tool, is needed to help the developer to ensure that, when a version of the complete product is compiled and linked, the appropriate versions of the component modules are used. Again, the user can be considerably helped by combining the version control and build tools with a configuration control database. A suitable query language can then be used to supply information about module versions, such as what versions of a given module are available and on what operating system/hardware configuration they run. In addition, management information can be provided such as fault reports for each version and where a given version of a module has been installed.

But management needs a broad variety of other tools. For example, managers need to know the status of every module. There are questions that require immediate answers, such as: Has module M_ABC been coded; has M_DEF been passed by the SQA group; did the UNIX version pass product testing? Management also needs to know how long it took the programmer to code a module so that the work may be billed appropriately, either directly to the client in instances where the client is charged by the hour or to the correct account when the module forms part of a larger product for which a lump sum is to be paid.

In addition to information appertaining to the components of the product, more general types of management information are needed. An example of this is critical path management (CPM), otherwise known as program evaluation review techniques (PERT).

12.1.1 PERT/CPM Tools

There are many hundreds of activities that have to be performed in the course of building a product, such as projecting cash flow or checking that the database manual is an accurate reflection of the structure of the database. Some activities have to precede others; for instance, a module cannot be coded until it has been designed. Other activities can be carried on in parallel. For example, suppose the product depicted in Figure 6.15 is developed top-down. After module A has been implemented, modules B, C, and D can be implemented in parallel.

Suppose that two activities are started at the same time and can be performed in parallel, and that both have to be completed before proceeding with the project as a whole. If the first takes 12 days, while the second needs only 3 days, then the first activity is *critical*. That is to say, any delay in the first activity will cause the project as a whole to be delayed. But the second activity can be delayed up to 9 days without adversely impacting the project; there is a *slack* of 9 days associated with the second activity. When using PERT/CPM [Moder, Phillips, and Davis, 1983], the manager inputs the activities, their estimated durations, and any precedence relations, that is, the activities which have to be completed before a specific activity can be started. The PERT/CPM package will then determine which of the hundreds of activities are critical, and it will also compute the slack for each of the noncritical activities. Most packages also print out a PERT chart showing the precedence relationships between the activities and highlighting the *critical path*, the path through the chart that consists of critical activities only. If any activity on the critical path is delayed, then so is the project as a whole.

A simple PERT chart is shown in Figure 12.1. There are 12 activities and 9 milestones. Starting with milestone A, activities AB, AC, and AD can be

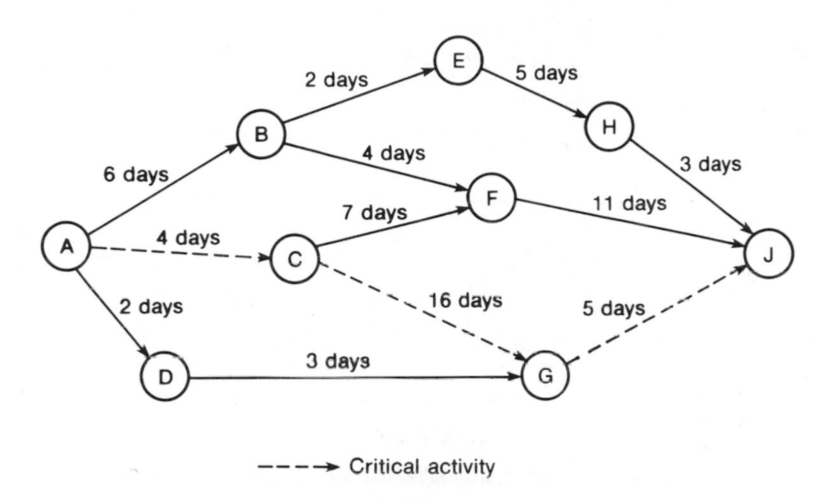

Figure 12.1 PERT chart showing estimated durations of activities and critical path.

started in parallel. Activity FJ cannot be started until both BF and CF are finished. The project as a whole is complete when activities HJ, FJ, and GJ are all complete. Completing the whole project will take at least 25 days. The critical path is ACGJ; if any of the critical activities, namely AC, CG, and GJ, is delayed in any way, the project as a whole will be delayed. On the other hand, if activity AD is delayed by up to 15 days, the project as a whole will not be delayed, because there is a slack of 15 days associated with activity AD.

Now suppose that activity AD is in fact delayed by 15 days. The situation at day 17 is shown in Figure 12.2. Actual durations of completed activities are italicized; activities that have been completed cannot be critical. There are now two critical paths, and activity DG has become critical. In other words, simply printing out a PERT chart showing the expected duration of each activity is in itself of little use. Data regarding actual durations must be continually input, and the PERT chart updated.

An important issue is: Who is responsible for the continual updating of the PERT data? After all, unless the information for the PERT chart is up to the minute, management will be unable to determine which activities are currently critical and to take appropriate action. What is needed is for all the software development tools to be integrated, and for information of all kinds, including source code, designs, documentation, contracts, and management information, to be stored in a product development database. Tools such as the PERT chart can then obtain their information from the database. In other words, what is needed is an integrated project support environment, or IPSE.

There is thus a natural progression within CASE. The simplest CASE device is a single *tool* such as an online interface checker or a build tool. Then

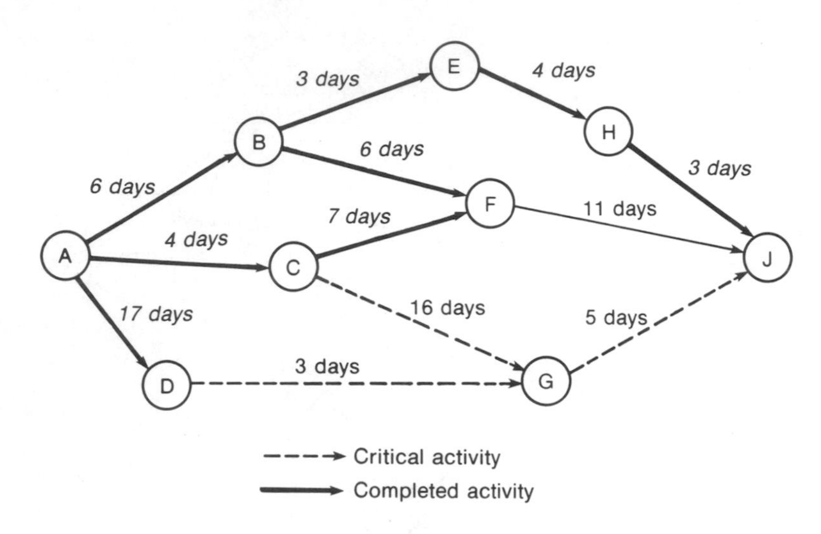

Figure 12.2 Updated PERT chart at day 17. Actual durations of completed activities are italicized.

tools can be combined, leading to an *environment* that can assist in the development of one or two phases within the software life cycle, such as specifications and design, or implementation. But such an environment will not necessarily provide management information even for the limited portion of the life cycle to which it is applicable, let alone for the project as a whole. Finally, an IPSE provides computer-aided support for every phase of the life cycle, as well as management information throughout.

Ideally, every software development organization should utilize an IPSE. But the cost of such an environment is large, not just the package itself but also the hardware on which to run it. For a smaller organization, a more restricted environment, or perhaps just a set of tools, will suffice. Thus, before examining a complete IPSE, other types of environments are described.

12.2 Software Development Environments

Dart et al. have introduced a taxonomy for classifying software development environments [Dart, Ellison, Feiler, and Habermann, 1987]. They identify four categories of environment, namely language-centered, structure-oriented, toolkit, and method-based environments.

12.2.1 Language-Centered Environments

Language-centered environments are built around a single language. Examples of such environments are Interlisp [Teitelman and Masinter, 1981] for Lisp and the Smalltalk environment [Goldberg, 1984] for Smalltalk. Language-centered environments are generally designed for rapid development, using tools such as incremental interpreters and compilers. When using an incremental compiler, if a small change is made to the source code then it is not necessary to reprocess the whole module; instead, the object code is modified to take the change into account.

Language-centered environments can be extended to programming-in-the-many (implementation by a team) by incorporating features needed to support multiple users at the same time. But such environments can have drawbacks. Because they are so closely bound up with a programming language, they generally support only the implementation phase and are difficult to extend to the other phases of the life cycle. Furthermore, current language-centered environments do not usually support project management information such as the status of each module or the time that has been devoted to each activity. These environments are good for prototyping, where their strength, namely rapid development of software, will outweigh their weakness, a general lack of support for programming-in-the-large.

A different sort of language-centered environment is the Ada programming support environment (APSE) [DoD, 1980]. The DoD requirements for the APSE stipulate that such environments must incorporate features for programming-in-the-large as well as programming-in-the-many. The problem of lack of support for the design phase is solved by using Ada Design Language (ADL) for the design [ANSI/IEEE 990, 1987]. ADL can be crudely described as "Ada + comments." Implementation then consists of translating the comments into valid Ada code. A commercially available example of this type of environment is the Rational Environment [Archer and Devlin, 1986].

12.2.2 Structure-Oriented Environments

A second type of environment is the structure-oriented environment. Here the environment is constructed around a structure editor (Section 10.10). Such an environment is therefore tied to the specific language whose structure is built into the editor. This class of environment is not at present much used in industry, which generally considers structure-oriented environments applicable only to the implementation phase and a tool for programming-in-the-small. One exception is the FLOW environment, which has been successfully used in industry for programming-in-the-many. It is built around a graphical design language and can therefore be used in the design phase and for implementation and maintenance. FLOW incorporates a limited management information database, as well as code generators for COBOL, FORTRAN, and Pascal [Dooley and Schach, 1985].

12.2.3 Toolkit Environments

The third type identified by Dart et al. [Dart, Ellison, Feiler, and Habermann, 1987] is the toolkit environment. This consists of a collection of tools put together to support mainly the implementation phase. The toolkit usually comprises an editor, compiler, linker, and debugger, together with version control tools. The difference between the tools of this category and the structure-oriented environment is that a toolkit is designed to handle any implementation language. Thus the editor cannot be structure-based. Probably the best-known example of a toolkit environment is the UNIX Programmer's Workbench [Dolotta, Haight, and Mashey, 1984]. Other commercial toolkits include the Information Engineering Facility, which includes tools for the whole life cycle, as does TekCASE. IDMS/Architect is a database design toolkit, while Life Cycle Management is a project management toolkit.

The strength of a toolkit is that any required tool can usually be added. For example, configuration control tools can be incorporated into a toolkit to help support programming-in-the-many. The weakness is that, while the tools may share a common user interface, the tools are usually not integrated. Thus it is difficult to ensure that the tools are used correctly or even used at all. For example, the automatic gathering of management information for later use by management tools such as a PERT package can be difficult to achieve.

12.2.4 Method-Based Environments

Finally, a method-based environment supports a specific method for developing software. Environments exist for a variety of the techniques discussed in this book, such as Gane and Sarson's structured systems analysis (Section 7.2), Jackson system development (Section 9.6), and Petri nets (Section 7.5). The majority of these environments provide graphical support for the specifications and design phases and incorporate a data dictionary. Some consistency checking is usually provided. Support for managing the development process is frequently incorporated into the environment. There are many commercially available environments of this type, including Excelerator, Analyst/Designer, The Design Machine, and HP Teamwork. Analyst/Designer is specific to Yourdon's method [Yourdon, 1989], while The Design Machine supports DSSD [Orr, 1981]. The HP Teamwork environment is useful for developing real-time software.

The emphasis in most method-based environments is on the support and formalization of the manual processes for software development laid down by the method. That is to say, these tools force users to utilize the method step by step in the way intended by its author, while assisting the user by providing graphical tools, a data dictionary, and consistency checking. This computerized framework is a strength of method-based environments in that users are forced to use a specific method and to use it correctly. But it can also be a weakness. The inherent emphasis on mechanization of programming-in-the-small frequently has the consequence that programming-in-the-large and programming-in-the-many are not adequately supported. For example, interface checking and version control are not often supported by this class of environment.

12.2.5 Comparison of Environment Types

No one environment is ideal for all products and all organizations, any more than one programming language can be considered to be "the best." Each of the four categories of environment described previously has its strengths and its weaknesses. Some are upperCASE environments, that is, they are used for the earlier stages of the software life cycle, namely for requirements, planning, specifications, and design. The lowerCASE environments are more appropriate for the later phases, that is, for implementation and for maintaining various versions of the product. Support for managing the software development process varies from environment to environment.

To date, no environment of any of the four classes described previously has been successfully extended to become an IPSE, an integrated project support environment supporting every phase of the software life cycle and providing the necessary support for programming-in-the-large and programming-in-the-many, including support for managing the entire process. Instead, the way to build an IPSE seems to be to start from scratch with the intention of producing a complete IPSE, rather than first constructing just an environment and adding on the additional features necessary for an IPSE. One reason why the incremental approach fails is that the basis of an IPSE is the software development database in which management information, all versions of all modules, contracts, plans, test cases, and the myriad other components of a software project are stored. Attempts to extend an environment into an IPSE will probably fail unless the database is designed from the very beginning to incorporate everything needed by the IPSE. This point is illustrated in the next section, where a specific IPSE is described.

12.3 ISTAR Integrated Project Support Environment

The ISTAR environment [Dowson, 1987a, 1987b, 1987c] was built by Imperial Software Technology (IST). From the start in 1983, the intention was to build an environment that would be language- and method-independent and would support every aspect and all phases of software production. Particular emphasis was paid to programming-in-the-large and programming-in-the-many in that project management is supported throughout, as is configuration control.

ISTAR does not fit into any of the four categories of environment described in the previous section. That it can handle any implementation language and any method precludes ISTAR from being language-centered, structure-oriented, or method-based. It also goes far beyond the toolkit type of environment. ISTAR is designed to permit compatible "foreign" tools, that is, tools constructed by other software organizations, to be integrated into the environment in such a way that they can make use of the ISTAR user interface and also access the project database. Thus, if a foreign tool is included in a particular configuration of ISTAR, users should be unaware that the tool is

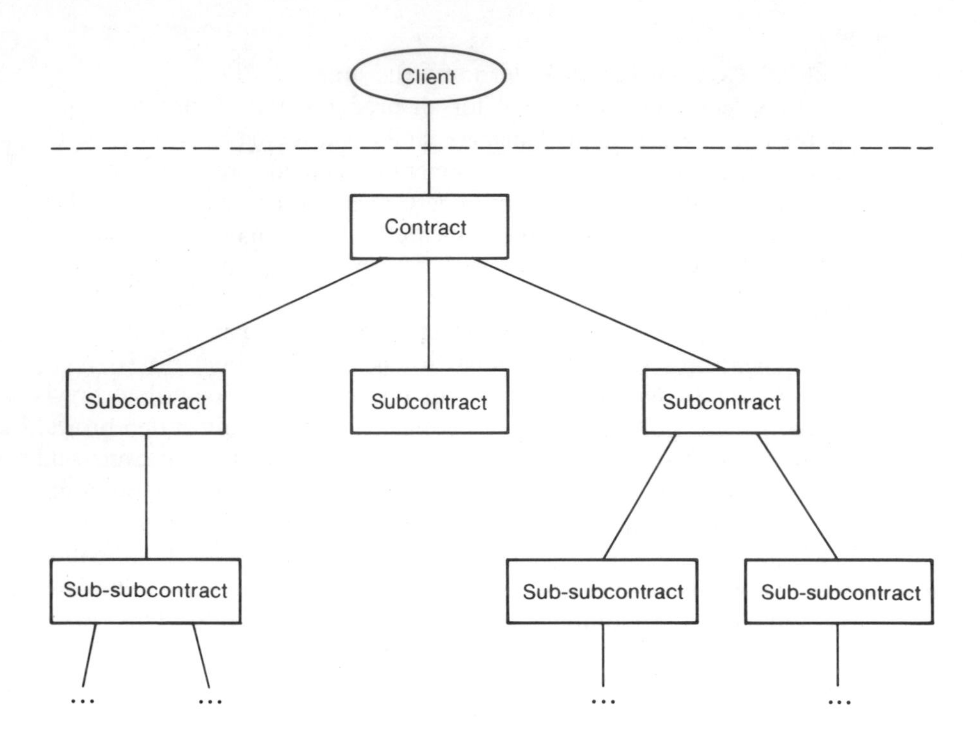

Figure 12.3 ISTAR contractual approach.

not intrinsic to ISTAR. Configurations of ISTAR have included C and Pascal programmer's workbenches as foreign tools. ISTAR also supports Ada; a validated Ada compiler has been incorporated.

The major concept on which ISTAR is based is the contractual approach. Every task in the development process is viewed as a contract. It is the contractor's responsibility to fulfill the contract as he or she thinks best, subject to any constraints in the contract such as deadlines, cost limits, timing, and storage constraints. Thus the project as a whole is a contract, but, unless explicitly forbidden to do so by the contract, the contractor may subcontract all or any part of the project. Each subcontractor in turn has the same right. A project may therefore be represented as shown in Figure 12.3. Each contract stipulates such items as acceptance criteria by which the successful completion of the contract may be judged, schedules for deliverables, reporting requirements, and standards to be used.

ISTAR does not force the developer to use any one specific type of project organization. For example, the project could be broken into subcontracts by phase. Thus one subcontract could be for drawing up the specifications, another for the design, a third for the implementation, and so on. This would be one way of proceeding if management decides to develop the product using the waterfall model. Conversely, management may decide to divide up a large product into separate pieces by functionality, assigning each to a different subcontractor, who would then be responsible for every aspect of developing that piece either directly or through the mechanism of further subcontracting.

All project information is stored in the contract database. ISTAR maintains a separate contract database for each subcontract, each database generally implemented on its own computer and incorporating its own workbench of tools. The various computers then communicate with one another, passing the necessary contract information. As milestones are reached, progress is recorded in the relevant database. When a subcontract is complete the resulting work is transferred either to the database of the contractor who awarded that contract or to the client, if the contract in question is the overall contract.

With regard to project management, the three main managerial activities supported are *planning*, *scheduling*, and *monitoring* of progress. Planning for the project as a whole requires planning on the part of each subcontractor. The necessary information is then passed from each subcontractor's database to that of the corresponding contractor, providing the contractor's planning tool with the information necessary to assist the contractor to perform planning at his or her level in the hierarchy. The scheduling tools make use of information from subcontractors to provide critical path information (Section 12.1.1), which can then be utilized by management to take corrective action if necessary. Information for both planning and scheduling tools is gathered by a monitoring tool.

ISTAR supports three views of an activity: the *task* view, reflecting the component tasks making up the activity; the *product* view, identifying the products resulting from the activity; and the *resource* view, indicating resource requirements for the activity. The various ISTAR management tools are integrated. For example, the scheduling tool communicates with the resource management tool to allocate resources to activities or, more accurately, to provide information to assist managers in allocating resources to activities—ISTAR provides tools, it does not automate the decision-making process. In this way, resources such as personnel, space, and computing resources can be appropriately allocated. Of course, resource conflicts may arise if managers try to schedule several activities simultaneously. ISTAR incorporates resource allocation tools that, in keeping with the ISTAR philosophy, detect such conflicts and inform management, but do not take any action themselves.

Even though ISTAR incorporates a wide range of management tools, it is not necessary to use most of them. For example, while the ISTAR estimating tool incorporates a full implementation of COCOMO (Section

4.5.3), it is unlikely to be used except perhaps at the highest levels of a contract hierarchy. Even reports from monitoring tools can be ignored; if management wishes to stipulate that all reports must be entered into the database in free-text format, it may do so. In addition, ISTAR incorporates office automation tools such as word processors for document preparation and production and electronic mail tools. These tools are naturally also accessed through the standard ISTAR user interface.

In late 1986, Motorola, Plessey, and British Telecom started medium-scale (5-person-year) commercial projects using ISTAR, with a view to evaluating it for large-scale projects. It is therefore too soon to have measurements of productivity and software quality improvements through use of ISTAR.

12.4 IPSE: A Universal Panacea?

On the one hand, the description of ISTAR sounds like a software developer's dream, providing support for every aspect of software development, with integrated tools, a uniform user interface, and management tools that cover the gamut of what is required for managing the development process. These, or similar features, are likely to be supported by future IPSEs, making software support at this level widely available to any organization that wants one. But there is a price to pay, namely the cost of the IPSE itself plus the hardware on which to run it. A version of an IPSE based on a personal computer seems very unlikely indeed. The database requirements alone are vast, including multiple versions of each source code module, the corresponding object code, test cases, reports of all kinds, and management information such as plans, schedules, timesheets, and cost accounting information. In short, for software organizations developing software on personal computers, purchasing an IPSE is like buying a solid gold howitzer to kill an ant.

On the other hand, an IPSE may well provide the necessary support for developers of large-scale software to produce high-quality software efficiently, within time and cost constraints. But it is too soon to tell whether IPSEs will prove to be cost-effective.

Chapter Review

CASE stands for computer-aided software engineering. CASE tools can assist with virtually every aspect of the software production life cycle (Section 12.1). There is a particular need for management information such as PERT/CPM charts, and computers can help in this respect as well (Section 12.1.1). Tools can be combined to form an environment that supports a complete phase such as design or implementation. In order to be able to select an appropriate environment for a specific purpose, it is useful to categorize environments as language-centered (Section 12.2.1), structure-oriented (Section 12.2.2), toolkit (Section 12.2.3), or method-based (Section 12.2.4). An integrated project support environment (IPSE) supports every phase of the life cycle; all project-related information is stored in its product development database. The ISTAR IPSE is described in some detail in Section 12.3. It is not clear yet whether use of IPSEs will prove to be cost-effective in the long run (Section 12.4).

For Further Reading

An introduction to PERT/CPM can be found in [Wiest and Levy, 1977] and [Moder, Phillips, and Davis, 1983].

Good starting points for information about CASE include [Hünke, 1980], [Howden, 1982], and also [Barstow, Shrobe, and Sandewall, 1984]. With regard to ensuring consistency between different views of the product, in advanced environments such as Pecan [Reiss, 1984], Magpie [Delisle, Menicosy, and Schwartz, 1984], and Cedar [Swinehart, Zellweger, and Hagmann, 1985], the problem is solved by having a single representation for the product within the computer and permitting multiple views of the same information.

Turning to specific environments, some language-centered environments are Interlisp [Teitelman and Masinter, 1981], Smalltalk [Goldberg, 1984], Cedar [Swinehart, Zellweger, and Hagmann, 1985], Eiffel [Meyer, 1988], and an implementation of an Ada APSE [Archer and Devlin, 1986]. Structure-oriented environments include FLOW [Dooley and Schach, 1985], Gandalf [Notkin, 1985; Habermann and Notkin, 1986], and Pecan [Reiss, 1984]. Toolkit environments include the UNIX Programmer's Workbench [Dolotta, Haight, and Mashey, 1984] and PCTE [Gallo, Minot, and Thomas, 1987]. There are many method-based environments, including PSL/PSA [Teichroew and Hershey, 1977] and Software Through Pictures [Wasserman and Pircher, 1987]. Information on ISTAR is to be found in [Dowson, 1987a, 1987b, 1987c].

In 1984, 1987, and 1989, ACM SIGSOFT and SIGPLAN jointly sponsored a Symposium on Practical Software Development Environments. The proceedings provide information on a broad spectrum of environments. Also useful are the proceedings of the annual International Workshops on Computer-Aided Software Engineering. Further papers on environments can be found in special issues of journals, including the November 1987 issue of *IEEE Computer*, the November 1987 issue of *IEEE Software*, and the June 1988 issue of *IEEE Transactions on Software Engineering*.

Problems

12.1 You are the owner and sole employee of One-Person Software Company. You decide that in order to be competitive you must buy CASE tools. You therefore apply for a bank loan for $12,000. Your bank manager asks you for a statement not more than one page in length (and preferably shorter) explaining in lay terms why you need CASE tools. How do you word the statement?

12.2 The newly appointed vice-president for software development of Old-Fashioned Software Corporation has hired you to help her change the way that the company develops software. There are about 220 employees, all writing COBOL code without the assistance of any CASE tools. Write a memo to the vice-president stating what sort of CASE equipment the company should purchase and carefully justifying your choices.

12.3 The vice-president responds to your memo of Problem 12.2 pointing out that there is a good possibility that the company will start developing software in languages other than COBOL in the near future. How do you respond? To what extent does this possibility invalidate your previous memo to her?

12.4 Three weeks later you receive another memo from the vice-president for software development stating that a decision has been taken to develop future products in C. What do you now advise with regard to CASE?

12.5 A day after you reply to the memo of Problem 12.4, a colleague mentions to you that the decision to select C for future products could be rescinded. There is a rumor in the office that Ada will be chosen instead of C, and another rumor that the COBOL diehards on the board of directors are fighting strongly for all software to be developed in COBOL, as in the past. Do you send another memo, and if so, what do you say? Or do you do nothing, and if so, why?

12.6 You now receive a memo from the vice-president for software maintenance pointing out that for the foreseeable future Old-Fashioned Software Corporation will have to maintain tens of millions of lines of COBOL code and asking you to advise with regard to CASE tools for supporting maintenance. What do you reply?

12.7 You and a friend decide to start Personal Computer Software Unlimited, developing software for personal computers on personal computers. Then an obscure cousin dies, leaving you $1 million on condition that you spend the money on an IPSE and the hardware needed to run it and that you keep the IPSE for at least 5 years. What do you do, and why?

12.8 You are a computer science professor at an excellent small liberal arts college. Programming assignments for computer science courses are done on a network of 35 personal computers. Your dean has asked you to advise the college with regard to purchasing CASE tools. The dean wishes to know whether CASE tools should be bought from the limited software budget, bearing in mind that, unless some sort of site license can be obtained, 35 copies of every CASE tool will have to be purchased. What do you advise?

12.9 (Readings in Software Engineering) Your instructor will distribute copies of [Dart, Ellison, Feiler, and Habermann, 1987]. Which category of environment do you consider to be the most useful CASE tool? Give reasons for your answer.

References

[ANSI/IEEE 990, 1987] "Recommended Practice for Ada as a Program Design Language," ANSI/IEEE 990-1987, American National Standards Institute, Inc., Institute of Electrical and Electronics Engineers, Inc., 1987.

[Archer and Devlin, 1986] J. E. ARCHER, JR., AND M. T. DEVLIN, "Rational's Experience Using Ada for Very Large Systems," *Proceedings of the First International Conference on Ada Programming Language Applications for the NASA Space Station*, NASA, 1986, pp. B2.5.1–B2.5.12.

[Barstow, Shrobe, and Sandewall, 1984] D. R. BARSTOW, H. E. SHROBE, AND E. SANDEWALL (Editors), *Interactive Programming Environments*, McGraw-Hill, New York, NY, 1984.

[Chen, 1983] P. P.-S. CHEN, "ER—A Historical Perspective and Future Directions," in: *Entity Relationship Approach to Software Engineering*, C. G. Davis (Editor), North-Holland, Amsterdam, The Netherlands, 1983.

[Dart, Ellison, Feiler, and Habermann, 1987] S. A. DART, R. J. ELLISON, P. H. FEILER, AND A. N. HABERMANN, "Software Development Environments," *IEEE Computer* **20** (November 1987), pp. 18–28.

[Delisle, Menicosy, and Schwartz, 1984] N. M. DELISLE, D. E. MENICOSY, AND M. D. SCHWARTZ, "Viewing a Programming Environment as a Single Tool," *Proceedings of the ACM SIGSOFT/SIGPLAN Software Engineering Symposium on Practical Software Development Environments, ACM SIGPLAN Notices* **19** (May 1984), pp. 49–56.

[DoD, 1980] "Requirements for the Programming Environment for the Common High-Order Language, STONEMAN," United States Department of Defense, February 1980.

[Dolotta, Haight, and Mashey, 1984] T. A. DOLOTTA, R. C. HAIGHT, AND J. R. MASHEY, "UNIX Time-Sharing System: The Programmer's Workbench," in: *Interactive Programming Environments*, D. R. Barstow, H. E. Shrobe, and E. Sandewall (Editors), McGraw-Hill, New York, NY, 1984, pp. 353–369.

[Dooley and Schach, 1985] J. W. M. DOOLEY AND S. R. SCHACH, "FLOW: A Software Development Environment Using Diagrams," *Journal of Systems and Software* **5** (August 1985), pp. 203–219.

[Dowson, 1987a] M. DOWSON, "ISTAR—An Integrated Project Support Environment," *Proceedings of the Second ACM SIGSOFT/SIGPLAN Software Engineering Symposium on Practical Development Support Environments*, *ACM SIGPLAN Notices* **22** (January 1987), pp. 27–33.

[Dowson, 1987b] M. DOWSON, "ISTAR and the Contractual Approach," *Proceedings of the Ninth International Conference on Software Engineering*, Monterey, CA, March 1987, pp. 287–288.

[Dowson, 1987c] M. DOWSON, "Integrated Project Support with ISTAR," *IEEE Software* **4** (November 1987), pp. 6–15.

[Gallo, Minot, and Thomas, 1987] F. GALLO, R. MINOT, AND I. THOMAS, "The Object Management System of PCTE as a Software Engineering Database Management System," *Proceedings of the Second ACM SIGSOFT/SIGPLAN Software Engineering Symposium on Practical Software Development Environments*, *ACM SIGPLAN Notices* **22** (January 1987), pp. 12–15.

[Goldberg, 1984] A. GOLDBERG, *Smalltalk-80: The Interactive Programming Environment*, Addison-Wesley, Reading, MA, 1984.

[Habermann and Notkin, 1986] A. N. HABERMANN AND D. NOTKIN, "Gandalf: Software Development Environments," *IEEE Transactions on Software Engineering* **SE-12** (December 1986), pp. 1117–1127.

[Howden, 1982] W. E. HOWDEN, "Contemporary Software Development Environments," *Communications of the ACM* **25** (May 1982), pp. 318–329.

[Hünke, 1980] H. HÜNKE (Editor), *Proceedings of the Symposium on Software Engineering Environments*, Lahnstein, West Germany, 1980.

[Meyer, 1988] B. MEYER, "Eiffel: A Language and Environment for Software Engineering," *Journal of Systems and Software* **8** (June 1988), pp. 199–246.

[Moder, Phillips, and Davis, 1983] J. J. MODER, C. R. PHILLIPS, AND E. W. DAVIS, *Project Management with CPM, PERT, and Precedence Diagramming*, Third Edition, Van Nostrand Reinhold, New York, NY, 1983.

[Nassi and Shneiderman, 1973] I. NASSI AND B. SHNEIDERMAN, "Flowchart Techniques for Structured Programs," *ACM SIGPLAN Notices* **8** (August 1973), pp. 12–26.

[Notkin, 1985] D. NOTKIN, "The GANDALF Project," *Journal of Systems and Software* **5** (May 1985), pp. 91–105.

[Orr, 1981] K. ORR, *Structured Requirements Definition*, Ken Orr and Associates, Topeka, KS, 1981.

[Reiss, 1984] S. P. REISS, "Graphical Program Development with PECAN Program Development Systems," *Proceedings of the ACM SIGSOFT/SIGPLAN Software Engineering Symposium on Practical

Software Development Environments, *ACM SIGPLAN Notices* **19** (May 1984), pp. 30–41.

[Swinehart, Zellweger, and Hagmann, 1985] D. C. SWINEHART, P. T. ZELLWEGER, AND R. B. HAGMANN, "The Structure of Cedar," *Proceedings of the ACM SIGPLAN 85 Symposium on Language Issues in Programming Environments*, *ACM SIGPLAN Notices* **20** (July 1985), pp. 230–244.

[Teichroew and Hershey, 1977] D. TEICHROEW AND E. A. HERSHEY, III, "PSL/PSA: A Computer-Aided Technique for Structured Documentation and Analysis of Information Processing Systems," *IEEE Transactions on Software Engineering* **SE-3** (January 1977), pp. 41–48.

[Teitelman and Masinter, 1981] W. TEITELMAN AND L. MASINTER, "The Interlisp Programming Environment," in: *Tutorial: Software Development Environments*, A. I. Wasserman (Editor), IEEE Computer Society Press, Washington, DC, 1981, pp. 73–81.

[Wasserman and Pircher, 1987] A. I. WASSERMAN AND P. A. PIRCHER, "A Graphical, Extensible Integrated Environment for Software Development," *Proceedings of the Second ACM SIGSOFT/SIGPLAN Software Engineering Symposium on Practical Software Development Environments*, *ACM SIGPLAN Notices* **22** (January 1987), pp. 131–142.

[Wiest and Levy, 1977] J. D. WIEST AND F. K. LEVY, *A Management Guide to PERT/CPM: With GERT/PDM/DCPM and other Networks*, Second Edition, Prentice-Hall, Englewood Cliffs, NJ, 1977.

[Yourdon, 1989] E. YOURDON, *Modern Structured Analysis*, Yourdon Press, Englewood Cliffs, NJ, 1989.

Chapter 13 Portability and Reusability

The ever-rising cost of software makes it imperative that some means be found for containing costs. One way is to ensure that the product as a whole can easily be adapted to run on a variety of different hardware/operating system combinations. Some of the costs of writing the product can then hopefully be recouped by selling versions that will run on other computers. Another reason for writing software that can easily be implemented on another computer is that the client organization may purchase new hardware, and all its software will then have to be converted to run on the new hardware. A product is considered to be *portable* if it is easier to adapt the product to run on the new computer than to write a new product from scratch.

Portability refers to the adaptation of a product to different hardware and operating systems without changing its functionality. A second way of cutting costs is to take pieces of one product and to use them, unchanged, in a different product; this is termed *reusability*. In this chapter first portability and then reusability are examined.

13.1 Portability

Portability may be defined as follows: Suppose a product P is compiled by compiler C and then runs on the *source* computer, namely hardware configuration H under operating system O. A product P′ is needed that functionally is equivalent to P, but must be compiled by compiler C′ and run on the *target* computer, namely hardware configuration H′ under operating system O′. If the cost of converting P into P′ is less than the cost of coding P′ from scratch, then P is defined to be *portable*.

For example, a numerical integration package has been implemented on a VAX 8800 running under the VAX/VMS operating system. It is written in FORTRAN and compiled by the VAX FORTRAN compiler. The package is now to be ported to a SUN 3/50 workstation running under the UNIX operating system. The compiler to be used will be the SUN FORTRAN compiler.

Offhand, this would seem to be a trivial task. After all, the package is written in a high-level language. But if the package contains any operating system calls such as to determine the CPU time used so far, then these require conversion. In addition, VAX FORTRAN and SUN FORTRAN are not identical languages; there are some features that are supported by one and not the other. Overall, the problem of porting software is a nontrivial one because of incompatibilities between different hardware configurations, operating systems, and compilers. Each of these aspects is now examined in turn.

13.1.1 Hardware Incompatibilities

Suppose that product P currently running on hardware configuration H is to be installed on hardware configuration H′. If H and H′ are both personal computers the obvious thing to do is to use H to copy P onto a floppy disk, and then to insert the disk in the disk drive of H′ and start computing. This may simply not be physically possible, because floppy disks come in different diameters, namely 8 inches and $5\frac{1}{4}$ inches, and there are also $3\frac{1}{2}$ inch diskettes. In addition, a double-sided disk cannot be read by a disk drive equipped to read only single-sided disks. The format of the disk is also critical; issues such as the density (number of bits per inch) affect readability.

For larger computers, magnetic tape would seem to be the way to port products or data from one computer to another. But seven-track tape and nine-track tape are incompatible. Information, that is, products and data, can be written at a number of different tape densities including 800 bpi (bytes per inch), 1600 bpi, and 6250 bpi. These densities are incompatible. The issue of odd or even parity can be a further complication.

Suppose now that the problem of physically copying product P to computer H′ has been solved. There is no guarantee that H′ can interpret the bit patterns created by H. A number of different character codes exist, the most popular of which are Extended Binary Coded Decimal Interchange Code (EBCDIC) and American Standard Code for Information Interchange (ASCII), the American version of the 7-bit ISO code [Mackenzie, 1980]. If H uses EBCDIC but H′ uses ASCII, then H′ will treat P as so much garbage.

While the original reason for these differences is historical, namely that researchers working independently for different manufacturers came up with different ways of doing the same thing, there are definite economic reasons for perpetuating them. To see this, consider the following imaginary situation. MCM Computer Manufacturers has sold hundreds of their MCM-1 computer. MCM now wishes to design, manufacture, and market a new computer, the MCM-2, that will be more powerful in every way than the

MCM-1, but will cost considerably less. Suppose further that the MCM-1 uses ASCII code, has 36-bit words consisting of four 9-bit bytes, uses seven-track tape at a density of 1600 bpi, and even parity. Now the chief computer architect of MCM decides that the MCM-2 should employ EBCDIC, have 16-bit words consisting of two 8-bit bytes, use nine-track tape at a density of 6250 bpi, and odd parity. The sales force will then have to tell current MCM-1 owners that the MCM-2 is going to cost them $35,000 less than any competitor's equivalent machine, but that it will cost them up to $200,000 to convert existing software and data from MCM-1 format to MCM-2 format. No matter how good the scientific reasons for designing the MCM-2 the new way may be, marketing considerations will ensure that the new computer will be compatible with the old one. A salesperson can then point out to an existing MCM-1 owner that not only is the MCM-2 computer $35,000 cheaper than any competitor's machine, but that if the customer is ill-advised enough to buy from a different manufacturer, then he or she will be spending $35,000 too much and will also have to pay some $200,000 to convert existing software and data to the format of the non-MCM machine.

Moving from the preceding imaginary situation to the real world, the most successful line of computers to date has been the IBM System/360–370 series [Gifford and Spector, 1987]. The success of this line of computers is due largely to full compatibility between machines; a product that runs on an IBM System/360 Model 30 built in 1964 will run unchanged on a IBM 3090 Model G built in 1989. But that same product that runs on the IBM System/360 Model 30 under OS/360 may require considerable modification before it can run on, say, a VAX 8800 under VMS. Part of the difficulty may be due to hardware incompatibilities. But part may be caused by operating system incompatibilities.

13.1.2 Operating System Incompatibilities

The job control languages (JCL) of any two computers are usually vastly different. Some of the difference is syntactic—the command for executing an executable load image might be @XEQ on one computer, //XQT on another, and .EXC on a third. When porting a product to a different operating system, syntactic differences are relatively straightforward to handle by simply translating commands from the one JCL into the other. But other differences can be more serious. For example, some operating systems support virtual memory. While the operating system may allow products to be of size, say, 8 megabytes, the actual area of main memory allocated to a particular product may be only, say, 64 kilobytes. What happens is that the user's product is partitioned into pages 4 kilobytes in size, and 16 of these pages can be in main memory at any one time. The rest of the pages are stored on disk and swapped in and out as needed by the virtual memory operating system. As a result, products can be written without any effective constraints as to size. But if a product that has been successfully implemented under a virtual memory operating system is to be ported to a operating system with physical constraints on product size, the entire product may have to be rewritten and then linked using overlay techniques to ensure that the size limit is not exceeded.

13.1.3 Numerical Software Incompatibilities

When a product is ported from one machine to another, or even compiled using a different compiler, the results of performing arithmetic may differ. On a 16-bit machine, that is, a computer with a word size of 16 bits, an integer will ordinarily be represented by one word (16 bits), and a double-precision integer by two adjacent words (32 bits). Unfortunately, some language implementations do not include double-precision integers. For example, standard Pascal does not include double-precision integers, and as a result, few Pascal implementations support that data type. Thus a product that functions perfectly on a compiler/hardware/operating system configuration in which Pascal integers are represented using 32 bits may fail to work correctly when ported to a computer in which integers are represented by only 16 bits. The obvious solution, namely representing integers larger than 2^{16} by floating-point numbers (**type** REAL), does not work, because integers are represented exactly, but floating-point numbers are in general only approximated using a mantissa (fraction) and exponent.

This problem can be solved in Ada, because in Ada it is possible to specify the range of an integer type and the precision (number of significant digits) of a floating-point type. The Ada–Europe Portability Working Group has produced a list of further recommendations for ensuring portability. These recommendations are specific to Ada and generally require a detailed understanding of the Ada Reference Manual [ANSI/MIL-STD-1815A, 1983]; the interested reader is referred to [Nissen and Wallis, 1984]. But where a numerical computation is performed in other languages, it is important, but often difficult, to ensure that numerical computations will be correctly performed on the target hardware/operating system.

13.1.4 Compiler Incompatibilities

Portability is difficult to achieve if a product is implemented in a language for which few compilers exist. If the product has been implemented in a specialized language such as CLU [Liskov, Snyder, Atkinson, and Schaffert, 1977], it may be necessary to rewrite it in a different language if the target computer does not have a compiler for that language. On the other hand, if a product is implemented in a popular language such as COBOL, FORTRAN, Lisp, Pascal, or C, the chances are good that a compiler or interpreter for that language can be found for a target computer.

Suppose that a product is written in a popular high-level language, standard FORTRAN, say. In theory, there should be no problem in porting the product from one machine to another—after all, standard FORTRAN is standard FORTRAN. Regrettably, that is not the case; in practice, there is no such thing as standard FORTRAN. Even though there is an ANSI FORTRAN standard, colloquially known as FORTRAN 77 [ANSI X3.9, 1977], there is no reason for a compiler writer to adhere to it. For example, a decision may be taken to support additional features not usually found in FORTRAN so that the marketing division can then tout a "new, extended FORTRAN compiler." Conversely, a compiler for a microcomputer may not be a full FORTRAN implementation. Also, if there is a deadline to produce a compiler, management may decide to bring out a less-than-complete implementation, intending to support the full standard in a later revision. Suppose that the compiler on the source computer supports a superset of ANSI FORTRAN. Suppose further that the target compiler is an implementation of standard ANSI FORTRAN. When a product implemented on that source computer is ported to the target, any portions of the product that make use of nonstandard ANSI FORTRAN constructs from the superset have to be recoded. Conversely, if the source compiler adheres to the standard, but the target compiler supports only a subset of the standard, then portability is again compromised.

There are several different Pascal standards. First came Jensen and Wirth's definition of the language [Jensen and Wirth, 1975]. There is the ANSI standard [ANSI/IEEE 770X3.97, 1983] and the ISO standard [ISO-7185, 1980]; an extended standard is under development [Joint X3J9/IEEE P770, 1986]. Notwithstanding this plethora of standards, subsets and supersets of Pascal abound. For instance, all Pascal standards specify that procedure and function names may be passed as parameters. But that feature is by no means universally supported by Pascal compilers. Conversely, many supersets of Pascal have been implemented. For example, some implementations of Pascal incorporate nonstandard bit manipulation operations such as bitwise **and** and **or**, perhaps in order to compete with C.

COBOL standards were developed by CODASYL (Committee on Data Systems Languages), a committee of American computer manufacturers and government and private users. Unfortunately, COBOL standards do not promote portability. A COBOL standard has an official life of 5 years, but each successive standard is not necessarily a superset of its predecessor. It is equally worrying that many features are left to the individual implementor, subsets may be termed "standard COBOL," and there is no restriction on extending the language to form a superset [Wallis, 1982].

At the time of writing both ANSI and ISO are nearing the end of their respective lengthy review processes for a joint standard for the programming language C. Most C compilers adhere quite closely to the original language specification [Kernighan and Ritchie, 1978]. The reason for this is that almost all C compiler writers use the standard front end of the portable C compiler, *pcc* [Johnson, 1979], and as a result the language accepted by the vast majority of compilers is identical. C products are, in general, easily ported from one implementation to another. An aid to C portability is the UNIX *lint* processor that can be used to determine implementation-dependent features, as well as constructs that may lead to difficulties when the product is ported to a target computer. Unfortunately, *lint* checks only the syntax and the static semantics, and is therefore not foolproof. But it can be of considerable help in reducing future problems. For example, in C it is legal to assign an integer value to a pointer and vice versa, but this is forbidden by *lint*. In some implementations, the size (number of bits) of an integer and a pointer will be the same, but the sizes may be different on other implementations; this sort of potential future portability problem can be flagged by *lint* and obviated by recoding the offending portions.

The only truly successful language standard is the Ada standard, embodied in the Ada Reference Manual [ANSI/MIL-STD-1815A, 1983]. Until the end of 1987, the name "Ada" was a registered trademark of the U.S. Government, Ada Joint Program Office (AJPO). As owner of the trademark, AJPO stipulated that the name Ada could legally be used only for language implementations that complied exactly with the standard; subsets and supersets were expressly forbidden. A mechanism was set up for validating Ada compilers, and only a compiler that successfully passed the validation process could be called an Ada compiler. Thus the trademark was used as a means of enforcing standards, and hence portability.

Now that the name Ada is no longer a trademark, enforcement of the standard is being achieved via a different mechanism. There is little or no market for an Ada compiler that has not been validated. Thus there are strong economic forces encouraging Ada compiler developers to have their compilers validated, and hence certified as conforming to the Ada standard.

A validation certificate is for a specific compiler running on specific hardware under a specific operating system; if an organization that has developed an Ada compiler wishes to port the compiler to another hardware and/or operating system configuration, revalidation is necessary. The reference manual for every Ada compiler must incorporate an appendix, namely Appendix F, in which the implementation-dependent characteristics of that Ada implementation are described. For example, it is possible to suppress type conversion and to treat a parameter simply as a bit pattern; an implementation may impose restrictions on the size of such an object. Technically, then, Ada products can be made fully portable, except with regard to features mentioned in Appendix F.

13.2 Why Portability?

In the light of the many barriers to porting software, the reader might well wonder if it is worthwhile to port software at all. One argument in favor of portability stated at the beginning of this chapter is that the cost of software can perhaps be partially recouped by porting the product to a different hardware/operating system configuration. But selling multiple variants of the software may not be possible. The application may be highly specialized, and no other client may need the software. For instance, the management information system software written for one major car rental corporation may simply be inapplicable to the way other car rental corporations are run. Alternatively, the software itself may give the client a competitive advantage, and selling copies of the product would then be tantamount to economic suicide. In the light of all this, is it not a waste of time and money to engineer portability into the product when it is designed?

The answer to this question is an emphatic NO. One major reason why portability is essential is that the life of a software product is generally longer than the life of the hardware for which it was first written. Good software products can have a life of 15 years, while hardware is frequently changed every 5 years or fewer. Thus good software can be implemented, over its lifetime, on three or more different hardware configurations.

One way to solve this problem is to buy upwardly compatible hardware. The only expense is the cost of the hardware; the software will not need to be changed. Nevertheless, in some cases it may be economically more sound to port the product to different hardware entirely. For example, the first version of a product may have been implemented 7 years ago on a mainframe. While it may be possible to buy a new mainframe on which the product will run without any changes being needed, it may be cheaper to implement multiple copies of the product on a network of personal computers, one on the desk of each employee who needs to use the product. In this instance, if the software has been written in such a manner as to promote portability, then porting the product to the personal computer network is financially a good solution.

But there are other kinds of software. Not all software is one of a kind. For example, many software organizations that write software for personal computers make their money by selling multiple copies. The profit on, say, a spreadsheet package is small and can in no way cover the expense of writing that package. In order to break even, 1000 (or perhaps 10,000) copies have to be sold. After this point, additional sales are pure profit. So if the product can be ported to additional types of hardware with ease, even more money can be made. Of course, as with all software, the product is not just the code; there is also documentation, including the manuals. Porting the spreadsheet package to other hardware means changing the documentation as well. Thus portability also means being able to change the documentation easily to reflect the target configuration, instead of having to write new documentation from scratch.

If the user organization has been using a particular product for a while, then the relevant personnel are familiar with it. Considerably less training is needed if the existing product is ported to a new computer, than if a completely new product were to be written. For this reason, too, portability is to be encouraged.

Ways of facilitating portability are now described.

13.3 Techniques for Achieving Portability

One way to try to achieve portability is not to allow programmers to use constructs that might cause problems when ported to another computer. For example, an obvious principle would seem to be: Write all software in a standard version of a high-level programming language. But how then is a portable operating system to be written? After all, it is inconceivable that an operating system could be written without at least some assembler code. Similarly, a compiler has to generate object code for a specific computer. Here, too, it is impossible not to have some implementation-dependent components.

13.3.1 Portable System Software

Instead of forbidding all implementation-dependent aspects, which would have the consequence of preventing most system software from being written, a better technique is to isolate any necessary implementation dependent pieces. An example of this technique is the way the UNIX operating system is constructed [Johnson and Ritchie, 1978]. About 9000 lines of the operating system are written in C. The remaining 1000 lines constitute the kernel. The kernel is written in assembler and must be rewritten for each implementation. About 1000 lines of the C code consist of device drivers; this code, too, must be rewritten each time. However, the remaining 8000 lines of C code remain largely unchanged from implementation to implementation.

The technique of isolating implementation-dependent pieces is also used in the portable P-compiler for Pascal. The compiler itself is written in Pascal. The P-compiler then translates the user's Pascal product into an intermediate language called P-code. In order to implement the P-compiler on a new machine, all that is needed is to write a P-code translator. To be more precise, a P-code version of the P-compiler is also needed in order to bootstrap the compiler on the target machine; for details, see [Nori, Ammann, Jensen, and Nägeli, 1981].

Another useful technique for increasing the portability of system software is to use levels of abstraction (Chapter 8). Consider, for example, graphical display routines for a workstation. A user inserts a command such as DRAW_LINE into his or her source code. The source code is compiled and then linked together with graphical routines. At run time, DRAW_LINE causes the workstation to draw a line on the screen as specified by the user. This can be implemented using two levels of abstraction. The upper level, written in a high-level language, interprets the user's command and calls the appropriate lower-level module to execute that command. If the graphical display routines are ported to a new type of workstation, then no changes need be made to the user's code or to the upper level of the graphical routines. However, the lower-level modules of the routines will have to be rewritten because they interface with the actual hardware, and the hardware of the new workstation is different from that of the workstation on which the package was previously implemented. This technique has also been successfully used for porting communications software that conforms to the seven levels of abstraction of the ISO–OSI model [Tanenbaum, 1981].

13.3.2 Portable Application Software

With regard to application software, rather than system software like operating systems and compilers, it is generally possible to write the product in a high-level language. In Chapter 10 it was pointed out that there is frequently no choice with regard to implementation language, but that where it is possible to select a language the choice should be made on the basis of cost–benefit analysis (Section 4.2.3). One of the factors that must enter into the cost–benefit analysis is the impact on portability.

At every stage in the development of a product, decisions can be taken that will result in a more portable resulting product. One potential problem can arise as a consequence of the fact that not all characters are supported by every computer. For example, comments in Pascal can be delimited by { ... } pairs or by (* ... *) pairs. Since some computers do not support brace brackets, it is preferable to use (* ... *) pairs. Also, some compilers distinguish between uppercase and lowercase letters. For such a compiler, variables This_Is_A_Name and this_is_a_name are two different variables. But other compilers treat the two names the same. A product that relies on differences between uppercase and lowercase letters can lead to hard-to-discover faults when the product is ported.

Just as there frequently is no choice of programming language, there may be no choice of operating system. However, if at all possible the operating system under which the product runs should be a popular one. This is an argument in favor of the UNIX operating system. UNIX has been implemented on a wide range of hardware. In addition, UNIX, or more precisely, a UNIX-like operating system, has been implemented on top of IBM VM/370 and VAX/VMS. An even more widely used operating system is MS-DOS. Just as use of a widely implemented programming language will promote portability, so too will use of a widely implemented operating system.

Language standards can play their part in achieving portability. If the coding standards of a development organization stipulate that only standard constructs may be used, then the resulting product is more likely to be portable. To this end, programmers must be provided with a list of nonstandard features supported by the compiler, but whose use is forbidden without prior managerial approval. Like other sensible programming standards, this one can be checked by machine.

Planning should also be done for potential future numerical incompatibilities. For example, if a product is being developed on 32-bit hardware but there is a possibility that in the future it may have to be ported to a 16-bit machine, then integers should be kept within the range $\pm 32{,}767$, and the modulus of real numbers should be within the range $\pm 10^{68}$. Also, no more than six decimal digits of precision should be assumed [Wallis, 1982].

It is necessary to plan for potential lack of compatibility between the operating system under which the product is being constructed and any future operating systems to which the product may be ported. If at all possible, operating system calls should be localized to one to two modules. In any event, every operating system call must be carefully documented. The documentation standard for operating system calls should assume that the next programmer to read the code will have no familiarity whatsoever with the current operating system, often a reasonable assumption.

Documentation should be provided to assist with future porting. An installation manual should be produced that will point out what parts of the product will have to be changed when porting the product, and what parts may have to be changed. In both instances, a careful explanation of what has to be done, and how to do it, must be provided. Also, lists of changes that will have be made in other manuals such as the user manual or the operator manual must also appear in the installation manual.

13.3.3 Portable Data

The problem of portability of data can be a vexing one. Problems of hardware incompatibilities were pointed out in Section 13.1.1. But even after such problems have been solved, software incompatibilities remain. Suppose that a data file is used by, say, a Pascal or COBOL product on the source computer. If that same product is now ported to another compiler/hardware/operating system configuration, it is unlikely that the data file can be read by the ported product; the format required for Pascal or COBOL data files in the new system will probably be totally different from that in the old system. For instance, the format of an indexed-sequential file is determined by the operating system; a different operating system generally implies a different format. Many files require headers containing information such as the format of the data in that file. The format of a header is almost always unique to the specific compiler and operating system under which that file was created. But bad as the situation is with regard to porting data files used by popular programming languages, the situation is far worse when database management systems are used.

The safest way of porting data is to construct an unstructured (sequential) file. This can then be ported with minimal difficulty to the target machine. From this unstructured file, the desired structured file, be it a Pascal file, a COBOL file, or a database, can be reconstructed. Two special conversion routines have to be written, one running on the source machine to convert the original structured file into sequential form, and one on the target machine to reconstruct the structured file from the ported sequential file. While this solution seems simple enough, the two routines will be nontrivial when conversions between complex database models have to be performed.

13.4 Reusability

A product is portable if it is easier to modify the product as a whole to run on another compiler/hardware/operating system configuration than to recode it from scratch. In contrast, reusability refers to reusing components of one product to facilitate the development of a different product with different functionality. A reusable component need not necessarily be a module or code fragment—it could be a design, a manual, or a set of test data.

When computers were first constructed, nothing was reused. Every time a product was developed, such items as multiplication routines, input/output routines, or routines for computing sines and cosines were constructed from scratch. Quite soon, however, it was realized that this was a considerable waste of effort, and subroutine libraries were constructed. Programmers could then simply invoke square root or sine functions whenever they wished. These subroutine libraries have become more and more sophisticated and have developed into run-time support routines. Thus, when a programmer calls a Pascal procedure, there is no need to write code to manage the stack or to pass the parameters; it is automatically handled by calling the appropriate run-time support routines. The concept of subroutine libraries has been also extended to large-scale statistical libraries such as SPSS [Norusis, 1982] and to numerical analysis libraries like NAG [Phillips, 1986].

The simplest form of reusability is to purchase a software package with the necessary functionality. From an economic viewpoint, this is an excellent scheme; the cost of buying a package is far less than having the same piece of software custom built. In practice, off-the-shelf packages may require customization to cater to a user's specific needs. Frequently, however, the source code is not available to the purchaser, and the package is then all but impossible to change in any way. Even if the source code is released, the problems of maintenance can make the cost of customization comparable to that of rewriting the software. One solution to this problem is an application generator This is a package that, when supplied with a list of parameters, constructs a customized software package within a specific limited domain such as accounts payable or payroll. A screen generator is another example of an application generator.

A different approach to reusability is to make sure that software components can easily be reused. Modularity in all its forms (Chapter 8) is the theoretical basis behind this technique. Modules of high cohesion and low coupling were employed in the 1970s. Information hiding of all types is now being tried, and objects in particular are receiving considerable interest in this regard. Related to this approach is the issue of module classification schemes and ways of describing the functionality of the module.

Suppose that a software organization has a large number of modules available for reuse. Some mechanism is needed to enable the software developer to know exactly what each module does. For example, suppose that a module is needed that will, say, traverse a tree in postorder, and that a module is available that does this. But before using the module, it is necessary to know exactly how the tree must be represented, and what happens if an attempt is made to traverse an empty tree. The knowledge needed before using a more complex module, such as one that performs file transfers between two computers, is correspondingly larger. A module interconnection language can be used to specify the interface of a module [Prieto-Díaz and Neighbors, 1986], but cannot specify the functionality of the module, that is, what the module does.

Although it will probably be some years before reusable software products become the norm in software development, much has already been achieved in practice. In the next section, three examples of reusability within the software industry are described.

13.5 Reusability Case Studies

13.5.1 Raytheon Missile Systems Division

In 1976, a study was undertaken at Raytheon's Missile Systems Division to determine whether reusability of designs and code was feasible [Lanergan and Grasso, 1984]. Over 5000 COBOL products in use were analyzed and classified. The researchers decided that there are only six basic functions performed in a business application product, and as a result between 40% and 60% of business application designs and modules could be standardized and reused. The basic functions were found to be: sort data; edit or manipulate data; combine data; explode data; update data; and report on data. For the next 6 years, a concerted attempt was made to reuse both design and code wherever possible.

The Raytheon approach uses reusability in two ways, namely what they termed functional modules and COBOL program logic structures. In Raytheon's terminology a *functional module* is a COBOL code fragment designed and coded for a specific purpose. A functional module can be purely data-related, such as a file description (**fd**) or a record description (level **01** in a **fd** or in **working-storage**). But the majority of functional modules include both data areas and procedure code, such as edit routines, database procedure division calls, tax computation routines, or date aging routines for accounts receivable. Use of the 3200 reusable modules has resulted in applications that average 60% reusable code. Functional modules were carefully designed, tested, and documented. Products that made use of these functional modules were found to be more reliable, and less testing of the product as a whole was needed.

The modules are stored in a standard copy library, and when needed, the **copy** verb is used. That is to say, the code is not physically present within the application product, but is included by the COBOL compiler at compilation time; the mechanism is similar to **#include** in C. The resulting source code is therefore shorter than if the copied code were physically present. As a consequence, maintenance is easier.

The Raytheon researchers also used what they termed a *COBOL program logic structure*. This is essentially a template, a framework that has to be fleshed out into a complete product. All four divisions of a COBOL product are present. But some data descriptions are incomplete, and some paragraphs consist only of the paragraph header. One example of a logic structure is the update logic structure. This is used for performing a sequential update, such as the case study in Chapter 5. Error handling is built in, as is sequence checking. The logic structure is 22 paragraphs in length. Many of the paragraphs can be filled in using by functional modules such as GET-TRANSACTION, SEQUENCE-CHECK-TRANSACTION, GET-MASTER, DELETE-A-RECORD, PRINT-PAGE-HEADING, and PRINT-CONTROL-TOTALS.

There are many advantages to the use of such templates. It makes the design and coding of a product quicker and easier because the framework of the product is already present; all that is needed is to fill in the details. Fault-prone areas such as end of file conditions are already built in and tested. In fact, testing as a whole is easier. But Raytheon believes that the major advantage that comes from the method occurs when the user requests modifications or enhancements. Once a maintenance programmer is familiar with the relevant logic structure, it is almost as if that maintenance programmer had been a member of the original development team.

By 1983, logic structures had been used over 5500 times in developing new products. About 60% of the code consisted of functional modules, that is, reusable code; this meant that design, coding, module testing, and documentation time could also be reduced by 60%, leading to an estimated 50% increase in productivity in software product development. But for Raytheon, the real benefit of the technique lies in the hope that the readability and understandability resulting from the consistent style will reduce the cost of maintenance by between 60% and 80%. While the arguments in Lanergan and Grasso's paper are strong [Lanergan and Grasso, 1984], no evidence of this hoped-for order-of-magnitude maintenance improvement has yet been published—perhaps it is still too soon.

It might seem that reusability is applicable only to business data-processing applications. But that is not so, as is demonstrated by the second case study, the Toshiba Software Factory.

13.5.2 Toshiba Software Factory

In 1977, the Toshiba Corporation started the Fuchu Software Factory at the Toshiba Fuchu Works in Tokyo, Japan. At the Fuchu Works, industrial process control systems are manufactured for, among other areas, electric power networks, nuclear power generators, factory automation, and traffic control; at the Software Factory, application software for the process control computers for those applications is manufactured [Matsumoto, 1984, 1987].

By 1985, a total of 2300 technical and managerial personnel were employed by the Software Factory. About 60% of the code is in FORTRAN augmented by real-time functions, 20% in an assembler-like language, and the rest in user-specified problem-oriented languages. The Software Factory measures productivity in lines of code. Because the effort to produce 1000 lines of FORTRAN is different from that needed to produce 1000 lines of assembler, the unit of productivity used is *equivalent assembler source lines*, or EASL; the usual conversion factor is that one line of high-level language is equivalent to four lines of assembler [Jones, 1978]. Using this measure, output from the Software Factory in 1985 was 7.2 million EASL. Products ranged in size from 1 to 21 million EASL, with an average size of 4 million EASL.

Software is developed using the waterfall model, with detailed reviews and inspections when each phase has been completed. Productivity, measured in EASL, is the driving force behind the software factory. It is monitored on both a projectwide basis, as well as on an individual basis. Annual productivity increases for the factory as a whole have been of the order of 8% to 9%. One of the items measured when appraising the performance of individuals is their fault rate. In the case of a programmer, for example, the number of faults per 1000 EASL is expected to decrease over time as a consequence of training and experience. Quality is an important aspect of the factory and is achieved through a number of different mechanisms, including use of reviews, inspections, and quality circles (groups of workers who meet on a regular basis to find ways of improving quality).

Software development is supported by a toolkit software development environment (Section 12.2.3), the software workbench (SWB). The environment consists of six subenvironments that interface with the software engineering database. Between them, four of the subenvironments support the various phases of the waterfall life cycle, and there are also subenvironments for project management and for quality assurance. About 750 workbenches (terminals) are linked into clusters, and thence to the central computer on which all SWB software is processed, although this is being replaced by a decentralized SWB.

Matsumoto attributes improvements in both productivity and quality to reuse of existing software [Matsumoto, 1987]. This reusable software includes not only modules, but also documentation of all kinds, such as designs, specifications, contracts, and manuals. A committee is responsible for deciding what parts should be placed in the *reusable software parts database*. After having been placed in the database, parts are indexed by keyword for later retrieval. Careful statistics are kept on the reuse rate of every part in the database. In 1985, the documentation reuse rate, that is, the number of reused pages divided by the total number of pages of documentation, was 32%. In the design phase, the reuse rate was 33%, while 48% of code was reused during the implementation phase. In addition, statistics are kept on the sizes of reused software parts: about 55% were 1 to 10K EASL in size, and 36% were in the 10K to 100K EASL range.

Corresponding statistics for 25 software products developed by NASA are now given.

13.5.3 NASA Software

Selby has characterized the reuse of software in a NASA software production environment in which software reuse is actively encouraged [Selby, 1988]. A total of 25 software products consisting of ground support software for unmanned spacecraft control were investigated. They ranged in size from 3000 to 112,000 source lines. The 7188 component modules were classified into four categories. Group 1 consisted of modules that were used without any changes. Group 2 were those modules reused with slight revisions, that is, less than 25% of the code was changed. Modules falling in group 3 were reused with major revisions; 25% or more of the code was changed. Group 4 modules were developed from scratch.

On average, 32% of the 7188 component modules were reused in modified or unmodified form. More specifically, 17% fell into group 1, 10% into group 2, and 5% into group 3. With regard to 2954 FORTRAN modules in the sample that were studied in detail, 45% of the modules were reused, with 28%, 10%, and 7% falling into groups 1, 2, and 3, respectively. In general, the reused modules were small, well documented, with simple interfaces and little input/output processing and tended to be terminal nodes in a module interconnection diagram, such as Figure 6.15.

These results are not really surprising. Small, well-documented modules are easier to comprehend than large modules with poor documentation, and are therefore more likely to be reused. In addition, a large module is likely to perform a number of functions, or perhaps one rather specialized function, and is therefore less likely to be reused than its smaller counterpart. A complex interface implies a large number of variables, and this, too, will tend to reduce the reusability of a module. Input/output processing can be somewhat application-specific, and this can lower its reusability. Finally, comparing terminal modules in a module interconnection diagram with modules higher up in the diagram, terminal modules are more likely to be functional modules, while modules higher up tend to be logic modules (Section 6.14.2). This further tends to increase the possibility of reusing a terminal module, rather than a nonterminal one.

A more constructive way of looking at Selby's results is to utilize them to ensure that modules can be reused in future products. Management should ensure that a specific design objective should be small modules with simple interfaces. Input/output processing should be localized to a few modules. All modules must be properly documented.

The overall lesson of all three case studies is that reusability is possible in practice. But the major push for reusability must come from management.

Chapter Review

A product is portable if it is easier to port to a different hardware/operating system configuration than to rewrite it from scratch (Section 13.1). Portability can be hampered by incompatibilities caused by hardware (Section 13.1.1), operating systems (Section 13.1.2), numerical software (Section 13.1.3), or compilers (Section 13.1.4). Nevertheless, it is important to try to make all products as portable as possible (Section 13.2). Ways of facilitating portability include using popular high-level languages, isolating the nonportable pieces of a product (Section 13.3.1), and adhering to language standards (Section 13.3.2). Reusability refers to reusing parts of a product in another product with different functionality (Section 13.4). Three very different reusability case studies are presented to show that, with the cooperation of management, it is possible to achieve a high level of reusability (Section 13.5).

For Further Reading

Three introductory texts on portability are [Wallis, 1982], [Wolberg, 1983], and [Lecarme and Gart, 1986]. Guidelines for achieving portability can be found in [Tanenbaum, Klint, and Bohm, 1978]. Portability of C and UNIX is discussed in [Johnson and Ritchie, 1978]. With regard to portability of Ada products, the reader should consult [Nissen and Wallis, 1984].

A review of the state of the art regarding reusability can be found in [Jones, 1984]. Further information on the reusability case studies in this chapter can be found in [Lanergan and Grasso, 1984], [Matsumoto, 1984, 1987], and [Selby, 1988]. [Ringland, 1984] is a description of five projects in which old software was reused. Some warnings regarding reuse are given in

[Tracz, 1988]. A classification scheme for module retrieval and reuse is described in [Prieto-Díaz and Freeman, 1987]. Reusability issues regarding Ada can be found in [Nissen and Wallis, 1984] and [Gargaro and Pappas, 1987]. Further papers on reusability are to be found in [Freeman, 1987] and in the September 1984 Special Issue on Reusability of *IEEE Transactions on Software Engineering.*

Problems

13.1 Explain how you would ensure that the library circulation system (Problem 7.4) is as portable as possible.

13.2 Explain how you would ensure that the product that checks whether a bank statement is correct (Problem 7.2) is as portable as possible.

13.3 Explain how you would ensure that the software for the automatic teller machine (ATM) of Problem 9.2 is as portable as possible.

13.4 Your organization is developing a real-time control system for a new type of laser that will be used in cancer therapy. You have been put in charge of writing two assembler modules. How will you instruct your team to ensure that the resulting code will be as portable as possible?

13.5 You are responsible for porting a 500,000-line FORTRAN product to a new computer that your company has just purchased. You manage to copy the source code to the new machine, but discover when you try to compile it that every one of the over 10,000 input/output statements has been written in a nonstandard FORTRAN syntax that the new compiler rejects. What do you do now?

13.6 Explain how you would ensure that as many modules as possible of the software for the library circulation system (Problem 7.4) can be reused in future products.

13.7 Explain how you would ensure that as many modules as possible of the product that checks whether a bank statement is correct (Problem 7.2) can be reused in future products.

13.8 Explain how you would ensure that as many modules as possible of the software for the automatic teller machine (ATM) of Problem 9.2 can be reused in future products.

13.9 You have just joined a large software organization that has hundreds of products comprising some 75,000 COBOL modules. You have been hired to come up with a plan for reusing as many of the modules as possible in future products. What do you do?

13.10 (Term Project) Port the product you implemented in Problem 10.12 to a different computer running under a different operating system.

13.11 (Readings in Software Engineering) Your instructor will distribute copies of [Prieto-Díaz and Freeman, 1987]. They describe their classification scheme as a "partial solution" to the software reuse problem. What else is needed?

References

[ANSI/MIL-STD-1815A, 1983] "Reference Manual for the Ada Programming Language," ANSI/MIL-STD-1815A, American National Standards Institute, Inc., United States Department of Defense, 1983.

[ANSI X3.9, 1977] "Programming Language FORTRAN," ANSI X3.9-1977, American National Standards Institute, Inc., 1977.

[ANSI/IEEE 770X3.97, 1983] "Pascal Computer Programming Language," ANSI/IEEE 770X3.97-1983, American National Standards Institute, Inc., Institute of Electrical and Electronic Engineers, Inc., 1983.

[Freeman, 1987] P. FREEMAN (Editor), *Tutorial: Software Reusability*, IEEE Computer Society Press, Washington, DC, 1987.

[Gargaro and Pappas, 1987] A. GARGARO AND T. L. PAPPAS, "Reusability Issues and Ada," *IEEE Software* **4** (July 1987), pp. 43–51.

[Gifford and Spector, 1987] D. GIFFORD AND A. SPECTOR, "Case Study: IBM's System/360–370 Architecture," *Communications of the ACM* **30** (April 1987), pp. 292–307.

[ISO-7185, 1980] "Specification for the Computer Programming Language Pascal," ISO-7185, International Standards Organization, 1980.

[Jensen and Wirth, 1975] K. JENSEN AND N. WIRTH, *Pascal User Manual and Report*, Second Edition, Springer-Verlag, New York, NY, 1975.

[Johnson, 1979] S. C. JOHNSON, "A Tour through the Portable C Compiler," Seventh Edition, UNIX Programmer's Manual, Bell Laboratories, January 1979.

[Johnson and Ritchie, 1978] S. C. JOHNSON AND D. M. RITCHIE, "Portability of C Programs and the UNIX System," *Bell System Technical Journal* **57** (No. 6, Part 2, 1978), pp. 2021–2048.

[Joint X3J9/IEEE P770, 1986] "Programming Language Extended Pascal," Working Draft, Joint X3J9/IEEE P770, Institute of Electrical and Electronic Engineers, Inc., 1986.

[Jones, 1978] T. C. JONES, "Measuring Programming Quality and Productivity," *IBM Systems Journal* **17** (No. 1, 1978), pp. 39–63.

[Jones, 1984] T. C. JONES, "Reusability in Programming: A Survey of the State of the Art," *IEEE Transactions on Software Engineering* **SE-10** (September 1984), pp. 488–494.

[Kernighan and Ritchie, 1978] B. W. KERNIGHAN AND D. M. RITCHIE, *The C Programming Language*, Prentice-Hall, Englewood Cliffs, NJ, 1978.

[Lanergan and Grasso, 1984] R. G. LANERGAN AND C. A. GRASSO, "Software Engineering with Reusable Designs and Code," *IEEE Transactions on Software Engineering* **SE-10** (September 1984), pp. 498–501.

[Lecarme and Gart, 1986] O. LECARME AND M. P. GART, *Software Portability*, McGraw-Hill, New York, NY, 1986.

[Liskov, Snyder, Atkinson, and Schaffert, 1977] B. LISKOV, A. SNYDER, R. ATKINSON, AND C. SCHAFFERT, "Abstraction Mechanisms in CLU,"

Communications of the ACM **20** (August 1977), pp. 564–576.

[Mackenzie, 1980] C. E. MACKENZIE, *Coded Character Sets: History and Development*, Addison-Wesley, Reading, MA, 1980.

[Matsumoto, 1984] Y. MATSUMOTO, "Management of Industrial Software Production," *IEEE Computer* **17** (February 1984), pp. 59–72.

[Matsumoto, 1987] Y. MATSUMOTO, "A Software Factory: An Overall Approach to Software Production," in: *Tutorial: Software Reusability*, P. Freeman (Editor), IEEE Computer Society Press, Washington, DC, 1987, pp. 155–178.

[Nissen and Wallis, 1984] J. NISSEN AND P. WALLIS (Editors), *Portability and Style in Ada*, Cambridge University Press, Cambridge, UK, 1984.

[Nori, Ammann, Jensen, and Nägeli, 1981] K. V. NORI, U. AMMANN, K. JENSEN, AND H. NÄGELI, "The Pascal (P) Compiler Implementation Notes," in: *Pascal—The Language and its Implementation*, D. W. Barron (Editor), John Wiley, Chichester, UK, 1981, pp. 125–170.

[Norusis, 1982] M. J. NORUSIS, *SPSS Introductory Guide: Basic Statistics and Operations*, McGraw-Hill, New York, NY, 1982.

[Phillips, 1986] J. PHILLIPS, *The NAG Library: A Beginner's Guide*, Clarendon Press, Oxford, UK, 1986.

[Prieto-Díaz and Freeman, 1987] R. PRIETO-DÍAZ AND P. FREEMAN, "Classifying Software for Reusability," *IEEE Software* **4** (January 1987), pp. 6–16.

[Prieto-Díaz and Neighbors, 1986] R. PRIETO-DÍAZ AND J. M. NEIGHBORS, "Module Interconnection Languages," *Journal of Systems and Software* **6** (November 1986), pp. 307–334.

[Ringland, 1984] J. RINGLAND, "Software Engineering in a Development Group," *Software—Practice and Experience* **14** (June 1984), pp. 533–559.

[Selby, 1988] R. W. SELBY, "Quantitative Studies of Software Reuse," Technical Report UCI TR-87-12, Version: March 22, 1988, Department of Information and Computer Science, University of California, Irvine, CA, 1988.

[Tanenbaum, 1981] A. S. TANENBAUM, "Network Protocols," *ACM Computing Surveys* **13** (December 1981), pp. 453–489.

[Tanenbaum, Klint, and Bohm, 1978] A. S. TANENBAUM, P. KLINT, AND W. BOHM, "Guidelines for Software Portability," *Software—Practice and Experience* **8** (November 1978), pp. 681–698.

[Tracz, 1988] W. TRACZ, "Software Reuse Myths," *ACM SIGSOFT Software Engineering Notes* **13** (January 1988), pp. 17–21.

[Wallis, 1982] P. J. L. WALLIS, *Portable Programming*, John Wiley and Sons, New York, NY, 1982.

[Wolberg, 1983] J. R. WOLBERG, *Conversion of Computer Software*, Prentice-Hall, Englewood Cliffs, NJ, 1983.

Chapter 14 Ada and Software Engineering

Around 1980, a major topic in both computer science research journals and computing magazines was the language Ada. Articles ranged from bitter invective to fulsome and unconditional praise of every aspect of Ada, and from carefully reasoned condemnation of Ada to well-balanced, generally favorable reviews. In short, every possible viewpoint for or against Ada was expressed, and at length.

Ada was developed for use in military embedded computer applications. An embedded computer is an integral part of a larger system whose primary purpose is not computation. Examples are a computer built into an intercontinental ballistic missile or a network of avionics computers on board a warplane. The primary function of an embedded system is to control the device in which it is embedded.

Since 1984, the Department of Defense (DoD) has required Ada to be used for the software for all new mission-critical applications. This means that embedded software in all new weapon systems, from handheld antitank devices to nuclear submarines, must be written in Ada. NATO issued a similar directive regarding its military software in 1986. Consequently, any discussion as to whether Ada "is a good thing" is irrelevant. Ada has arrived on the software scene, for better or for worse. DoD possesses more hardware than any other organization and funds the writing of more software than perhaps the next 20 largest organizations put together; more than half of that software is embedded. No matter what the critics of Ada may feel, and even Ada's strongest supporters grudgingly admit that Ada has *some* imperfections, Ada has been adopted by DoD. As a consequence, the very nature of software engineering will be impacted by this new language.

In this chapter Ada is examined not as a language, but rather from the viewpoint of the interaction between Ada and software engineering. In order to appreciate the impact of Ada on software engineering both now and in the future, it is necessary to know why Ada was developed, and how much is at stake in the Ada endeavor. As with the rest of this book, a previous knowledge of Ada is not required.

14.1 History of Ada

During the early 1970s, the Department of Defense became acutely aware of problems with its software. One of the more worrisome facts was that at least 450 different languages were being used in DoD products. A major implication of this proliferation was that maintenance, difficult enough under normal circumstances, was made all but impossible by the need to find competent maintenance programmers for this babel of languages. In addition, tools to support software development and maintenance were rudimentary, mainly because of the enormous cost of buying or building an adequate toolkit for each of those languages.

The situation with regard to embedded software was particularly bad. The problem was partly caused by the fact that embedded software, being almost exclusively real-time in nature and frequently distributed as well, is more difficult to develop and maintain than, say, batch-mode data-processing applications. Another reason was because of the scale of the problem. In 1973, embedded software consumed 56% of the total DoD software budget [Fisher, 1976]. Scientific software, taking about 5% of the DoD software budget, was almost all FORTRAN code. About 19% of software expenditure went to data-processing software, written mostly in COBOL. The remaining 20% of the DoD software budget was taken up with other types of software costs. Embedded software was not merely the major focus of the DoD software initiative, but it was in this area that the largest variety of languages was employed. Each branch of the Armed Forces had its own favorite real-time language; the Army supported TACPOL, the Navy promoted CMS-2, while the choice of the Air Force was JOVIAL. In short, the situation regarding embedded software was bad, and, bearing the maintenance implications in mind, could only get worse.

DoD embedded software tends to be large, namely of the order of hundreds of thousands or millions of lines of code, to have a lifetime of 10 to 15 years, and to change frequently over that period as requirements change [Fisher, 1976]. In addition, embedded software is almost always subject to space constraints in that the size of the computer embedded within, say, a tank, a drone, or a helicopter, is generally restricted. Furthermore, the time constraints of real-time software are ever present. Finally, embedded software must be highly reliable. To put it bluntly, once a missile has been launched from a nuclear submarine, it is too late to make any changes to its software in the event of a failure of any sort being detected.

In 1975, the High Order Language Working Group (HOLWG) was set up within DoD in order to spearhead a move towards a common high-order language for all the armed services. The intention was that HOLWG should come up with three items: a list of requirements for a common DoD high-order language, an evaluation of existing languages against this list, and a recommendation regarding the adoption of one language or perhaps a small set of programming languages that met the requirements [Whitaker, 1978]. In order to determine the requirements for a DoD language, a series of requirements documents was produced by HOLWG and circulated to DoD, industry, and academia for comments. The first requirements document was termed STRAWMAN [DoD, 1975a]. A sequence of documents followed, each a refinement of its predecessor incorporating comments and suggestions received from the various review panels. The documents were named WOODENMAN [DoD, 1975b], TINMAN [DoD, 1976], IRONMAN [DoD, 1977a], Revised IRONMAN [DoD, 1977b], and STEELMAN [DoD, 1978a].

During 1976, a total of 23 existing languages were evaluated against the TINMAN specifications, which included support for the principles of software engineering such as portability and abstraction, and the inclusion of real-time constructs. The conclusion was that no one existing language was adequate. The Defense Advanced Research Projects Agency (DARPA) was then given the task of managing the development of a common high-order language, DoD-1, that would satisfy IRONMAN. Rather than have the language designed by a small, private, DoD-sponsored committee, DARPA set up a worldwide public competition. Altogether 17 language proposals were received from all over the world, designed by teams which included academics, industry professionals, and military experts in their ranks.

Four of the 17 submissions were chosen for further development. In order to ensure that reviewers would not know the identities of the competing organizations, the language proposals were given the names Blue, Green, Red, and Yellow. The four languages were evaluated against the Revised IRONMAN specifications. Red and Green won this phase, and during March and April 1979, further development of both took place, followed by evaluation against STEELMAN. The winner was Green, a mainly European team led by Jean Ichbiah of Honeywell Bull, France. That is why the Ada Reference Manual [ANSI/MIL-STD-1815A, 1983] is bound with a green cover, as are issues of *Ada Letters*, a major Ada research journal published by the Special Interest Group on Ada of the Association for Computing Machinery (ACM SIGAda).

But DoD was as unhappy with the name Green as it was with the earlier name DoD-1. Jack Cooper of the U.S. Navy suggested the name Ada, after Ada, Countess of Lovelace, daughter of the poet Lord Byron. She had programmed Babbage's Analytic Engine, the first computer, in the first half of the nineteenth century. Although Babbage's design was correct, the Analytic Engine could not be constructed because of the limitations of nineteenth-century technology. The Deputy Secretary of Defense obtained permission from the Countess of Lovelace's heir, the Earl of Lytton, to use the name Ada [Carlson, Druffel, Fisher, and Whitaker, 1980]. The language manual and a rationale for the design were then circulated worldwide through the medium of *ACM SIGPLAN Notices* [Preliminary Ada Reference Manual, 1979; Ichbiah, Barnes, Heliard, Krieg-Brückner, Roubine, and Wichmann, 1979]. This elicited still further comment, resulting in the final version of the language manual being handed over to DoD in 1980. The language was approved by DoD and assigned military standard MIL-STD-1815. The number 1815 is significant, being the year of birth of Ada Lovelace.

The 1980 version of the language was then submitted to the American National Standards Institute (ANSI) for adoption as an American national standard. Public review of Ada elicited a considerable volume of comments and criticism, centering mainly on the complexity of the language. Although the reference manual was drastically changed, the language itself underwent only minor alterations. In 1983, the revised language standard was approved by both DoD and ANSI [ANSI/MIL-STD-1815A, 1983].

Once the language, and a name for that language, had been approved by DoD, but before any compilers had been written, a validation procedure for Ada compilers was set up. A validation suite of over 2500 tests was drawn up, the so-called Ada Compiler Validation Capability (ACVC). In order to be validated, a compiler must successfully compile instances of valid Ada, but must flag illegal constructs at compilation time and detect run-time faults. A compiler developer can obtain a tape of the ACVC and test its compiler; running the complete validation suite can take 1 week. Once the developer is satisfied that the compiler passes the tests, the compiler is independently validated by the Ada Validation Organization. If the compiler passes this scrutiny it may legally be termed a "validated Ada compiler." Validation has no implications regarding the speed of the compiler or the efficiency or compactness of the compiled code. All that validation means is that the language handled by the compiler is standard Ada and not a subset or a superset. Additional tests are continually added to the suite, and a compiler therefore has to be revalidated each year against the current version of the validation suite.

In parallel with the language effort, the specifications for the Ada programming support environment (APSE) progressed through a series of documents. The first, named SANDMAN, was circulated to a restricted group in June, 1978. Thereafter followed PEBBLEMAN [DoD, 1978b], and STONEMAN [DoD, 1980]. In 1980, HOLWG was dissolved, and control of the Ada effort was transferred to the newly created Ada Joint Projects Office (AJPO). To date, Ada compilers have been produced and validated for a wide variety of different machines. In addition, a number of Ada development environments are currently available; a complete APSE, however, has not yet been built.

14.2 Software Engineering Implications of Ada

When choosing an implementation language for a product, one of the factors that is sometimes taken into account is that choice of language may have some impact on the implementation phase and could also affect integration, product testing, and maintenance. That is to say, the impact that a language could have on the life cycle should be considered, but it is usually unlikely to be a major factor. Not so with Ada. Once the specifications have been drawn up, Ada can have a major impact on all the remaining phases of the life cycle. This impact could be for the better or for the worse. After all, Ada is merely a tool, and like other tools can be correctly used or misused. In this section Ada is analyzed from the viewpoint of how it can be used to engineer better software. A theme of this section is that, in order to make use of the power of Ada, it is not enough merely to be a programmer; software engineering skills are needed to use Ada correctly. The reason is that, as will be shown, fundamental principles of software engineering underlie much of Ada, and proper use of Ada requires an understanding not just of the language itself, but also of the principles that underlie it.

14.2.1 Ada and Modularity

The fact that Ada was used to illustrate most of the modularity concepts of Chapter 8 indicates that Ada supports the partitioning of a large product into modules. Unlike standard Pascal, which does not permit separate compilation [ANSI/IEEE 770X3.97, 1983], Ada allows modules to be compiled separately. But Ada goes beyond separate compilation of modules. Ada *compilation units* include procedures, functions, packages, tasks (for concurrent programming), and generics, which are similar to macros. A compilation unit consists of two parts, the *specification* and the *body*. Consider an Ada package. As explained in Chapter 8, entities of the package that are to be visible outside the package and therefore can be utilized by other compilation units are declared in the package specification. These may include types, constants, procedures, and names of *exceptions* (run-time faults such as table overflow or array index out of range). The pieces that are needed to make the package work appear in the package body, including declarations of types and variables that are local to the package body, implementation of the procedures declared in the specification, and details of how exceptions are to be handled. The specification and the body of any compilation unit can be compiled separately. However, a body cannot be compiled until the corresponding specification has been compiled, because declarations needed for compiling the body are contained in the specification.

Ada supports all types of information hiding. With regard to data abstraction, it is a principle of software engineering that knowledge as to how a type is implemented should not be made use of outside the compilation unit in which that type is defined. Packages satisfying this principle are self-contained and hence reusable. Furthermore, if programmers write code that does not depend on the way that a type is implemented, such as the number of words it occupies, the resulting code will be portable. But Ada goes further than this. Through the use of private types Ada can actually ensure that information as to how a data structure is implemented is kept secret. Private types thus prevent any operations, other than those specifically mentioned in the package specification and invoked via the package interface, from being performed on data structures. This results in more reliable code and code that is easier to prove correct. In Ada, objects (Section 8.7) are frequently implemented as packages with private types.

Ada also supports procedural abstraction (Section 8.4.1) through the medium of *generic procedures*, essentially templates that can be instantiated for a variety of different parameters. For example, a generic procedure can be written that interchanges two items (Figure 14.1). This generic procedure, INTERCHANGE, can then be instantiated to yield a procedure INTEGER_INTERCHANGE that interchanges two integers, or a different procedure BOOLEAN_INTERCHANGE that interchanges two Boolean variables. In each case, the programmer must specify the type of the items to be interchanged. For example, the ITEM => INTEGER clause in Figure 14.1(c) causes every instance of ITEM to be replaced by INTEGER when the procedure is instantiated. A generic procedure is therefore on a higher level of abstraction than an ordinary procedure that can be applied to only one type of parameter. Generic procedures promote reusability in that they can be reused in a broader range of products than ordinary procedures.

Ada is not just a programming language. Ada Design Language (ADL) is frequently used to design products that are to be implemented in Ada [ANSI/IEEE 990, 1987]. An ADL design essentially consists of a framework of Ada control statements, with the details of each block informally expressed as comments. Implementation then consists of translating the comments into valid Ada code. There are a number of advantages to using ADL. First, the entire product can be developed in one language; there is no need to have to translate from a design language into an implementation language, and the problem of faults induced by mistranslations of design into code is thus bypassed. The product can be compiled and executed at all times since ADL has essentially the same syntax as Ada, so the transition from design to implementation is smooth. The power of Ada can be used in the

```
generic
   type ITEM is private;
procedure INTERCHANGE (FIRST, SECOND : in out ITEM);
```

(a)

```
procedure INTERCHANGE (FIRST, SECOND : in out ITEM) is
   TEMP : ITEM;
begin
   TEMP := FIRST;
   FIRST := SECOND;
   SECOND := TEMP;
end INTERCHANGE;
```

(b)

```
procedure INTEGER_INTERCHANGE is new INTERCHANGE (ITEM => INTEGER);
   . . .
   . . .
procedure BOOLEAN_INTERCHANGE is new INTERCHANGE (ITEM => BOOLEAN);
```

(c)

Figure 14.1 Generic procedure for interchanging two items. (a) Generic procedure specification. (b) Generic procedure body. (c) Instantiations of procedure for items of type INTEGER or BOOLEAN.

design phase, as can the software tools available to support Ada implementation.

Does the use of ADL constrain the designer? A detailed answer to that question is given in [Rajlich, 1985b]; here, only top-down and bottom-up design and implementation are examined. ADL is executable code because comments are ignored by a compiler, and the end product of design using ADL is Ada code. As a result, ADL supports an incremental life cycle in which design and implementation proceed in parallel. When a compilation unit has been designed, the design team can hand it over to the implementation team and then proceed with designing the next compilation unit.

In bottom-up development, the lowest-level compilation units are designed first. These will generally be packages, procedures, and functions. The next highest level of compilation units can then invoke the procedures and functions and make use of the packages by means of **with** clauses. This is shown in Figure 14.2. First, packages PACKAGE_1 and PACKAGE_2 are designed and implemented. Then compilation unit MAIN is designed. The presence of the **with** clause in MAIN ensures that any entities defined in a

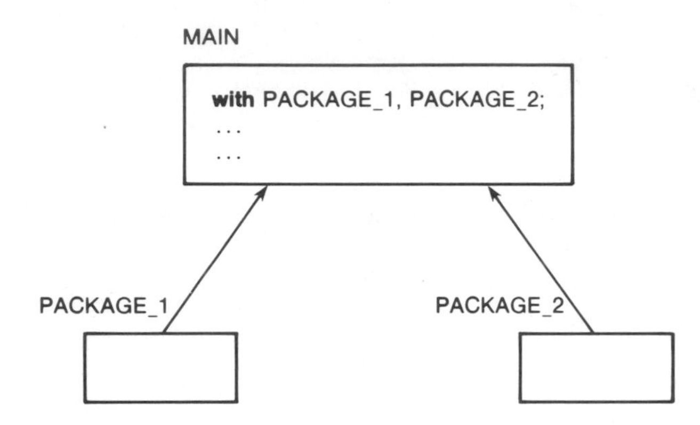

Figure 14.2 Compilation unit MAIN utilizes packages PACKAGE_1 and PACKAGE_2.

package specification such as a type, a procedure, or a function can be used within MAIN in the same way as if the package itself were incorporated within MAIN.

With regard to top-down development, a module can be designed with the procedures and functions that it calls implemented as stubs as explained in Section 6.14.2. But packages present a problem. Suppose that the product shown in Figure 14.2 is to be developed top-down using the incremental life cycle. First, compilation unit MAIN is designed. At this stage, neither PACKAGE_1 nor PACKAGE_2 has been designed, and thus there is no way of writing the **with** clause; the resulting ADL design therefore cannot be compiled. Top-down incremental development is fine from the viewpoint of procedures and functions. But packages have to be designed before the compilation units that employ them. This is one of the reasons why object-oriented design (OOD) is so popular among Ada software developers. In OOD (Section 9.8) first the objects are extracted, and then the design is built in terms of those objects. As a result, both top-down and bottom-up incremental development can be performed in conjunction with OOD, because in both instances all the objects (packages) are designed first and implemented, and then the procedures and functions are designed and implemented top-down or bottom-up, as the case may be.

There is another Ada mechanism that promotes top-down development. In a package, it is possible to define and then to compile the body of a compilation unit separately from the package in which it is defined. Consider the package shown in Figure 14.3. The two procedures FAHRENHEIT_TO_CELSIUS and CELSIUS_TO_FAHRENHEIT can be compiled independently of the package in which are they declared. This promotes top-down development. First, a package specification can be developed, then its body, and, finally, the bodies of the procedures in that package.

```
package CONVERT_TEMPERATURE is
   type CELSIUS is digits 6;
   type FAHRENHEIT is digits 6;
     -- CELSIUS and FAHRENHEIT are defined to be floating point numbers
     -- with 6 significant decimal digits.

   procedure FAHRENHEIT_TO_CELSIUS (F : in FAHRENHEIT; C : out CELSIUS);
   procedure CELSIUS_TO_FAHRENHEIT (C : in CELSIUS; F : out FAHRENHEIT);
end CONVERT_TEMPERATURE;
```

(a)

```
package body CONVERT_TEMPERATURE is
   SCALE_FACTOR : constant CELSIUS := 5.0/9.0;
     -- SCALE_FACTOR is a constant of type CELSIUS with value 5/9.
   FREEZING_POINT_FAHRENHEIT : constant FAHRENHEIT := 32.0;

     -- FREEZING_POINT_FAHRENHEIT is a constant of type FAHRENHEIT with value 32.

   procedure FAHRENHEIT_TO_CELSIUS (F : in FAHRENHEIT; C : out CELSIUS) is separate;
   procedure CELSIUS_TO_FAHRENHEIT (C : in CELSIUS; F : out FAHRENHEIT) is separate;
end CONVERT_TEMPERATURE;
```

(b)

```
separate (CONVERT_TEMPERATURE)
procedure FAHRENHEIT_TO_CELSIUS (F : in FAHRENHEIT; C : out CELSIUS) is
begin
   C := CELSIUS (F – FREEZING_POINT_FAHRENHEIT) * SCALE_FACTOR;
     -- The expression (F – FREEZING_POINT_FAHRENHEIT) is of type FAHRENHEIT.
     -- In order to ensure that the strong typing rules of Ada are not violated, it is
     -- necessary to convert ("coerce") this expression to type CELSIUS.
     -- This is achieved by the construction CELSIUS (. . .)
end FAHRENHEIT_TO_CELSIUS;
```

(c)

```
separate (CONVERT_TEMPERATURE)
procedure CELSIUS_TO_FAHRENHEIT (C : in CELSIUS; F : out FAHRENHEIT) is
begin
   F := FAHRENHEIT (C/SCALE_FACTOR) + FREEZING_POINT_FAHRENHEIT;
     -- The expression C/SCALE_FACTOR is of type CELSIUS.
     -- In order to ensure that the strong typing rules of Ada are not violated, it is
     -- necessary to convert ("coerce") this expression to type FAHRENHEIT.
     -- This is achieved by the construction FAHRENHEIT (. . .)
end CELSIUS_TO_FAHRENHEIT;
```

(d)

Figure 14.3 Package for converting temperatures. (a) Package specification. (b) Package body. (c) Separate procedure FAHRENHEIT_TO_CELSIUS. (d) Separate procedure CELSIUS_TO_FAHRENHEIT.

14.2.2 Ada and Reliability

Ada supports strong type checking. This means that every variable must be of a specific type, a variable can take on only those values that are permitted for that type, and the only operations that can be performed on that variable are operations permitted for variables of that type. Ada does not permit its typing rules to be broken. For example, suppose that B is a variable of **type** INTEGER. Then the assignment statement

```
B := A + 5;
```

is invalid unless A is also of **type** INTEGER.

But if A is of a real type, such as fixed point, floating point, or a type derived from them, then it is necessary to *coerce* A to **type** INTEGER. The preceding assignment statement would have to be replaced by

```
B := INTEGER(A) + 5;
```

Thus, if A has the value 3.2, converting it to **type** INTEGER causes its value to be truncated to 3, and the result of the addition operation is that the value 8 is assigned to B.

There are a number of advantages of strong typing. When a variable is declared, the range of values it can assume is implicitly or explicitly declared at the same time. If this range is violated, then either the programmer declared the variable incorrectly or made some other sort of programming fault. Either way, the requirement of strong typing will cause the fault to be detected by the Ada compiler if the fault is in the syntax or static semantics of the code, or otherwise by the Ada run-time routines. Second, strong typing can reduce faults in parameter lists. Consider the Ada code shown in Figure 14.4. While **type** X_ANGLE and **type** Y_ANGLE are both specified to be integers in the range 0 through 90, the fact that they have different names means that they are different Ada types. An Ada compiler would detect a fault in **procedure** TYPE_CHECK because the types of the actual parameters in the call COMPUTE_ANGLES (ANGLE_2, ANGLE _1) do not agree with the types of the formal parameters in the declaration of COMPUTE_ANGLES. If **procedure** COMPUTE_ANGLES had been defined in one compilation unit but invoked in another, then the Ada linker would have detected the strong typing violation. Enforcement of strong typing across interfaces is then a further useful way of detecting faults. The fault in Figure 14.4 might have been difficult to detect had ANGLE_1 and ANGLE_2 simply both been defined to be of **type** INTEGER, or even if both had been defined to be of the same type. It is important that users of Ada be aware not just of the features of Ada, but why they are important, and how they can be used to improve the quality of software. That is to say, Ada merely provides tools; it is up to the individual

```
procedure TYPE_CHECK is

  type X_ANGLE is range 0 .. 90;
  type Y_ANGLE is range 0 .. 90;

  ANGLE_1 : X_ANGLE;
  ANGLE_2 : Y_ANGLE;

  procedure COMPUTE_ANGLES (X : in X_ANGLE; Y : in Y_ANGLE) is
  begin
    . . .
  end COMPUTE_ANGLES;

begin
  COMPUTE_ANGLES (ANGLE_2, ANGLE_1);
end TYPE_CHECK;
```

Figure 14.4 Ada code showing strong typing violation.

designer to use Ada correctly. But it is unreasonable to expect that subranges and other Ada devices be utilized unless the designer has previously been trained in both Ada itself and software engineering.

Strong typing is not the only mechanism that assists with achieving reliable code. Another is range checking. Consider the code fragment shown in Figure 14.5. If the value of D supplied to the input statement GET(D) is outside the range 0 .. 9, then a failure is detected. But Ada goes beyond reliability in that it supports fault tolerance, that is, the ability to try to recover in the event of a run-time failure. If the value of D is out of range, it is not very helpful if the product simply aborts. Instead, Ada supports a mechanism for recovery. In the example, **procedure** RESUBMIT_DATA is invoked when there is a constraint violation, thereby allowing the product to try to recover from the violation. While it is clearly not possible to foresee every conceivable

```
   type DIGIT is range 0 .. 9;
   D : DIGIT;
begin
   . . .
   GET(D);
   . . .
   exception
      when CONSTRAINT_ERROR =>
         RESUBMIT_DATA;
end;
```

Figure 14.5 Code fragment to illustrate range checking and recovery.

type of exception that might arise, Ada allows the designer to do his or her best to anticipate failure modes and to incorporate appropriate exception handlers to deal with them. Such an Ada product is more reliable than an equivalent Pascal product, say, where a range violation results in immediate and irrevocable failure termination. Ada was designed specifically for embedded products, an application area in which both reliability and fault tolerance are vital. It is therefore not surprising that Ada incorporates mechanisms for supporting both these features.

14.2.3 Ada and Structured Programming

It should go without saying that Ada supports structured programming in both the classical and broader senses as explained in Section 10.3.2. It also supports **goto** statements. It is up to the individual to write readable code; use of Ada in itself does not guarantee good quality code. What is needed is judgement. For instance, Ada supports the **exit** statement, which can be used to terminate a loop. An example of the use of **exit** statements is shown in Figure 14.6. The loop will be exited either if a RAW_ITEM is of an INVALID_TYPE or if the corresponding PROCESSED_ITEM is of an INVALID_TYPE. There are thus three ways of leaving the loop: the two **exit** statements and the bottom of the loop. If no invalid item is encountered before all NUMBER_OF_CASES have been processed, then the loop terminates in the conventional way. But if an invalid item comes up, then one of the two **exit** statements will terminate the loop prematurely.

The code of Figure 14.6 is not structured, at least not in the classical sense, because the **loop** block has one entry, but three exits. But when the code is classically structured as shown in Figure 14.7, it is far less readable, even though the resulting three blocks, namely the **loop** block and the two **if-then-else** blocks, have only one entry and one exit each. Again, a programmer needs to understand structured programming and its implications fully before coding in Ada, and this sort of knowledge is often restricted to software engineers.

```
for I in 1 .. NUMBER_OF_CASES loop
    GET_NEXT_RAW_ITEM (RAW_ITEM);
    exit when TYPE_OF (RAW_ITEM) = INVALID_TYPE;
    PROCESS_ITEM (RAW_ITEM, PROCESSED_ITEM);
    exit when TYPE_OF (PROCESSED_ITEM) = INVALID_TYPE;
end loop;
```

Figure 14.6 Use of **exit** statement.

```
I := 0;
NO_ERROR := TRUE;
while I < NUMBER_OF_CASES and NO_ERROR loop
   GET_NEXT_RAW_ITEM (RAW_ITEM);
   if TYPE_OF (RAW_ITEM) = INVALID_TYPE then
      NO_ERROR := FALSE;
   else
      PROCESS_ITEM (RAW_ITEM, PROCESSED_ITEM);
      if TYPE_OF (PROCESSED_ITEM) = INVALID_TYPE then
         NO_ERROR := FALSE;
      else
         I := I + 1;
      end if;
   end if;
end loop;
```

Figure 14.7 Classically structured version of Figure 14.6.

14.2.4 Ada and Maintenance

The most important ingredient needed for maintainability is readability, and a number of features have been incorporated into the language to achieve this. One example is *named association of parameters*. Suppose that a procedure has been defined as follows:

```
procedure COMPUTE_SPEED (OLD_X_COORD, OLD_Y_COORD,
   OLD_Z_COORD, NEW_X_COORD, NEW_Y_COORD,
   NEW_Z_COORD: in REAL; SPEED: out REAL);
```

One particular invocation of COMPUTE_SPEED is

```
COMPUTE_SPEED (13.459, -18.634, 28.775, 24.762, -98.628,
   45.350, VALUE);
```

Such a call is not easy to understand without looking at the definition of the procedure, and even then it is easy to confuse parameters. But Ada allows the name of each formal parameter to be associated with the corresponding actual parameter, thus:

```
COMPUTE_SPEED (OLD_X_COORD => 13.459,
   OLD_Y_COORD => -18.634, OLD_Z_COORD => 28.775,
   NEW_X_COORD => 24.762, NEW_Y_COORD => -98.628,
   NEW_Z_COORD => 45.350, SPEED => VALUE);
```

This version of the call is easy to read. While it is true that named association of parameters is nothing more than a clever syntactic trick, it is indicative of the care that has gone into the design of Ada to ensure that a broad variety of facilities for achieving high-quality code is made available to Ada programmers.

Another Ada mechanism for promoting readability is *overloading*. Consider a language like Pascal that allows the programmer to define new types, such as complex numbers. While the operator * is used in Pascal to denote multiplication of both real numbers and of integers, it cannot also be used for multiplication of complex numbers; for this purpose a procedure, MULTIPLY_COMPLEX, say, must be declared. But in Ada, operators such as * may be redefined in new contexts; this is termed *overloading*. If a complex-number package is written, the specificaton will probably include a declaration such as

```
function "*" (C_1, C_2: in COMPLEX_NUMBER)
    return COMPLEX_NUMBER;
```

That is, the function named * takes two arguments, both complex numbers, and the value it returns is a complex number. The implementation of this new use of * then appears in the package body. Code that uses the complex-numbers package can use the operator * for multiplying integers and real numbers, but also for multiplying complex numbers. The resulting code will include statements like

```
C_DIR := C_STREAM * C_PAUSE;
```

This is easier to comprehend than the equivalent statement

```
MULTIPLY_COMPLEX (C_STREAM, C_PAUSE, C_DIR);
```

that uses a procedure rather than operator overloading. Again, this is essentially "syntactic sugar." But tricks like this can make a considerable difference to the maintainability of a product. Incidentally, there is a price to pay for overloading in that it adds to the complexity of the Ada compiler.

What happens when a module is changed during maintenance? To be more specific, suppose that the specification of a package is changed in some way. Obviously, both the specification and body of that package must be recompiled, but so must any compilation units that make use of that package via a **with** clause. The compiler can easily determine the interdependencies between modules and disable the relevant compilation units until they have been recompiled.

14.2.5 Ada and Reusability

Many features of Ada promote reusability. For example, the fact that types and subprograms can be passed as parameters allows compilation units to be employed in different contexts. Another feature that encourages module

```
const SIZE = 17;
type MATRIX = array [1 .. SIZE, 1 .. SIZE] of CHAR;
procedure TRANSPOSE_MATRIX (IN_MATRIX : MATRIX; var OUT_MATRIX : MATRIX);
```

(a)

```
type MATRIX is array (INTEGER range <> , INTEGER range <> ) of CHARACTER;
procedure TRANSPOSE_MATRIX (IN_MATRIX : in MATRIX; OUT_MATRIX : out MATRIX);
```

(b)

Figure 14.8 Declarations for matrix transposition procedure using (a) Pascal and (b) Ada.

reuse is the generic, essentially a parametrized template. An example of a generic procedure is given in Figure 14.1. But Ada also supports generic packages. Just as a generic procedure can be instantiated to provide an implementation of an algorithm operating on a particular data structure, so a generic package can be instantiated to provide a package tailored to a specific purpose such as the implementation of an object.

In a language like Pascal, the dimensions of an array passed as a parameter must be specified. For example, Figure 14.8(a) shows a series of Pascal declarations for a procedure to transpose a 17 × 17 array of characters. If the procedure is now to be used for transposing, say, a 6 × 6 array, the **const** declaration must be changed, and the product recompiled. It is not possible to write a routine that will transpose a matrix of a size to be specified at run time. In contrast, Ada allows unconstrained arrays. The declaration of **type** MATRIX in Figure 14.8(b) merely stipulates that a MATRIX is a two-dimensional array of characters; its size is not specified. The procedure can be used to transpose matrices of any size.

The support of unconstrained arrays as formal parameters, as well as other features such as dynamic arrays, allows Ada to be used in subroutine libraries. Efforts are currently underway to translate numerical analysis libraries written in FORTRAN into Ada so that developers of Ada software can benefit from the thousands, if not millions, of person-hours that have gone into the development of those libraries. Other subroutine libraries such as statistical routines are also being translated. The resulting libraries indeed constitute reusable code.

But not every feature of Ada promotes reusability. Consider the package shown in Figure 14.3. The two procedures FAHRENHEIT_TO_CELSIUS and CELSIUS_TO_FAHRENHEIT both use the **constant** SCALE_FACTOR and the **constant** FREEZING_POINT_FAHRENHEIT defined in the body of package CONVERT_TEMPERATURE. This means that these two procedures cannot be reused on their own, that is, without **package**

CONVERT_TEMPERATURE. This is reflected by the first clause of both procedures, namely **separate** (CONVERT_TEMPERATURE). Thus, while the **separate** feature promotes top-down development, it hampers reusability.

14.2.6 Ada and Real-Time Software

Since Ada was specifically designed for embedded systems, it embodies a number of features for real-time products. The Ada **task** is a compilation unit that operates in parallel with other compilation units. A high-level intertask communication mechanism, the *rendezvous*, supports communication between tasks and synchronization of tasks. The topic of tasking is somewhat specialized and requires a knowledge of Ada beyond that needed for the rest of this book. Accordingly, the interested reader should refer to one of the many excellent Ada textbooks currently available, such as [Gehani, 1983] or [Cohen, 1986].

Nevertheless, there is one aspect of Ada tasking that should be mentioned. Tasks can be set up as concurrent processes. That is to say, there may be a single CPU, with a number of tasks that have to share that CPU. For example, five tasks may measure the temperature at particular points in a steel mill, while seven other tasks control various aspects of the steel production process. The scheduling of tasks, that is, which task gets the next time quantum of the CPU and how long that quantum will last, is controlled by the Ada run-time scheduler; it is not possible to implement arbitrary scheduling algorithms [Cornhill, Sha, Lehoczky, Rajkumar, and Tokuda, 1987; Ardö, 1987; McCormick, 1987]. Despite the fact that Ada was specifically designed for embedded software, complaints regarding the overall Ada tasking mechanism are often aired.

Two features of Ada intended to assist in real-time control are the priority mechanism, which allows a designer to specify that one task is more urgent than another, and timing facilities. Both have limited utility [Burns and Wellings, 1987]. In fact, the real-time facilities of Ada as a whole are woefully inadequate from the viewpoint of being able to handle real-time constraints.

14.3 Economics of Ada

Detractors of Ada sometimes refer to it as "PL/2." By comparing Ada to PL/I, they are hitting at the size, complexity, and subtleties of Ada, and at the consequent difficulty in understanding all aspects of Ada in sufficient depth to use it correctly under all circumstances. If name calling is deemed to be necessary, then a more appropriate name would be COBOL/2. Like Ada, COBOL (COmmon Business Oriented Language) was a product of DoD. Developed under the direction of Captain (later Rear Admiral) Grace Murray Hopper, COBOL was approved by DoD in 1960. Thereafter, DoD effectively would not buy hardware for running data-processing applications unless that hardware had a COBOL compiler [Sammet, 1978]. DoD was, and still is, the world's largest purchaser of computer hardware, and, in the 1960s, a considerable proportion of DoD software was written for data-processing purposes; embedded software is a comparatively recent development. As a result, COBOL compilers were written as a matter of urgency for virtually every computer. Also, commercial software for DoD had to be written in COBOL.

The result is that COBOL is now the world's most popular programming language. There is more operations mode software written in COBOL than in all other programming languages put together. While it is true that this might simply be a reflection of the fact that most software is written for data-processing purposes, it does not negate the fact that COBOL is the world's most widely used programming language. Although fourth-generation languages (4GLs) are undoubtedly growing in popularity for new applications, maintenance is still the major software activity, and that maintenance is being performed on existing COBOL software. In short, DoD put its stamp onto the world's software via its first major programming language, COBOL.

It is likely that Ada will have a similar impact on the world software scene. In terms of DoD requirements, Ada is being used for embedded software. The translation of scientific subroutine libraries into Ada means that Ada will soon be used for DoD scientific software as well. While Ada was specifically designed for embedded software, as early as 1984 half a million lines of Ada code had been written for various data-processing applications by Softplan, a Finnish software organization [Norokorpi, 1984]. Thus there is no need to continue with the use of COBOL for DoD data-processing software. Even though the transition to Ada in this area will take longer, it is probably inevitable.

With DoD paying billions of dollars annually for software and hardware, Ada compilers, tools, and environments are being constructed by a wide variety of organizations who are forced, willingly or unwillingly, to jump on the Ada bandwagon. Organizations that write embedded software for DoD are purchasing these Ada products. In addition, they are investing millions, if not billions, of dollars in training Ada developers and maintainers. This training is far more detailed and lengthy than for ordinary programming languages that do not require a knowledge of software engineering before they can be meaningfully used. The money for all this hardware, software, and training is effectively coming from DoD, because the cost is built into the price of the Ada products that DoD commissions.

While Ada is undeniably an expensive programming language paid for by U.S. taxpayers, the payoff comes in the future. As explained previously, Ada was developed in order to rationalize the DoD software effort. While the up-front cost of Ada is vast, the hope is that this cost can be recouped at the maintenance stage. After all, DoD embedded software generally has a lifetime of 10 to 15 years. By using one standard language, maintenance costs should be reduced. If Ada is as good a language as anticipated, maintenance should prove to be easier and cheaper than with other languages. In addition to the maintenance implications, the Ada initiative should reduce the number of military hardware projects that have reportedly been abandoned [Kulik, 1987] or significantly delayed [Lake, 1987] because of the poor quality of their embedded software.

The Ada endeavor was started around 1975. The first compilers that were fast enough for practical development of Ada software appeared around 1985. Significant figures on maintenance of Ada products are not likely to be available much before 1995. Thus it will be at least 20 years from the start of the Ada initiative before it will be known whether the project is a success or not. A consoling thought is that the efforts of the best computer experts in the world went into the design of Ada. If the Ada initiative fails, at least we have done the very best we could.

Chapter Review

The historical background leading to the development of Ada is presented in Section 14.1. Ada is then analyzed from the viewpoint of software engineering. Specifically, it is shown in Sections 14.2.1 through 14.2.5 that Ada supports modularity, reliability, structured programming, maintenance, and reusability. However, Ada does not seem ideal for real-time systems, the very area for which Ada was originally designed (Section 14.2.6). Finally, the economic implications of Ada are described in Section 14.3.

For Further Reading

There are numerous excellent introductions to Ada, including [Gehani, 1983] and [Cohen, 1986]. The Ada language reference manual is [ANSI/MIL-STD-1815A, 1983]. Balanced appraisals of Ada are presented in [Wegner, 1984] and [Sammet, 1986], while [Lieblein, 1986] is a report on the status of Ada. Useful insights into the design of Ada can be found in [Ichbiah, Barnes, Heliard, Krieg-Brückner, Roubine, and Wichmann, 1979]. A detailed explanation of the Ada compiler validation suite can be found in [Goodenough, 1986].

Ada Design Language (ADL) is described in [ANSI/IEEE 990, 1987]. The way that Ada supports programming-in-the-large is discussed in [Wolf, Clarke, and Wileden, 1985]. The issue of Ada and reusability is presented in [Burton, Aragon, Bailey, Koehler, and Mayes, 1987]. [Buzzard and Mudge, 1985] is an analysis of Ada as an object-oriented language. [Nielsen and Shumate, 1987] contains an overview of Ada design methods, together with a description of a design method for real-time systems in Ada. For information on Ada for real-time systems, the reader should consult the proceedings of the

International Workshops on Real-Time Ada Issues which appear in *ACM SIGAda Ada Letters*.

While the earlier Ada articles appeared in *ACM SIGPLAN Notices*, since 1981 a primary journal for Ada has been *ACM SIGAda Ada Letters*. There are a number of annual Ada conferences, including the Annual Conference on Ada Technology, ACM SIGAda meeting, Washington Ada Symposium, and Ada–Europe Conference. Conference summaries frequently appear in *ACM SIGAda Ada Letters*.

Problems

14.1 A year ago, you and a friend started a two-person software production company, developing software for personal computers. You have had a number of clients and have successfully developed a variety of financial software implemented in COBOL and BASIC. Today, you see an advertisement offering one-third off the price of an Ada compiler, together with the additional memory needed to run Ada on your personal computer. Your partner urges you to buy it, saying that the compiler will enable your company to enter the burgeoning Ada market. List the potential costs and benefits of buying the compiler. (Do not give dollar amounts). Also state the possible risks involved in the venture.

14.2 You are the vice-president for education of Old-Fashioned Software Corporation. It has finally been decided that 50 of the company's computer professionals will be given the training to enable them to develop future products in Ada. You have the job of selecting the 50 from the 220 professionals at Old-Fashioned Software. How would you choose them?

14.3 Having chosen the 50 computer professionals of Problem 14.2, what topics would you teach them in order to turn them into productive Ada developers?

14.4 You have been hired by Acme Defense Contractor Software Producers, a large organization that develops and maintains software mainly for weapons systems. It is the first day of your first job. You specifically joined Acme because you want to develop complex real-time embedded systems in Ada, your favorite programming language. To your dismay, you are assigned to a team responsible for writing a payroll product in COBOL. Your manager says that your software engineering skills will ensure that the payroll product will be the finest that Acme has ever developed. What do you reply, bearing in mind that resigning on day 1 will not look good on your résumé?

14.5 You have just designed Beb, a new programming language that you think has all the strengths of Ada, but none of its weaknesses. How would you publicize Beb? What do you think the reaction of DoD, NATO, and other large Ada users will be?

14.6 (Term Project) Reimplement in Ada the product you implemented in Problem 10.12.

14.7 (Readings in Software Engineering) Your instructor will distribute copies of [Sammet, 1986]. Which features of Ada are truly unique, that is to say, have not been derived or copied from earlier programming languages?

References

[ANSI/IEEE 770X3.97, 1983] "Pascal Computer Programming Language," ANSI/IEEE 770X3.97-1983, American National Standards Institute, Inc., Institute of Electrical and Electronic Engineers, Inc., 1983.

[ANSI/IEEE 990, 1987] "Recommended Practice for Ada as a Program Design Language," ANSI/IEEE 990-1987, American National Standards Institute, Inc., Institute of Electrical and Electronics Engineers, Inc., 1987.

[ANSI/MIL-STD-1815A, 1983] "Reference Manual for the Ada Programming Language," ANSI/MIL-STD-1815A, American National Standards Institute, Inc., United States Department of Defense, 1983.

[Ardö, 1987] A. Ardö, "Real-Time Efficiency of Ada in a Multiprocessor Environment," *Proceedings of the International Workshop on Real-Time Ada Issues, ACM SIGAda Ada Letters* **VII** (No. 6, 1987), pp. 40–42.

[Burns and Wellings, 1987] A. Burns and A. J. Wellings, "Real-Time Ada Issues," *Proceedings of the International Workshop on Real-Time Ada Issues, ACM SIGAda Ada Letters* **VII** (No. 6, 1987), pp. 43–46.

[Burton, Aragon, Bailey, Koehler, and Mayes, 1987] B. A. Burton, R. W. Aragon, S. A. Bailey, I. D. Koehler, and L. A. Mayes, "The Reusable Software Library," *IEEE Software* **4** (July 1987), pp. 25–33.

[Buzzard and Mudge, 1985] G. D. Buzzard and T. N. Mudge, "Object-Based Computing and the Ada Programming Language," *IEEE Computer* **18** (March 1985), pp. 11–19.

[Carlson, Druffel, Fisher, and Whitaker, 1980] W. E. Carlson, L. E. Druffel, D. A. Fisher, and W. A. Whitaker, "Introducing Ada," *Proceedings of the ACM Annual Conference, ACM 80*, Nashville, TN, 1980, pp. 263–271.

[Cohen, 1986] N. H. Cohen, *Ada as a Second Language*, McGraw-Hill, New York, NY, 1986.

[Cornhill, Sha, Lehoczky, Rajkumar, and Tokuda, 1987] D. Cornhill, L. Sha, J. P. Lehoczky, R. Rajkumar, and H. Tokuda, "Limitations of Ada for Real-Time Scheduling," *Proceedings of the International Workshop on Real-Time Ada Issues, ACM SIGAda Ada Letters* **VII** (No. 6, 1987), pp. 33–39.

[DoD, 1975a] "Requirements for High Order Programming Languages, STRAWMAN," United States Department of Defense, July 1975.

[DoD, 1975b] "Requirements for High Order Programming Languages, WOODENMAN," United States Department of Defense, August 1975.

[DoD, 1976] "Requirements for High Order Programming Languages, TINMAN," United States Department of Defense, June 1976.

[DoD, 1977a] "Requirements for High Order Programming Languages, IRONMAN," United States Department of Defense, January 1977.

[DoD, 1977b] "Requirements for High Order Programming Languages, Revised IRONMAN," United States Department of Defense, July 1977.

[DoD, 1978a] "Requirements for High Order Programming Languages, STEELMAN," United States Department of Defense, June 1978.

[DoD, 1978b] "Requirements for the Programming Environment for the Common High-Order Language, PEBBLEMAN," United States Department of Defense, July 1978.

[DoD, 1980] "Requirements for the Programming Environment for the Common High-Order Language, STONEMAN," United States Department of Defense, February 1980.

[Fisher, 1976] D. A. FISHER, "A Common Programming Language for the Department of Defense—Background and Technical Requirements," Institute for Defense Analyses, Report P-1191, 1976.

[Gehani, 1983] N. GEHANI, *Ada, An Advanced Introduction*, Prentice-Hall, Englewood Cliffs, NJ, 1983.

[Goodenough, 1986] J. B. GOODENOUGH, "Ada Compiler Validation: An Example of Software Testing Theory and Practice," *Proceedings of the CRAI Workshop on Software Factories and Ada*, Capri, Italy, May 1986, *Lecture Notes in Computer Science 275*, Springer-Verlag, Berlin, West Germany, pp. 195–221.

[Ichbiah, Barnes, Heliard, Krieg-Brückner, Roubine, and Wichmann, 1979] J. D. ICHBIAH, J. G. P. BARNES, J. C. HELIARD, B. KRIEG-BRÜCKNER, O. ROUBINE, AND B. A. WICHMANN, "Rationale for the Design of the Ada Programming Language," *ACM SIGPLAN Notices* **14** (June 1979), Part B.

[Kulik, 1987] S. KULIK, "Sophisticated Weaponry: How Reliable?" *Defense Science and Electronics* **6** (August 1987), p. 37.

[Lake, 1987] J. S. LAKE, "The Beleaguered B-1B," *Defense Science and Electronics* **6** (April 1987), pp. 31–34.

[Lieblein, 1986] E. LIEBLEIN, "The Department of Defense Software Initiative —A Status Report," *Communications of the ACM* **29** (August 1986), pp. 734–744.

[McCormick, 1987] F. McCormick, "Scheduling Difficulties of Ada in the Hard Real-Time Environment," *Proceedings of the International Workshop on Real-Time Ada Issues, ACM SIGAda Ada Letters* **VII** (No. 6, 1987), pp. 49–50.

[Nielsen and Shumate, 1987] K. W. Nielsen and K. Shumate, "Designing Large Real-Time Systems in Ada," *Communications of the ACM* **30** (August 1987), pp. 695–715.

[Norokorpi, 1984] J. Norokorpi, Informal Report, ACM SIGAda Meeting, Washington, DC, November 1984.

[Preliminary Ada Reference Manual, 1979] "Preliminary Ada Reference Manual," *ACM SIGPLAN Notices* **14** (June 1979), Part A.

[Rajlich, 1985b] V. Rajlich, "Paradigms for Design and Implementation in Ada," *Communications of the ACM* **28** (July 1985), pp. 718–727.

[Sammet, 1978] J. E. Sammet, "The Early History of COBOL," *Proceedings of the History of Programming Languages Conference*, Los Angeles, CA, 1978, pp. 199–276.

[Sammet, 1986] J. E. Sammet, "Why Ada is Not Just Another Programming Language," *Communications of the ACM* **29** (August 1986), pp. 722–732.

[Wegner, 1984] P. Wegner, "Capital-Intensive Software Technology. Part 4: Accomplishments and Deficiencies of Ada," *IEEE Software* **1** (July 1984), pp. 39–42.

[Whitaker, 1978] W. A. Whitaker, "The U.S. Department of Defense Common High Order Language Effort," *ACM SIGPLAN Notices* **13** (February 1978), pp. 19–29.

[Wolf, Clarke, and Wileden, 1985] A. L. Wolf, L. A. Clarke, and J. C. Wileden, "Ada-Based Support for Programming-in-the-Large," *IEEE Software* **2** (March 1985), pp. 58–71.

Chapter 15 Experimentation in Software Engineering

Throughout the book, techniques are analyzed and compared using several criteria, but frequently on the basis of experimentation. The purpose of this chapter is to equip the reader to evaluate the validity of such experiments in software engineering.

The reader with a background in the biological sciences or medicine may be reading this book with a growing sense of disbelief. In biology, an experimental result cannot be claimed to be true on the basis of a single sample—an adequate amount of data has to be presented, using the appropriate statistical techniques. Furthermore, a result published by one laboratory is not accepted as valid until it has been successfully repeated at an independent laboratory. In the field of medicine, the Food and Drug Administration (FDA) will not allow a new drug to be marketed until it has first passed a long and careful series of trials performed on hundreds of animals, and then an even longer and more careful series of trials on thousands of human beings. Yet in this book, claims have been made on the basis of experiments that not merely have been performed only once, but where the number of samples is equal to 1! For example, a case was made for the chief programmer team based on one single experiment using one sample, namely *The New York Times* clipping file (Section 10.8.1). In other disciplines, a similar claim would be laughed out of court, yet in software engineering a single datum seems to be adequate to prove anything. In this chapter software engineering experimentation is examined to see if that charge is true.

15.1 Why Experimentation on Methods Is Needed

As evinced by earlier chapters of this book, a wide range of methods has been developed for every phase of software development, from cost estimation to design, from implementation to product integration. While some methods are mutually compatible, other methods are diametrically opposed—consider, for example, democratic team organization (Section 10.7) versus chief programmer team organization (Section 10.8). The differences between competing methods are not always as marked, but they are generally sufficiently large for management to have to decide between them. For example, if data-oriented design is to be used, management must then decide between a number of competing methods, including the methods of Jackson, Warnier, and Orr (Sections 9.6 and 9.7). Decisions like these have to be taken during the planning stage of every software project. The life-cycle model for the project must be chosen, as well as the methods to be used in each phase of that life cycle. Unfortunately, there is little hard evidence to help management to take such decisions for selecting methods for programming-in-the-many.

In this book, *experimentation-in-the-small* and *experimentation-in-the-many* will be used to denote experimentation in the areas of programming-in-the-small and in programming-in-the-many, respectively. The distinction is by no means an artificial one: Experimentation-in-the-small is an acceptable scientific technique for determining the validity of a variety of software engineering techniques for programming-in-the-small. But as will be shown, experimentation-in-the-many cannot be used to compare methods for programming-in-the-many [Schach, 1982].

15.2 Experimentation-in-the-Small

Suppose that a software engineer has come up with a new technique for coding products small enough to be developed by a single individual. The software engineer believes that this technique is extremely easy to use and will result in code containing significantly fewer faults than when competing techniques are used. One way of subjecting this new technique to controlled experimentation is for the software engineer to approach a professor of computer science at a nearby college and ask him or her to set up an experiment with 50 students in a computer science class as subjects. A programming assignment that the professor is about to hand out to the class is used for the experiment. Half the students, chosen at random, are told to do the assignment using the technique taught in class, and the other 25 are to use the new coding technique proposed by the software engineer. The professor then grades the students' assignments, and by applying the appropriate statistics [John, 1971] can determine whether this new technique is indeed as promising as the software engineer believes.

There are a number of problems with this experimental procedure. There is a well-known charge that "psychology is the psychology of 19-year-old college sophomores." What is alleged is that, when a psychology professor undertakes research, what sometimes happens is that a group of students are paid an hourly rate to take part in an experiment. The results of that experiment are then extrapolated to the population as a whole, notwithstanding the fact that college students are by no means representative of the general population, let alone those students who volunteer for psychology experiments. It is somewhat different when a classroom programming assignment is turned into an experiment, because the students are not volunteers. Nevertheless, there are a number of differences between students doing a class assignment and professional programmers developing a piece of software. First, their motivations are different. College students are usually motivated by the grade they hope to achieve for their assignment. On the other hand, the motivation for professional programmers-in-the-small is almost always fame and fortune. They develop software, either on a full-time or part-time basis, in return for a salary, an hourly wage, or a lump sum payment on completion of the product. In a few cases a programmer will devote many hours of his or her free time to developing software that hopefully will eventually be sold to a major software organization and earn the programmer millions of dollars.

A second difference is that of scale. Students are given perhaps 3 weeks to complete an average classroom programming assignment. (In practice, the entire assignment is usually done the night before it is due, but that is another story!) A major undergraduate programming project may, at best, be some 10 weeks in length; after all, the instructor has to spend some time at the beginning of the semester or quarter explaining the theory behind the project and how it is to be done and some time at the end grading and evaluating the project. But students have other courses in addition to their computer science project course, and also a hectic social life and/or a part-time job, so the scale of an undergraduate project can never be as large as that of a product developed by a professional programmer-in-the-small. One way to remedy this is to use graduate students, because graduate courses are usually more demanding timewise than undergraduate courses, and it is therefore reasonable to assign larger projects to such a class. But it is not clear that graduate students are representative of professional programmers-in-the-small, and therefore that results so obtained can be extrapolated beyond the walls of the graduate school. Admittedly, there is no proof that undergraduates are representative either, but since the majority of professional programmers do not have a graduate degree in computer science, undergraduate computer science students are probably more representative than their graduate counterparts.

At the same time, it is also possible to conduct experiments on programming-in-the-small using software professionals as subjects. An example of this is the research performed by Myers comparing strategies for module testing described in Section 6.12 [Myers, 1978a]. In his paper Myers described the subjects of his experiment as 59 *highly experienced* programmers (author's italics), presumably in order to make it clear that his research was not conducted on sophomores. Myers' research is one of many projects that have used software professionals as subjects, and the results of such research are more difficult to challenge than when the identical experiments have been performed on college students.

While programming-in-the-small is an important area of software engineering, many more people are employed in programming-in-the-many. Experimentation in this area for the purpose of comparing competing methods is now considered.

15.3 Experimentation-in-the-Many

There are a number of difficulties when attempting to compare two methods for programming-in-the-many by means of controlled experimentation.

15.3.1 Use of Students

When performing experimentation-in-the-many, students cannot be used as subjects. It was pointed out previously that the difference of scale between classroom projects and real-life projects precludes the use of students in experimentation-in-the-small. The situation is considerably worse with regard to experimentation-in-the-many. The largest practical classroom group project is generally a team of three students working together for 10 weeks; when teams are larger, the actual work is usually done by only two or three members of the team. Bearing in mind that an undergraduate can probably devote at most 10 hours a week to a single course, this means that a team of three students can put in at most 2 person-months of effort. But a project that can be completed in 2 person-months can hardly be considered to be programming-in-the-many. A larger effort is likely to be possible only from a graduate class, and even then 3 person-months is probably the upper limit. Furthermore, in order to be able to make valid statistical inferences, a minimum of 20 teams is needed. This means that the class size must be at least 60, which is not common at the graduate level in computer science. The experiment of Boehm, Gray, and Seewaldt in the field of prototyping described in Section 3.3.4 was run with only seven teams, four of size 3 and three of size 2, a total of 18 graduate students [Boehm, Gray, and Seewaldt, 1984]. The average team effort was 2.7 person-months. There are a number of reasons why this experiment has been attacked, including the argument that the product was hardly large enough to constitute programming-in-the-many and that the subjects were computer science graduate students and not computer professionals. In addition, the experiment has not been repeated by an independent group.

15.3.2 Comparison of Methods

Suppose that a new design method has been proposed for programming-in-the-many. One experiment that might be used to compare the new

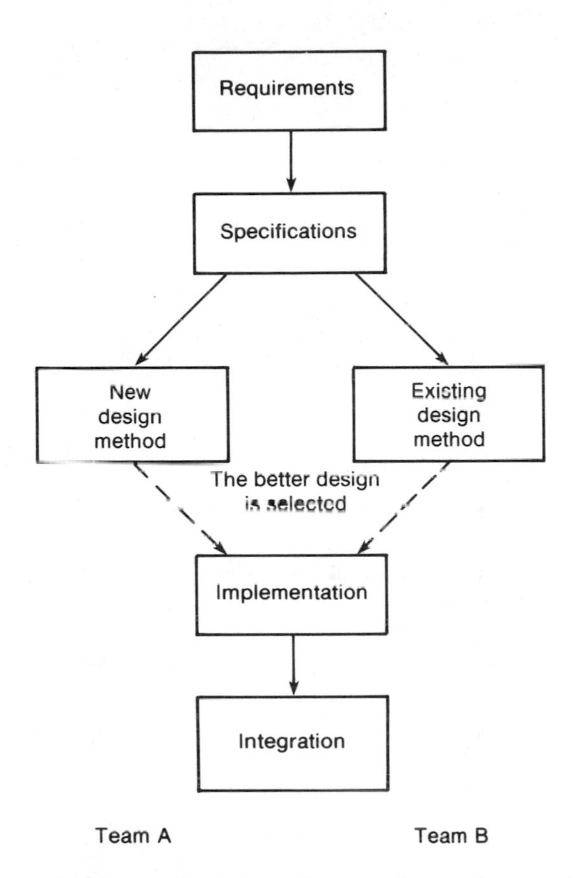

Figure 15.1 Impractical experiment for comparing two design methods.

design method with an existing method is shown in Figure 15.1. Once the requirements have been determined and the specifications have been drawn up, two teams get to work. Team A uses the new design method, while team B uses the existing design method. The resulting designs are then compared to see which method results in a better design, and this design is then used for the remainder of the life cycle.

The problem with this scheme is: How are two designs to be compared? While a bad design is easy to detect at any stage, the problem here is to distinguish between a good design and an excellent design. Put in other terms, if an organization is satisfied with its existing design method but is investigating a new, and apparently superior, method with a view to possible adoption, an experiment such as that shown in Figure 15.1 will not be adequate. The reason is that since there is no objective way of evaluating designs, the best

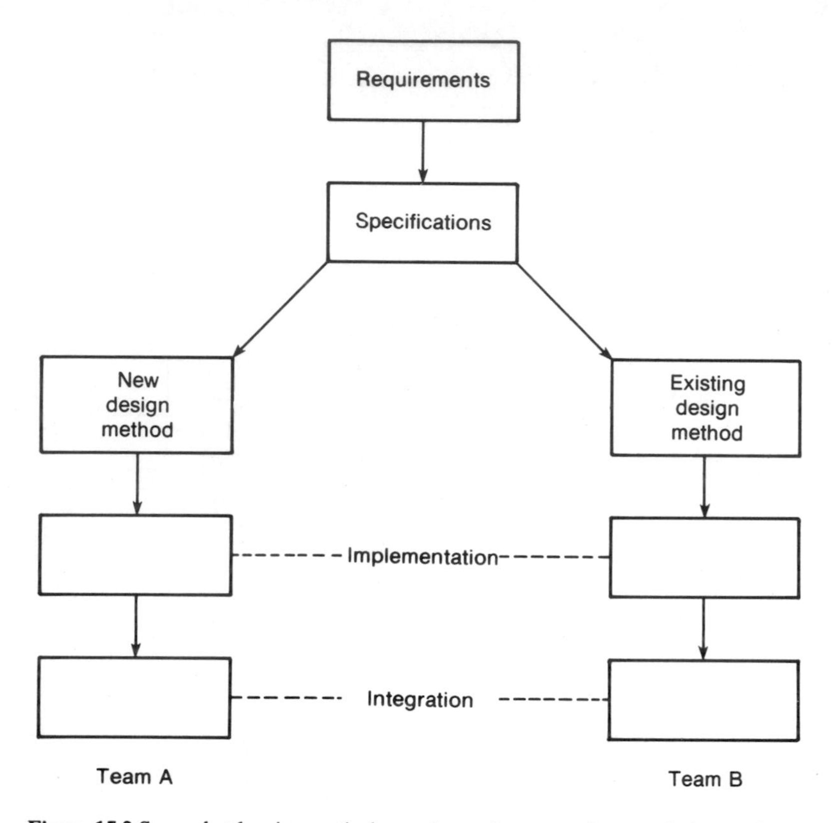

Figure 15.2 Somewhat less impractical experiment for comparing two design methods.

way, and practically speaking, the *only* way of comparing two good designs is to implement them both, and then compare the resulting products and the effort needed to build those products.

This is shown in Figure 15.2. After the specifications have been drawn up, team A uses the new design method to design the product, and team B uses the existing design method. Thereafter, both teams use the same implementation and integration methods, and all the way through the same testing techniques are used. At the end, the two products are compared. Since everything except the design methods is identical, any differences between the two products would be due to differences in the design methods. There are just two things wrong with this experiment—it costs too much to perform, and even if it were performed as described, the results would be meaningless.

15.3.3 Cost

The first problem is that of cost. Suppose that the cost of the product as a whole is $200,000. Perhaps the client can be charged as much as $275,000.

But now the design, implementation, and integration phases are to be done in duplicate. The cost to the software developer of the work of the second team will be at least $125,000. This means that the project will now cost the developer $325,000, while the client is paying only $275,000. No software development organization that is run for profit can afford to perform such an experiment. But the reality is worse than that. The programming-in-the-small experiment proposed in Section 15.2 used 50 undergraduates, because if each technique had been used by only one or two students, then the results could have been ascribed to chance or to differences between individual students. By using a larger number of subjects, individual fluctuations are averaged out, and statistical methods can be used to measure differences between methods. But the experiment of Figure 15.2 uses only two teams. No sensible conclusions could be drawn from an experiment based on such a small sample.

To get around this problem, the experiment must incorporate many more teams. At least 10 teams should use each design method, and the resulting 20 products should be compared. But the cost of this larger experiment would be $200,000 for a single version of the product plus $125,000 for each of the 19 additional teams, making a total of $2,575,000. Clearly the prohibitive cost factor makes this experiment impractical. But this proposed experiment compares merely two of perhaps hundreds of design methods that have been put forward. Suppose that only eight methods are to be compared. The cost of performing the experiment with 10 teams for each method will be over $10,000,000. In short, experimentation-in-the-many is simply too expensive a way of comparing methods.

Maintenance is almost always the most important phase of software production, and this holds even when comparing design methods. Thus, when comparing two designs, what should happen is that the products of team A and team B should be separately maintained until the maintenance implications of the new design method can be compared with those of the existing one. The cost implications of building two versions of the same product have been pointed out. How much more impractical is the idea of maintaining two parallel versions for the next 10 or 15 years, let alone the 20, or preferably more, parallel versions needed for statistical proof. The question of cost means that continuing an impractical experiment through the maintenance phase is even more impractical. Nevertheless, the critical difference when comparing two design methods is how the resulting designs behave during the maintenance phase.

15.3.4 Differences between Individual Software Professionals

Returning to the experiment depicted in Figure 15.2, suppose that team A completes the product in only 20 person-months, while team B takes 29 person-months. At first sight, this seems to mean that the new design method used by team A promotes faster product development than the existing design method used by team B. But that is not necessarily the case. It is quite conceivable that the members of team A are simply better software engineers than those of team B, and that the observed differences are due solely to differences between individuals in the two teams.

Is it conceivable that such large differences can exist between individual software engineers? As described in Section 4.5, Sackman performed a series of experiments comparing the abilities of computer professionals and observed differences between pairs of programmers of up to 28 to 1, with respect to items such as coding and debugging time and product size [Sackman, Erikson, and Grant, 1968]. Superficially, this is easy to explain: An experienced programmer will almost always outperform an entry-level programmer. But that is *not* what Sackman measured. He worked with matched pairs of computer professionals, comparing, say, two individuals with 12 years of experience implementing operating systems in assembler. Another pair of his subjects might be two entry-level programmers, both trained at the same institute and both with only 2 months of programming experience in implementing data capture products. What is most alarming about Sackman's results is that his biggest observed differences, such as the figure of 28 to 1 quoted previously, were between pairs of *experienced* programmers. Thus the difference between the times needed for the two teams of Figure 15.2, namely 20 person-months for team A and 29 person-months for team B, could easily be explained by differences between abilities of individual team members.

This objection can be overcome by repeating the experiment on a different project, but rearranging the members of team A and team B into two new teams. A number of experiments, each lasting probably about a year, would be run, interchanging team members each time. Then statistical methods such as factorial design can be used to separate out the effect of individual abilities [John, 1971]. The trouble with this is that there is no guarantee that all members of both teams will stay at the same organization long enough to complete the experiment. If just one team member resigns, retires, is fired, drops dead, or is transferred to another division, then the experiment will be invalidated. In other words, in a police state it may be possible to ensure that participants do not change jobs for the 5 years or so needed to run a complete set of trials, but in a free country there is virtually no way of ensuring that an experiment can be run long enough with the same subjects to be able to subtract out the effects of individual abilities.

Taking all these difficulties into account, there is apparently no way of conducting acceptable experimental trials to compare directly two methods for programming-in-the-many. For every project, decisions regarding choice of method have to be made by management. These decisions generally have to be made on the basis of inadequate information. But they nevertheless have to be made. Since direct experimentation does not seem to be the answer, alternative ways of gaining information about software development methods are discussed in the next section.

15.4 Alternatives to Controlled Experimentation

One alternative to controlled experimentation is to base decisions regarding choice of method on theoretical, rather than experimental, grounds. For example, the author has observed that four methods are consistently reported in the literature to work well in conjunction with one another [Schach, 1982]. The four methods are composite/structured design (Section 8.1), structured programming (Section 10.3), top-down development (Section 6.14.2), and chief programmer team organization (Section 10.8). He found that there were several common properties underlying all four methods. For example, all four are hierarchical in nature. On theoretical grounds it was concluded that, when developing a product, compatible methods should be chosen. That is to say, a set of methods should be selected that exhibit similar theoretical properties.

Rather than totally abandoning experimentation, an alternative approach is to conduct carefully controlled experiments not on complete methods, but rather on those techniques of software engineering that can be tested in this way. For example, Curtis, Sheppard, and Milliman investigated the extent to which software complexity metrics could be used to predict the time needed to find a fault in each of three different FORTRAN products 25 to 225 lines long. The experiment was conducted on 54 professional programmers. It was shown that both a metric of Software Science and McCabe's cyclomatic number (Section 6.11.2) are better predictors of the difficulty of finding a fault than lines of code [Curtis, Sheppard, and Milliman, 1979]. Another example of experimentation into techniques using professional programmers is the work of Fagan described in Chapter 6. Fagan showed that design inspections are a better testing technique than design walkthroughs [Fagan, 1976].

It is also possible to conduct experiments on code, rather than on human beings. An example of this is the work of Bailey and Basili [Bailey and Basili, 1981]. Eighteen large NASA software products were analyzed in order to be able to predict the cost of developing future products. Twenty-one factors, both objective and subjective, were measured for each product and used to develop a cost prediction metric. In addition to developing a metric accurate to within 1 standard deviation, another result of the experiment was that the cost per line to NASA of reusing existing code was shown to be one-fifth of the cost of writing new code.

The preceding examples are a small fraction of the experimentation that has been carried out. The interested reader is urged to read [Basili, Selby, and Hutchens, 1986], where over 100 experiments in software engineering are referenced. Many of those experiments can be challenged on various grounds.

For example, very few of them have been validated through repetition by a different group of experimenters. In addition, the majority of the experiments were conducted on students, rather than on computer professionals.

These criticisms notwithstanding, it is clear that experimental software engineering is a thriving field of research. Many useful results have been obtained that can indeed assist management in directing the production of software. But much still remains to be done, both in proving beyond all doubt results that are currently open to criticism and in conducting experimentation in those areas of software engineering where the current state of the art can best be described as "intuition."

Chapter Review

The need for experimentation on methods is explained in Section 15.1. While experimentation with respect to programming-in-the-small is an accepted scientific technique (Section 15.2), controlled experimentation on methods for programming-in-the-many is effectively impossible because of factors such as cost (Section 15.3.3) and individual differences between software professionals (Section 15.3.4). In Section 15.4 some alternatives to such experimentation are put forward.

For Further Reading

The paper by Basili, Selby, and Hutchens categorizes more than 100 experiments in software engineering [Basili, Selby, and Hutchens, 1986]. In addition, a brief description is given of many of them. This paper is therefore a fund of information regarding experiments that have been performed in software engineering.

Many of those experiments were described in *IEEE Transactions on Software Engineering*. Others were described in the proceedings of the various IEEE International Conferences on Software Engineering. Both these sources should be consulted when researching software engineering experiments.

Problems

15.1 A professor of software engineering suggests that an alternative to controlled experimentation is careful examination of estimates of time, cost, and resource requirements, and the reasons given for deviating from those estimates. He feels that if this procedure is carried out consistently for a number of years and a sufficiently large number of projects is tracked, then statistical techniques can be used to deduce the principles of software engineering that are too expensive to test by means of experimentation-in-the-many. Will this work? Give careful reasons for your answer.

15.2 Design an experiment to test whether knowledge of software engineering is needed to develop Ada software. Estimate the cost of running your experiment.

15.3 (Readings in Software Engineering) Your instructor will distribute copies of [Basili, Selby, and Hutchens, 1986]. Decide for yourself which of the experimental results presented there were proved to your satisfaction, giving reasons for your conclusions.

References

[Bailey and Basili, 1981] J. W. BAILEY AND V. R. BASILI, "A Meta-Model for Software Development Resource Expenditures," *Proceedings of the Fifth International Conference on Software Engineering*, San Diego, CA, 1981, pp. 107–116.

[Basili, Selby, and Hutchens, 1986] V. R. BASILI, R. W. SELBY, AND D. H. HUTCHENS, "Experimentation in Software Engineering," *IEEE Transactions on Software Engineering* **SE-12** (July 1986) pp. 733–743.

[Boehm, Gray, and Seewaldt, 1984] B. W. BOEHM, T. E. GRAY, AND T. SEEWALDT, "Prototyping Versus Specifying: A Multi-Project Experiment," *IEEE Transactions on Software Engineering* **SE-10** (May 1984), pp. 290–303.

[Curtis, Sheppard, and Milliman, 1979] B. CURTIS, S. B. SHEPPARD, AND P. MILLIMAN, "Third Time Charm: Stronger Prediction of Programmer Performance by Software Complexity Metrics," *Proceedings of the Fourth International Conference on Software Engineering*, Munich, West Germany, 1979, pp. 356–360.

[Fagan, 1976] M. E. FAGAN, "Design and Code Inspections to Reduce Errors in Program Development," *IBM Systems Journal* **15** (No. 3, 1976), pp. 182–211.

[John, 1971] P. W. M. JOHN, *Statistical Design and Analysis of Experiments*, Macmillan, New York, NY, 1971.

[Myers, 1978a] G. J. MYERS, "A Controlled Experiment in Program Testing and Code Walkthroughs/Inspections," *Communications of the ACM* **21** (September 1978), pp. 760–768.

[Sackman, Erikson, and Grant, 1968] H. SACKMAN, W. J. ERIKSON, AND E. E. GRANT, "Exploratory Experimental Studies Comparing Online and Offline Programming Performance," *Communications of the ACM* **11** (January 1968), pp. 3–11.

[Schach, 1982] S. R. SCHACH, "A Unified Theory for Software Production," *Software—Practice and Experience* **12** (July 1982), pp. 683–689.

Chapter 16 Automatic Programming

For many software engineers, the future lies in *automatic programming*. That is to say, instead of software being developed by software engineers, the entire process will be automated. If a user wants a software product, then that user will sit at a terminal and answer a series of questions posed by an automatic programming tool. A typical question might be: What do you want the product to do? The user might reply: keep track of the inventory in 17 warehouses, or compute my tax return, or compute the energy levels for U^{238} using the liquid drop model, or generate a radar interference pattern. Each answer is not the answer of a software professional, but rather that of a user who knows little or nothing about computers. Furthermore, each answer is couched in the user's terms and assumes a technical knowledge on the part of the automatic programming tool with regard to the application area, be it warehouses, income tax, nuclear physics, or radar. Compared to the way a software professional would respond, the user's answer is incomplete and imprecise. Also, details have been omitted. In response to such an answer, the tool will ask a further question in order to obtain clarification. A series of questions and answers follows, until eventually the automatic programming tool has determined the user's requirements. It then outputs a software product that does exactly what the user wants.

Is this scenario fact or fantasy? There is unanimous agreement that, as far as the twentieth century is concerned, the idea of a user-friendly, automatic programming tool that can output products in any domain, be it complex scientific programming or real-time embedded software, is nothing more than idle speculation. But some computer scientists believe that current research in artificial intelligence could lead to such a scenario being possible, albeit some time in the future. In their opinion, in 20 or 50 years time, say, there will be no need for software engineers—once we have an automatic programming tool, it can be used to generate all other software, including more advanced automatic programming tools. At the same time, other computer researchers are equally convinced that such an automatic programming tool can never be built. Just as the alchemists of the Middle Ages strove in vain to find the philosophers' stone that could turn base metals into gold, so, they believe, the concept of a tool that can generate any type of software product to its user's total satisfaction is a futile quest.

While most people have 20/20 hindsight, attempting to determine the future is an inexact science. Instead of trying to predict what software engineering will be like in the year 2020, a description will be given of some of the research that has led people to conclude that automatic programming is, or is not, a viable proposition.

16.1 Proposed Automated Life Cycle

The term "automatic programming" was used previously without any sort of definition. The term has a long history. When software was written exclusively in machine code, assembler was considered to be automatic programming. Then the term was used to describe early FORTRAN compilers. It has been suggested that automatic programming is always "the next step," that is, whatever is just beyond the current state of the art.

One view of automatic programming is as part of a future automated life cycle [Balzer, Cheatham, and Green, 1983]. This view will be illustrated by comparison with the current way of producing software.

When software is developed, requirements are obtained, and the specifications are drawn up. Using these specifications, a design is produced. The process of constructing the design is largely informal, unless a method-based environment (Section 12.2.4) is used that restricts the designer to following the steps of a particular method. But even when a method-based environment is used, the design process itself is virtually always undocumented. That is to say, the starting point, namely the specifications, and the finishing point, namely the design, are indeed documents, but the process of going from specifications to design is rarely recorded in any form. This lack of documentation of the development process can adversely affect maintenance. If perfective or adaptive maintenance has to be performed, a change to the requirements is made. This change is then implemented via a change to the specifications followed by a change to the design. But changing the design can be a problem if there is no record of the various design decisions, nor any explanation as to why they were made when the original specifications were used to develop the original design.

A similar problem arises with regard to implementation decisions. A programmer is given the design of a module and is told to implement it. What frequently happens is that there are a number of different ways of implementing that module, and the programmer has to decide which would be the best way. Very rarely indeed is any record kept of the alternatives that the programmer considered, and why he or she opted for one alternative rather than another. Again, when a module is maintained, at the very least a record of these decisions could help to steer the maintenance programmer away from poor choices of implementation strategy.

A further problem with source code is that it frequently has been optimized in some way. For example, the original implementation may be correct, but it runs too slowly. In order to speed it up, changes are made to the product as a whole. Such changes in one module may make use of information in another module. The resulting code is hard to understand, because reading just the one optimized module will not clarify why it was implemented the way it was. At the same time, the original programmer is exceedingly unlikely to have provided a list of the other modules that have to be understood before the optimizations performed on the one module make any sense.

In fact, a maintenance programmer is lucky to have any documentation at all. What happens all too often in practice is that the source code constitutes the sole documentation. There is then no record for the maintenance programmer of any decisions, let alone an explanation of why they were made. This exacerbates the maintenance problems described in Chapter 11.

The way software is currently produced is shown in Figure 16.1. During the requirements phase the client's needs are determined. The next phase is the specifications phase; the client's needs are expressed in the form of the specifications. The design phase consists of producing the design using the specifications as a basis. During implementation, the design document of each module is used as the basis for coding that module. The basis for maintenance is the source code. During the maintenance phase, the specifications and design, if they still exist, may be changed, but the objective of the maintenance phase is to change the source code. Contrast this with the proposed automated life cycle shown in Figure 16.2. Here the output from the specifications phase is the *formal* specifications. These formal specifications are then used by the automatic tool to create the desired product. Maintenance is performed on the specifications and not the source code; the new source code is created from the modified formal specifications. This process is sometimes termed *maintenance by replay*.

There is still one problem that this life cycle does not handle. The automatically created product may not be efficient, and some sort of optimization may be needed to speed up the product and/or reduce its size. For example, a more efficient algorithm or a specific data structure may be needed in one module. However, if this optimization is applied to the source code, then maintenance would also have to be applied to the source code in order to retain the optimization information in the new version. This would have the effect of nullifying the greatest strength of automatic programming, the fact

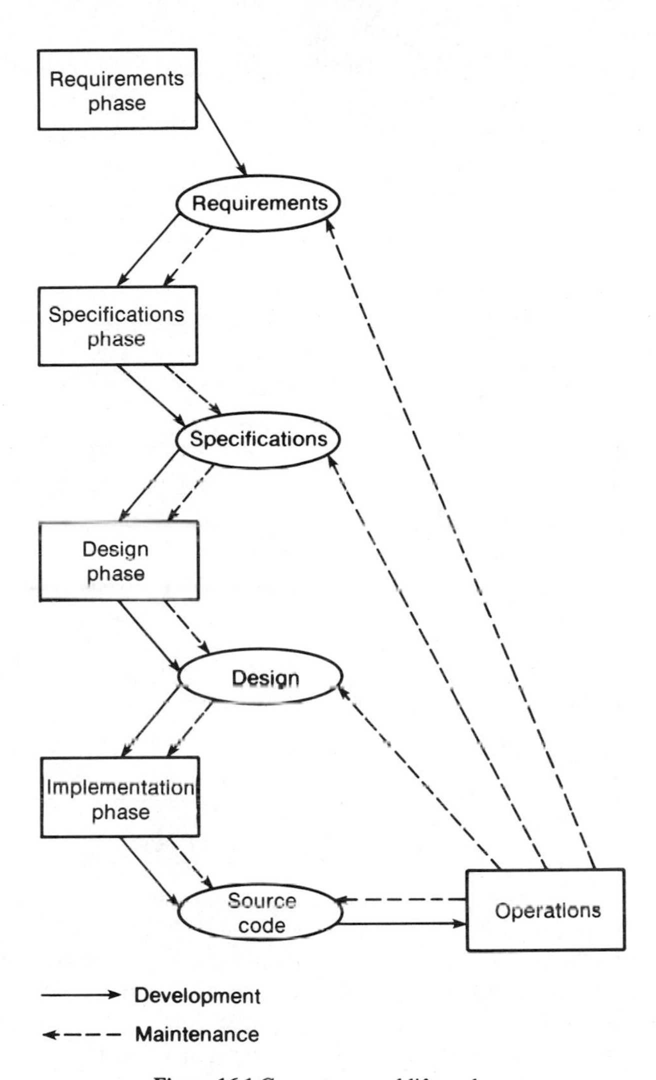

Figure 16.1 Current manual life cycle.

that maintenance is applied to the specifications, not the source code. The solution to this is to set up a file of optimization strategies, as shown in Figure 16.3. When the automatic tool is called upon to create a new version of the source code, it first consults the optimization file so that the necessary strategies, such as algorithms and/or data structures, can be incorporated in every version of the code. It may also be necessary during the maintenance phase to make changes to the file of optimization strategies; this is also reflected in Figure 16.3.

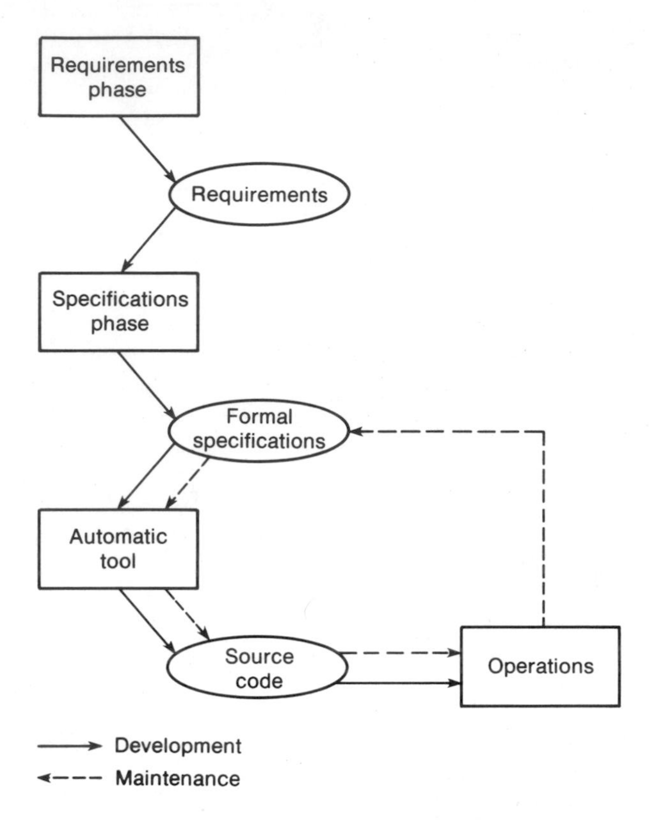

Figure 16.2 Proposed automated life cycle.

This scheme has a number of advantages. First, the difficulties caused by maintaining the source code will disappear. There is no problem with missing design decisions or implementation decisions, because both the design and the implementation are automatically created from the formal specifications, using the optimization strategies. There is no problem with missing documentation, because the formal specifications and the optimization strategies are all the documentation that is needed. Maintenance is performed on the highest conceptual level, namely the formal specifications, rather than the lowest conceptual level, the source code. This has the effect of simplifying maintenance. Also, the problem of trying to understand optimized code falls away. First, the code is never examined. Second, the optimization strategies, instead of being implemented in the code, are explicitly stated in a separate file.

Of course, this scheme also has its disadvantages. This proposed life cycle is constructed around the formal specifications, which have to be

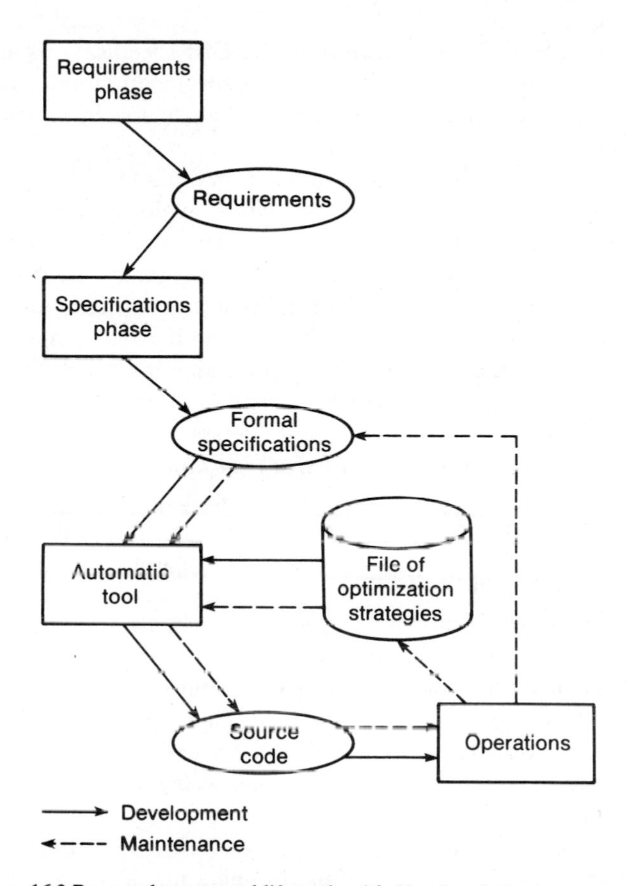

Figure 16.3 Proposed automated life cycle with file of optimization strategies.

expressed in some formal specification language. In Chapter 7 a case was made for formal specifications, but several difficulties of formal specifications were also pointed out. Furthermore, in the previous paragraph it was stated that maintenance is simplified because it is applied to the highest conceptual level, the specifications. But if the specifications are difficult to understand as is frequently the case with formal specifications, this is a mixed blessing.

16.2 Towards an Automated Life Cycle

For the past 20 years, Balzer has been working towards implementing automatic programming [Balzer, 1985]. The formal specification language for his system is called Gist [Swartout, 1983]. The idea behind Gist is that users should be able to write formal specifications that are conceptually similar to the product that they wish to describe, and that the resulting specifications should therefore be easy to read. In practice, Gist specifications are as difficult to read as those written in any other formal specification language. As Balzer explains it, a formal language excludes mechanisms such as summaries and overviews, alternative points of view, diagrams, and examples [Balzer, 1985]. To try to overcome this problem, a Gist paraphraser was constructed to translate Gist specifications into English. Not only does this paraphraser make the specifications readable, but it helps uncover errors in specifications. The paraphraser is not merely a translator, but can reorganize the material to make it more understandable. Use of a paraphraser may help in the future to overcome some of the problems with formal specification languages.

The state of the art in automatic programming has not yet reached the point where researchers are able to construct a tool that takes a formal specification as input and creates executable source code as output. Instead, one of the mechanisms used is *successive transformations*. The formal specifications undergo a series of transformations applied stepwise to specific constructs within the product. Eventually, the specifications are completely transformed into executable source code. The transformation process is by no means fully automated; human intervention is needed in deciding which transformation to apply next, and precisely where to apply that transformation. Balzer's group has addressed not just the issue of successively transforming the specifications into the desired product, but also of describing the series of transformations performed. Using their language Paddle [Wile, 1983], a record can be kept of the sequence of transformations that the specifications have undergone in the course of transformation from Gist to executable code (Lisp). Changes to the transformation sequence can be achieved by modifying the Paddle record of formal development and then replaying it. This represents a first step towards the idea of modifying a file of optimization strategies as shown in Figure 16.3.

Balzer's work is an attempt to achieve automatic programming, that is, an automated tool that can create products in any application domain. Although progress has been made, there is still a considerable way to go.

16.3 Domain-Specific Automatic Programming

Instead of trying to achieve full automatic programming, other researchers have aimed at intermediate goals, such as automatic programming in a restricted domain. The more restricted the domain, the easier it is to produce an automatic programming tool. For example, report generators and screen generators have been available for well over a decade. A report generator generates the code to produce a printed report. All that the user needs to provide are details like the format of the report, the heading, the data to be processed, the primary control break, secondary control break, and so on. The principle behind a report generator is that one data-processing report is very much like any other report. If the user provides the relevant information, it is relatively straightforward to generate a module that will print the desired report. Similar observations can be made about screen generators, tools that can be used to generate screens for inputting data in business data-processing applications.

A report generator is an automatic programming tool in a very small domain. An example of automatic programming in a somewhat larger domain would be a tool that automatically creates business data-processing software products. A promising line of research in a different domain is the ΦNIX ("phi-nix") project [Barstow, 1985]. The application area is oil-well logging. An important oil-prospecting technique is to drill a hole into a rock formation where oil and/or natural gas are suspected. Then an instrument at the end of a long cable, a logging tool, is lowered into the hole. The logging tool can measure petrophysical properties such as the resistivity of the rock surrounding the borehole or the time for sound to travel through the rock (sonic transit time). As the logging tool is raised to the surface, a series of measurements is made and plotted against the depth of the tool in the borehole. The resulting graph is called an oil-well log. A geological or petroleum engineer then analyzes the log in order to determine the types of minerals in the formation, both solid and fluid, and their relative proportions.

An example of a product that could be generated when the ΦNIX project has been successfully completed is one which takes as input logs of resistivity and sonic transit time and produces as output an estimate of the materials in the formation. In order to do this, the generator must have at its disposal information regarding physical laws, such as resistivity is inversely proportional to the square of the water saturation of the rock. It is also necessary to supply a list of assumptions that are to be made by the generator, such as the rock surrounding the borehole is sandstone, and the fluid through which the logging tool passes is a mixture of hydrocarbons and water. Another example of a possible product within the ΦNIX domain of oil-well logging is the embedded software to control the logging tool that measures the sonic transit time.

16.4 Programming Assistants

Rather than restrict the domain, an alternative approach is to restrict the power of the tool. That is, instead of building an automatic programming tool, a programmer's assistant is built. The assistant then handles as much of the drudge work as possible, leaving the creative work to the human software engineer.

A number of assistants of this kind have been constructed. The Programmer's Apprentice is an intelligent editor [Waters, 1982]. The software engineer is expected to decide the hard aspects, such as what should be done and what algorithms should be used, while the Programmer's Apprentice assists in any way it can. A major strength of the system is that it incorporates a number of templates, termed *clichés*, and it understands their semantics. A cliché is essentially a standard technique of dealing with a task, such as computing a square root or determining whether two numbers are equal within some small quantity 10^{-5}, say. From the data-processing viewpoint, clichés include enumerating the elements of a file or printing a simple report.

Suppose a report is to be printed. One piece of information that must be supplied to the cliché is the title that will appear at the top of every page. In this respect, a cliché is merely a macro. But another required piece of information is the enumerator, that is, the mechanism for enumerating the elements that will be printed in the report. This enumerator can itself be a cliché; for example, it can be the cliché for enumerating the elements of a file. Whereas macros simply implement substitution of a dummy variable by an actual variable, such as replacing dummy variable REPORT_TITLE by a specific character string, the Programmer's Apprentice allows clichés to be combined. For this to be feasible, the Programmer's Apprentice has to understand the semantics of each cliché, and hence goes far beyond what a macro generator can do. For example, the line limit in a report is initialized to 64 lines per page, based on the assumption that there are 66 lines on a page, with the title occupying the first line and the second line blank. If the user specifies a long REPORT_TITLE that will occupy two lines, then the line limit in the body of the report is automatically changed to 63 lines per page.

The Programmer's Apprentice is large and is implemented in Lisp. For example, one component of the Programmer's Apprentice is KBEmacs (knowledge-based apprentice in Emacs), which consists of some 40,000 lines of Lisp [Waters, 1985]. As a result it is slow, too slow to be used in the practical software production environment. In addition, it incorporates far fewer than the 1000 or so clichés that are estimated to be necessary for the Programmer's Apprentice to be of use outside the research environment.

Other forms of experimental assistant have been constructed. For example, a testing assistant has been built that helps a programmer to select, execute, and modify test cases [Chapman, 1982]. Assistants have also been constructed for product development and maintenance [Ambras and O'Day, 1988; Kaiser, Feiler, and Popovich, 1988]. Overall, research into assistants could eventually lead to an automatic programming tool if a way could be found to reduce the human component and increase the role played by the computer.

16.5 Case against Automatic Programming

All the types of automatic programming tool described previously make use of artificial intelligence (AI) techniques. All might be described as

```
if
    the animal is a mammal and
    it has pointed teeth and
    it has claws and
    its eyes point forward
then
    it is a carnivore
```

Figure 16.4 Example of a rule.

expert systems, or *knowledge-based systems*, or *rule-based systems*, three more-or-less synonymous terms. A brief introduction to rule-based systems is now given.

16.5.1 Rule-Based Systems

A *rule-based system* is a software product that incorporates a set of rules, the *rule base*, plus a software system for manipulating the rule base, the *inference engine*. An example of a rule is shown in Figure 16.4. A rule consists of two parts: the precondition and the postcondition. In the example, the precondition consists of four separate conditions, all of which have to hold for the precondition to be true; the postcondition, "it is a carnivore," holds if the precondition is true. In general, there will be a number of rules stored in the rule base. The inference engine finds a rule whose precondition is true, and that rule then *fires*. The state of the rule-based system is then modified to reflect the fact that the postcondition is now true. In this instance, if the animal is a mammal with pointed teeth, claws, and forward-pointing eyes, then the fact that the animal is a carnivore will now be associated with that animal.

There is some opposition to the use of the term "expert system." Consider the example of driving a car. A novice uses rules like "when the speed of the car reaches 15 miles per hour, change from first gear into second gear." The novice follows such rules blindly. In contrast, an expert driver reacts instinctively. He or she does not break down a situation into components to see which rule can be applied. Instead, a complete situation is recalled. For example, when an expert driver rounds a corner at speed and the road is wet, the decision as to whether to brake, change to a lower gear, or merely take the foot off the accelerator is done instinctively, and not on the basis of rules.

Similar arguments hold for the game of chess. A novice has a set of rules such as "never exchange your queen for an opponent's pawn, because a queen is more valuable than a pawn." A novice adheres strictly to such rules—a queen sacrifice is out of the question for a novice. A chess expert, on the other hand, sees the position as a whole, not in terms of its components. As a result, an expert simply does not use rules. Incidentally, even the best chess-playing machines cannot beat the best human players.

Because almost every expert system uses rules but human experts do not, Hubert and Stuart Dreyfus prefer to use the term *novice system*, rather than *expert system* [Dreyfus and Dreyfus, 1986b]. To bolster this terminology, they point out that in those areas where human know-how is needed to deliver judgement, human experts outperform rule-based systems. The one exception seems to be R1 (now called XCON), a rule-based system, which aids in the configuration of VAX computers. It determines from the customer's purchase order what substitutions and additions are needed to make the order consistent and complete. It also produces a set of diagrams for organizing the components such as CPUs, memory, and disks. The product, which comprises over 10,000 rules [Barker and O'Connor, 1989], reduces the time needed to configure an order and produces accurate configurations. The point is that one does not have to be an expert to configure VAX computers; it can be done on the basis of a set of rules. The reason that XCON works so well is that there are so many possible combinations that even humans use rule-based techniques to decide on the configuration of an order. Thus all that XCON is doing is copying a human activity, and not surprisingly it carries out its tasks much more quickly and more accurately than a human could. In this respect, XCON is like a computerized accounts receivable product. Such a product bills debtors and credits their payments exactly the same way a human clerk would, but, given the correct input data, the computer does it faster and more accurately than a human.

Not all expert systems are necessarily rule-based. For example, a current research thrust in AI is neural networks, and a future expert system based on neural networks rather than rules cannot prima facie be termed a "novice system." Nevertheless, the current state of the art in knowledge-based software engineering is such that it is difficult to contradict those who would claim that the goal of an automatic programming tool that can interact with a novice user and then output a product in any application domain can never be achieved.

Whether or not the goal can be reached, the struggle to develop such an automatic tool has been a fruitful one in that a number of exciting spin-offs have resulted from the research. Most areas of software engineering have been enriched as a consequence of researchers' efforts. For example, we now understand aspects of specifications, implementation, testing, and even maintenance that we might not have discovered had it not been for research into automatic programming. It is true that if an automatic programming tool were developed, it would instantly put every software engineer permanently out of work. Nevertheless, the struggle towards automatic programming, even if the goal can never be attained, can only enrich the field of software engineering.

Chapter Review

In this chapter an automatic programming life cycle is proposed (Section 16.1). While it is clearly a future scenario (Section 16.2), automatic programming restricted to a single domain (Section 16.3) or an intelligent programming assistant (Section 16.4) seem to be more reasonable initial goals. Finally, the viability of automatic programming is discussed in Section 16.5.

For Further Reading

The paper [Frenkel, 1985] provides a somewhat nontechnical overview of automatic programming. An evaluation of expert systems is to be found in [Bobrow, Mittal, and Stefik, 1986].

The work of Balzer is described in [Balzer, Cheatham, and Green, 1983] and [Balzer, 1985]. Details about Gist appear in [Swartout, 1983], and Paddle is described in [Wile, 1983]. A detailed description of XCON is to be found in [Barker and O'Connor, 1989]. The ΦNIX project is described in [Barstow, 1985]. Various aspects of the programmer's apprentice project can be found in [Waters, 1982], [Waters, 1985] and in the November 1988 issue of *IEEE Computer*. Other types of assistants are described in [Singer, Ledgard, and Hueras, 1981], [Chapman, 1982], [Ambras and O'Day, 1988], [Kaiser, Feiler, and Popovich, 1988], and [Karimi and Konsynski, 1988].

The case against expert systems is clearly made in [Dreyfus and Dreyfus, 1986a], as well as in their book, *Mind Over Machine* [Dreyfus and Dreyfus, 1986b]. The interplay between software engineering and artificial intelligence is described in [Simon, 1986]. In the November 1985 issue of *IEEE Transactions on Software Engineering* a variety of other papers regarding the interaction between software engineering and artificial intelligence can be found.

Problems

16.1 Suppose that an automatic programming tool has been developed that can work in only one domain, namely to generate further automatic programming tools. How could such a tool be used?

16.2 Do you believe that automatic programming will be achieved in your lifetime? Give careful reasons for your answer.

16.3 Once automatic programming tools have been developed, what will software engineers do?

16.4 (Readings in Software Engineering) Your instructor will distribute copies of [Dreyfus and Dreyfus, 1986a]. Do you agree with the Dreyfus brothers' views? Justify your answer.

References

[Ambras and O'Day, 1988] J. AMBRAS AND V. O'DAY, "MicroScope: A Knowledge-Based Programming Environment," *IEEE Software* **5** (May 1988), pp. 50–58.

[Balzer, 1985] R. BALZER, "A 15 Year Perspective on Automatic Programming," *IEEE Transactions on Software Engineering* **SE-11** (November 1985), pp. 1257–1268.

[Balzer, Cheatham, and Green, 1983] R. BALZER, T. E. CHEATHAM, JR., AND C. GREEN, "Software Technology in the 1990's: Using a New Paradigm," *IEEE Computer* **16** (November 1983), pp. 39–45.

[Barker and O'Connor, 1989] V. E. BARKER AND D. E. O'CONNOR, "Expert Systems for Configuration at Digital: XCON and Beyond," *Communications of the ACM* **32** (March 1989), pp. 298–318.

[Barstow, 1985] D. R. BARSTOW, "Domain-Specific Automatic Programming," *IEEE Transactions on Software Engineering* **SE-11** (November 1985), pp. 1321–1336.

[Bobrow, Mittal, and Stefik, 1986] D. G. BOBROW, S. MITTAL, AND M. J. STEFIK, "Expert Systems: Perils and Promise," *Communications of the ACM* **29** (September 1986), pp. 880–894.

[Chapman, 1982] D. CHAPMAN, "A Program Test Assistant," *Communications of the ACM* **25** (September 1982), pp. 625–634.

[Dreyfus and Dreyfus, 1986a] H. DREYFUS AND S. DREYFUS, "Why Computers May Never Think Like People," *Technology Review* **89** (January 1986), pp. 43–61.

[Dreyfus and Dreyfus, 1986b] H. L. DREYFUS AND S. E. DREYFUS, *Mind Over Machine*, The Free Press, New York, 1986.

[Frenkel, 1985] K. A. FRENKEL, "Toward Automating the Software-Development Cycle," *Communications of the ACM* **28** (June 1985), pp. 578–589.

[Kaiser, Feiler, and Popovich, 1988] G. E. KAISER, P. H. FEILER, AND S. S. POPOVICH, "Intelligent Assistance for Software Development and Maintenance," *IEEE Software* **5** (May 1988), pp. 40–49.

[Karimi and Konsynski, 1988] J. KARIMI AND B. R. KONSYNSKI, "An Automated Design Assistant," *IEEE Transactions on Software Engineering* **14** (February 1988), pp. 194–210.

[Simon, 1986] H. A. SIMON, "Whether Software Engineering Needs to be Artificially Intelligent," *IEEE Transactions on Software Engineering* **SE-12** (July 1986), pp. 726–732.

[Singer, Ledgard, and Hueras, 1981] A. SINGER, H. LEDGARD, AND J. F. HUERAS, "The Annotated Assistant: A Step Towards Human Engineering," *IEEE Transactions on Software Engineering* **SE-7** (July 1981), pp. 353–374.

[Swartout, 1983] W. SWARTOUT, "The Gist Behavior Explainer," *Proceedings of the National Conference on Artificial Intelligence AAAI-83*, 1983, pp. 402–407.

[Waters, 1982] R. C. WATERS, "The Programmer's Apprentice: Knowledge-Based Program Editing," *IEEE Transactions on Software Engineering* **SE-8** (January 1982), pp. 1–12.

[Waters, 1985] R. C. WATERS, "The Programmer's Apprentice: A Session with KBEmacs," *IEEE Transactions on Software Engineering* **SE-11** (November 1985), pp. 1296–1320.

[Wile, 1983] D. S. WILE, "Program Developments: Formal Explanations of Implementations," *Communications of the ACM* **26** (November 1983), pp. 902–911.

Appendix

Plain Vanilla Ice Cream Corporation

Plain Vanilla Ice Cream Corporation (PVIC) manufactures and sells only one product, namely the most delicious vanilla-flavored ice cream imaginable. The ice cream is made in the PVIC factory in Roomys, Wisconsin, and sold exclusively at the 23 PVIC stores, 7 in Wisconsin, 10 in Illinois, and 6 in Michigan. PVIC stores sell nothing but vanilla-flavored ice cream. Notwithstanding the fact that Plain Vanilla Ice Cream is an outstanding product, PVIC profits have steadily decreased over the past 4 years, so much so that management and shareholders alike are worried about the long-term viability of the corporation. A management consultant has been called in, and she recommends that PVIC implement *management by objectives*.

Each store is assigned an objective, a sales target for each month. For example, Blanche Weiss, the store manager at the PVIC store in Gleeda, Michigan, has been told that her target for January is to sell 40,000 gallons (40 kgal) of ice cream. For February, her sales target is 50 kgal. The target figures for the complete year appear in Figure A.1.

The actual sales for January at the Gleeda store are only 22 kgal. *Effectiveness*, a measure of how closely the objective has been attained, is defined as follows:

$$\text{Effectiveness} = \frac{\text{actual sales}}{\text{target sales}} \times 100\%$$

Thus the effectiveness of the Gleeda store for the month of January is $22/40 \times 100$, or 55%.

In February, actual sales rise to 40 kgal. Since the February target is 50 kgal, February effectiveness is 80%. Blanche Weiss receives a visit from the

	Jan	Feb	Mar	Apr	May	Jun	Jul	Aug	Sep	Oct	Nov	Dec
Target sales (kgal)	40	50	50	60	60	65	70	75	80	80	85	90
Actual sales (kgal)	22	40	75									
Effectiveness	55%	80%	150%									
Shortfall	45%	20%	—									
Objective achieved	No	Yes	Yes									

Figure A.1 Data for management by objectives through March for Plain Vanilla Ice Cream Store at Gleeda, Michigan.

regional sales manager for Michigan. He is furious. For 2 months in a row, Blanche has failed to make her sales target. Blanche protests, pointing out that her effectiveness has increased from 55% to 80%. The regional sales manager suddenly realizes that effectiveness alone is not an adequate measure of performance; improvement from month to month must also be taken into account.

PVIC management then decides on the following policy. If a store fails to achieve 100% efficiency in a given month, the efficiency *shortfall* for that month is defined to be the difference between 100% efficiency and the actual efficiency. Thus the January shortfall for the Gleeda store is (100% − 55%), or 45%, while the February shortfall is 20%. A store manager is deemed to have achieved his or her objective for the month if an efficiency of at least 95% is attained, or if the shortfall for the month is less than half of the previous month's shortfall.

Thus, in the case of the PVIC store in Gleeda, Blanche did not achieve her objective in January because her efficiency was only 55%. But even though her efficiency in February was only 80%, her February shortfall, namely 20%, was less than half of her January shortfall of 45%. Because of this improvement, in terms of company criteria she has achieved her February objective. In March, her sales of 75 kgal exceed her target of 50 kgal. Her efficiency is 150%, and again she achieves her objective.

The management by objectives plan works so well in all three states that PVIC management decides that the system should be computerized. Each month the following reports must be printed.

1. Store Report

A one-page report containing the following information must be printed each month for each store. The report must be printed on three-part paper. One copy is to be sent to the store manager, one to the relevant regional sales manager, and one to the vice-president for sales at the company's headquarters in Roomys.

Store number (3 digits)
Store street name and number (up to 25 characters)
Store city (up to 25 characters)
Store state (2 characters)
Store ZIP code (9 digits)
Store telephone number, including area code (10 digits)
Store manager's name
Store region (state) (2 letters)
Sales target, in kgal, for January through December (4 digits)
Actual sales, in kgal, for January through current month (4 digits)
Efficiency, for January through current month (3 digits plus % sign)
Shortfall, for January through current month (2 digits plus % sign), or "—" if efficiency $\geq$ 100%
Whether objective has been achieved, for January through current month (3 characters)
If objective has not been achieved this month, a one-line message stating this

2. Report for Regional Sales Manager

PVIC stores are organized into regions, corresponding to the states in which PVIC sells ice cream. A one-page report containing the following information must be printed each month for each region (state). The report must be printed on two-part paper, one copy for the regional sales manager, and the other for the vice-president for sales in Roomys.

Name of region (state) (2 letters)
Name of regional sales manager
Address of regional sales manager (2 lines)
Telephone number of regional sales manager, including area code (10 digits)
For each store in the region (state), one line containing:
- Store number (3 digits)
- Store street name and number (up to 25 characters)
- Store city (up to 25 characters)
- Target for this month, in kgal (4 digits)
- Actual sales for this month, in kgal (4 digits)
- Efficiency for this month (3 digits plus % sign)
- Whether or not objective was achieved this month (3 characters)

Number of stores in that region (3 digits)
Number of stores achieving their objective this month (3 digits)
Number of stores not achieving their objective this month (3 digits)
Percentage of stores not achieving their objective this month (3 digits plus % sign)

3. Report for Vice-President for Sales

A one-page report containing the following information must be printed each month.

For each region (state) in which PVIC sells ice cream, one line containing:
Name of region (state) (2 letters)
Number of stores in that region (3 digits)
Number of stores achieving their objective this month (3 digits)
Number of stores not achieving their objective this month (3 digits)
Percentage of stores not achieving their objective this month (3 digits plus % sign)
One line containing the following information for all stores combined:
Total number of PVIC stores (3 digits)
Total number of stores achieving their objective this month (3 digits)
Total number of stores not achieving their objective this month (3 digits)
Percentage of stores not achieving their objectives this month (3 digits plus % sign)

Input Routines

The following input routines must be written. Initially, information regarding stores and their managers must be input. This includes the number, address, and telephone number of each store, as well as the name of its manager. A list of region names, as well as the name, address, and telephone number of the relevant regional sales managers, must also be provided. Then, at the beginning of each year, the sales target for each store for each month must be provided. At the beginning of every month, sales figures for the previous month for each store must be input.

Update Routines

The preceding routines are required as a matter of urgency. Once the product is working, update routines will have to be written for changing any of the information stored in the files. The update routines must incorporate authorization checks to ensure that information is changed only by managers who are entitled to do so, and a record of any such changes must be kept. Once a week, a report listing all changes, and who authorized them, must be delivered to the vice-president for sales.

Bibliography

The chapter number in parentheses denotes the chapter in which the item has been referenced.

[Abbott, 1983] R. J. Abbott, "Program Design by Informal English Descriptions," *Communications of the ACM* **26** (November 1983), pp. 882–894. (Chapter 9)

[Alavi, 1984] M. Alavi, "An Assessment of the Prototyping Approach to Information Systems Development," *Communications of the ACM* **27** (June 1984), pp. 556–563. (Chapter 3)

[Albrecht, 1979] A. J. Albrecht, "Measuring Application Development Productivity," *Proceedings of the IBM SHARE/GUIDE Applications Development Symposium*, Monterey, CA, October 1979, pp. 83–92. (Chapter 4)

[Albrecht and Gaffney, 1983] A. J. Albrecht and J. E. Gaffney, Jr., "Software Function, Source Lines of Code, and Development Effort Prediction: A Software Science Validation," *IEEE Transactions on Software Engineering* **SE-9** (November 1983), pp. 639–648. (Chapter 4)

[Alford, 1985] M. Alford, "SREM at the Age of Eight; The Distributed Computing Design System," *IEEE Computer* **18** (April 1985), pp. 36–46. (Chapter 7)

[Ambras and O'Day, 1988] J. Ambras and V. O'Day, "MicroScope: A Knowledge-Based Programming Environment," *IEEE Software* **5** (May 1988), pp. 50–58. (Chapter 16)

[ANSI X3.9, 1977] "Programming Language FORTRAN," ANSI X3.9-1977, American National Standards Institute, Inc., 1977. (Chapter 13)

[ANSI/IEEE 729, 1983] "A Standard Glossary of Software Engineering Terminology," ANSI/IEEE 729-1983, American National Standards Institute, Inc., Institute of Electrical and Electronic Engineers, Inc., 1983. (Chapter 6)

[ANSI/IEEE 770X3.97, 1983] "Pascal Computer Programming Language," ANSI/IEEE 770X3.97-1983, American National Standards Institute, Inc., Institute of Electrical and Electronic Engineers, Inc., 1983. (Chapters 13 and 14)

[ANSI/IEEE 829, 1983] "Software Test Documentation," ANSI/IEEE 829-1983, American National Standards Institute, Inc., Institute of Electrical and Electronic Engineers, Inc., 1983. (Chapter 4)

[ANSI/IEEE 990, 1987] "Recommended Practice for Ada as a Program Design Language," ANSI/IEEE 990-1987, American National Standards Institute, Inc., Institute of Electrical and Electronics Engineers, Inc., 1987. (Chapters 12 and 14)

[ANSI/MIL-STD-1815A, 1983] "Reference Manual for the Ada Programming Language," ANSI/MIL-STD-1815A, American National Standards Institute, Inc., United States Department of Defense, 1983. (Chapters 8, 13, and 14)

[Apple, 1984] *Macintosh Pascal User's Guide*, Apple Computer, Inc., Cupertino, CA, 1984. (Chapter 10)

[Archer and Devlin, 1986] J. E. ARCHER, JR., AND M. T. DEVLIN, "Rational's Experience Using Ada for Very Large Systems," *Proceedings of the First International Conference on Ada Programming Language Applications for the NASA Space Station*, NASA, 1986, pp. B2.5.1–B2.5.12. (Chapter 12)

[Ardö, 1987] A. ARDÖ, "Real-Time Efficiency of Ada in a Multiprocessor Environment," *Proceedings of the International Workshop on Real-Time Ada Issues, ACM SIGAda Ada Letters* **VII** (No. 6, 1987), pp. 40–42. (Chapter 14)

[Aron, 1983] J. D. ARON, *The Program Development Process. Part II. The Programming Team*, Addison-Wesley, Reading, MA, 1983. (Chapter 10)

[Babich, 1986] W. A. BABICH, *Software Configuration Management: Coordination for Team Productivity*, Addison-Wesley, Reading, MA, 1986. (Chapter 11)

[Bailey and Basili, 1981] J. W. BAILEY AND V. R. BASILI, "A Meta-Model for Software Development Resource Expenditures," *Proceedings of the Fifth International Conference on Software Engineering*, San Diego, CA, 1981, pp. 107–116. (Chapter 15)

[Baker, 1972] F. T. BAKER, "Chief Programmer Team Management of Production Programming," *IBM Systems Journal* **11** (No. 1, 1972), pp. 56–73. (Chapter 10)

[Balzer, 1985] R. BALZER, "A 15 Year Perspective on Automatic Programming," *IEEE Transactions on Software Engineering* **SE-11** (November 1985), pp. 1257–1268. (Chapters 3, 7, and 16)

[Balzer, Cheatham, and Green, 1983] R. BALZER, T. E. CHEATHAM, JR., AND C. GREEN, "Software Technology in the 1990's: Using a New Paradigm," *IEEE Computer* **16** (November 1983), pp. 39–45. (Chapter 16)

[Barker and O'Connor, 1989] V. E. BARKER AND D. E. O'CONNOR, "Expert Systems for Configuration at Digital: XCON and Beyond," *Communications of the ACM* **32** (March 1989), pp. 298–318. (Chapter 16)

[Barstow, 1985] D. R. BARSTOW, "Domain-Specific Automatic Programming," *IEEE Transactions on Software Engineering* **SE-11** (November 1985), pp. 1321–1336. (Chapter 16)

[Barstow, Shrobe, and Sandewall, 1984] D. R. BARSTOW, H. E. SHROBE, AND E. SANDEWALL (Editors), *Interactive Programming Environments*, McGraw-Hill, New York, NY, 1984. (Chapter 12)

[Basili and Hutchens, 1983] V. R. BASILI AND D. H. HUTCHENS, "An Empirical Study of a Syntactic Complexity Family," *IEEE Transactions on Software Engineering* **SE-9** (November 1983), pp. 664–672. (Chapter 6)

[Basili and Selby, 1987] V. R. BASILI AND R. W. SELBY, "Comparing the Effectiveness of Software Testing Strategies," *IEEE Transactions on Software Engineering* **SE-13** (December 1987), pp. 1278–1296. (Chapter 6)

[Basili and Weiss, 1984] V. R. BASILI AND D. M. WEISS, "A Methodology for Collecting Valid Software Engineering Data," *IEEE Transactions on Software Engineering* **SE-10** (November 1984), pp. 728–738. (Chapter 6)

[Basili, Selby, and Hutchens, 1986] V. R. BASILI, R. W. SELBY, AND D. H. HUTCHENS, "Experimentation in Software Engineering," *IEEE Transactions on Software Engineering* **SE-12** (July 1986) pp. 733–743. (Chapter 15)

[Behrens, 1983] C. A. BEHRENS, "Measuring the Productivity of Computer Systems Development Activities with Function Points," *IEEE Transactions on Software Engineering* **SE-9** (November 1983), pp. 648–652. (Chapter 4)

[Beizer, 1984] B. BEIZER, *Software System Testing and Quality Assurance*, Van Nostrand Reinhold, New York, NY, 1984. (Chapters 4 and 6)

[Bentley, 1986] J. BENTLEY, *Programming Pearls*, Addison-Wesley, Reading, MA, 1986. (Chapter 10)

[Berliner and Zave, 1987] E. F. BERLINER AND P. ZAVE, "An Experiment in Technology Transfer: PAISLey Specification of Requirements for an Undersea Lightwave Cable System," *Proceedings of the Ninth International Conference on Software Engineering*, Monterey, CA, March 1987, pp. 42–50. (Chapter 7)

[Berry, 1978] D. M. BERRY, Personal Communication, 1978. (Chapter 8)

[Berry, 1985] D. M. BERRY, "On the Application of Ada and its Tools to the Information Hiding Decomposition Methodology for the Design of Software Systems," in: *Methodologies for Computer System Design*, W. K. Giloi and B. D. Shriver (Editors), Elsevier North-Holland, Amsterdam, The Netherlands, 1985, pp. 308–321. (Chapter 8)

[Berry and Wing, 1985] D. M. BERRY AND J. M. WING, "Specifying and Prototyping: Some Thoughts on Why They Are Successful," in: *Formal Methods and Software Development*, *Proceedings of the International Joint Conference on Theory and Practice of Software Development*, *Volume 2*, Springer-Verlag, Berlin, West Germany, 1985, pp. 117–128. (Chapter 6)

[Berzins, Gray, and Naumann, 1986] V. BERZINS, M. GRAY, AND D. NAUMANN, "Abstraction-Based Software Developments," *Communications of the ACM* **29** (May 1986), pp. 402–415. (Chapter 8)

[Bjørner, 1987] D. BJØRNER, "On the Use of Formal Methods in Software Development," *Proceedings of the Ninth International Conference on Software Engineering*, Monterey, CA, March 1987, pp. 17–29. (Chapter 7)

[Blair, Murphy, Schach, and McDonald, 1988] J. A. BLAIR, L. C. MURPHY, S. R. SCHACH, AND C. W. MCDONALD, "Rapid Prototyping, Bottom-Up Design, and Reusable Modules: A Case Study," *ACM Mid-Southeast Summer Meeting*, Nashville, TN, May 1988. (Chapter 3)

[Boar, 1984] B. H. BOAR, *Application Prototyping: A Requirements Definition Strategy for the '80s*, John Wiley and Sons, New York, NY, 1984. (Chapter 3)

[Bobrow and Stefik, 1983] D. G. BOBROW AND M. STEFIK, *The Loops Manual*, Intelligent Systems Laboratory, Xerox Corporation, Palo Alto, CA, 1983. (Chapters 8 and 9)

[Bobrow, Mittal, and Stefik, 1986] D. G. BOBROW, S. MITTAL, AND M. J. STEFIK, "Expert Systems: Perils and Promise," *Communications of the ACM* **29** (September 1986), pp. 880–894. (Chapter 16)

[Boehm, 1973] B. W. BOEHM, "Software and its Impact: A Quantitative Assessment," *Datamation* **19** (May 1973), pp. 48–59. (Chapter 1)

[Boehm, 1976] B. W. BOEHM, "Software Engineering," *IEEE Transactions on Computers* **C-25** (December 1976), pp. 1226–1241. (Chapters 1 and 2)

[Boehm, 1979] B. W. BOEHM, "Software Engineering, R & D Trends and Defense Needs," in: *Research Directions in Software Technology*, P. Wegner (Editor), The MIT Press, Cambridge, MA, 1979. (Chapter 1)

[Boehm, 1980] B. W. BOEHM, "Developing Small-Scale Application Software Products: Some Experimental Results," *Proceedings of the Eighth IFIP World Computer Congress*, October 1980, pp. 321–326. (Chapter 1)

[Boehm, 1981] B. W. BOEHM, *Software Engineering Economics*, Prentice-Hall, Englewood Cliffs, NJ, 1981. (Chapters 1, 3, and 4)

[Boehm, 1983] B. W. BOEHM, "The Hardware/Software Cost Ratio: Is it a Myth?" *IEEE Computer* **16** (March 1983), pp. 78–80. (Chapter 1)

[Boehm, 1984a] B. W. BOEHM, "Software Engineering Economics," *IEEE Transactions on Software Engineering* **SE-10** (January 1984), pp. 4–21. (Chapter 4)

[Boehm, 1984b] B. W. BOEHM, "Verifying and Validating Software Requirements and Design Specifications," *IEEE Software* **1** (January 1984), pp. 75–88. (Chapter 6)

[Boehm, 1988] B. W. BOEHM, "A Spiral Model of Software Development and Enhancement," *IEEE Computer* **21** (May 1988), pp. 61–72. (Chapter 3)

[Boehm and Papaccio, 1988] B. W. BOEHM AND P. N. PAPACCIO, "Understanding and Controlling Software Costs," *IEEE Transactions on Software Engineering* **14** (October 1988), pp. 1462–1477. (Chapters 1 and 2)

[Boehm, Gray, and Seewaldt, 1984] B. W. BOEHM, T. E. GRAY, AND T. SEEWALDT, "Prototyping Versus Specifying: A Multi-Project Experiment," *IEEE Transactions on Software Engineering* **SE-10** (May 1984), pp. 290–303. (Chapters 3 and 15)

[Boehm, Penedo, Stuckle, Williams, and Pyster, 1984] B. W. BOEHM, M. H. PENEDO, E. D. STUCKLE, R. D. WILLIAMS, AND A. B. PYSTER, "A Software Development Environment for Improving Productivity," *IEEE Computer* **17** (June 1984), pp. 30–44. (Chapters 3 and 4)

[Böhm and Jacopini, 1966] C. BÖHM AND G. JACOPINI, "Flow Diagrams, Turing Machines, and Languages with Only Two Formation Rules," *Communications of the ACM* **9** (May 1966), pp. 366–371. (Chapter 10)

[Bologna, Quirk, and Taylor, 1985] S. BOLOGNA, W. J. QUIRK, AND J. R. TAYLOR, "Simulation and System Validation," in: *Verification and Validation of Real-Time Software*, W. J. Quirk (Editor), Springer-Verlag, Berlin, West Germany, 1985, pp. 179–201. (Chapter 6)

[Booch, 1983] G. BOOCH, *Software Engineering with Ada*, Second Edition, Benjamin/Cummings, Menlo Park, CA, 1983. (Chapter 9)

[Booch, 1986] G. BOOCH, "Object-Oriented Development," *IEEE Transactions on Software Engineering* **SE-12** (February 1986), pp. 211–221. (Chapters 8 and 9)

[Brady, 1977] J. M. BRADY, *The Theory of Computer Science*, Chapman and Hall, London, UK, 1977. (Chapter 7)

[Branstad and Powell, 1984] M. BRANSTAD AND P. B. POWELL, "Software Engineering Project Standards," *IEEE Transactions on Software Engineering* **SE-10** (January 1984), pp. 73–78. (Chapter 4)

[Bratley, Fox, and Schrage, 1987] P. BRATLEY, B. L. FOX, AND L. E. SCHRAGE, *Guide to Simulation*, Second Edition, Springer-Verlag, New York, NY, 1987. (Chapter 4)

[Brooks, 1975] F. P. BROOKS, JR., *The Mythical Man-Month: Essays on Software Engineering*, Addison-Wesley, Reading, MA, 1975. (Chapters 1, 2, 3, and 10)

[Brooks, 1986] F. P. BROOKS, JR., "No Silver Bullet," in: *Information Processing '86*, H.-J. Kugler (Editor), Elsevier North-Holland, New York, NY, 1986. Reprinted in: *IEEE Computer* **20** (April 1987), pp. 10–19. (Chapter 2)

[Bruno and Marchetto, 1986] G. BRUNO AND G. MARCHETTO, "Process-Translatable Petri Nets for the Rapid Prototyping of Process Control Systems," *IEEE Transactions on Software Engineering* **SE-12** (February 1986), pp. 346–357. (Chapter 7)

[Budde, Kühlenkamp, Mathiassen, and Züllighoven, 1984] R. BUDDE, K. KÜHLENKAMP, L. MATHIASSEN, AND H. ZÜLLIGHOVEN (Editors), *Approaches to Prototyping: Proceedings of the Working Conference on Prototyping*, Namur, Belgium, October 1983, Springer-Verlag, Berlin, West Germany, 1984. (Chapter 3)

[Buhr, 1984] R. J. A. BUHR, *System Design in Ada*, Prentice-Hall, Englewood Cliffs, NJ, 1984. (Chapter 9)

[Burns and Wellings, 1987] A. BURNS AND A. J. WELLINGS, "Real-Time Ada Issues," *Proceedings of the International Workshop on Real-Time Ada Issues, ACM SIGAda Ada Letters* **VII** (No. 6, 1987), pp. 43–46. (Chapter 14)

[Burton, Aragon, Bailey, Koehler, and Mayes, 1987] B. A. BURTON, R. W. ARAGON, S. A. BAILEY, I. D. KOEHLER, AND L. A. MAYES, "The Reusable Software Library," *IEEE Software* **4** (July 1987), pp. 25–33. (Chapter 14)

[Buzzard and Mudge, 1985] G. D. BUZZARD AND T. N. MUDGE, "Object-Based Computing and the Ada Programming Language," *IEEE Computer* **18** (March 1985), pp. 11–19. (Chapters 8 and 14)

[Cameron, 1986] J. R. CAMERON, "An Overview of JSD," *IEEE Transactions on Software Engineering* **SE-12** (February 1986), pp. 222–240. (Chapter 9)

[Cameron, 1988] J. CAMERON, *JSP & JSD: The Jackson Approach to Software Development*, Second Edition, IEEE Computer Society Press, Washington, DC, 1988. (Chapter 9)

[Card, McGarry, and Page, 1987] D. N. CARD, F. E. MCGARRY, AND G. T. PAGE, "Evaluating Software Engineering Technologies," *IEEE Transactions on Software Engineering* **SE-13** (July 1987), pp. 845–851. (Chapters 1 and 2)

[Carlson, Druffel, Fisher, and Whitaker, 1980] W. E. CARLSON, L. E. DRUFFEL, D. A. FISHER, AND W. A. WHITAKER, "Introducing Ada," *Proceedings of the ACM Annual Conference, ACM 80*, Nashville, TN, 1980, pp. 263–271. (Chapter 14)

[Chandrasekharan, Dasarathy, and Kishimoto, 1985] M. CHANDRASEKHARAN, B. DASARATHY, AND Z. KISHIMOTO, "Requirements-Based Testing of Real-Time Systems: Modeling for Testability," *IEEE Computer* **18** (April 1985), pp. 71–80. (Chapter 7)

[Chapman, 1982] D. CHAPMAN, "A Program Test Assistant," *Communications of the ACM* **25** (September 1982), pp. 625–634. (Chapter 16)

[Chen, 1983] P. P.-S. CHEN, "ER—A Historical Perspective and Future Directions," in: *Entity Relationship Approach to Software Engineering*, C. G. Davis (Editor), North-Holland, Amsterdam, The Netherlands, 1983. (Chapter 12)

[Cherry, 1986] G. CHERRY, *The PAMELA Designer's Handbook*, Thought Tools, Reston, VA, 1986. (Chapter 9)

[Chow, 1985] T. S. CHOW (Editor), *Tutorial: Software Quality Assurance: A Practical Approach*, IEEE Computer Society Press, Washington, DC, 1985. (Chapter 4)

[Clarke, Podgurski, Richardson, and Zeil, 1985] L. A. CLARKE, A. PODGURSKI, D. J. RICHARDSON, AND S. J. ZEIL, "A Comparison of Data Flow Path Selection Criteria," *Proceedings of the Eighth International Conference on Software Engineering*, London, UK, August 1985, pp. 244–251. (Chapter 6)

[Cobb, 1985] R. H. COBB, "In Praise of 4GLs," *Datamation* **31** (July 15, 1985), pp. 90–96. (Chapter 10)

[Cohen, 1986] N. H. COHEN, *Ada as a Second Language*, McGraw-Hill, New York, NY, 1986. (Chapter 14)

[Conte, Dunsmore, and Shen, 1986] S. D. CONTE, H. E. DUNSMORE, AND V. Y. SHEN, *Software Engineering Metrics and Models*, Benjamin/Cummings, Menlo Park, CA, 1986. (Chapter 4)

[Coolahan and Roussopoulos, 1983] J. E. COOLAHAN, JR., AND N. ROUSSOPOULOS, "Timing Requirements for Time-Driven Systems Using Augmented Petri Nets," *IEEE Transactions on Software Engineering* **SE-9** (September 1983), pp. 603–616. (Chapter 7)

[Cornhill, Sha, Lehoczky, Rajkumar, and Tokuda, 1987] D. CORNHILL, L. SHA, J. P. LEHOCZKY, R. RAJKUMAR, AND H. TOKUDA, "Limitations of Ada for Real-Time Scheduling," *Proceedings of the International Workshop on Real-Time Ada Issues, ACM SIGAda Ada Letters* **VII** (No. 6, 1987), pp. 33–39. (Chapter 14)

[Coulter, 1983] N. S. COULTER, "Software Science and Cognitive Psychology," *IEEE Transactions on Software Engineering* **SE-9** (March 1983), pp. 166–171. (Chapters 4 and 5)

[Côté, Bourque, Oligny, and Rivard, 1988] V. CÔTÉ, P. BOURQUE, S. OLIGNY, AND N. RIVARD, "Software Metrics: An Overview of Recent Results," *Journal of Systems and Software* **8** (March 1988), pp. 121–131. (Chapter 4)

[Cox, 1986] B. J. COX, *Object-Oriented Programming: An Evolutionary Approach*, Addison-Wesley, Reading, MA, 1986. (Chapters 8 and 9)

[Cragon, 1982] H. G. CRAGON, "The Myth of the Hardware/Software Cost Ratio," *IEEE Computer* **15** (December 1982), pp. 100–101. (Chapter 1)

[Crossman, 1982] T. D. CROSSMAN, "Inspection Teams, Are They Worth It?" *Proceedings of the Second National Symposium on EDP Quality Assurance*, Chicago, IL, November 1982. (Chapter 6)

[Currit, Dyer, and Mills, 1986] P. A. CURRIT, M. DYER, AND H. D. MILLS, "Certifying the Reliability of Software," *IEEE Transactions on Software Engineering* **SE-12** (January 1986), pp. 3–11. (Chapters 3 and 6)

[Curtis, Sheppard, and Milliman, 1979] B. CURTIS, S. B. SHEPPARD, AND P. MILLIMAN, "Third Time Charm: Stronger Prediction of Programmer Performance by Software Complexity Metrics," *Proceedings of Fourth International Conference on Software Engineering*, Munich, West Germany, 1979, pp. 356–360. (Chapter 15)

[Dahl, Dijkstra, and Hoare, 1972] O.-J. DAHL, E. W. DIJKSTRA, AND C. A. R. HOARE, *Structured Programming*, Academic Press, New York, NY, 1972. (Chapter 6)

[Dahl, Myrhaug, and Nygaard, 1973] O.-J. DAHL, B. MYRHAUG, AND K. NYGAARD, *SIMULA begin*, Auerbach, Philadelphia, PA, 1973. (Chapters 8 and 9)

[Daly, 1977] E. B. DALY, "Management of Software Development," *IEEE Transactions on Software Engineering* **SE-3** (May 1977), pp. 229–242. (Chapter 1)

[Dart, Ellison, Feiler, and Habermann, 1987] S. A. DART, R. J. ELLISON, P. H. FEILER, AND A. N. HABERMANN, "Software Development Environments," *IEEE Computer* **20** (November 1987), pp. 18–28. (Chapters 7 and 12)

[Dasarathy, 1985] B. DASARATHY, "Timing Constraints of Real-Time Systems: Constructs for Expressing Them, Methods of Validating Them," *IEEE Transactions on Software Engineering* **SE-11** (January 1985), pp. 80–86. (Chapter 6)

[Dasgupta and Pearce, 1972] A. K. DASGUPTA AND D. W. PEARCE, *Cost–Benefit Analysis*, Macmillan, London, UK, 1972. (Chapter 4)

[Date, 1986] C. J. DATE, *An Introduction to Database Systems*, Fourth Edition, Addison-Wesley, Reading, MA, 1986. (Chapter 10)

[Deitel, 1983] H. M. DEITEL, *An Introduction to Operating Systems*, Addison-Wesley, Reading, MA, 1983. (Chapter 9)

[Delisle and Schwartz, 1987] N. DELISLE AND M. SCHWARTZ, "A Programming Environment for CSP," *Proceedings of the Second ACM SIGSOFT/SIGPLAN Software Engineering Symposium on Practical Software Development Environments*, *ACM SIGPLAN Notices* **22** (January 1987), pp. 34–41. (Chapter 7)

[Delisle, Menicosy, and Schwartz, 1984] N. M. DELISLE, D. E. MENICOSY, AND M. D. SCHWARTZ, "Viewing a Programming Environment as a Single Tool," *Proceedings of the ACM SIGPLAN/SIGSOFT Software Engineering Symposium on Practical Software Development Environments*, *ACM SIGPLAN Notices* **19** (May 1984), pp. 49–56. (Chapter 12)

[DeMarco, 1978] T. DEMARCO, *Structured Analysis and System Specification*, Yourdon Press, New York, NY, 1978. (Chapter 7)

[DeMarco, 1982] T. DEMARCO, *Controlling Software Projects: Management, Measurement, and Estimation*, Yourdon Press, New York, NY, 1982. (Chapter 4)

[DeMarco and Lister, 1988] T. DEMARCO AND T. LISTER, *Peopleware: Productive Projects and Teams*, Dorset House, New York, NY, 1988. (Chapter 10)

[De Millo, Lipton, and Perlis, 1979] R. A. DE MILLO, R. J. LIPTON, AND A. J. PERLIS, "Social Processes and Proofs of Theorems and Programs," *Communications of the ACM* **22** (May 1979), pp. 271–280. (Chapter 6)

[DeRemer and Kron, 1976] F. DEREMER AND H. H. KRON, "Programming-in-the-Large Versus Programming-in-the-Small," *IEEE Transactions on Software Engineering* **SE-2** (June 1976), pp. 80–86. (Chapter 10)

[Dijkstra, 1968a] E. W. DIJKSTRA, "Go To Statement Considered Harmful," *Communications of the ACM* **11** (March 1968), pp. 147–148. (Chapter 10)

[Dijkstra, 1968b] E. W. DIJKSTRA, "The Structure of the 'THE' Multiprogramming System," *Communications of the ACM* **11** (May 1968), pp. 341–346. (Chapter 8)

[Dijkstra, 1968c] E. W. DIJKSTRA, "A Constructive Approach to the Problem of Program Correctness," *BIT* **8** (No. 3, 1968), pp. 174–186. (Chapter 6)

[Dijkstra, 1972] E. W. DIJKSTRA, "The Humble Programmer," *Communications of the ACM* **15** (October 1972), pp. 859–866. (Chapter 6)

[Dijkstra, 1976] E. W. DIJKSTRA, *A Discipline of Programming*, Prentice-Hall, Englewood Cliffs, NJ, 1976. (Chapters 5 and 6)

[DoD, 1975a] "Requirements for High Order Programming Languages, STRAWMAN," United States Department of Defense, July 1975. (Chapter 14)

[DoD, 1975b] "Requirements for High Order Programming Languages, WOODENMAN," United States Department of Defense, August 1975. (Chapter 14)

[DoD, 1976] "Requirements for High Order Programming Languages, TINMAN," United States Department of Defense, June 1976. (Chapter 14)

[DoD, 1977a] "Requirements for High Order Programming Languages, IRONMAN," United States Department of Defense, January 1977. (Chapter 14)

[DoD, 1977b] "Requirements for High Order Programming Languages, Revised IRONMAN," United States Department of Defense, July 1977. (Chapter 14)

[DoD, 1978a] "Requirements for High Order Programming Languages, STEELMAN," United States Department of Defense, June 1978. (Chapter 14)

[DoD, 1978b] "Requirements for the Programming Environment for the Common High-Order Language, PEBBLEMAN," United States Department of Defense, July 1978. (Chapter 14)

[DoD, 1980] "Requirements for the Programming Environment for the Common High-Order Language, STONEMAN," United States Department of Defense, February 1980. (Chapters 12 and 14)

[Dolotta, Haight, and Mashey, 1984] T. A. DOLOTTA, R. C. HAIGHT, AND J. R. MASHEY, "UNIX Time-Sharing System: The Programmer's Workbench," in: *Interactive Programming Environments*, D. R. Barstow, H. E. Shrobe, and E. Sandewall (Editors), McGraw-Hill, New York, NY, 1984, pp. 353–369. (Chapter 12)

[Dooley and Schach, 1985] J. W. M. DOOLEY AND S. R. SCHACH, "FLOW: A Software Development Environment Using Diagrams," *Journal of Systems and Software* **5** (August 1985), pp. 203–219. (Chapters 10 and 12)

[Dowson, 1987a] M. DOWSON, "ISTAR—An Integrated Project Support Environment," *Proceedings of the Second ACM SIGSOFT/SIGPLAN Software Engineering Symposium on Practical Development Support Environments, ACM SIGPLAN Notices* **22** (January 1987), pp. 27–33. (Chapter 12)

[Dowson, 1987b] M. DOWSON, "ISTAR and the Contractual Approach," *Proceedings of the Ninth International Conference on Software Engineering*, Monterey, CA, March 1987, pp. 287–288. (Chapter 12)

[Dowson, 1987c] M. DOWSON, "Integrated Project Support with ISTAR, *IEEE Software* **4** (November 1987), pp. 6–15. (Chapter 12)

[Dreyfus and Dreyfus, 1986a] H. DREYFUS AND S. DREYFUS, "Why Computers May Never Think like People," *Technology Review* **89** (January 1986), pp. 43–61. (Chapter 16)

[Dreyfus and Dreyfus, 1986b] H. L. DREYFUS AND S. E. DREYFUS, *Mind Over Machine*, The Free Press, New York, NY, 1986. (Chapter 16)

[Dunn, 1984] R. H. DUNN, *Software Defect Removal*, McGraw-Hill, New York, NY, 1984. (Chapter 6)

[Dunn and Ullman, 1982] R. DUNN AND R. ULLMAN, *Quality Assurance for Computer Software*, McGraw-Hill, New York, NY, 1982. (Chapter 4)

[Elshoff, 1976] J. L. ELSHOFF, "An Analysis of Some Commercial PL/I Programs," *IEEE Transactions on Software Engineering* **SE-2** (June 1976), pp. 113–120. (Chapter 1)

[Emery, 1974] J. C. EMERY, "Cost/Benefit Analysis of Information Systems," in: *System Analysis Techniques*, J. D. Cougar and R. W. Knapp (Editors), John Wiley and Sons, New York, NY, 1974, pp. 395–425. (Chapter 1)

[Endres, 1975] A. ENDRES, "An Analysis of Errors and their Causes in System Programs," *IEEE Transactions on Software Engineering* **SE-1** (June 1975), pp. 140–149. (Chapter 6)

[Fagan, 1974] M. E. FAGAN, "Design and Code Inspections and Process Control in the Development of Programs," Technical Report IBM-SSD TR 21.572, IBM Corporation, December 1974. (Chapter 1)

[Fagan, 1976] M. E. FAGAN, "Design and Code Inspections to Reduce Errors in Program Development," *IBM Systems Journal* **15** (No. 3, 1976), pp. 182–211. (Chapters 6 and 15)

[Fagan, 1986] M. E. FAGAN, "Advances in Software Inspections," *IEEE Transactions on Software Engineering* **SE-12** (July 1986), pp. 744–751. (Chapter 6)

[Faulk and Parnas, 1988] S. R. FAULK AND D. L. PARNAS, "On Synchronization in Hard-Real-Time Systems," *Communications of the ACM* **31** (March 1988), pp. 274–287. (Chapter 9)

[Feldman, 1979] S. I. FELDMAN, "Make—A Program for Maintaining Computer Programs," *Software—Practice and Experience* **9** (April 1979), pp. 225–265. (Chapter 11)

[Feldman, 1981] M. B. FELDMAN, "Data Abstraction, Structured Programming, and the Practicing Programmer," *Software—Practice and Experience* **11** (July 1981), pp. 697–710. (Chapter 8)

[Ferrentino and Mills, 1977] A. B. FERRENTINO AND H. D. MILLS, "State Machines and Their Semantics in Software Engineering," *Proceedings of the First International Computer Software and Applications Conference, COMPSAC '77*, Chicago, IL, 1977, pp. 242–251. (Chapter 7)

[Fisher, 1976] D. A. FISHER, "A Common Programming Language for the Department of Defense—Background and Technical Requirements," Institute for Defense Analyses, Report P-1191, 1976. (Chapter 14)

[Freeman, 1987] P. FREEMAN (Editor), *Tutorial: Software Reusability*, IEEE Computer Society Press, Washington, DC, 1987. (Chapter 13)

[Freeman and Wasserman, 1983] P. FREEMAN AND A. I. WASSERMAN, *Tutorial: Software Design Techniques*, Fourth Edition, IEEE Computer Society Press, Washington, DC, 1983. (Chapter 9)

[Freiman and Park, 1979] F. R. FREIMAN AND R. E. PARK, "PRICE Software Model—Version 3: An Overview," *Proceedings of the IEEE-PINY Workshop on Quantitative Software Models*, October 1979, pp. 32–41. (Chapter 4)

[Frenkel, 1985] K. A. FRENKEL, "Toward Automating the Software-Development Cycle," *Communications of the ACM* **28** (June 1985), pp. 578–589. (Chapter 16)

[Gallo, Minot, and Thomas, 1987] F. GALLO, R. MINOT, AND I. THOMAS, "The Object Management System of PCTE as a Software Engineering Database Management System," *Proceedings of the Second ACM SIGSOFT/SIGPLAN Software Engineering Symposium on Practical Software Development Environments, ACM SIGPLAN Notices* **22** (January 1987), pp. 12–15. (Chapter 12)

[Gane and Sarsen, 1979] C. GANE AND T. SARSEN, *Structured Systems Analysis: Tools and Techniques*, Prentice-Hall, Englewood Cliffs, NJ, 1979. (Chapters 7 and 9)

[Garcia-Molina, Germano, and Kohler, 1984] H. GARCIA-MOLINA, F. GERMANO, JR., AND W. H. KOHLER, "Debugging a Distributed Computer System," *IEEE Transactions on Software Engineering* **SE-10** (March 1984), pp. 210–219. (Chapter 6)

[Gargaro and Pappas, 1987] A. GARGARO AND T. L. PAPPAS, "Reusability Issues and Ada," *IEEE Software* **4** (July 1987), pp. 43–51. (Chapter 13)

[Garman, 1981] J. R. GARMAN, "The 'Bug' Heard 'Round the World," *ACM SIGSOFT Software Engineering Notes* **6** (October 1981), pp. 3–10. (Chapter 6)

[Gehani, 1982] N. GEHANI, "Specifications: Formal and Informal—A Case Study," *Software—Practice and Experience* **12** (May 1982), pp. 433–444. (Chapter 7)

[Gehani, 1983] N. GEHANI, *Ada, An Advanced Introduction*, Prentice-Hall, Englewood Cliffs, NJ, 1983. (Chapter 14)

[Gehani and McGettrick, 1986] N. GEHANI AND A. MCGETTRICK (Editors), *Software Specification Techniques*, Addison-Wesley, Reading, MA, 1986. (Chapter 7)

[Gelperin and Hetzel, 1988] D. GELPERIN AND B. HETZEL, "The Growth of Software Testing," *Communications of the ACM* **31** (June 1988), pp. 687–695. (Chapter 6)

[Ghezzi and Mandrioli, 1987] C. GHEZZI AND D. MANDRIOLI, "On Eclecticism in Specifications: A Case Study Centered around Petri Nets," *Proceedings of the Fourth International Workshop on Software Specification and Design*, Monterey, CA, April 1987, pp. 216–224. (Chapter 7)

[Gifford and Spector, 1987] D. GIFFORD AND A. SPECTOR, "Case Study: IBM's System/360–370 Architecture," *Communications of the ACM* **30** (April 1987), pp. 292–307. (Chapter 13)

[Gilb, 1985] T. GILB, "Evolutionary Delivery versus the 'Waterfall Model'," *ACM SIGSOFT Software Engineering Notes* **10** (July 1985), pp. 49–61. (Chapter 3)

[Gilb, 1988] T. GILB, *Principles of Software Engineering Management*, Addison-Wesley, Wokingham, UK, 1988. (Chapter 3)

[Glass, 1982] R. L. GLASS, "Real-Time Checkout: The 'Source Error First' Approach," *Software—Practice and Experience* **12** (January 1982), pp. 77–83. (Chapter 6)

[Glass, 1983] R. L. GLASS (Editor), *Real-Time Software*, Prentice-Hall, Englewood Cliffs, NJ, 1983. (Chapter 6)

[Glass and Noiseux, 1981] R. L. GLASS AND R. A. NOISEUX, *Software Maintenance Guidebook*, Prentice-Hall, Englewood Cliffs, NJ, 1981. (Chapter 11)

[Goel, 1985] A. L. GOEL, "Software Reliability Models: Assumptions, Limitations, and Applicability," *IEEE Transactions on Software Engineering* **SE-11** (December 1985), pp. 1411–1423. (Chapter 6)

[Goldberg, 1984] A. GOLDBERG, *Smalltalk-80: The Interactive Programming Environment*, Addison-Wesley, Reading, MA, 1984. (Chapters 8, 9, and 12)

[Goldberg, 1986] R. GOLDBERG, "Software Engineering: An Emerging Discipline," *IBM Systems Journal* **25** (No. 3/4, 1986), pp. 334–353. (Chapter 9)

[Gomaa, 1986] H. GOMAA, "Software Development of Real-Time Systems," *Communications of the ACM* **29** (July 1986), pp. 657–668. (Chapter 9)

[Good, 1983] D. I. GOOD, "The Proof of a Distributed System in GYPSY," Technical Report ICSCA-CMP-30, University of Texas at Austin, 1983. (Chapter 6)

[Goodenough, 1979] J. B. GOODENOUGH, "A Survey of Program Testing Issues," in: *Research Directions in Software Technology*, P. Wegner

(Editor), The MIT Press, Cambridge, MA, 1979, pp. 316–340. (Chapter 6)

[Goodenough, 1986] J. B. GOODENOUGH, "Ada Compiler Validation: An Example of Software Testing Theory and Practice," *Proceedings of the CRAI Workshop on Software Factories and Ada*, Capri, Italy, May 1986, *Lecture Notes in Computer Science 275*, Springer-Verlag, Berlin, West Germany, pp. 195–221. (Chapter 14)

[Goodenough and Gerhart, 1975] J. B. GOODENOUGH AND S. L. GERHART, "Toward a Theory of Test Data Selection," *Proceedings of the Third International Conference on Reliable Software*, Los Angeles, CA, 1975, pp. 493–510. Also published in: *IEEE Transactions on Software Engineering* **SE-1** (June 1975), pp. 156–173. Revised version: J. B. Goodenough and S. L. Gerhart, "Toward a Theory of Test Data Selection: Data Selection Criteria," in: *Current Trends in Programming Methodology, Volume 2*, R. T. Yeh (Editor), Prentice-Hall, Englewood Cliffs, NJ, 1977, pp. 44–79. (Chapters 6 and 7)

[Gordon, 1979] M. J. C. GORDON, *The Denotational Description of Programming Languages, An Introduction*, Springer-Verlag, New York, NY, 1979. (Chapter 7)

[Grant, 1985] F. J. GRANT, "The Downside of 4GLs," *Datamation* **31** (July 15, 1985), pp. 99–104. (Chapter 10)

[Gremillion, 1984] L. L. GREMILLION, "Determinants of Program Repair Maintenance Requirements," *Communications of the ACM* **27** (August 1984), pp. 826–832. (Chapter 6)

[Guha, Lang, and Bassiouni, 1987] R. K. GUHA, S. D. LANG, AND M. BASSIOUNI, "Software Specification and Design Using Petri Nets," *Proceedings of the Fourth International Workshop on Software Specification and Design*, Monterey, CA, April 1987, pp. 225–230. (Chapter 7)

[Guimaraes, 1983] T. GUIMARAES, "Managing Application Program Maintenance Expenditures," *Communications of the ACM* **26** (October 1983), pp. 739–746. (Chapter 11)

[Guimaraes, 1985] T. GUIMARAES, "A Study of Application Program Development Techniques," *Communications of the ACM* **28** (May 1985), pp. 494–499. (Chapter 10)

[Gustafson and Kerr, 1982] G. G. GUSTAFSON AND R. J. KERR, "Some Practical Experience with a Software Quality Assurance Program," *Communications of the ACM* **25** (January 1982), pp. 4–12. (Chapter 4)

[Guttag, 1977] J. GUTTAG, "Abstract Data Types and the Development of Data Structures," *Communications of the ACM* **20** (June 1977), pp. 396–404. (Chapter 8)

[Habermann and Notkin, 1986] A. N. HABERMANN AND D. NOTKIN, "Gandalf: Software Development Environments," *IEEE Transactions on*

Software Engineering **SE-12** (December 1986), pp. 1117–1127. (Chapter 12)

[Halbert and O'Brien, 1987] D. C. HALBERT AND P. D. O'BRIEN, "Using Types and Inheritance in Object-Oriented Programming," *IEEE Software* **4** (September 1987), pp. 71–79. (Chapter 8)

[Halstead, 1977] M. H. HALSTEAD, *Elements of Software Science*, Elsevier North-Holland, New York, NY, 1977. (Chapters 4 and 6)

[Hamer and Frewin, 1982] P. G. HAMER AND G. D. FREWIN, "M. H. Halstead's Software Science—A Critical Examination," *Proceedings of the IEEE Sixth International Conference on Software Engineering*, Tokyo, Japan, 1982, pp. 197–205. (Chapter 4)

[Hammons and Dobbs, 1985] C. HAMMONS AND P. DOBBS, "Coupling, Cohesion and Package Unity in Ada," *ACM SIGAda Ada Letters* **IV** (No. 6, 1985), pp. 49–59. (Chapter 8)

[Hansen, 1971] W. HANSEN, "Creation of Hierarchic Text with a Computer Display," Ph.D. Thesis, Computer Science Department, Stanford University, Stanford, CA, 1971. (Chapter 10)

[Hansen, 1983] K. HANSEN, "Data Structured Program Design," Ken Orr and Associates, Topeka, KS, 1983. (Chapter 9)

[Hayes, 1985] I. J. HAYES, "Applying Formal Specification to Software Development in Industry," *IEEE Transactions on Software Engineering* **SE-11** (February 1985), pp. 169–178. (Chapter 7)

[Helmer-Hirschberg, 1966] O. HELMER-HIRSCHBERG, *Social Technology*, Basic Books, New York, 1966. (Chapter 4)

[Hoare, 1969] C. A. R. HOARE, "An Axiomatic Basis for Computer Programming," *Communications of the ACM* **12** (October 1969), pp. 576–583. (Chapter 6)

[Hoare, 1985] C. A. R. HOARE, *Communicating Sequential Processes*, Prentice-Hall International, Englewood Cliffs, NJ, 1985. (Chapter 7)

[Hoare, 1987] C. A. R. HOARE, "An Overview of Some Formal Methods for Program Design," *IEEE Computer* **20** (September 1987), pp. 85–91. (Chapter 9)

[Horowitz, Kemper, and Narasimhan, 1985] E. HOROWITZ, A. KEMPER, AND B. NARASIMHAN, "A Survey of Application Generators," *IEEE Software* **2** (January 1985), pp. 40–54. (Chapter 10)

[Houghton, 1984] R. C. HOUGHTON, JR., "Online Help Systems: A Conspectus," *Communications of the ACM* **27** (February 1984), pp. 126–133. (Chapter 10)

[Howden, 1982] W. E. HOWDEN, "Contemporary Software Development Environments," *Communications of the ACM* **25** (May 1982), pp. 318–329. (Chapter 12)

[Howden, 1987] W. E. HOWDEN, *Functional Program Testing and Analysis*, McGraw-Hill, New York, NY, 1987. (Chapter 6)

[Hünke, 1980] H. HÜNKE (Editor), *Proceedings of the Symposium on Software Engineering Environments*, Lahnstein, West Germany, 1980. (Chapter 12)

[Hwang, 1981] S.-S. V. HWANG, "An Empirical Study in Functional Testing, Structural Testing, and Code Reading Inspection," Scholarly Paper 362, Department of Computer Science, University of Maryland, College Park, MD, 1981. (Chapter 6)

[Ichbiah, Barnes, Heliard, Krieg-Brückner, Roubine, and Wichmann, 1979] J. D. ICHBIAH, J. G. P. BARNES, J. C. HELIARD, B. KRIEG-BRÜCKNER, O. ROUBINE, AND B. A. WICHMANN, "Rationale for the Design of the Ada Programming Language," *ACM SIGPLAN Notices* **14** (June 1979), Part B. (Chapter 14)

[IEEE 1028, 1986] "Draft Standard for Software Reviews and Audits," IEEE P1028, Institute of Electrical and Electronic Engineers, Inc., 1986. (Chapter 6)

[IEEE 1058, 1987] "Draft Standard for Software Project Management Plans," IEEE P1058, Institute of Electrical and Electronic Engineers, Inc., 1987. (Chapter 4)

[ISO-7185, 1980] "Specification for the Computer Programming Language Pascal," ISO-7185, International Standards Organization, 1980. (Chapter 13)

[IWSSD, 1986] Call for Papers, Fourth International Workshop on Software Specification and Design, *ACM SIGSOFT Software Engineering Notes* **11** (April 1986), pp. 94–96. (Chapter 7)

[Jackson, 1975] M. A. JACKSON, *Principles of Program Design*, Academic Press, New York, NY, 1975. (Chapter 9)

[Jackson, 1983] M. A. JACKSON, *System Development*, Prentice-Hall, Englewood Cliffs, NJ, 1983. (Chapters 7 and 9)

[Jacob, 1985] R. J. K. JACOB, "A State Transition Diagram Language for Visual Programming," *IEEE Computer* **18** (August 1985), pp. 51–59. (Chapter 7)

[Jensen and Wirth, 1975] K. JENSEN AND N. WIRTH, *Pascal User Manual and Report*, Second Edition, Springer-Verlag, New York, NY, 1975. (Chapter 13)

[John, 1971] P. W. M. JOHN, *Statistical Design and Analysis of Experiments*, Macmillan, New York, NY, 1971. (Chapter 15)

[Johnson, 1979] S. C. JOHNSON, "A Tour through the Portable C Compiler," Seventh Edition, UNIX Programmer's Manual, Bell Laboratories, January 1979. (Chapter 13)

[Johnson and Ritchie, 1978] S. C. JOHNSON AND D. M. RITCHIE, "Portability of C Programs and the UNIX System," *Bell System Technical Journal* **57** (No. 6, Part 2, 1978), pp. 2021–2048. (Chapter 13)

[Joint X3J9/IEEE P770, 1986] "Programming Language Extended Pascal," Working Draft, Joint X3J9/IEEE P770, Institute of Electrical and Electronic Engineers, Inc., 1986. (Chapter 13)

[Jones, 1978] T. C. JONES, "Measuring Programming Quality and Productivity," *IBM Systems Journal* **17** (No. 1, 1978), pp. 39–63. (Chapters 6 and 13)

[Jones, 1984] T. C. JONES, "Reusability in Programming: A Survey of the State of the Art," *IEEE Transactions on Software Engineering* **SE-10** (September 1984), pp. 488–494. (Chapter 13)

[Jones, 1986] C. JONES, *Programming Productivity*, McGraw-Hill, New York, NY, 1986. (Chapters 4 and 6)

[Jones, 1987] C. JONES, Letter to the Editor, *IEEE Computer* **20** (December 1987), p. 4. (Chapter 4)

[Kaiser, Feiler, and Popovich, 1988] G. E. KAISER, P. H. FEILER, AND S. S. POPOVICH, "Intelligent Assistance for Software Development and Maintenance," *IEEE Software* **5** (May 1988), pp. 40–49. (Chapter 16)

[Kampen, 1987] G. R. KAMPEN, "An Eclectic Approach to Specification," *Proceedings of the Fourth International Workshop on Software Specification and Design*, Monterey, CA, April 1987, pp. 178–182. (Chapter 7)

[Karimi and Konsynski, 1988] J. KARIMI AND B. R. KONSYNSKI, "An Automated Design Assistant," *IEEE Transactions on Software Engineering* **14** (February 1988), pp. 194–210. (Chapter 16)

[Kelly, 1987] J. C. KELLY, "A Comparison of Four Design Methods for Real-Time Systems," *Proceedings of the Ninth International Conference on Software Engineering*, Monterey, CA, March 1987, pp. 238–252. (Chapter 9)

[Kemerer, 1987] C. F. KEMERER, "An Empirical Validation of Software Cost Estimation Models," *Communications of the ACM* **30** (May 1987), pp. 416–429. (Chapter 4)

[Kernighan and Plauger, 1974] B. W. KERNIGHAN AND P. J. PLAUGER, *The Elements of Programming Style*, McGraw-Hill, New York, NY, 1974. (Chapter 10)

[Kernighan and Ritchie, 1978] B. W. KERNIGHAN AND D. M. RITCHIE, *The C Programming Language*, Prentice-Hall, Englewood Cliffs, NJ, 1978. (Chapter 13)

[King and Schrems, 1978] J. L. KING AND E. L. SCHREMS, "Cost–Benefit Analysis in Information Systems Development and Operation," *ACM Computing Surveys* **10** (March 1978), pp. 19–34. (Chapter 4)

[Kmielcik et al., 1984] J. KMIELCIK ET AL., "SCR Methodology User's Manual," Grumman Aerospace Corporation, Report SRSR-A6-84-002, 1984. (Chapter 9)

[Knuth, 1974] D. E. KNUTH, "Structured Programming with **go to** Statements," *ACM Computing Surveys* **6** (December 1974), pp. 261–301. (Chapter 10)

[Kulik, 1987] S. KULIK, "Sophisticated Weaponry: How Reliable?" *Defense Science and Electronics* **6** (August 1987), p. 37. (Chapter 14)

[Lake, 1987] J. S. LAKE, "The Beleaguered B-1B," *Defense Science and Electronics* **6** (April 1987), pp. 31–34. (Chapter 14)

[Lamport, 1980] L. LAMPORT, "'Sometime' is Sometimes 'Not Never': On the Temporal Logic of Programs," *Proceedings of the Seventh Annual ACM Symposium on Principles of Programming Languages*, Las Vegas, NV, 1980, pp. 174–185. (Chapter 6)

[Landwehr, 1983] C. E. LANDWEHR, "The Best Available Technologies for Computer Security," *IEEE Computer* **16** (July 1983), pp. 86–100. (Chapter 6)

[Lanergan and Grasso, 1984] R. G. LANERGAN AND C. A. GRASSO, "Software Engineering with Reusable Designs and Code," *IEEE Transactions on Software Engineering* **SE-10** (September 1984), pp. 498–501. (Chapter 13)

[Lantz, 1985] K. E. LANTZ, *The Prototyping Methodology*, Prentice-Hall, Englewood Cliffs, NJ, 1985. (Chapter 3)

[Laughery and Laughery, 1985] K. R. LAUGHERY, JR., AND K. R. LAUGHERY, SR., "Human Factors in Software Engineering: A Review of the Literature," *Journal of Systems and Software* **5** (February 1985), pp. 3–14. (Chapter 1)

[Leavenworth, 1970] B. LEAVENWORTH, Review #19420, *Computing Reviews* **11** (July 1970), pp. 396–397. (Chapters 6 and 7)

[Lecarme and Gart, 1986] O. LECARME AND M. P. GART, *Software Portability*, McGraw-Hill, New York, NY, 1986. (Chapter 13)

[Ledgard, 1975] H. LEDGARD, *Programming Proverbs*, Hayden Books, Rochelle Park, NJ, 1975. (Chapter 10)

[Lehman and Belady, 1985] M. M. LEHMAN AND L. A. BELADY (Editors), *Program Evolution, Processes of Software Change*, Academic Press, London, UK, 1985. (Chapter 3)

[Licker, 1985] P. S. LICKER, *The Art of Managing Software Development People*, John Wiley and Sons, New York, NY, 1985. (Chapter 10)

[Lieblein, 1986] E. LIEBLEIN, "The Department of Defense Software Initiative —A Status Report," *Communications of the ACM* **29** (August 1986), pp. 734–744. (Chapter 14)

[Lientz and Swanson, 1980] B. P. LIENTZ AND E. B. SWANSON, *Software Maintenance Management: A Study of the Maintenance of Computer Applications Software in 487 Data Processing Organizations*, Addison-Wesley, Reading, MA, 1980. (Chapter 11)

[Lientz, Swanson, and Tompkins, 1978] B. P. LIENTZ, E. B. SWANSON, AND G. E. TOMPKINS, "Characteristics of Application Software Maintenance," *Communications of the ACM* **21** (June 1978), pp. 466–471. (Chapters 1 and 11)

[Linger, 1980] R. C. LINGER, "The Management of Software Engineering. Part III. Software Design Practices," *IBM Systems Journal* **19** (No. 4, 1980), pp. 432–450. (Chapter 7)

[Linger, Mills, and Witt, 1979] R. C. LINGER, H. D. MILLS, AND B. I. WITT, *Structured Programming: Theory and Practice*, Addison-Wesley, Reading, MA, 1979. (Chapter 6)

[Liskov and Guttag, 1986] B. LISKOV AND J. GUTTAG, *Abstraction and Specification in Program Development*, The MIT Press, Cambridge, MA, 1986. (Chapter 8)

[Liskov and Zilles, 1974] B. LISKOV AND S. ZILLES, "Programming with Abstract Data Types," *ACM SIGPLAN Notices* **9** (April 1974), pp. 50–59. (Chapter 8)

[Liskov, Snyder, Atkinson, and Schaffert, 1977] B. LISKOV, A. SNYDER, R. ATKINSON, AND C. SCHAFFERT, "Abstraction Mechanisms in CLU," *Communications of the ACM* **20** (August 1977), pp. 564–576. (Chapters 8 and 13)

[London, 1971] R. L. LONDON, "Software Reliability through Proving Programs Correct," in: *Proceedings of the IEEE International Symposium on Fault-Tolerant Computing*, March 1971. (Chapters 6 and 7)

[Luckham and von Henke, 1985] D. C. LUCKHAM AND F. W. VON HENKE, "An Overview of Anna, a Specification Language for Ada," *IEEE Software* **2** (March 1985), pp. 9–22. (Chapter 7)

[Mackenzie, 1980] C. E. MACKENZIE, *Coded Character Sets: History and Development*, Addison-Wesley, Reading, MA, 1980. (Chapter 13)

[MacNair and Sauer, 1985] E. A. MACNAIR AND C. H. SAUER, *Elements of Practical Performance Modeling*, Prentice-Hall, Englewood Cliffs, NJ, 1985. (Chapter 4)

[Manna, 1974] Z. MANNA, *Mathematical Theory of Computation*, McGraw-Hill, New York, NY, 1974. (Chapter 6)

[Manna and Pnueli, 1981] Z. MANNA AND A. PNUELI, "Verification of Concurrent Programs, Part I: The Temporal Framework," Report No. STAN-CS-81-836, Department of Computer Science, Stanford University, Stanford, CA, June 1981. (Chapter 6)

[Manna and Waldinger, 1978] Z. MANNA AND R. WALDINGER, "The Logic of Computer Programming," *IEEE Transactions on Software Engineering* **SE-4** (1978), pp. 199–229. (Chapter 6)

[Mantei, 1981] M. MANTEI, "The Effect of Programming Team Structures on Programming Tasks," *Communications of the ACM* **24** (March 1981), pp. 106–113. (Chapter 10)

[Mantei and Teorey, 1988] M. M. MANTEI AND T. J. TEOREY, "Cost/Benefit Analysis for Incorporating Human Factors in the Software Development Lifecycle," *Communications of the ACM* **31** (April 1988), pp. 428–439. (Chapter 3)

[Martin, 1985] J. MARTIN, *Fourth-Generation Languages, Volumes I, II, and III*, Prentice-Hall, Englewood Cliffs, NJ, 1985. (Chapter 10)

[Martin and McClure, 1985] J. P. MARTIN AND C. MCCLURE, *Diagramming Techniques for Analysts and Programmers*, Prentice-Hall, Englewood Cliffs, NJ, 1985. (Chapters 7 and 9)

[Mason and Carey, 1983] R. E. A. MASON AND T. T. CAREY, "Prototyping Interactive Information Systems," *Communications of the ACM* **26** (May 1983), pp. 347–354. (Chapter 3)

[Mathis, 1986] R. F. MATHIS, "The Last 10 Percent," *IEEE Transactions on Software Engineering* **SE-12** (June 1986), pp. 705–712. (Chapters 1 and 2)

[Matsumoto, 1984] Y. MATSUMOTO, "Management of Industrial Software Production," *IEEE Computer* **17** (February 1984), pp. 59–72. (Chapter 13)

[Matsumoto, 1987] Y. MATSUMOTO, "A Software Factory: An Overall Approach to Software Production," in: *Tutorial: Software Reusability*, P. Freeman (Editor), IEEE Computer Society Press, Washington, DC, 1987, pp. 155–178. (Chapter 13)

[McCabe, 1976] T. J. MCCABE, "A Complexity Measure," *IEEE Transactions on Software Engineering* **SE-2** (December 1976), pp. 308–320. (Chapter 6)

[McCormick, 1987] F. MCCORMICK, "Scheduling Difficulties of Ada in the Hard Real-Time Environment," *Proceedings of the International Workshop on Real-Time Ada Issues, ACM SIGAda Ada Letters* **VII** (No. 6, 1987), pp. 49–50. (Chapter 14)

[Merlin, 1974] P. MERLIN, "A Study of the Recoverability of Computing Systems," Ph.D. Dissertation, University of California, Irvine, CA, 1974. (Chapter 7)

[Metcalfe and Boggs, 1976] R. M. METCALFE AND D. R. BOGGS, "Ethernet: Distributed Packet Switching for Local Computer Networks," *Communications of the ACM* **19** (July 1976), pp. 395–404. (Chapter 6)

[Metzger, 1981] P. W. METZGER, *Managing a Programming Project*, Second Edition, Prentice-Hall, Englewood Cliffs, NJ, 1981. (Chapter 10)

[Meyer, 1985] B. MEYER, "On Formalism in Specifications," *IEEE Software* **2** (January 1985), pp. 6–26. (Chapter 7)

[Meyer, 1986] B. MEYER, "Genericity versus Inheritance," *Proceedings of the Conference on Object-Oriented Programming Systems, Languages and Applications, ACM SIGPLAN Notices* **21** (November 1986), pp. 391–405. (Chapter 8)

[Meyer, 1987] B. MEYER, "Reusability: The Case for Object-Oriented Design," *IEEE Software* **4** (March 1987), pp. 50–64. (Chapter 8)

[Meyer, 1988] B. MEYER, "Eiffel: A Language and Environment for Software Engineering," *Journal of Systems and Software* **8** (June 1988), pp. 199–246. (Chapter 12)

[Miller, 1956] G. A. MILLER, "The Magical Number Seven, Plus or Minus Two: Some Limits on our Capacity for Processing Information," *The Psychological Review* **63** (March 1956), pp. 81–97. (Chapter 5)

[Mills, 1976] H. D. MILLS, "Software Development," *IEEE Transactions on Software Engineering* **SE-2** (September 1976), pp. 265–273. (Chapters 1 and 2)

[Mills, 1988] H. D. MILLS, "Stepwise Refinement and Verification in Box-Structured Systems," *IEEE Computer* **21** (June 1988), pp. 23–36. (Chapter 5)

[Mills, Basili, Gannon, and Hamlet, 1987] H. D. MILLS, V. R. BASILI, J. D. GANNON, AND R. G. HAMLET, *Principles of Computer Programming: A Mathematical Approach*, Allyn and Bacon, Newton, MA, 1987. (Chapter 6)

[Mills, Dyer, and Linger, 1987] H. D. MILLS, M. DYER, AND R. C. LINGER, "Cleanroom Software Engineering," *IEEE Software* **4** (September 1987), pp. 19–25. (Chapter 6)

[Mills, Linger, and Hevner, 1987] H. D. MILLS, R. C. LINGER, AND A. R. HEVNER, "Box Structured Information Systems," *IBM Systems Journal* **26** (No. 4, 1987), pp. 395–413. (Chapter 5)

[Mishan, 1982] E. J. MISHAN, *Cost–Benefit Analysis: An Informal Introduction*, Third Edition, George Allen & Unwin, London, UK, 1982. (Chapter 4)

[Moder, Phillips, and Davis, 1983] J. J. MODER, C. R. PHILLIPS, AND E. W. DAVIS, *Project Management with CPM, PERT, and Precedence Diagramming*, Third Edition, Van Nostrand Reinhold, New York, NY, 1983. (Chapter 12)

[Moon, 1986] D. A. MOON, "Object-Oriented Programming with *Flavors*," *Proceedings of the Conference on Object-Oriented Programming Systems, Languages and Applications, ACM SIGPLAN Notices* **21** (November 1986), pp. 1–8. (Chapters 8 and 9)

[Moran, 1981] T. P. MORAN (Editor), Special Issue: The Psychology of Human–Computer Interaction, *ACM Computing Surveys* **13** (March 1981). (Chapter 5)

[Munoz, 1988] C. U. MUNOZ, "An Approach to Software Product Testing," *IEEE Transactions on Software Engineering* **14** (November 1988), pp. 1589–1596. (Chapter 6)

[Musa, Iannino, and Okumoto, 1987] J. D. MUSA, A. IANNINO, AND K. OKUMOTO, *Software Reliability: Measurement, Prediction, Application*, McGraw-Hill, New York, NY, 1987. (Chapters 4 and 6)

[Musser, 1980] D. R. MUSSER, "Abstract Data Type Specification in the AFFIRM System," *IEEE Transactions on Software Engineering* **SE-6** (January 1980), pp. 24–32. (Chapter 6)

[Myers, 1975] G. J. MEYERS, *Reliable Software Through Composite Design*, Petrocelli/Charter, New York, NY, 1975. (Chapter 8)

[Myers, 1976] G. J. MYERS, *Software Reliability: Principles and Practices*, Wiley-Interscience, New York, NY, 1976. (Chapter 6)

[Myers, 1978a] G. J. MYERS, "A Controlled Experiment in Program Testing and Code Walkthroughs/Inspections," *Communications of the ACM* **21** (September 1978), pp. 760–768. (Chapters 6 and 15)

[Myers, 1978b] G. J. MYERS, *Composite/Structured Design*, Van Nostrand Reinhold, New York, NY, 1978. (Chapter 8)

[Myers, 1979] G. J. MYERS, *The Art of Software Testing*, John Wiley and Sons, New York, NY, 1979. (Chapters 2 and 6)

[Nassi and Shneiderman, 1973] I. NASSI AND B. SHNEIDERMAN, "Flowchart Techniques for Structured Programs," *ACM SIGPLAN Notices* **8** (August 1973), pp. 12–26. (Chapter 12)

[Naur, 1964] P. NAUR, "The Design of the GIER ALGOL Compiler," in: *Annual Review in Automatic Programming, Volume 4*, Pergamon Press, Oxford, UK, 1964, pp. 49–85. (Chapter 7)

[Naur, 1969] P. NAUR, "Programming by Action Clusters," *BIT* **9** (No. 3, 1969), pp. 250–258. (Chapters 6 and 7)

[Nelson, Haibt, and Sheridan, 1983] R. A. NELSON, L. M. HAIBT, AND P. B. SHERIDAN, "Casting Petri Nets into Diagrams," *IEEE Transactions on Software Engineering* **SE-9** (September 1983), pp. 590–602. (Chapter 7)

[Neumann, 1980] P. G. NEUMANN, Letter from the Editor, *ACM SIGSOFT Software Engineering Notes* **5** (July 1980), p. 2. (Chapter 1)

[Neumann, 1986] P. G. NEUMANN, "On Hierarchical Design of Computer Systems for Critical Applications," *IEEE Transactions on Software Engineering* **SE-12** (September 1986), pp. 905–920. (Chapter 8)

[Nielsen and Shumate, 1987] K. W. NIELSEN AND K. SHUMATE, "Designing Large Real-Time Systems in Ada," *Communications of the ACM* **30** (August 1987), pp. 695–715. (Chapters 9 and 14)

[Nissen and Wallis, 1984] J. NISSEN AND P. WALLIS (Editors), *Portability and Style in Ada*, Cambridge University Press, Cambridge, UK, 1984. (Chapter 13)

[Norden, 1958] P. V. NORDEN, "Curve Fitting for a Model of Applied Research and Development Scheduling," *IBM Journal of Research and Development* **2** (July 1958), pp. 232–248. (Chapter 4)

[Nori, Ammann, Jensen, and Nägeli, 1981] K. V. Nori, U. Ammann, K. Jensen, and H. Nägeli, "The Pascal (P) Compiler Implementation Notes," in: *Pascal—The Language and its Implementation*, D. W. Barron (Editor), John Wiley, Chichester, UK, 1981, pp. 125–170. (Chapter 13)

[Norman and Draper, 1986] D. A. Norman and S. W. Draper (Editors), *User Centered System Design: New Perspectives on Human–Computer Interaction*, Lawrence Erlbaum Associates, Hillsdale, NJ, 1984. (Chapter 1)

[Norokorpi, 1984] J. Norokorpi, Informal Report, ACM SIGAda Meeting, Washington, DC, November 1984. (Chapter 14)

[Norusis, 1982] M. J. Norusis, *SPSS Introductory Guide: Basic Statistics and Operations*, McGraw-Hill, New York, NY, 1982. (Chapter 13)

[Notkin, 1985] D. Notkin, "The GANDALF Project," *Journal of Systems and Software* **5** (May 1985), pp. 91–105. (Chapter 12)

[Oest, 1986] O. N. Oest, "VDM from Research to Practice," *Proceedings of IFIP Congress, Information Processing '86*, 1986, pp. 527–533. (Chapter 7)

[Orr, 1981] K. Orr, *Structured Requirements Definition*, Ken Orr and Associates, Inc., Topeka, KS, 1981. (Chapters 7, 9, and 12)

[Ottenstein, 1979] L. M. Ottenstein, "Quantitative Estimates of Debugging Requirements," *IEEE Transactions on Software Engineering* **SE-5** (September 1979), pp. 504–514. (Chapter 6)

[Parikh, 1988] G. Parikh (Editor), *Techniques of Program and System Maintenance*, Second Edition, QED Information Sciences, Wellesley, MA, 1981. (Chapter 11)

[Parikh and Zvegintzov, 1983] G. Parikh and N. Zvegintzov (Editors), *Tutorial: Software Maintenance*, IEEE Computer Society Press, Washington, DC, 1983. (Chapter 11)

[Parnas, 1971] D. L. Parnas, "Information Distribution Aspects of Design Methodology," *Proceedings of the IFIP Congress*, Ljubljana, Yugoslavia, 1971, pp. 339–344. (Chapters 8 and 9)

[Parnas, 1972a] D. L. Parnas, "A Technique for Software Module Specification with Examples," *Communications of the ACM* **15** (May 1972), pp. 330–336. (Chapters 8 and 9)

[Parnas, 1972b] D. L. Parnas, "On the Criteria to be Used in Decomposing Systems into Modules," *Communications of the ACM* **15** (December 1972), pp. 1053–1058. (Chapters 8 and 9)

[Parnas, 1979] D. L. Parnas, "Designing Software for Ease of Extension and Contraction," *IEEE Transactions on Software Engineering* **SE-5** (March 1979), pp. 128–138. (Chapter 2)

[Parnas and Weiss, 1987] D. L. Parnas and D. M. Weiss, "Active Design Reviews: Principles and Practices," *Journal of Systems and Software* **7** (December 1987), pp. 259–265. (Chapter 6)

[Parnas, Clements, and Weiss, 1985] D. L. PARNAS, P. C. CLEMENTS, AND D. M. WEISS, "The Modular Structure of Complex Systems," *IEEE Transactions on Software Engineering* **SE-11** (March 1985), pp. 259–266. (Chapter 8)

[Pearce, 1971] D. W. PEARCE, *Cost–Benefit Analysis*, Macmillan, London, UK, 1971. (Chapter 4)

[Perez, 1988] E. P. PEREZ, "Simulating Inheritance with Ada," *ACM SIGAda Ada Letters* **VIII** (No. 5, 1988), pp. 37–46. (Chapter 8)

[Perry, 1983] W. E. PERRY, *A Structured Approach to Systems Testing*, Prentice-Hall, Englewood Cliffs, NJ, 1983. (Chapter 6)

[Peterson, 1981] J. L. PETERSON, *Petri Net Theory and the Modeling of Systems*, Prentice-Hall, Englewood Cliffs, NJ, 1981. (Chapter 7)

[Peterson, 1988] G. E. PETERSON (Editor), *Tutorial: Object-Oriented Computing, Volume 2: Implementations*, IEEE Computer Society Press, Washington, DC, 1988. (Chapter 9)

[Petri, 1962] C. A. PETRI, "Kommunikation mit Automaten," Ph.D. Dissertation, University of Bonn, West Germany, 1962. (In German). (Chapter 7)

[Petschenik, 1985] N. H. PETSCHENIK, "Practical Priorities in System Testing," *IEEE Software* **2** (September 1985), pp. 18–23. (Chapter 6)

[Phillips, 1986] J. PHILLIPS, *The NAG Library: A Beginner's Guide*, Clarendon Press, Oxford, UK, 1986. (Chapter 13)

[Pollack, Hicks, and Harrison, 1971] S. L. POLLACK, H. T. HICKS, JR., AND W. J. HARRISON, *Decision Tables: Theory and Practice*, Wiley-Interscience, New York, NY, 1971. (Chapter 7)

[Poston and Bruen, 1987] R. M. POSTON AND M. W. BRUEN, "Counting Down to Zero Software Failures," *IEEE Software* **4** (September 1987), pp. 54–61. (Chapter 4)

[Preliminary Ada Reference Manual, 1979] "Preliminary Ada Reference Manual," *ACM SIGPLAN Notices* **14** (June 1979), Part A. (Chapter 14)

[Prieto-Díaz and Freeman, 1987] R. PRIETO-DÍAZ AND P. FREEMAN, "Classifying Software for Reusability," *IEEE Software* **4** (January 1987), pp. 6–16. (Chapter 13)

[Prieto-Díaz and Neighbors, 1986] R. PRIETO-DÍAZ AND J. M. NEIGHBORS, "Module Interconnection Languages," *Journal of Systems and Software* **6** (November 1986), pp. 307–334. (Chapter 13)

[Putnam, 1978] L. N. PUTNAM, "A General Empirical Solution to the Macro Software Sizing and Estimating Problem," *IEEE Transactions on Software Engineering* **SE-4** (July 1978), pp. 345–361. (Chapter 4)

[Quirk, 1983] W. J. QUIRK, "Recent Developments in the SPECK Specification System," Report CSS.146, Harwell, UK, 1983. (Chapter 6)

[Quirk, 1985] W. J. QUIRK (Editor), *Verification and Validation of Real-Time Software*, Springer-Verlag, Berlin, West Germany, 1985. (Chapter 6)

[Rajlich, 1985a] V. RAJLICH, "Stepwise Refinement Revisited," *Journal of Systems and Software* **5** (February 1985), pp. 81–88. (Chapter 5)

[Rajlich, 1985b] V. RAJLICH, "Paradigms for Design and Implementation in Ada," *Communications of the ACM* **28** (July 1985), pp. 718–727. (Chapter 14)

[Ramamoorthy, Prakash, Tsai, and Usuda, 1984] C. V. RAMAMOORTHY, A. PRAKASH, W.-T. TSAI, AND Y. USUDA, "Software Engineering: Problems and Perspectives," *IEEE Computer* **17** (October 1984), pp. 191–209. (Chapters 1 and 2)

[Rapps and Weyuker, 1985] S. RAPPS AND E. J. WEYUKER, "Selecting Software Test Data Using Data Flow Information," *IEEE Transactions on Software Engineering* **SE-11** (April 1985), pp. 367–375. (Chapter 6)

[Reifer, 1986] D. J. REIFER (Editor), *Tutorial: Software Management*, Third Edition, IEEE Computer Society Press, Washington, DC, 1986. (Chapter 4)

[Reiss, 1984] S. P. REISS, "Graphical Program Development with PECAN Program Development Systems," *Proceedings of the ACM SIGSOFT/SIGPLAN Software Engineering Symposium on Practical Software Development Environments, ACM SIGPLAN Notices* **19** (May 1984), pp. 30–41. (Chapter 12)

[Ringland, 1984] J. RINGLAND, "Software Engineering in a Development Group," *Software—Practice and Experience* **14** (June 1984), pp. 533–559. (Chapter 13)

[Rochkind, 1975] M. J. ROCHKIND, "The Source Code Control System," *IEEE Transactions on Software Engineering* **SE-1** (October 1975), pp. 255–265. (Chapter 11)

[Ross, 1985] D. T. ROSS, "Applications and Extensions of SADT," *IEEE Computer* **18** (April 1985), pp. 25–34. (Chapter 7)

[Royce, 1970] W. W. ROYCE, "Managing the Development of Large Software Systems: Concepts and Techniques," *Proceedings of WestCon*, August 1970. (Chapter 3)

[Rubinstein and Hersh, 1984] R. RUBINSTEIN AND H. M. HERSH, *The Human Factor: Designing Computer Systems for People*, Digital Press, Burlington, MA, 1984. (Chapter 1)

[Sackman, 1970] H. SACKMAN, *Man–Computer Problem Solving: Experimental Evaluation of Time-Sharing and Batch Processing*, Auerbach, Princeton, NJ, 1970. (Chapter 4)

[Sackman, Erikson, and Grant, 1968] H. SACKMAN, W. J. ERIKSON, AND E. E. GRANT, "Exploratory Experimental Studies Comparing Online and Offline Programming Performance," *Communications of the ACM* **11** (January 1968), pp. 3–11. (Chapters 4 and 15)

[Sammet, 1978] J. E. Sammet, "The Early History of COBOL," *Proceedings of the History of Programming Languages Conference*, Los Angeles, CA, 1978, pp. 199–276. (Chapter 14)

[Sammet, 1986] J. E. Sammet, "Why Ada is Not Just Another Programming Language," *Communications of the ACM* **29** (August 1986), pp. 722–732. (Chapter 14)

[Schach, 1982] S. R. Schach, "A Unified Theory for Software Production," *Software—Practice and Experience* **12** (July 1982), pp. 683–689. (Chapter 15)

[Schach and Stevens-Guille, 1979] S. R. Schach and P. D. Stevens-Guille, "Two Aspects of Computer-Aided Design," *Transactions of the Royal Society of South Africa* **44** (Part 1, 1979), pp. 123–126. (Chapter 8)

[Scheffer, Stone, and Rzepka, 1985] P. A. Scheffer, A. H. Stone, III, and W. E. Rzepka, "A Case Study of SREM," *IEEE Computer* **18** (April 1985), pp. 47–54. (Chapter 7)

[Schneidewind, 1987] N. F. Schneidewind, "The State of Software Maintenance," *IEEE Transactions on Software Engineering* **SE-13** (March 1987), pp. 303–310. (Chapter 11)

[Schulmeyer and McManus, 1987] G. G. Schulmeyer and J. I. McManus (Editors), *Handbook of Software Quality Assurance*, Van Nostrand Reinhold, New York, NY, 1987. (Chapter 4)

[Schwartz and Delisle, 1987] M. D. Schwartz and N. M. Delisle, "Specifying a Lift Control System with CSP," *Proceedings of the Fourth International Workshop on Software Specification and Design*, Monterey, CA, April 1987, pp. 21–27. (Chapter 7)

[Seitz, 1985] C. L. Seitz, "The Cosmic Cube," *Communications of the ACM* **28** (January 1985), pp. 22–33. (Chapter 6)

[Selby, 1988] R. W. Selby, "Quantitative Studies of Software Reuse," Technical Report UCI TR-87-12, Version: March 22, 1988, Department of Information and Computer Science, University of California, Irvine, CA, 1988. (Chapter 13)

[Selby, Basili, and Baker, 1987] R. W. Selby, V. R. Basili, and F. T. Baker, "Cleanroom Software Development: An Empirical Evaluation," *IEEE Transactions on Software Engineering* **SE-13** (September 1987), pp. 1027–1037. (Chapters 3 and 6)

[Shaw, 1984] M. Shaw, "Abstraction Techniques in Modern Programming Languages," *IEEE Software* **1** (October 1984), pp. 10–26. (Chapter 8)

[Shen, Conte, and Dunsmore, 1983] V. Y. Shen, S. D. Conte, and H. E. Dunsmore, "Software Science Revisited: A Critical Analysis of the Theory and its Empirical Support," *IEEE Transactions on Software Engineering* **SE-9** (March 1983), pp. 155–165. (Chapter 4)

[Shneiderman, 1980] B. SHNEIDERMAN, *Software Psychology: Human Factors in Computer and Information Systems*, Winthrop Publishers, Cambridge, MA, 1980. (Chapter 1)

[Shneiderman and Mayer, 1975] B. SHNEIDERMAN AND R. MAYER, "Towards a Cognitive Model of Programmer Behavior," Technical Report TR-37, Indiana University, Bloomington, IN, 1975. (Chapter 8)

[Shooman, 1983] M. L. SHOOMAN, *Software Engineering: Design, Reliability, and Management*, McGraw-Hill, New York, NY, 1983. (Chapter 6)

[Shriver and Wegner, 1987] B. SHRIVER AND P. WEGNER (Editors), *Research Directions in Object-Oriented Programming*, The MIT Press, Cambridge, MA, 1987. (Chapter 8)

[Simon, 1986] H. A. SIMON, "Whether Software Engineering Needs to be Artificially Intelligent," *IEEE Transactions on Software Engineering* **SE-12** (July 1986), pp. 726–732. (Chapter 16)

[Singer, Ledgard, and Hueras, 1981] A. SINGER, H. LEDGARD, AND J. F. HUERAS, "The Annotated Assistant: A Step Towards Human Engineering," *IEEE Transactions on Software Engineering* **SE-7** (July 1981), pp. 353–374. (Chapter 16)

[Smith, Kotik, and Westfold, 1985] D. R. SMITH, G. B. KOTIK, AND S. J. WESTFOLD, "Research on Knowledge-Based Software Environments at the Kestrel Institute," *IEEE Transactions on Software Engineering* **SE-11** (November 1985), pp. 1278–1295. (Chapter 7)

[Sobell, 1985] M. G. SOBELL, *A Practical Guide to UNIX System V*, Benjamin/Cummings, Menlo Park, CA, 1985. (Chapter 10)

[Squires, 1982] S. L. SQUIRES, *Working Papers from the ACM SIGSOFT Rapid Prototyping Workshop, ACM SIGSOFT Software Engineering Notes* **7** (December 1982). (Chapter 3)

[Stefik and Bobrow, 1986] M. STEFIK AND D. G. BOBROW, "Object-Oriented Programming: Themes and Variations," *The AI Magazine* **6** (No. 4, 1986), pp. 40–62. (Chapters 8 and 9)

[Stephenson, 1976] W. E. STEPHENSON, "An Analysis of the Resources Used in Safeguard System Software Development," Bell Laboratories, Draft Paper, August 1976. (Chapter 1)

[Stevens, Myers, and Constantine, 1974] W. P. STEVENS, G. J. MYERS, AND L. L. CONSTANTINE, "Structured Design," *IBM Systems Journal* **13** (No. 2, 1974), pp. 115–139. (Chapter 8)

[Stroustrup, 1986] B. STROUSTRUP, *The C++ Programming Language*, Addison-Wesley, Reading, MA, 1986. (Chapters 8 and 9)

[Swartout, 1983] W. SWARTOUT, "The Gist Behavior Explainer," *Proceedings of the National Conference on Artificial Intelligence AAAI-83*, 1983, pp. 402–407. (Chapter 16)

[Swinehart, Zellweger, and Hagmann, 1985] D. C. SWINEHART, P. T. ZELLWEGER, AND R. B. HAGMANN, "The Structure of Cedar," *Proceedings of the ACM SIGPLAN 85 Symposium on Language Issues in Programming Environments*, *ACM SIGPLAN Notices* **20** (July 1985), pp. 230–244. (Chapter 12)

[Symons, 1988] C. R. SYMONS, "Function Point Analysis: Difficulties and Improvements," *IEEE Transactions on Software Engineering* **14** (January 1988), pp. 2–11. (Chapter 4)

[Takahashi and Kamayachi, 1985] M. TAKAHASHI AND Y. KAMAYACHI, "An Empirical Study of a Model for Program Error Prediction," *Proceedings of the Eighth International Conference on Software Engineering*, London, UK, 1985, pp. 330–336. (Chapter 6)

[Tamir, 1980] M. TAMIR, "ADI: Automatic Derivation of Invariants," *IEEE Transactions on Software Engineering* **SE-6** (January 1980), pp. 40–48. (Chapter 6)

[Tanenbaum, 1981] A. S. TANENBAUM, "Network Protocols," *ACM Computing Surveys* **13** (December 1981), pp. 453–489. (Chapter 13)

[Tanenbaum, 1984] A. S. TANENBAUM, *Structured Computer Organization*, Second Edition, Prentice-Hall, Englewood Cliffs, NJ, 1984. (Chapter 8)

[Tanenbaum, Klint, and Bohm, 1978] A. S. TANENBAUM, P. KLINT, AND W. BOHM, "Guidelines for Software Portability," *Software—Practice and Experience* **8** (November 1978), pp. 681–698. (Chapter 13)

[Teichroew and Hershey, 1977] D. TEICHROEW AND E. A. HERSHEY, III, "PSL/PSA: A Computer-Aided Technique for Structured Documentation and Analysis of Information Processing Systems," *IEEE Transactions on Software Engineering* **SE-3** (January 1977), pp. 41–48. (Chapters 7 and 12)

[Teitelbaum and Reps, 1981] T. TEITELBAUM AND T. REPS, "The Cornell Program Synthesizer: A Syntax-Directed Programming Environment," *Communications of the ACM* **24** (September 1981), pp. 563–573. (Chapter 10)

[Teitelman and Masinter, 1981] W. TEITELMAN AND L. MASINTER, "The Interlisp Programming Environment," in: *Tutorial: Software Development Environments*, A. I. Wasserman (Editor), IEEE Computer Society Press, Washington, DC, 1981, pp. 73–81. (Chapter 12)

[Thayer, 1988] R. H. THAYER (Editor), *Tutorial: Software Engineering Management*, IEEE Computer Society Press, Washington, DC, 1988. (Chapter 4)

[Thayer, Pyster, and Wood, 1981] R. H. THAYER, A. B. PYSTER, AND R. C. WOOD, "Major Issues in Software Engineering Project Management," *IEEE Transactions on Software Engineering* **SE-7** (July 1981), pp. 333–342. (Chapter 2)

[Tichy, 1985] W. F. Tichy, "RCS—A System for Version Control," *Software—Practice and Experience* **15** (July 1985), pp. 637–654. (Chapter 11)

[Tracz, 1979] W. J. Tracz, "Computer Programming and the Human Thought Process," *Software—Practice and Experience* **9** (February 1979), pp. 127–137. (Chapter 5)

[Tracz, 1988] W. Tracz, "Software Reuse Myths," *ACM SIGSOFT Software Engineering Notes* **13** (January 1988), pp. 17–21. (Chapter 13)

[van der Poel and Schach, 1983] K. G. van der Poel and S. R. Schach, "A Software Metric for Cost Estimation and Efficiency Measurement in Data Processing System Development," *Journal of Systems and Software* **3** (September 1983), pp. 187–191. (Chapter 4)

[van Hoeve and Engmann, 1987] F. van Hoeve and R. Engmann, "An Object-Oriented Approach to Application Generation," *Software—Practice and Experience* **17** (September 1987), pp. 623–645. (Chapter 8)

[Vick and Ramamoorthy, 1984] C. R. Vick and C. V. Ramamoorthy (Editors), *Handbook of Software Enginering*, Van Nostrand Reinhold, New York, NY, 1984. (Chapters 1 and 2)

[Wahl and Schach, 1988] N. J. Wahl and S. R. Schach, "A Methodology and Distributed Tool for Debugging Dataflow Programs," *Proceedings of the Second Workshop on Software Testing, Verification, and Analysis*, Banff, Canada, July 1988, pp. 98–105. (Chapter 6)

[Wallis, 1982] P. J. L. Wallis, *Portable Programming*, John Wiley and Sons, New York, NY, 1982. (Chapter 13)

[Walsh, 1979] T. J. Walsh, "A Software Reliability Study Using a Complexity Measure," *Proceedings of the AFIPS National Computer Conference*, New York, NY, 1979, pp. 761–768. (Chapter 6)

[Ward and Mellor, 1985] P. T. Ward and S. J. Mellor, *Structured Development for Real-Time Systems, Volumes 1, 2 and 3*, Yourdon Press, New York, NY, 1985. (Chapter 9)

[Warnier, 1976] J. D. Warnier, *Logical Construction of Programs*, Van Nostrand Reinhold, New York, NY, 1976. (Chapter 9)

[Warnier, 1981] J. D. Warnier, *Logical Construction of Systems*, Van Nostrand Reinhold, New York, NY, 1981. (Chapters 7 and 9)

[Wasserman, 1985] A. I. Wasserman, "Extending State Transition Diagrams for the Specification of Human–Computer Interaction," *IEEE Transactions on Software Engineering* **SE-11** (August 1985), pp. 699–713. (Chapter 7)

[Wasserman and Pircher, 1987] A. I. Wasserman and P. A. Pircher, "A Graphical, Extensible Integrated Environment for Software Development," *Proceedings of the Second ACM SIGSOFT/SIGPLAN Software Engineering Symposium on Practical Software Development Envi-

ronments, *ACM SIGPLAN Notices* **22** (January 1987), pp. 131–142. (Chapter 12)

[Waters, 1982] R. C. WATERS, "The Programmer's Apprentice: Knowledge-Based Program Editing," *IEEE Transactions on Software Engineering* **SE-8** (January 1982), pp. 1–12. (Chapter 16)

[Waters, 1985] R. C. WATERS, "The Programmer's Apprentice: A Session with KBEmacs," *IEEE Transactions on Software Engineering* **SE-11** (November 1985), pp. 1296–1320. (Chapter 16)

[Wegner, 1984] P. WEGNER, "Capital-Intensive Software Technology. Part 4: Accomplishments and Deficiencies of Ada," *IEEE Software* **1** (July 1984), pp. 39–42. (Chapter 14)

[Weinberg, 1971] G. M. WEINBERG, *The Psychology of Computer Programming*, Van Nostrand Reinhold, New York, NY, 1971. (Chapters 1 and 10)

[Weinberg and Freedman, 1984] G. M. WEINBERG AND D. P. FREEDMAN, "Reviews, Walkthroughs, and Inspections," *IEEE Transactions on Software Engineering* **SE-10** (January 1984), pp. 68–72. (Chapter 6)

[Weyuker, 1988] E. J. WEYUKER, "An Empirical Study of the Complexity of Data Flow Testing," *Proceedings of the Second Workshop on Software Testing, Verification, and Analysis*, Banff, Canada, July 1988, pp. 188–195. (Chapter 6)

[Whitaker, 1978] W. A. WHITAKER, "The U.S. Department of Defense Common High Order Language Effort," *ACM SIGPLAN Notices* **13** (February 1978), pp. 19–29. (Chapter 14)

[Wiest and Levy, 1977] J. D. WIEST AND F. K. LEVY, *A Management Guide to PERT/CPM: With GERT/PDM/DCPM and other Networks*, Second Edition, Prentice-Hall, Englewood Cliffs, NJ, 1977. (Chapter 12)

[Wile, 1983] D. S. WILE, "Program Developments: Formal Explanations of Implementations," *Communications of the ACM* **26** (November 1983), pp. 902–911. (Chapter 16)

[Wirth, 1971] N. WIRTH, "Program Development by Stepwise Refinement," *Communications of the ACM* **14** (April 1971), pp. 221–227. (Chapters 5 and 6)

[Wirth, 1975] N. WIRTH, *Algorithms + Data Structures = Programs*, Prentice-Hall, Englewood Cliffs, NJ, 1975. (Chapter 5)

[Wirth, 1985] N. WIRTH, *Programming in Modula-2*, Third Corrected Edition, Springer-Verlag, Berlin, West Germany, 1985. (Chapter 8)

[Wolberg, 1983] J. R. WOLBERG, *Conversion of Computer Software*, Prentice-Hall, Englewood Cliffs, NJ, 1983. (Chapter 13)

[Wolf, Clarke, and Wileden, 1985] A. L. WOLF, L. A. CLARKE, AND J. C. WILEDEN, "Ada-Based Support for Programming-in-the Large," *IEEE Software* **2** (March 1985), pp. 58–71. (Chapter 14)

[Wong, 1984] C. WONG, "A Successful Software Development," *IEEE Transactions on Software Engineering* **SE-10** (November 1984), pp. 714–727. (Chapter 3)

[Woodward, Hedley, and Hennell, 1980] M. R. WOODWARD, D. HEDLEY, AND M. A. HENNELL, "Experience with Path Analysis and Testing of Programs," *IEEE Transactions on Software Engineering* **SE-6** (May 1980), pp. 278–286. (Chapter 6)

[World Book Encyclopedia, 1982] *World Book Encyclopedia*, World Book-Childcraft International, Inc., Chicago, IL, 1982, Volume N, p. 430. (Chapter 9)

[Yau and Tsai, 1986] S. S. YAU AND J. J.-P. TSAI, "A Survey of Software Design Techniques," *IEEE Transactions on Software Engineering* **SE-12** (June 1986), pp. 713–721. (Chapter 9)

[Young and Gregory, 1972] D. M. YOUNG AND R. T. GREGORY, *A Survey of Numerical Mathematics, Volume I*, Addison-Wesley, Reading, MA, 1972. (Chapter 8)

[Yourdon, 1989] E. YOURDON, *Modern Structured Analysis*, Yourdon Press, Englewood Cliffs, NJ, 1989. (Chapters 7 and 12)

[Yourdon and Constantine, 1979] E. YOURDON AND L. L. CONSTANTINE, *Structured Design: Fundamentals of a Discipline of Computer Program and Systems Design*, Prentice-Hall, Englewood Cliffs, NJ, 1979. (Chapters 8 and 9)

[Zave, 1984] P. ZAVE, "The Operational versus the Conventional Approach to Software Development," *Communications of the ACM* **27** (February 1984), pp. 104–118. (Chapter 7)

[Zave, 1986] P. ZAVE, "Case Study: The PAISLey Approach Applied to its Own Software Tools," *Computer Languages* **11** (January 1986), pp. 15–28. (Chapter 7)

[Zave, 1988] P. ZAVE, "PAISLey User Documentation, Volumes 1, 2, and 3," AT & T Bell Laboratories, 1988. (Chapter 7)

[Zave and Cole, 1983] P. ZAVE AND G. E. COLE, JR., "A Quantitative Evaluation of the Feasibility of, and Suitable Hardware Architectures for, an Adaptive, Parallel Finite-Element System," *ACM Transactions on Mathematical Software* **9** (September 1983), pp. 271–292. (Chapter 7)

[Zave and Schell, 1986] P. ZAVE AND W. SCHELL, "Salient Features of an Executable Specification Language and its Environment," *IEEE Transactions on Software Engineering* **SE-12** (February 1986), pp. 312–325. (Chapter 7)

[Zelkowitz, Shaw, and Gannon, 1979] M. V. ZELKOWITZ, A. C. SHAW, AND J. D. GANNON, *Principles of Software Engineering and Design*, Prentice-Hall, Englewood Cliffs, NJ, 1979. (Chapters 1 and 11)

Name Index

This index includes only names and references cited in the actual text.

Name Index

Subject Index